Mammalian Toxicology of Insecticides

Issues in Toxicology

Series Editors:
Professor Diana Anderson, *University of Bradford, UK*
Dr Michael D Waters, *Integrated Laboratory Systems, Inc, N Carolina, USA*
Dr Martin F Wilks, *University of Basel, Switzerland*
Dr Timothy C Marrs, *Edentox Associates, Kent, UK*

Titles in the Series:
1: Hair in Toxicology: An Important Bio-Monitor
2: Male-mediated Developmental Toxicity
3: Cytochrome P450: Role in the Metabolism and Toxicity of Drugs and other Xenobiotics
4: Bile Acids: Toxicology and Bioactivity
5: The Comet Assay in Toxicology
6: Silver in Healthcare
7: *In Silico* Toxicology: Principles and Applications
8: Environmental Cardiology
9: Biomarkers and Human Biomonitoring, Volume 1: Ongoing Programs and Exposures
10: Biomarkers and Human Biomonitoring, Volume 2: Selected Biomarkers of Current Interest
11: Hormone-Disruptive Chemical Contaminants in Food
12: Mammalian Toxicology of Insecticides

How to obtain future titles on publication:
A standing order plan is available for this series. A standing order will bring delivery of each new volume immediately on publication.

For further information please contact:
Book Sales Department, Royal Society of Chemistry, Thomas Graham House, Science Park, Milton Road, Cambridge, CB4 0WF, UK
Telephone: +44 (0)1223 420066, Fax: +44 (0)1223 420247, Email: books@rsc.org
Visit our website at http://www.rsc.org/Shop/Books/

Mammalian Toxicology of Insecticides

Edited by

Timothy C. Marrs OBE MD DSc FRCP FRCPath FBTS MRSC
Edentox Associates, Edenbridge, Kent TN8 6AQ, UK and
National Poisons Information Service (Birmingham Unit), UK and
City Hospital, Birmingham B18 7QH, UK

RSC Publishing

Issues in Toxicology No. 12

ISBN: 978-1-84973-191-1
ISSN: 1757-7179

A catalogue record for this book is available from the British Library

Published by The Royal Society of Chemistry,
Thomas Graham House, Science Park, Milton Road,
Cambridge CB4 0WF, UK

Registered Charity Number 207890

For further information see our web site at www.rsc.org

Preface

Mammals and insects are much more similar in structure than are mammals and fungi or mammals and plants (target organisms of other pesticides). This produces the problem that insecticides are frequently toxic to mammals by mechanisms directly related to their toxicity to the target organisms, *i.e.* insects. This is particularly the case with neurotoxicity, many insecticides targeting the insect nervous system. However, some insecticides are targeted at structures or hormonal systems specific to insects (insect growth regulators, chitin synthesis inhibitors and ecdysone agonists) and these are generally of low mammalian toxicity. For some reason, many of these insecticides are mildly haematotoxic.

There are thus issues specific to insecticides that do not occur with other pesticides, and this is the reason for the need for a book specifically on insecticide toxicology in mammals (books are available on toxicity to insects).

As old groups of insecticides are discarded (*e.g.* organochlorines), new ones come on the market. Some of these are similar to older insecticides, targeting the nervous system (*e.g.* the neonicotinoids). Others target structures and systems unique to insects. A relatively new approach is that of biological insecticides.

In view of the economic damage done by insects as well as the damage to human and animal health by disease vectors, I hope this book will fill a niche in an important area of toxicology.

Tim Marrs
Edenbridge

Issues in Toxicology No. 12
Mammalian Toxicology of Insecticides
Edited by Timothy C. Marrs

Published by the Royal Society of Chemistry, www.rsc.org

Contents

Issues in Toxicology No. 12
Mammalian Toxicology of Insecticides
Edited by Timothy C. Marrs

Published by the Royal Society of Chemistry, www.rsc.org

CHAPTER 1

Toxicology of Insecticides—Introductory Considerations

TIMOTHY C MARRS

Edentox Associates, Pinehurst, Four Elms Road, Edenbridge, Kent, UK and UK National Poisons Information Service (Birmingham Centre), City Hospital, Birmingham, UK. Email: timothymarrs05@aol.com

1.1 Introduction

Insecticides are a group of substances with heterogeneous toxicity, whose desired activity is the killing of unwanted insects. Closely allied are acaricides or miticides, terms used for substances that kill mites. Indeed, many acaricides are also insecticides. Many insecticides have mammalian toxicity that is related to their toxicity to the target organism, especially those insecticides that target the insect nervous system. Notable exceptions to this are insecticides that target systems present in insects but not mammals, such as the juvenile hormone analogues and chitin-synthesis inhibitors. Even those insecticides that target systems present in both insects and mammals may have target organism specificity conferred by physical or metabolic differences between insects and mammals.

To be weighed against their mammalian toxicity are the facts that insects and are important sources of agricultural loss of food and other crops such as cotton and can give rise to damage to buildings, where construction is of wood. A very important role of insecticides in public health is vector control. Many insects carry diseases such as malaria, yellow fever, Lyme disease, dengue and sleeping sickness, all of which cause considerable mortality and/or morbidity.

Issues in Toxicology No. 12
Mammalian Toxicology of Insecticides
Edited by Timothy C. Marrs

Published by the Royal Society of Chemistry, www.rsc.org

Probably the most serious of these is malaria,[1] where the use of insecticides is a very important part of disease control.

As with many uses of chemicals, the key to a successful insecticide is selective toxicity against target insects and away from non-target insects and mammals.[2,3] An ideal insecticide will interfere with a biological system in the insect that has no counterpart in non-target species: this is the advantage of the juvenile hormone analogue and chitin-synthesis inhibiting insecticides. In the case of agricultural insecticides, the insecticide should be toxic to insects, but less toxic to plants, to humans and to other non-target organisms. In the case of insecticides targeting the insect neurological system, insecticides often exploit the relative accessibility of the insect nervous system to xenobiotics, when compared to humans, or different distribution of neurotransmission systems, together with the lack of a nervous system in plants.

In recent years, there has been some concern as to the possibility of deleterious effects from exposure to multiple pesticides, *inter alia* insecticide exposure, either as residues in food and water or at the workplace, home or environment, or indeed a combination of these pathways; this is further discussed below. Another recent development is the use of microbial insecticides: these are discussed in Chapter 10.

1.2 Classification

The insecticides can be classified in several ways, for example into those of natural or synthetic origin (see Table 1.1). This division, while perhaps useful for organic farmers, tell us little about toxicology. Insecticides of natural origin—*e.g.* pyrethrum, nicotine, rotenone (Derris), the 'mectins—have little in common from the point of view of mammalian toxicology and, of course, the synthetic insecticides can be divided into many groups, depending on their mode of action and/or structure. Indeed some natural insecticides (*e.g.* pyrethrum) have more in common with synthetic insecticides (the synthetic pyrethroids) than they do with other natural insecticides.

Another way of classifying insecticides is by chemical structure (Table 1.2).

This classification is of much more use to toxicologists, partly because it provides some guide to effects in mammals. However, it is important to remember that this classification is not a complete guide to mammalian toxicology. The organophosphate esters are commonly anticholinesterase

Table 1.1 Examples of insecticides and synthetic and natural origin.

Synthetic insecticides	*Natural insecticides*
Organochlorines	Pyrethrum
Organophosphates	Nicotine
Carbamates	Rotenone (Derris)
Synthetic pyrethroids	Avermectins
Neonicotinoids	

Table 1.2 Classification of insecticides by chemical structure.

Group by chemical structure	*Examples*
Organochlorines	DDT
Organophosphates	Malathion
Carbamates	Aldicarb
Pyrethrins/synthetic pyrethroids	Pyrethrum, permethrin
Nicotine/neonicotinoids	Nicotine, imidacloprid

insecticides (exceptionally, pyrazophos is a fungicide), but other organic compounds of phosphorus have different actions both in target and non-target species *e.g.* glyphosate, which is a herbicide. Similarly, most carbamate insecticides are *N*-methyl carbamates: other carbamates have fungicidal and herbicidal action and are not cholinesterase inhibitors.

The most useful classification for toxicologists is probably by mode of action in the target species and a recent classification on that basis has been proposed by the Insecticide Resistance Action Committee, although it should be noted that the mode of action in target species is not always completely known.[4] Nevertheless, in many cases where the mode of action in target insects is known, it may provide some guide to the toxicological action in mammals. Thus, many insecticides act on the insect nervous system and their effects in mammals are often related to this. Action on the insect nervous system may be on enzymes involved in neurotransmission (anticholinesterases—see Chapter 4) or directly on receptors (nicotine and the neonicotinoids [see Chapter 6] and fipronil [see Chapters 9 and 12]). Other neuronal structures may be targeted: the avermectins stimulate the release and binding of γ-aminobutyric acid (GABA), a neurotransmitter, at nerve endings (see Chapter 8). Organochlorines such as DDT have a more general action on sodium channels in neurons (see Chapter 3), and pyrethrins and synthetic pyrethroids also act on sodium channels by keeping them open (see Chapter 5). The reason for the utility of this classification is that the mammalian toxicology of these potentially neurotoxic insecticides is in large part related to the neurotoxicity in target organisms. With insecticides that target the insect nervous system, specificity towards insects and away from mammals may be achieved by exploiting the greater accessibility of the insect nervous system, metabolic differences between insects and mammals, or differences in distribution of receptors and other components of neurotransmission systems. Alternatively, specificity can be achieved by differences in binding affinity for receptors or by a combination of means.

Another group of insecticides target structures in insects that are not found in mammals, including the insect growth regulators (juvenile hormone analogues), chitin-synthesis inhibitors and ecdysone agonists. The toxicity of these compounds to mammals is generally low and, strangely, often referable to the mammalian haematological system (see Chapter 7).

Yet another way of classifying insecticides is by their use, in that this may decide how they are regulated (see below).

1.3 Regulation of Insecticides

In the European Union (EU), including the UK, agricultural, horticultural and home garden insecticides are regulated under the Plant Protection Products Directive[5] and subsequent regulations,[6] whereas insecticides used in public hygiene and biocides are regulated under the Biocidal Products Directive.[7] A major change in regulation of plant protection products is the replacement of a largely risk-based regulatory system under the Plant Protection Products Directive by a hazard- and risk-based system. There is little experience of such a system in either the EU or the rest of the world, and it remains to be seen how the new system will work.

Insecticides or acaricides used on animals to control ectoparasites are classified in the EU as veterinary medicines (this is not the case in the USA). Some of these veterinary medicines have a high toxicological profile, as those that are used in food-producing animals can give rise to residues in food in the same way that pesticides used in plant protection can. Furthermore, there has been some concern that ectoparasiticides, used for example on dogs and cats, may give rise to significant exposure of children through the hand-to-mouth pathway. Another consideration is that, because of the scale of their use, some veterinary medicines (*e.g.* sheep dips) may have a significant environmental impact. In the UK, mostly during the early 1990s, there was concern about the safety of organophosphorus (OP) sheep dips to farmers and farm workers.[8] In the EU, Directive 65/65[9] provided the initial basis for the regulation of veterinary pharmaceuticals, and this has been amended many times. Other legislation also impinges on veterinary drug regulation, notably Regulation EC 726/2004,[10] which established the European Medicines Agency (EMEA) in London. Previously called the European Medicines Evaluation Agency, it has retained the acronym EMEA. Also established was the Committee for Medicinal Products for Veterinary Use (CVMP). A few insecticides are used in human medicine, notably malathion for headlouse treatment and trichlorfon (metrifonate), used in tropical medicine: in the EU, these are covered under similar legislation to veterinary drugs, but the relevant committee at the EMEA is the Committee for Proprietary Medicinal Products (CPMP). Another use for insecticides is vector control of insect-borne diseases, notably malaria. In the UK, endemic malaria in East Anglia and north Kent was largely eradicated by the time of the First World War, while in Italy malaria was eradicated after the Second World War and Spain was declared malaria-free in 1964. Insecticide vector control is therefore not a major issue in the EU and certainly not in the UK. However, in other countries where malaria is endemic, mosquito control is important in public health and the use of DDT in this role continues to provoke controversy.[11] Nor should it be forgotten that there are many other diseases spread by insects; with them, insecticides make a major contribution to public health. Lice and fleas (wingless insects) and ticks (mites) can all carry disease, some of which are very serious, such as typhus and bubonic plague.

Methods of risk assessment used by authorities that regulate plant protection products, biocides and veterinary medicines have until recently not differed much, but the relatively few insecticides used in human medicine are in general regulated on a risk-benefit basis. Logically, insecticides used in vector control should be regulated in the same way. The other compounds are subjected to risk assessment during the authorization process, on the general assumption that no individual benefit accrues from exposure. The regulatory authorities endeavour to ensure that exposure is below reference doses/health-based guidance values such as the *acceptable daily intake* (ADI). For insecticides with appreciable acute toxicity another reference dose, the *acute reference dose* (ARfD) is also elaborated, while for operators there is a third reference dose, the *acceptable operator exposure level* (AOEL). The ADI is normally calculated from the most sensitive study in the most sensitive animal species, additionally incorporating a safety factor of 100 or more.[12] The ADI represents a mean maximum intake over time, but to avoid acute toxicity from insecticides, the concept of the ARfD was developed. The calculation of the ARfD also involves a safety factor, most often of 100, from those studies appropriate for assessing acute toxicity.[13] In the risk assessment process for food exposure the calculated exposure of a 'high-level' consumer is compared with the reference doses.

1.4 Exposure

There are a number of possible ways in which humans can be exposed to insecticides. Thus food consumers may be exposed to insecticides used on food crops or food-producing animals, and of course farmers and other applicators may be exposed. Insecticides used domestically in wood preservation, as household insecticides or on pet animals may be important sources of exposure of the general public. Moreover, insecticides may get into the water supply. It should also be remembered that the more acutely toxic insecticides have been used for suicide and murder.

The effects of insecticide residues in food and water probably cause the greatest concern to the general public. In fact, reports of definitively diagnosed clinical poisoning by residues of pesticides in food seems to be comparatively rare, certainly in the EU and USA, and in comparison with deliberate and accidental poisonings and occupational intoxication; nevertheless, cases are reported from time to time (see below). The reasons for the paucity of reports of poisoning from residues are not clear: it may reflect a true rarity of poisoning or a lack of recognition of food-borne poisoning, which might, for example, be mistaken for bacterial food poisoning. Of course, occupational poisonings are far more easily identified, the proximity of cause and effect making diagnosis easier. Acute poisoning, where insecticides had been used in accordance with regulations, seems likely to be a very uncommon occurrence, if it occurs at all, at least in developed countries, simply because of the complexity of the regulatory process and the strictness of enforcement of those regulations. The problem of consumers who consume high levels of some foodstuffs is taken into account when *maximum residue levels* (MRLs) are pronounced to be toxicologically acceptable by comparison with the appropriate ADI. It is probable that residues of many multiples of the MRL would probably be necessary to produce acute poisoning. Thus it seems highly unlikely that extreme consumption, by itself, could give rise to insecticide poisoning from approved uses. In fact, analysis of reported consumer poisonings by insecticides shows that most reported instances occur from:

- spillage of insecticides on to food during storage or transport
- eating grain or seed potatoes treated with insecticides, where the food article was not intended for human consumption
- improper application of insecticides or failure to observe harvest intervals.[14–16]

The last may have been responsible for clinical poisoning with methamidophos reported in Taiwan by Wu and colleagues.[17] Also, aldicarb, a carbamate anticholinesterase of high acute toxicity, has caused outbreaks of poisoning in the USA and Canada[18] and in Ireland.[19] The insecticides responsible have often been ones with low LD_{50} (<20 mg/kg bw).

For risk assessment purposes, exposure is calculated from field trials, where pesticide residues are measured, and from dietary surveys of food consumption. Also from field trial data an MRL is established to ensure proper usage of the insecticide by the farmer and this MRL must also be toxicologically acceptable. For older pesticides, there may be data from residue surveillance programs showing the frequency and magnitude of residues.

1.4.1 Exposure to Mixtures of Insecticides

Exposure to insecticides from all sources and simultaneous exposure to mixtures of insecticides has always been a matter of concern. However, risk assessment of pesticides has, until recently, been route-specific and by individual insecticide. This started to change with the enactment in 1996 of the US Food Quality Protection Act.[20] This act introduced the terms *aggregate risk assessment* (assessment of pesticides from all sources of exposure) and *cumulative risk assessment* (risk assessment of more than one pesticide at a time). The first of these, aggregate risk assessment, is comparatively uncontroversial and simply a matter of data gathering, but the second presents many practical and theoretical problems. The term cumulative risk assessment has been criticised as inviting confusion with pharmacological cumulation, but the use of the term has become widespread and will be used here. Cumulative risk assessment is based upon simple and well-recognised toxicological principles that pesticides with a similar mechanism of toxicological action will act similarly, while pesticides with a different mechanism of toxicological action will act independently. Unfortunately, there are some problems with definitions of combined actions of chemicals, *inter alia* pesticides. The UK Committee on Toxicity of Chemicals in Food, Consumer Products and the Environment (COT) produced a report on toxicity of mixtures of chemicals with special reference to pesticides[21] and the definitions used in that report are given in Table 1.3.

Table 1.3 Combined effects of pesticides—terminology.

Types of combined effect	*Subtypes*	*Synonyms*	*Effects observed*
Non-interactive	Simple similar action	Additivity	Dose additivity
	Simple dissimilar action	Independent action	Effect additivity
Interactive	Potentiation	Synergy, synergism	Greater than dose-additive effect
	Antagonism		Less than dose-additive effect

Adapted from WiGRAMP final report.[21]

In cumulative risk assessment the default assumptions are that pesticides with the same toxic action will produce, in combination, simple similar action and will exhibit dose additivity, whereas those with different toxic actions will show simple dissimilar action, which will produce effect additivity and no interaction will occur (no potentiation or antagonism). This means that before doing cumulative risk assessment, it is necessary to take a view on whether compounds possess a similar mechanism of action. That is to say, whether a group of compounds constitute a common mechanism group (CMG). This is easiest with groups such as anticholinesterase OPs when they are all known to act at a single macromolecule (acetylcholinesterase), but much more difficult where compounds have similar effects but possibly by multiple mechanisms, *e.g.* endocrine disruptors. Even with as homogeneous a group of compounds as the OPs, there are compounds that are atypical in target organism, for example pyrazophos or only have weak anticholinesterase activity (ethephon, tolclofos-methyl). Factors that need to be considered in defining CMGs have been reviewed.[22,23] Carrying out a cumulative risk assessment is easy in theory; it consists of adding the residues of the compounds together having allowed for potency and comparing the result with some figure to toxicological acceptability. In practice there are many problems and more than one way of doing cumulative risk assessment,[24] but these are by no means insurmountable. In the USA, much progress has been made, with various groups of compounds having been examined, *e.g.* OPs,[25] *N*-methyl carbamates[26] and some non-insecticidal pesticides. In the EU, the European Food Safety Authority (EFSA) organized a colloquium in 2006.[27] The EFSA panel on plant protection products and their residues (PPR panel) has proposed a tiered approach to cumulative risk assessment.[28] One concern addressed by the COT was concern about the 'cocktail effect': that is, the possibility that harmful effects that may be produced by exposure to food residues of many different pesticides and other chemicals, not necessarily toxicologically similar.[21] Further, there is concern in Europe that exposure to chemicals from a variety of sources, even at low doses, may be deleterious. There is no clear evidence of the occurrence of such effects from pesticides and the importance of any such combined actions in humans is unclear. In fact it may simply be that the question of the existence of the 'cocktail effect' is unanswerable.

1.5 Newer Developments

Toxicology is a continually developing science, and a number of changes have occurred over the last few years. A major change in the regulation of plant protection products has been discussed above, in relation to hazard and risk. Two other developments are

- in hazard identification—developmental neurotoxicity
- in exposure assessment—probabilistic exposure assessment.

1.5.1 Developmental Neurotoxicity

Developmental neurotoxicity is a major public health problem with neuroactive substances, notably ethanol,[29] but also some other substances.[30] Since many insecticides are neurotoxic, the possibility has had to be considered that they might be developmental neurotoxicants. A major meeting was held at Williamsburg, Virginia, USA in 1989 to consider how this problem might be approached.[31] While many aspects of developmental neurotoxicity are undoubtedly tested in the conventional pesticide data package, there was concern that some aspects might be being missed. Consequently the United States Environmental Protection Agency (EPA) elaborated a guideline for a test of developmental neurotoxicity in the rat.[32] There is also an Organisation for Economic Co-operation and Development (OECD) guideline.[33] These guidelines describe a test that is not mechanism based but rather intended to detect as broad a range of developmental neurotoxicants as possible.

The test substance is normally administered orally to several groups of pregnant animals during gestation and lactation (day 6 of gestation to post-natal day 10) and offspring are selected for evaluation of neurotoxicity. The end points that are evaluated are gross neurological and behavioural abnormality, motor activity, auditory startle response and assessment of learning. At termination, the offspring brains are weighed and neuropathology is carried out (normally day 60 after birth). The EPA is evaluating these studies and accumulating a database.[34,35] Developmental neurotoxicity tests (DNTs) done for the EPA have been submitted to European regulatory authorities as part of packages when available even when nothing adverse occurred, but there is no general requirement to do DNTs for the EU (see Hass, 2003).[36]

1.5.2 Probabilistic Exposure Assessment

This is sometimes incorrectly called probabilistic risk assessment (no serious attempt has been made to apply probabilistic methodology to toxicology studies). Probabilistic exposure assessment uses distributions of data, *e.g.* food item intake, pesticide residue level and other factors, to calculate the distribution of pesticide residue intake.[37,38] There is evidence that it gives a better guide to pesticide intake than deterministic exposure calculations and its use becomes becomes essential when cumulative risk assessment is carried out, to avoid 'compounded conservatism'.[39]

1.6 Economic Benefits of Insecticides

It must be recognised that insects are important sources of agricultural loss and can damage property, for example where construction is of wood. Also, as discussed above, insects are vectors of disease in humans and other animals and failure to control disease vectors may render land unusable for agriculture or even uninhabitable. Weighed against the disadvantages of insecticides that accrue from their toxic effects on non-target species, including humans.

1.7 Nomenclature of Insecticides

In addition to chemical names, insecticides have national common names (*e.g.* British Standards Institute [BSI], American National Standards Institute [ANSI]) as well as ISO (International Organization for Standardization) common names.[40,41] The list of ISO names is continually expanding and is updated regularly.[42] ISO names are allocated in English, French and Russian, and the English-language ISO name is not always the same as the BSI or ANSI common name. An example is iodofenphos, which is the BSI name, where the English-language ISO name is jodfenphos (curiously, the French language ISO name is iodofenphos!).

A few insecticides are used as drugs either in human or veterinary medicine. As such they have international non-proprietary names (INNs), as well as United States Adopted Names (USANs) and British Approved Names (BANs). INNs are bestowed in a number of languages (English, Latin, French, Russian, Spanish, Arabic, Chinese, and Latin) by the World Health Organization.[43] In some cases the INN differs from the ISO pesticide name: for example, trichlorfon, an insecticide, is the same as metrifonate, a pharmaceutical used in human medicine. Insecticides are commonly sold under trade names given by their manufacturers. It is customary to capitalize the initial letter of trade names but not of ISO or other approved names.

1.8 Purity of Insecticides

It is very important that in tests on pesticides the product used is identified precisely. Insecticides will have a technical specification, which gives the minimum content of the pure active ingredient and the identity and maximum content of impurities present at or greater than a certain level (normally 0.1% of the total content of the active substance). Pesticides are sold as a formulation which comprises the active substance and other substances. The formulation recipe consists of the nominal target content for the pure active substance, with acceptable tolerance limits, and the name(s) and quantity of all other components.[44] It should be noted that substances other than the active ingredient may contribute to acute toxicity: an example is the contamination of diazinon, an OP, with sulfotep, a more acutely toxic anticholinesterase.[45] Sometimes impurities will produce qualitatively different toxicity from the active substance; thus several trialkylphosphorothioates, contaminants of OP insecticides, produce lung injury.[46,47]

References

1. J. G. Breman, Eradicating malaria, *Sci. Prog.*, 2009, **92**, 1–38.
2. A. Albert, *Selective Toxicity: The Physico-Chemical Basis of Therapy*, Chapman & Hall, London, 7th edn, 1985.
3. T. Narahashi, X. Zhao, T. Ikeda, K. Nagata and J. Z. Yeh, Differential actions of insecticides on target sites: basis for selective toxicity, *Hum. Exp. Toxicol.*, 2007, **26**, 361–366.
4. Insecticide Resistance Action Committee. September 2010. http://www.irac-online.org/wp-content/uploads/2009/09/MoA-classification_v7.0.4–5Oct10.pdf
5. Council of the European Communities Directive 91/414/EEC of 15 July 1991 concerning the placing of plant protection products on the market. *Official Journal of the European Union*, **L 230**, 19/08/1991, pp. 1–32.
6. European Parliament and Council Regulation (EC) No. 1107/2009 of 21 October 2009 concerning the placing of plant protection products on the market and repealing Council Directives 79/117/EEC and 91/414/EEC. *Official Journal of the European Union*, **L 309**, 24.11.2009, pp. 1–46.
7. European Parliament and of the Council Directive 98/8/EC of the of 16 February 1998 concerning the placing of biocidal products on the market. *Official Journal of the European Union*, **L 123**, 24/4/98, pp. 1–63.
8. T. C. Marrs and P. Edwards, Medicines used to control and treat external parasites of sheep—toxicology and the phenomenon of reported adverse human responses to organophosphorus sheep dips, in *Veterinary Pharmacovigilance*, ed. K. N. Woodward, Wiley-Blackwell, Chichester, UK, 2009, pp. 517–528.
9. Council of the European Economic Community Directive 65/65/EEC of 26 January 1965 on the approximation of provisions laid down by law, regulation or administrative action relating to medicinal products. *Official Journal of the European Union,* **L 22**, 9/2/1965 p. 369 (as amended by Directives 66/454/EEC, 75/319/EEC, 83/570/EEC, 87/21/EEC, 89/341/EEC, 89/342/EEC, 89/343/EEC, 92/27/EEC, 92/73/EEC, 93/39/EEC).
10. European Parliament and Council Regulation EC 726/2004 of 31 March 2004 laying down Community procedures for the authorisation and supervision of medicinal products for human and veterinary use and establishing a European Medicines Agency. *Official Journal of the European Union*, **L 136**, 30/4/2004, pp. 1–32.
11. K. Raghavendra, T. K. Barik, B. P. Reddy, P. Sharma, A. P. Dash, Malaria vector control: from past to future. *Parasitol. Res.* 2011, **108**, 757–779.

12. *Environmental Health Criteria 104. Principles for the Toxicological Assessment of Pesticide Residues in Food.* World Health Organization, Geneva, 1990.
13. R. Solecki, L. Davies, V. Dellarco, I. Dewhurst, M. Raaij and A. Tritscher, Guidance on setting of acute reference dose (ARfD) for pesticides, *Food Chem. Toxicol.*, 2005, **43**, 1569–1593.
14. A. Ferrer and J. P. Cabral, Epidemics due to pesticide contamination of food. *Food Addit. Contam.*, 1989, **6** Suppl 1, S95–98.
15. A. Ferrer and R. Cabral, Toxic epidemics caused by alimentary exposure to pesticides: a review. *Food Addit. Contam.*, 1991, **8**, 755–775.
16. A. Ferrer and R. Cabral, Recent epidemics of poisoning by pesticides, *Toxicol. Lett.*, 1995, **82–83**, 55–63.
17. M. L. Wu, J. F. Deng, W. J. Tsai, J. Ger, S. S. Wong and H. P. Li, Food poisoning due to methamidophos-contaminated vegetables, *J. Toxicol. Clin. Toxicol.*, 2001, **39**, 333–336.
18. A. H. Hall and B. H. Rumack, Incidence, presentation and therapeutic attitudes to anticholinesterase poisoning in the USA, in *Clinical and Experimental Toxicology of Organophosphates and Carbamates*, ed. B. Ballantyne and T. C. Marrs, Butterworth-Heinemann, Oxford, 1992, pp. 471–493.
19. Department of Agriculture and Food, Press release 117/92. Government Information Services, Dublin, 1992.
20. 104th United States Congress, *Food Quality Protection Act of 1996*. August 3, 1996, P.L. 104–170.
21. Committee on Toxicity of Chemicals in Food, Consumer Products and the Environment, *Review of Risk Assessment of Mixtures of Pesticides and Similar Substances.* WiGRAMP final report. Food Standards Agency, London, 2002. http://cot.food.gov.uk/cotwg/wigramp/wigrampfinalreport
22. U.S. Environmental Protection Agency, *Guidance for Identifying Pesticide Chemicals and Other Substances That Have a Common Mechanism of Toxicity*. Office Of Pesticide Programs, Office of Prevention, Pesticides, and Toxic Substances, Washington, D.C., January 29, 1999. http://www.epa.gov/fedrgstr/EPA-PEST/1999/February/Day-05/
23. P. A. Fenner-Crisp, FQPA science issues: common mechanism of toxicity and cumulative risk assessment, *Regul. Toxicol. Pharmacol.*, 2000, **31**, 308–310.
24. C. F. Wilkinson, G. R. Christoph, E. Julien, J. M. Kelley, J. Kronenberg, J. McCarthy and R. Reiss, Assessing the risks of exposures to multiple chemicals with a common mechanism of toxicity: how to cumulate? *Regul. Toxicol. Pharmacol.*, 2000, **31**(1), 30–43.
25. U.S. Environmental Protection Agency, *Organophosphate Pesticides: Revised Cumulative Risk Assessment*. U.S. Environmental Protection Agency, Washington, DC, June 10, 2002. http://www.epa.gov/opp00001/cumulative/rra-op/
26. U.S. Environmental Protection Agency, *N-methyl Carbamate Revised Cumulative Risk Assessment*, U.S. Environmental Protection Agency, Washington, DC, September 26, 2007. http://www.epa.gov/oppsrrd1/cumulative/common_mech_groups.htm#carbamate

27. European Food Safety Authority, *EFSA's 7th Scientific Colloquium—Cumulative Risk Assessment of Pesticides to Human Health: the Way Forward.* 28 and 29 November 2006, European Food Safety Authority, Parma, Italy, 2006. http://www.efsa.europa.eu/en/events/event/colloque061128.htm
28. EFSA Panel on Plant Protection Products and their Residues (PPR panel). Scientific opinion on risk assessment for a selected group of pesticides from the triazole group to test possible methodologies to assess cumulative effects from exposure through food from these pesticides on human health. *EFSA Journal*, 2009, **7**, 1167–1170.
29. B. B. Bruce, V. Biousse, A. L. Dean and N. J. Newman, Neurologic and ophthalmic manifestations of fetal alcohol syndrome, *Rev. Neurol. Dis.*, 2009, **6**, 13–20.
30. J. Julvez and P. Grandjean, Neurodevelopmental toxicity risks due to occupational exposure to industrial chemicals during pregnancy, *Ind. Health.*, 2009, **47**, 459–468.
31. Qualitative and quantitative comparability of human and animal developmental neurotoxicity. Williamsburg, Virginia, April 11–13, 1989. Proceedings. *Neurotoxicol. Teratol.*, 1990, **12**, 173–292.
32. U.S. Environmental Protection Agency, *Health Effects Guidelines OPPTS 870.6300 Developmental Neurotoxicity Study.* U.S. Environmental Protection Agency, Washington, D.C., 1998.
33. *OECD Guidelines for the Testing of Chemicals, Section 4: Health Effects. Test No. 426: Developmental Neurotoxicity Study.* Organisation for Economic Co-operation and Development, Paris, 15 Oct 2007. http://www.oecdbookshop.org/oecd/display.asp?K=5L4FG25MNKXS&DS=Test-No.-426-Developmental-Neurotoxicity-Study, accessed 6 July 2011.
34. S. Makris, K. Raffaele, W. Sette, J. Seed, A retrospective analysis of twelve developmental neurotoxicity studies submitted to the US EPA office of prevention, pesticides and toxic substances. U.S. Environmental Protection Agency, Washington, D.C., 1998. http://www.epa.gov/scipoly/sap/1999/may/10xtx428.pdf
35. S. L. Makris, K. Raffaele, S. Allen, W. J. Bowers, U. Hass, E. Alleva, G. Calamandrei, L. Sheets, P. Amcoff, N. Delrue and K. M. Crofton, *et al.*, A retrospective performance assessment of the developmental neurotoxicity study in support of OECD test guideline 426, *Environ. Health Perspect.*, 2009, **117**, 17–25.
36. U. Hass, Current status of developmental neurotoxicity: regulatory view, *Toxicol. Lett.*, 2003, **140–141**, 155–159.
37. J. D. van Klaveren and P. E. Boon, Probabilistic risk assessment of dietary exposure to single and multiple pesticide residues or contaminants: summary of the work performed within the SAFE FOODS project, *Food Chem. Toxicol.*, 2009, **47**(12), 2879–82.
38. A. R. Boobis, B. C. Ossendorp, U. Banasiak, P. Y. Hamey, I. Sebestyen and A. Moretto, Cumulative risk assessment of pesticide residues in food, *Toxicol. Lett.*, 2008, **180**, 137–150.

39. A. C. Cullen, Measures of compounding conservatism in probabilistic risk assessment, *Risk Anal.*, 1994, **14**(4), 389–393.
40. International Organization for Standardization, *Progress in Standardization. Annex 1 ISO Common Names for Pesticides. Annex 3 ISO Common Names for Pesticides.* International Organization for Standardization, Geneva, 1965.
41. International Organization for Standardization, *International Standard 1750.* International Organization for Standardization, Geneva, 1981.
42. International Organization for Standardization, *International Standard 1750.* ISO 1750:1981/Amendment: 2008 International Organization for Standardization, Geneva, 2008. http://www.iso.org/iso/iso_catalogue/catalogue_tc/catalogue_detail.htm?csnumber=44984
43. World Health Organization, *International Non-Proprietary Names Guidance.* World Health Organization, Geneva. http://www.who.int/medicines/services/inn/en/index.html
44. Health and Safety Executive, Chemicals Regulation Directorate, *Data Requirements Handbook.* Health and Safety Executive, Bootle.
45. E. P. Meier, W. H. Dennis, A. B. Rosencrance, W. F. Randall, W. J. Cooper and M. C. Warner, Sulfotepp, a toxic impurity in formulations of diazinon, *Bull. Environ. Contam. Toxicol.*, 1979, **23**, 158–164.
46. T. Imamura, J. Gandy and L. Hasegawa, Development of tolerance to a pneumotoxic impurity of malathion, *J. Toxicol. Environ. Health*, 1985, **15**, 279–291.
47. D. Dinsdale. Pulmonary toxicity of anticholinesterases, in *Clinical and Experimental Toxicology of Organophosphate and Carbamate Compounds*, ed. B. Ballantyne and T. C. Marrs, Butterworth-Heinemann, Oxford, 1992, pp. 156–166.

CHAPTER 2

Mammalian Metabolism of Insecticides

JANICE E CHAMBERS* AND EDWARD C MEEK

Mississippi State University, Center for Environmental Health Sciences, College of Veterinary Medicine, 240 Wise Center Drive, Mississippi State, MS, USA. *Email: chambers@cvm.msstate.edu

2.1 Introduction

Synthetic insecticides have been used in crop protection since the 1950s and have been critical components of successful modern agriculture. However, there has been a transition in these last 60+ years from the compounds that are more metabolically and chemically stable, and therefore are more likely to be persistent and bioaccumulative in the environment, to compounds that are metabolically and chemically unstable and therefore are not persistent and bioaccumulative in the environment. Shifts in usage in developed nations from the stable organochlorine insecticides to the unstable classes such as organophosphates, *N*-methyl carbamates and synthetic pyrethroids has occurred over the years because of the bioaccumulative character of the organochlorine insecticides. In addition, shifts from compounds that are more toxic to people and other mammals to those that are less toxic have occurred to provide a greater margin of safety to farm workers and the general public and their commodity animals as well as animals in the ecosystems. Both of these types of usage shifts have been dependent, at least in part, on metabolism. Clearly the legacy organochlorine insecticides were not well metabolized, which led to their bioaccumulation, and, in fact, many of these legacy organochlorine insecticides

Issues in Toxicology No. 12
Mammalian Toxicology of Insecticides
Edited by Timothy C. Marrs

Published by the Royal Society of Chemistry, www.rsc.org

(or a few of their stable metabolites) are frequently found in people and animals today. Some of the insecticides used to replace the organochlorine insecticides had high mammalian toxicity levels, such as some of the early organophosphates, and were in turn replaced by insecticides with lower mammalian toxicity levels, and the lower toxicities were frequently the result of rapid and effective detoxication in mammals.

This chapter will describe the metabolism of the major groups of synthetic insecticides which have been developed and used over the last 6–7 decades. The major xenobiotic metabolizing enzymes involved in insecticide metabolism will be described. Examples of the metabolism of representative examples of each of the major use classes will be provided. Many of these metabolic pathways were elucidated many decades ago when the initial insecticides in the use classes were developed. The reader is referred to three foundational references which detail the metabolic pathways of many of the insecticides: Aizawa[1,2] and Dikshith.[3]

2.2 Overview of Major Xenobiotic Metabolizing Enzymes Important in the Mammalian Metabolism of Insecticides

As is true for most xenobiotics, the insecticides, even the modern ones, need to have some lipophilic character in order to readily cross biological membranes and to be absorbed into the body from any exposure route. If these insecticides are suitable substrates for the xenobiotic metabolizing enzymes, these enzymes can convert the lipophilic (and potentially bioaccumulative insecticides) into polar entities that can then be eliminated through the urinary or biliary (faecal) routes.

The xenobiotic metabolizing enzymes are generally divided into two overall classes, phase 1 and phase 2. The majority of phase 1 reactions are oxidations, reductions or hydrolyses, with oxidations and hydrolyses being the most important two reaction categories in mammals (reductions are infrequent in mammalian tissues because of the need for a reducing environment to carry them out). In general most phase 1 reactions add or uncover a reactive group on the parent insecticide, making the phase 1 metabolite more polar and therefore more likely to be excreted, although the metabolite may not be hydrophilic enough to partition effectively into the water of the urine or faeces for appreciable excretion. The presence of this new reactive moiety can make the phase 1 metabolite amenable to further metabolism, but this moiety may also make the phase 1 metabolite more biologically reactive, creating the reactive metabolite which induces the acute toxicity characteristic of the insecticide. The phase 2 reactions add an endogenous ligand to their substrates at a site containing a suitable reactive moiety. In the majority of cases the phase 2 metabolite is water soluble and can be excreted in the urine or bile (to enter the faeces). The substrates for the phase 2 reactions must contain a moiety that allows reaction with the endogenous ligand. In many cases this moiety is provided by the phase 1 reactions, so a sequence of phase 1 and phase 2 reactions is needed for the parent insecticide and its metabolites to be cleared from the body. In some cases the parent insecticide contains such a moiety and can be a substrate for the phase 2 reactions without undergoing a phase 1 reaction. Highest activities of the xenobiotic metabolizing enzymes occur in the liver in mammals. The reader is referred to the following references for a more detailed overview of xenobiotic metabolism and for more detailed discussion of the mechanism of toxicity of the various insecticide classes: Klaassen[4] and Smart and Hodgson.[5] Acute toxicity levels are taken from Meister.[6]

To be effective in pest control, insecticides must act quickly for fast and effective insect kill, so the common insecticides induce a neurotoxic effect that rapidly incapacitates the insect. In many cases the same neurochemical target (*i.e.* enzyme, receptor, channel) that mediates the toxic action in insects is also present in mammals and can be impacted in much the same way. The majority of common insecticides are either acutely toxic as the parent insecticide or have a reactive metabolite from a single phase 1 reaction (and these are primarily a subset of the organophosphate insecticides). Therefore the majority of the phase 1 reactions (with the exception of some bioactivations on organophosphates) are detoxications and the phase 2 reactions are detoxications. In the majority of cases the phase 1 reactions are the important ones from a toxicological standpoint in that they either are the initial detoxication step or they provide the reactive metabolite that induces the primary acute toxicity. The phase 2 reactions can make the insecticides/metabolites more readily excretable, but usually do not mediate the initial detoxication.

2.2.1 Phase 1 Reactions

Of primary importance among the phase 1 reactions are the oxidations catalysed by cytochromes P450 (CYPs). The CYPs constitute a family of enzymes that accommodate a wide variety of exogenous and endogenous substrates. They are monooxygenases and catalyse the addition of one atom of molecular oxygen into the substrate followed by rearrangements to yield several types of metabolites, depending upon the structure of the substrate. Among the most common types of reactions relevant to insecticides catalysed by the CYPs are: hydroxylations of aromatic and aliphatic moieties; *N*-, *O*-, and *S*-dealkylations; epoxidations of carbon-carbon double bonds; desulfuration, where a P=S is converted to P=O; and sulfoxidations of sulfur ethers to sulfoxides and subsequently to sulfones. The latter three reactions are usually bioactivations. The CYPs require reduced nicotine adenine dinucleotide phosphate (NADPH) and molecular oxygen.

Oxidations are also catalysed by the flavin monooxygenases (FMOs), though on a more limited scale than by the CYPs. The FMOs also require NADPH and molecular oxygen but they have more limited substrate specificity than the CYPs.

Another important group of phase 1 reactions are hydrolyses which are usually detoxication reactions. Carboxylesterases catalyse the hydrolysis of carboxylic esters, yielding a carboxylic acid and an alcohol as products. A-esterases (frequently called paraoxonase, PON) catalyse the hydrolysis of phosphotriesters, yielding a phosphoric acid diester and an alcohol as products. Epoxide hydratases (hydrolases) catalyse the hydration of epoxides, yielding a dihydrodiol as a product. Water is the only cofactor required by hydrolases, so a cofactor requiring the input of metabolic energy is not needed for this group of reactions, in contrast to the other phase 1 reactions and the phase 2 reactions.

A rather specialized phase 1 reaction is that catalysed by DDT dehydrochlorinase, which removes hydrogen chloride from DDT, creating a carbon-carbon double bond, and yielding DDE as a product (discussed further below). The reaction requires glutathione which is not depleted in the reaction and is not a glutathione-conjugating phase 2 reaction, as described below.

2.2.2 Phase 2 Reactions

Conjugation reactions whereby an endogenous ligand is added to the substrate constitute the phase 2 reactions. As indicated above, the phase 2 enzymes can add the endogenous ligand to a reactive group when such occurs in the parent insecticide, or it can react with a phase 1 metabolite that has had a reactive group produced by the phase 1 enzymes. Two important phase 2 reactions are sulfation, where the substrate is conjugated to sulfate from the cofactor phosphoadenosylphosphosulfate (PAPS), catalysed by sulfotransferases, and glucuronidation, where the substrate is conjugated to glucuronic acid from the cofactor uridine diphosphoglucuronic acid (UDPGA), catalysed by glucuronosyl transferases. A third important phase 2 reaction is mercapturic acid formation, which involves conjugation of the substrate to the tripeptide glutathione through cysteine's sulfhydryl group catalysed by glutathione *S*-transferases, followed by the removal of glutamic acid by a γ-glutamyltranspeptidase, removal of glycine by a glycinase, and acetylation of cysteine by an acetylase, yielding the final metabolite which is a mercapturic acid. All of the above metabolites, *i.e.* sulfates, glucuronides and mercapturic acids, are water soluble and readily excreted.

In the following sections, metabolism within the major classes of synthetic insecticides will be discussed, along with examples of the structures within these classes. Only phase 1 reactions will be shown, because these are the most toxicologically relevant reactions.

2.3 Organochlorine Insecticides

There are two very important major classes within the organochlorine insecticides, the chlorinated diphenylethane class which includes the well-known insecticide DDT that has been very widely used, and the chlorinated cyclodiene class which includes a number of compounds (*e.g.* chlordane, dieldrin) that have also been extensively used. All insecticides in these two classes are lipophilic and generally poorly metabolized, which has led to global bioaccumulation of some of these insecticides or their stable metabolites. These compounds were developed early in the history of synthetic insecticides, used widely because of their efficacy and frequently their low cost, and most were restricted or phased out by developed countries because of the ability of the compounds to bioaccumulate and because of concerns about their toxicity. While a number of metabolites have been identified for these compounds, in general the production of metabolites is slow and ineffective compared to the other classes of insecticides that will be discussed.

2.3.1 Chlorinated Diphenylethane Insecticides

The chlorinated diphenylethane class comprises a relatively small group of insecticides, but contains what is probably the most famous of all synthetic insecticides, DDT (1,1,1-trichloro-2,2-bis(*p*-chlorophenyl)ethane, sometimes unofficially called dichlorodiphenyltrichloroethane from which the term DDT arose). The class also includes dicofol (which is a metabolite of DDT, but also a pesticide in its own right) and methoxychlor. The acute neurotoxic action of DDT results from the prolonged opening of neuronal sodium channels, which leads to hyperexcitability within the nervous system.

2.3.1.1 DDT

The commercial product called DDT included two major isomers, *p,p′*-DDT and *o,p′*-DDT, in an 80%/20% mixture. *p,p′*-DDT (abbreviated as DDT throughout this discussion) is more neurotoxic than *o,p′*-DDT and also more neurotoxic than its metabolites; since it is also more abundant, its metabolism will be discussed. Metabolism seems to be restricted to the ethane and not the rings, and was elucidated in the mouse.[7,8] The metabolic scheme for DDT is illustrated in Figure 2.1.

One of the most important metabolic reactions of DDT is the conversion of DDT to DDE through the removal of hydrogen chloride by DDT dehydrochlorinase, a glutathione-dependent reaction that occurs quite readily, in contrast to most of the other reactions occurring with organochlorine insecticides. DDE is more lipophilic than DDT and is the main bioaccumulative compound in the DDT family (DDT and its metabolites), being found frequently in blood and fat stores of people and animals throughout the world. An additional metabolite TDE can be formed from either DDT itself through dechlorination, or from DDE through reduction of the double bond. From TDE an additional dechlorination can occur, followed by oxidative dechlorination to the acid DDA, which is readily excreted. An additional pathway of metabolism involves hydroxylation to dicofol, which has itself been used as an acaricide. However, dicofol is considerably less toxic than DDT (rat oral LD_{50}s of 820–960 and 113 mg kg^{-1}, respectively), so this pathway is also a detoxication.

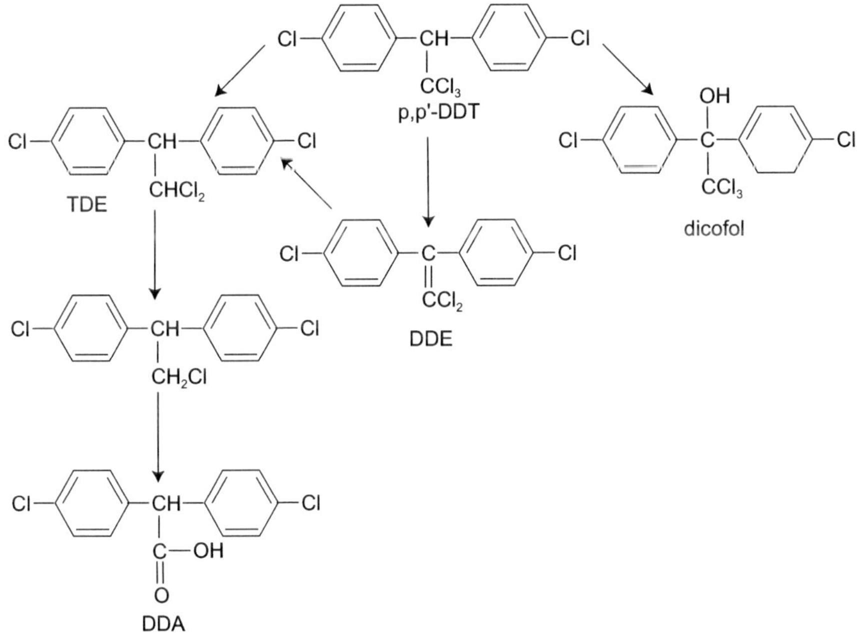

Figure 2.1 Metabolism of DDT.

2.3.1.2 Methoxychlor

Methoxychlor has methoxy groups substituted for the chlorines on the two phenyl rings of DDT. These methoxy substitutions make methoxychlor readily metabolized[9] and therefore relatively non-toxic to mammals (rat oral LD_{50} 6000 mg kg^{-1}) and not likely to bioaccumulate. These two methoxy groups can be sequentially removed by *O*-demethylation to yield a metabolite hydroxyphenyltrichloroethane (HPTE) which has been shown to resemble 17β-estradiol and therefore displays oestrogenic properties. HPTE can be dehydrochlorinated, followed by oxidative dechlorination, similar to what is observed with DDT. Lastly a decarboxylation, followed by oxidation of the carbon bridging the two rings can occur, yielding *p,p′*-dihydroxybenzophenone (Figure 2.2).

2.3.2 Chlorinated Cyclodiene Insecticides

The chlorinated cyclodienes are a relatively large group of compounds that are also highly lipophilic, refractory to most metabolism, and therefore bioaccumulative. This group of compounds is characterized by a fully chlorinated cyclohexane ring with one double bond and a chlorinated endomethylene bridge. A second ring is bonded to the cyclohexane ring, and it may contain oxygen, chlorine, sulfur or another endomethylene bridge. These compounds are neurotoxic and antagonize the γ-aminobutyric acid (GABA) receptor so suppress chloride flux through the GABA receptor ionophore.[10] Through

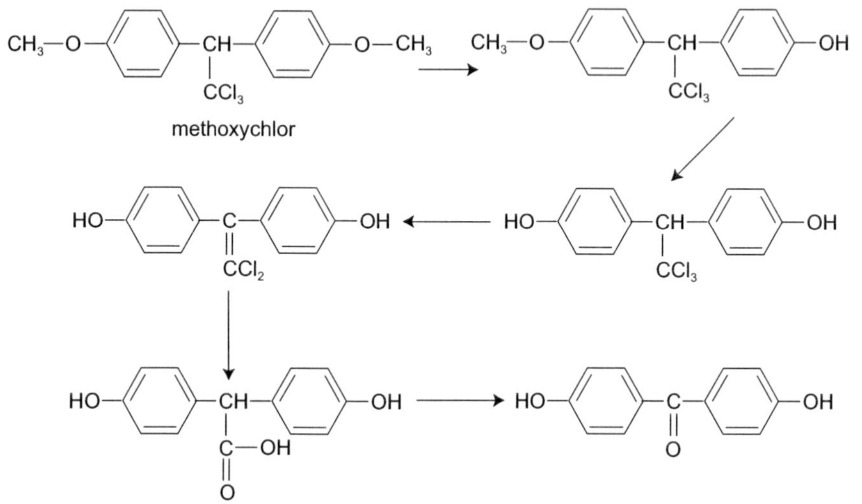

Figure 2.2 Metabolism of methoxychlor.

inhibition of the action of the inhibitory neurotransmitter GABA, excitatory input in the nervous system predominates and hyperexcitability occurs. The metabolism of a few important representatives of the chlorinated cyclodiene class of insecticides is provided.

2.3.2.1 Aldrin/Dieldrin

Aldrin ((1*R*,4*S*,4a*S*,5*S*,8*R*,8a*R*)-1,2,3,4,10,10-hexachloro-1,4,4a,5,8a-hexahydro-1,4:5,8-dimethanonaphthalene) is epoxidized to dieldrin ((1*R*,4*S*,4a*S*,5*R*,6*R*,7*S*,8*S*,8a*R*)-1,2,3,4,10,10-hexachloro-1,4,4a,5,6,7,8,8a-octahydro-6,7-epoxy-1,4:5,8-dimethanonaphthalene) (Figure 2.3).[11,12] Dieldrin was also used as a commercial insecticide. The rat oral LD_{50}s for aldrin and dieldrin are very similar (38–67 mg kg^{-1} and 37–87 mg kg^{-1}, respectively), which is somewhat surprising since epoxides are usually more biologically reactive than their parent compounds that possess a carbon-carbon double bond, but it is possible that the aldrin epoxidation reaction is facile and the *in vivo* toxicity tests reflect dieldrin toxicity for both aldrin and dieldrin. Dieldrin can be hydroxylated at the methylene bridge,[13] followed by epoxide hydration.[14,15] Photodieldrin can be formed in mammals by further cyclization of dieldrin.[12] Lastly, oxidation at the methylene bridge of dieldrin can lead to cyclization to dieldrin pentachloroketone.[16]

2.3.2.2 Endrin

Endrin is an isomer of dieldrin, but considerably more toxic (rat oral LD_{50} 7–15 mg kg^{-1}). Hydroxylation can occur at three locations (Figure 2.4).[17] A further oxidation product, 12-ketoendrin, can be formed which is very toxic

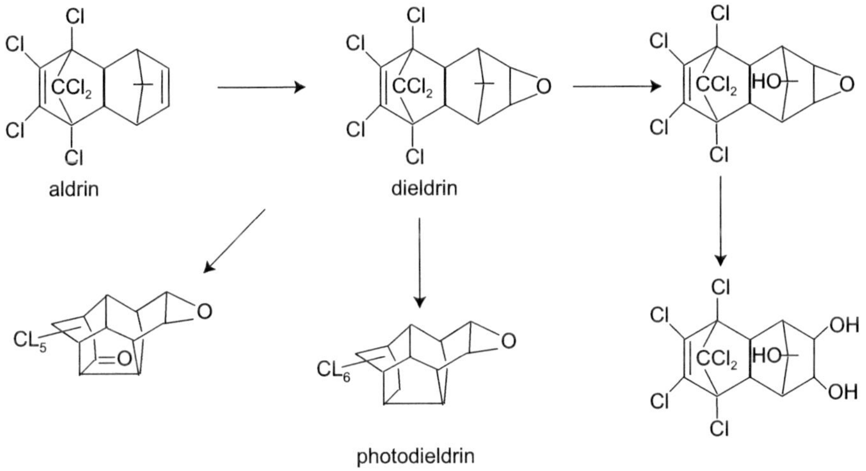

Figure 2.3 Metabolism of aldrin and dieldrin.

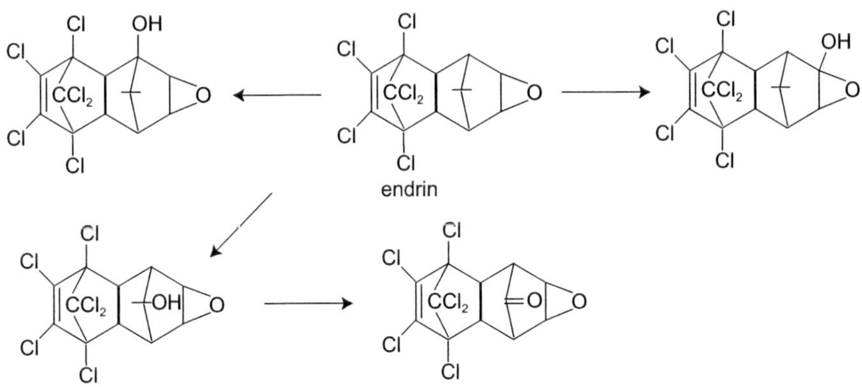

Figure 2.4 Metabolism of endrin.

(LD_{50}, 1 mg kg^{-1}) and which is a very potent GABA receptor antagonist,[10,18,19] so even though 12-ketoendrin is a minor metabolite, its formation might be extremely important in the overall high toxicity level of endrin.

2.3.2.3 Chlordane

Chlordane (1,2,4,5,6,7,8,8-octachloro-2,3,3a,4,7,7a-hexahydro-4,7-methanoindane) exists in *cis* and *trans* isomers, which have slightly different metabolic pathways (Figure 2.5). The most likely metabolism of *cis*-chlordane is through two hydroxylations in which two of the chlorines are removed, and probably involves some unknown intermediates. The most likely metabolism of *trans*-chlordane is

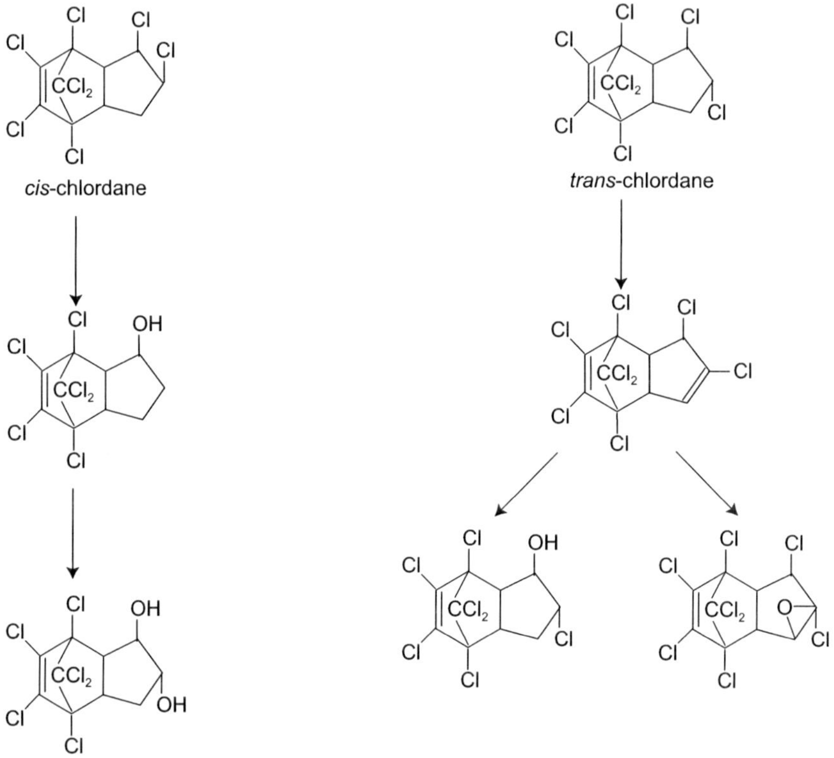

Figure 2.5 Metabolism of chlordane.

dehydrogenation followed by epoxidation of the resultant double bond.[20] Lastly, both the *cis* and *trans* isomers can be dehydrochlorinated to heptachlor, which has been used as an insecticide in its own right (Figure 2.6).[20] Since heptachlor is more toxic than technical chlordane (rat oral LD_{50}s, 147–220 mg kg^{-1} and 367–515 mg kg^{-1}, respectively), conversion of chlordane to heptachlor could be considered bioactivation.

2.3.2.4 Heptachlor

Epoxidation and hydroxylation are the primary metabolic pathways for heptachlor (1,4,5,6,7,8,8-heptachloro-3a,4,7,7a-tetrahydro-4,7-methanoindene) (Figure 2.6).[21] Because epoxides are typically more biologically reactive than the parent carbon-carbon double bonds, the formation of heptachlor epoxide may be the reason that heptachlor is more toxic than chlordane.

2.4 Organophosphorus Insecticides

As environmental and health concerns arose about the organochlorine insecticides and decisions were made to phase these out, the organophosphorus (OP)

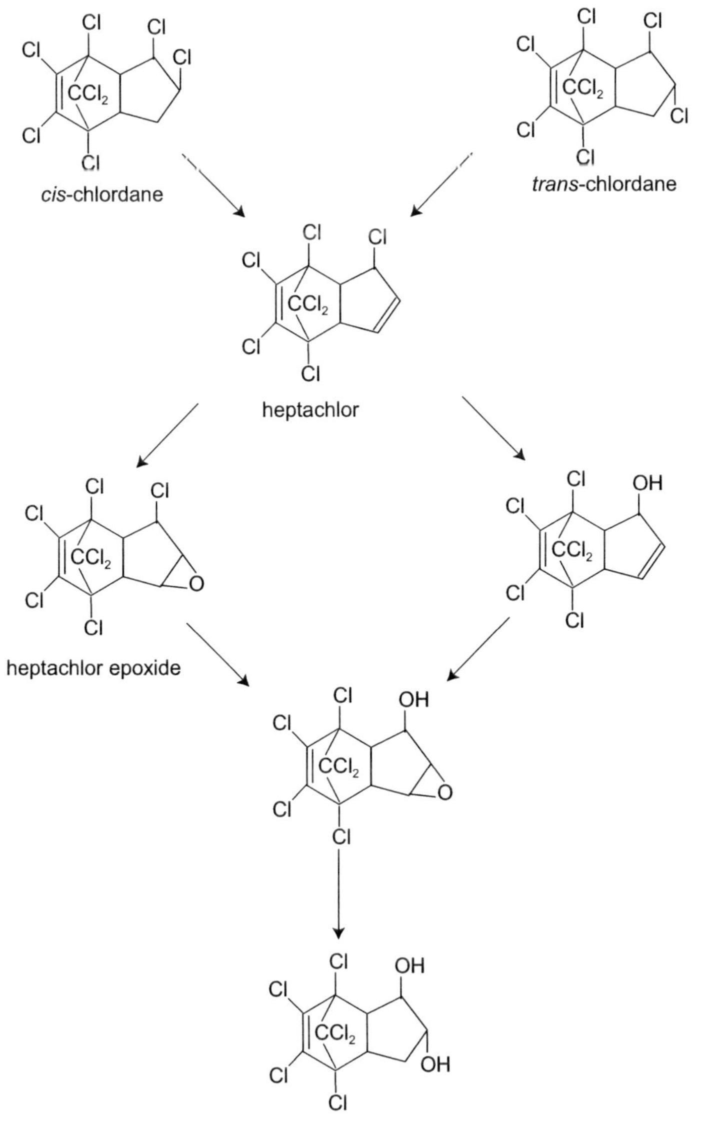

Figure 2.6 Metabolism of heptachlor.

insecticides arose as a prominent class for replacement. The organophosphorus insecticides (frequently called OPs or organophosphates for convenience, although technically organophosphate refers to only a subset of the class) arose from the chemical technology of the Second World War in which nerve agents (gases) were developed as chemical warfare agents. The OP insecticides are, as parent compounds or active metabolites, potent and persistent inhibitors of acetylcholinesterase, the critical nervous system enzyme that degrades the neurotransmitter acetylcholine. The subsequent accumulation of acetylcholine leads to hyperexcitability in the widely distributed cholinergic pathways of the central and peripheral nervous systems, and results in a wide variety of signs of poisoning. The OP insecticides are metabolically labile, so even those that are quite lipophilic do not bioaccumulate. One serious unintended consequence of the replacement of some of the relatively safe organochlorine insecticides with some of the early highly toxic OP insecticides was the accidental poisoning of agricultural workers who were not aware of the hazard of these compounds. Such chemical accidents led to the development of less toxic OP insecticides, with the lower toxicity frequently the result of very efficient mammalian detoxication mechanisms. The OP insecticide class is large and diverse, representing a variety of aliphatic, aromatic and heterocyclic structures bonded to a central phosphorus atom, and a variety of toxicity levels exist within the class.[22]

2.4.1 Parathion

Parathion (*O,O*-diethyl *O*-4-nitrophenyl phosphorothionate) is one of the earliest of the OP insecticides developed, and is one of the most toxic (rat oral LD_{50} 2–22 mg kg^{-1}). It is a phosphorothionate characterized by a P=S bond, and is a relatively weak anticholinesterase. However it is readily metabolized by a desulfuration reaction to the active oxon metabolite, paraoxon, characterized by a P=O bond. This reaction is mediated by CYPs through an unstable phosphooxythiiran intermediate which breaks down non-enzymatically to yield paraoxon (Figure 2.7).[23] The resultant oxon metabolite is a potent anticholinesterase and mediates the acetylcholinesterase inhibition induced by parathion. The unstable phosphooxythiiran intermediate may also degrade through a second pathway, termed oxidative dearylation (oxidative cleavage), to yield two detoxified products, diethyl phosphoric acid/diethyl phosphorothioic acid plus 4-nitrophenol. An additional, but less important, CYP-mediated reaction with parathion is *O*-dealkylation to yield desethylparathion and acetaldehyde,[24] and this is a detoxication. Lastly the nitro substitution on the ring of parathion can be reduced, yielding aminoparathion,[25] which is also a detoxication reaction. However, despite the several detoxication options, the high acute toxicity level of parathion suggests that a large fraction of parathion is converted to paraoxon. A detoxication pathway for paraoxon exists, through the calcium-dependent hydrolysis catalysed by A-esterases (paraoxonase), to yield diethyl phosphate plus 4-nitrophenol; however, despite its name paraoxonase is not as efficient in hydrolysing paraoxon as it is for some other oxons (such as chlorpyrifos-oxon and diazoxon, the active metabolites of chlorpyrifos and diazinon, respectively), contributing to the high acute toxicity of parathion (and the lower toxicity of both chlorpyrifos and diazinon compared to parathion).

To illustrate some of the diversity of structures within the OP insecticide class and the bioactivation of some of the more prominent of the phosphorothionates, Figure 2.8 shows the desulfuration reactions of parathion and parathion-methyl (both highly toxic), malathion (very low toxicity) and chlorpyrifos and diazinon (moderate toxicity levels).

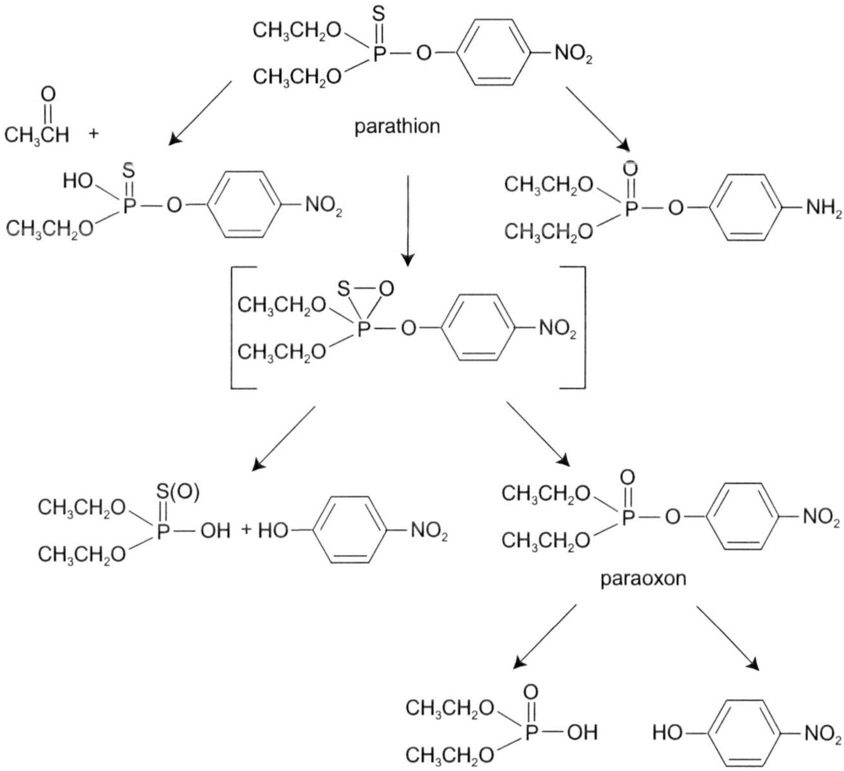

Figure 2.7 Metabolism of parathion.

2.4.2 Diazinon

Another common OP insecticide which displays a moderate acute toxicity level (rat oral LD_{50} 300–400 mg kg^{-1}) is diazinon (*O,O*-diethyl *O*-(2-isopropyl-4-methyl-6-pyrimidinyl)phosphorothioate). It can undergo bioactivation through desulfuration and detoxication through oxidative cleavage. The parent insecticide can also be detoxified through hydroxylation of the aliphatic constituents on the pyrimidine ring, and the active metabolite diazoxon can be hydrolysed by A-esterases (Figure 2.9).[26–28]

2.4.3 Malathion

One of the least toxic synthetic insecticides of any chemical class is malathion (rat oral LD_{50} 1000–1375 mg kg^{-1}) (*O,O*-dimethyl phosphorodithioate of diethyl mercaptosuccinate). Malathion possesses two carboxylic acid esters which are highly amenable to hydrolysis by carboxylesterases. Malathion can be bioactivated to its oxon, malaoxon; either the parent malathion or malaoxon can be hydrolysed, and therefore detoxified, to the α and β

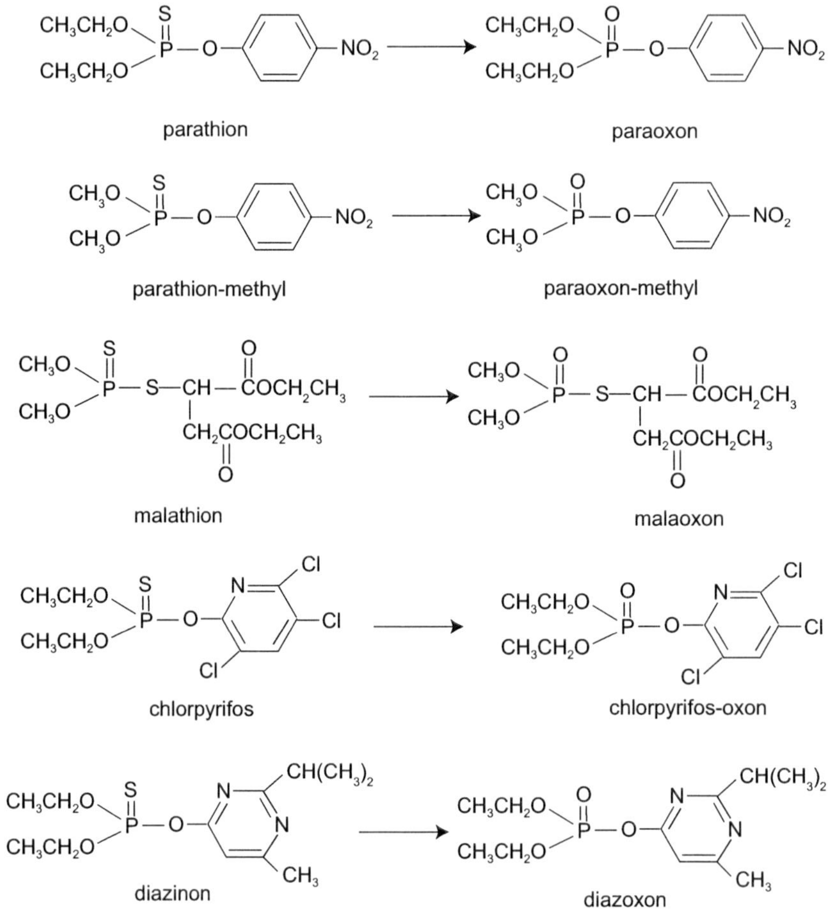

Figure 2.8 Desulfuration of parathion, methyl parathion, malathion, chlorpyrifos and diazinon.

monoacids and subsequently to the diacid (Figure 2.10).[29,30] Because mammals have very efficient carboxylesterases and insects do not, malathion is a highly selective insecticide with high toxicity to insects and very low acute toxicity to mammals. In addition to oxidative cleavage similar to the compounds discussed above (and not shown in the figure for simplicity), an additional reaction is glutathione transferase-mediated demethylation, also a detoxication.

2.5 *N*-Methylcarbamate Insecticides

The *N*-methylcarbamates are also a group of easily metabolized anti-cholinesterases, with a diversity of structures and a range of acute toxicity levels.

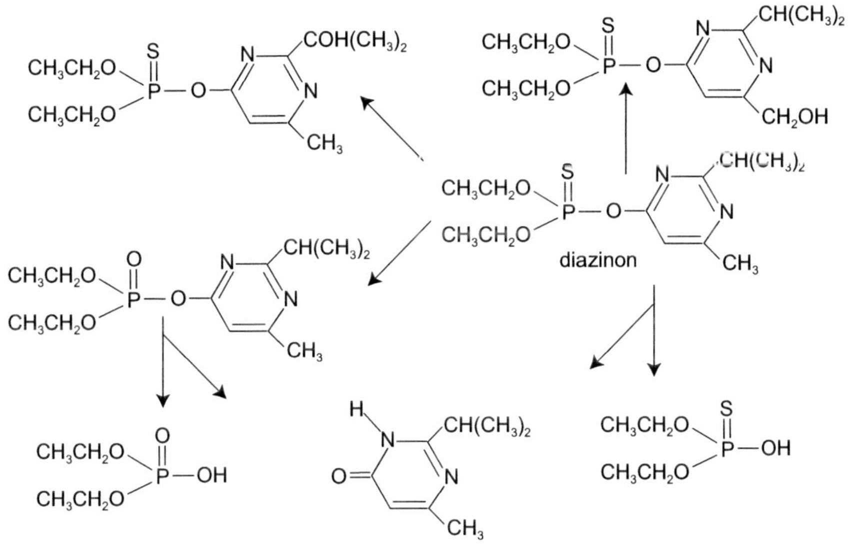

Figure 2.9 Metabolism of diazinon.

They elicit similar signs of toxicity to the organophosphorus insecticides, but in contrast to the OP insecticides which induce a persistent inhibition of acetylcholinesterase, the carbamates inhibit acetylcholinesterase only transiently. Because of their metabolic lability, the carbamates do not bioaccumulate. The parent insecticides are the anticholinesterases, so metabolism leads to detoxified products.

2.5.1 Carbaryl

Carbaryl (1-naphthyl *N*-methylcarbamate) is a widely used insecticide of moderate mammalian toxicity (rat oral LD_{50} 246–283 mg kg^{-1}). Aromatic ring hydroxylation (directly or following ring epoxidation), aliphatic hydroxylation and hydrolysis of the carbamate bond all occur leading to a variety of metabolites that are all less toxic than the parent insecticide (Figure 2.11).[31,32]

2.5.2 Carbofuran

Carbofuran (2,3-dihydro-2,2-dimethyl-7-benzofuranyl methylcarbamate) is a highly toxic insecticide (rat oral LD_{50} 11 mg kg^{-1}). It is metabolized to a variety of metabolites through furan ring oxidation, aliphatic oxidation and carbamate hydrolysis although it does not seem to experience aromatic ring hydroxylation (Figure 2.12).[2,31,33] Even though there are numerous detoxified metabolites identified, carbofuran displays a very high level of acute toxicity.

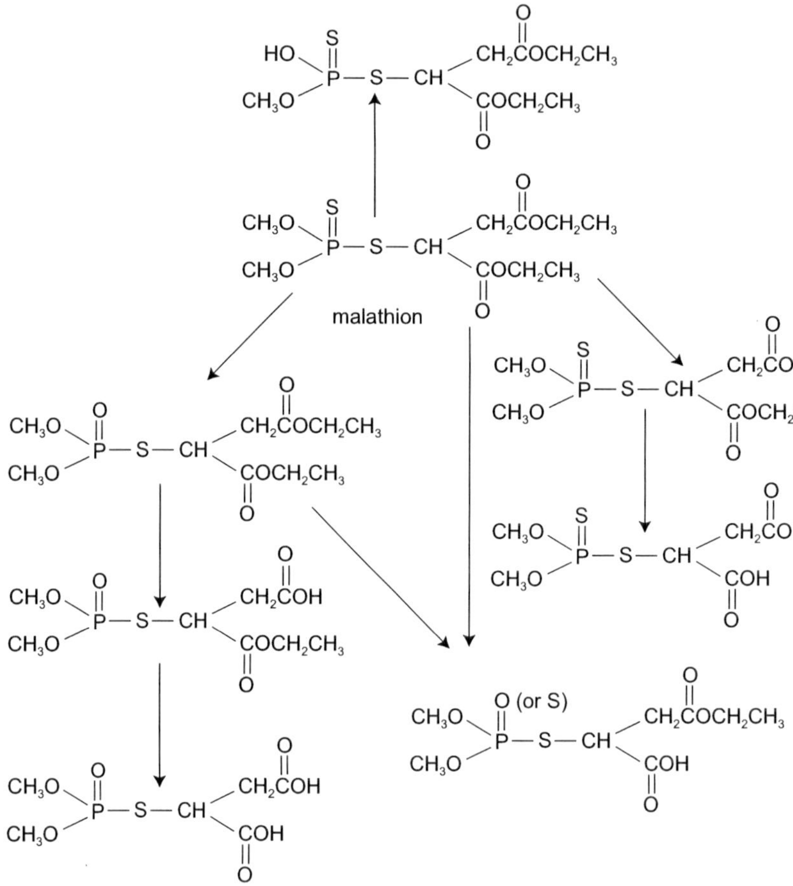

Figure 2.10 Metabolism of malathion.

2.6 Pyrethroid Insecticides

The pyrethroids are a class of insecticides that have both natural components, derived from the flowers of *Chrysanthemum cinerariaefolium*, and synthetic derivatives whose chemistries are based only very loosely on the chemistries of the natural products. The natural pyrethroids have been used for many years in a flower extract called pyrethrum, which has six active principles: pyrethrins I and II, cinerins I and II, and jasmolins I and II. Pyrethrum has a low mammalian toxicity (rat oral LD_{50} 1500 mg kg^{-1}). Because pyrethrum components are so environmentally labile, and therefore the residual efficacy of pyrethrum is so short, the synthetic pyrethroids were synthesized to provide more environmental stability and therefore a more prolonged field efficacy. The pyrethroids have the same acute toxicity mechanism of action as DDT, the prolonged opening of neuronal sodium channels, despite having structures that do not resemble DDT at all.

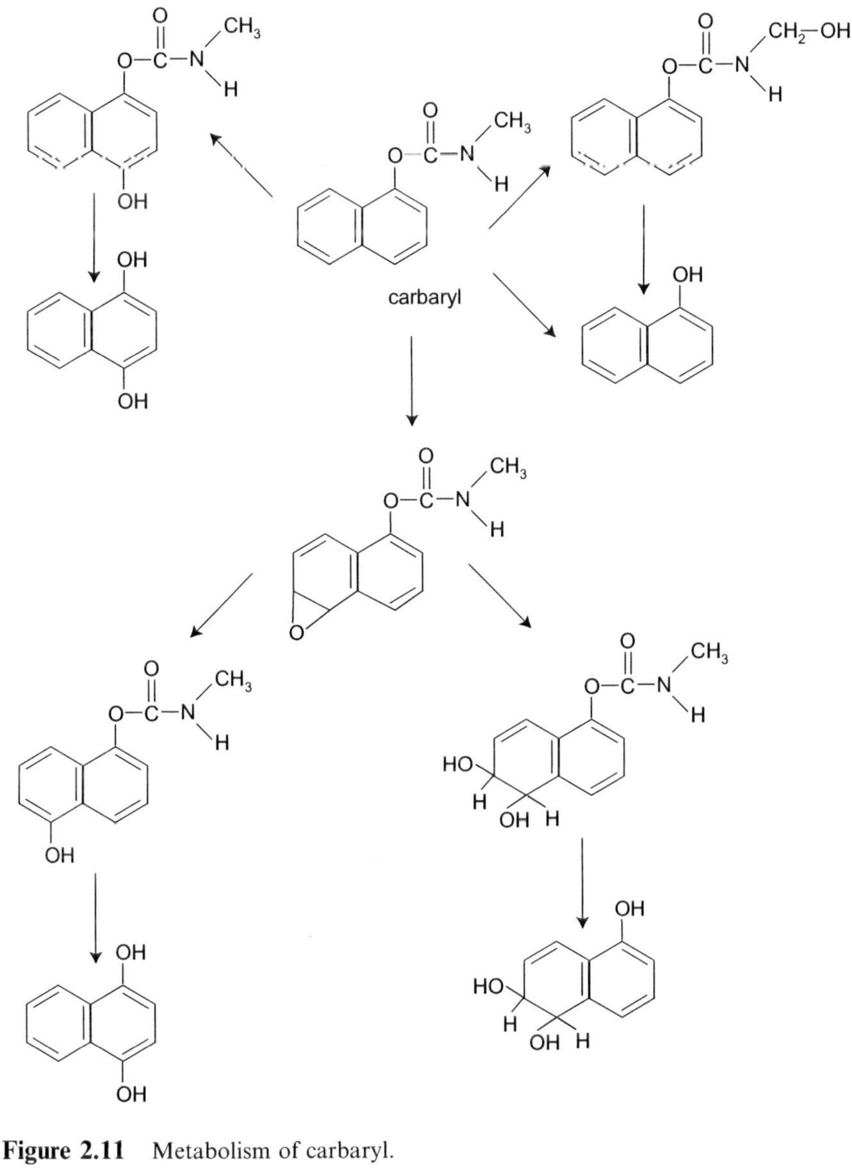

Figure 2.11 Metabolism of carbaryl.

2.6.1 Pyrethrins I and II

Pyrethrins I and II are derived from the esterification of pyrethrolone with chrysanthemic acid and pyrethroic acid, respectively. Therefore the pyrethrins have an ester bond in the interior of the molecules, but curiously it is not readily hydrolysed (Figure 2.13).[34] However the exterior ester bond of pyrethrin II is hydrolysed and several aliphatic oxidations of pyrethrin I and pyrethrin II occur.

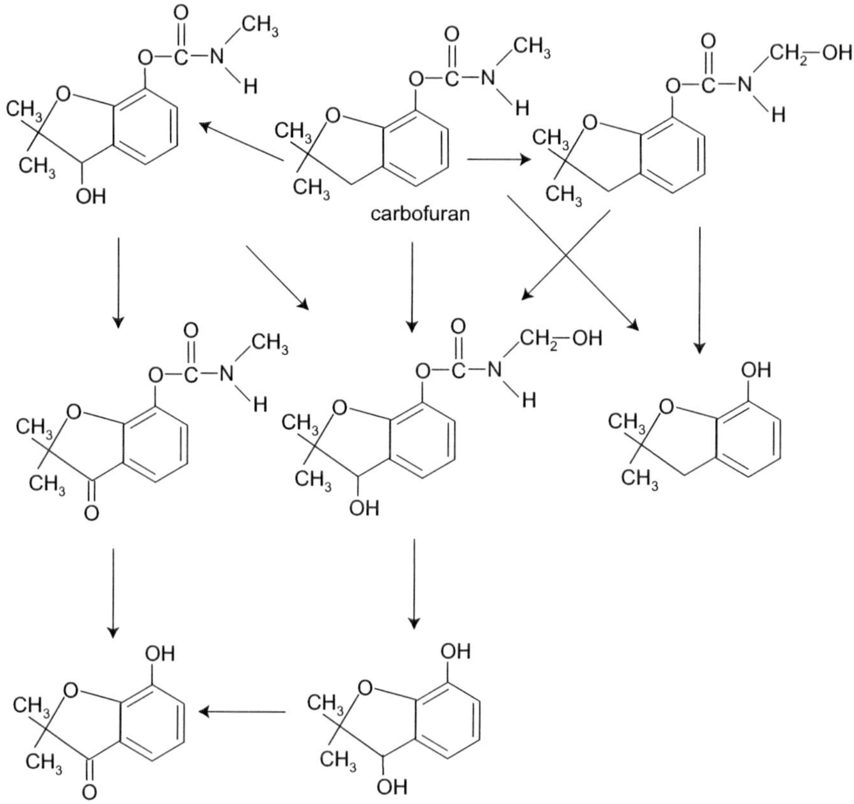

Figure 2.12 Metabolism of carbofuran.

2.6.2 Permethrin

One example of a synthetic pyrethroid is permethrin (3-phenoxybenzyl (1*RS*,3*RS*; 1*RS*,3*SR*)-3-(2,2-dichlorovinyl)-2,2-dimethylcyclopropanecarboxylate), whose technical mixture includes 25–40% of the *cis* isomer and 60–75% of the *trans* isomer. Permethrin displays a low acute toxicity level (rat oral LD_{50} 430–4000 mg kg^{-1}). The metabolism of the major isomer, *trans*-permethrin, includes hydrolysis of the internal ester bond (in contrast to pyrethrins I and II), and several aliphatic and aromatic oxidations, making permethrin highly labile and contributing to the low mammalian toxicity (Figure 2.14).[35]

2.7 Neonicotinoid Insecticides

While the above classes of insecticides have been in use for many years, a more recent class of insecticides has been developed, termed the neonicotinoids, to distinguish them from nicotine, a natural highly toxic compound from tobacco with historical but not modern usage. Neonicotinoids stimulate the postsynaptic

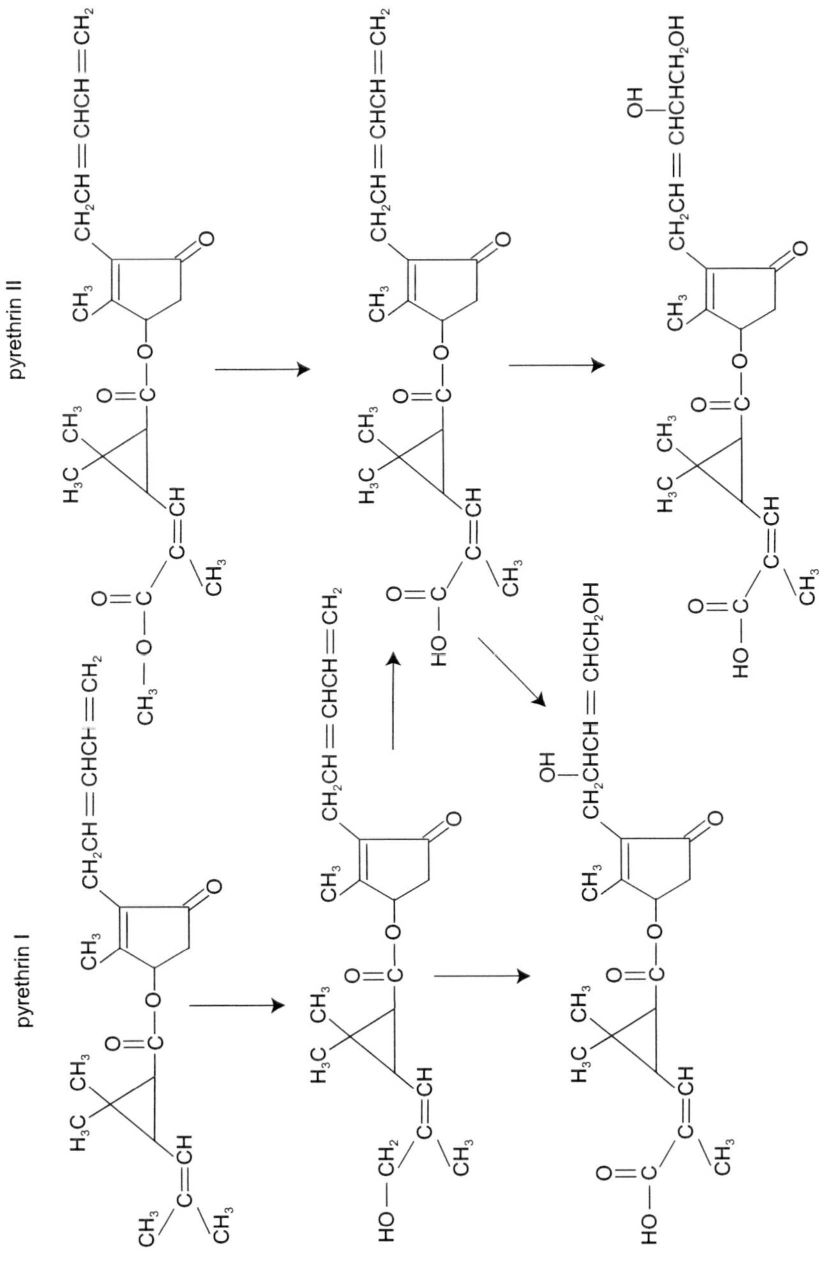

Figure 2.13 Metabolism of pyrethrins I and II.

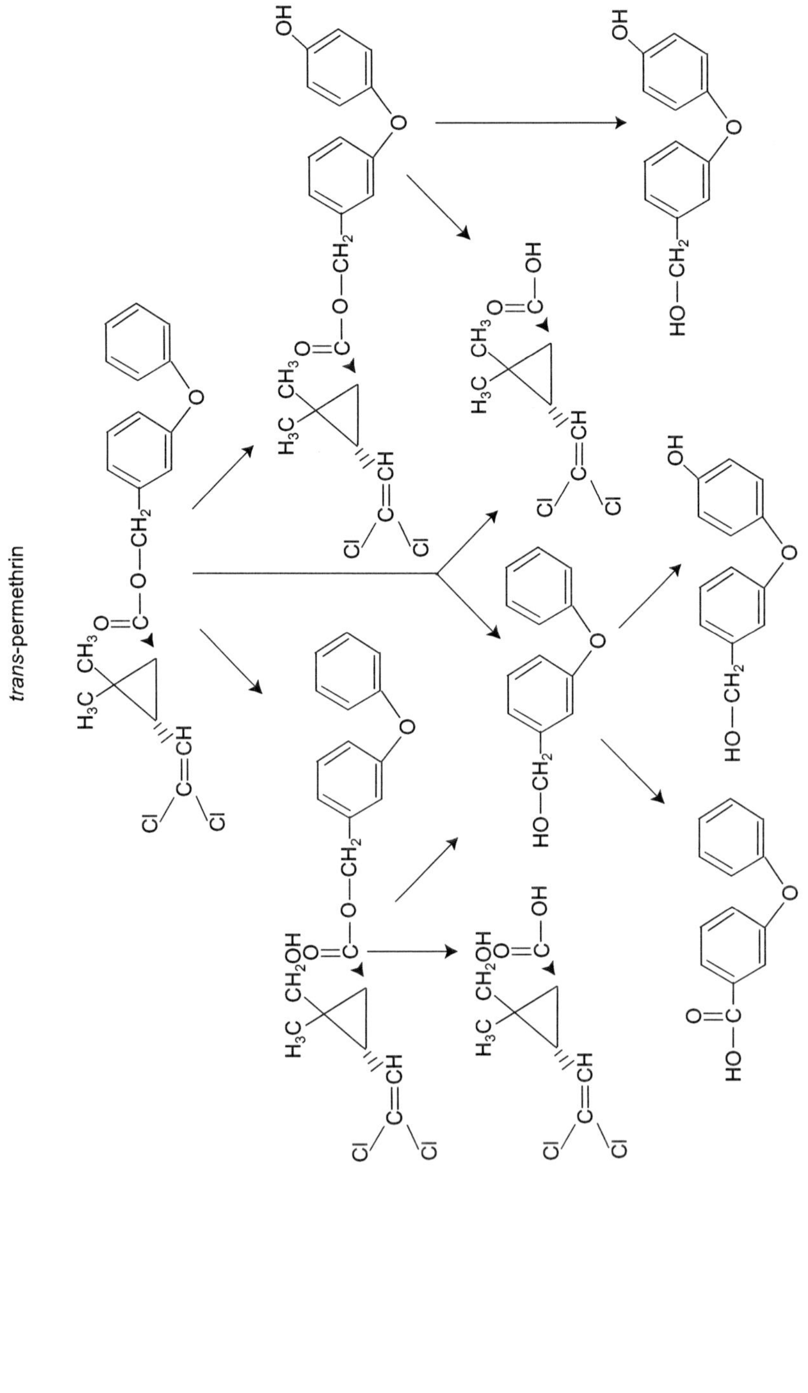

Figure 2.14 Metabolism of *trans*-permethrin.

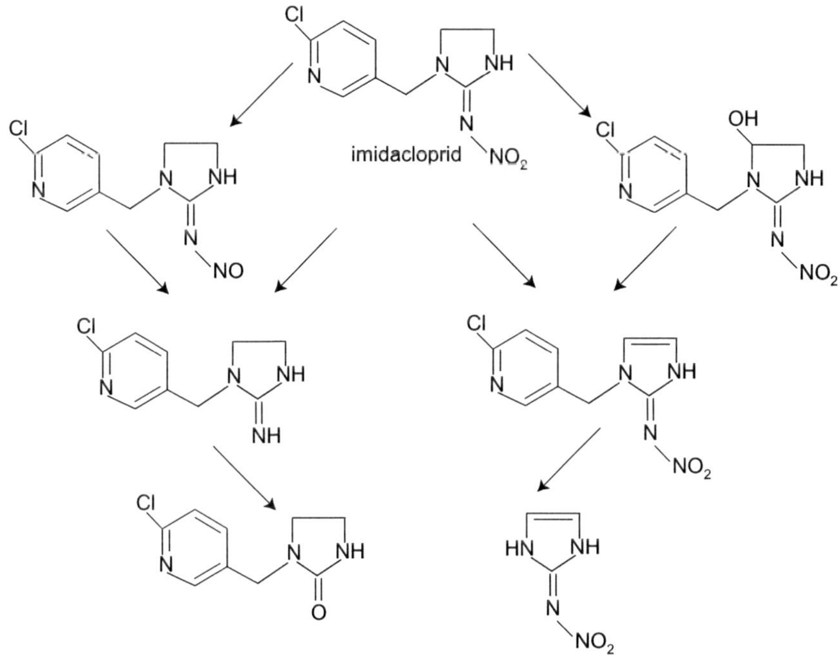

Figure 2.15 Metabolism of imidacloprid.

nicotinic acetylcholine receptors in the insect central nervous system, and are highly toxic to insects but not to mammals, and therefore are effective at low application rates. The neonicotinoids are at present heavily used globally.

2.7.1 Imidacloprid

Imidacloprid (1-(6-chloro-3-pyridylmethyl)-*N*-nitroimidazolidin-2-ylideneamine) is a widely used representative of the neonicotinoid class and has a relatively low mammalian toxicity (rat oral LD_{50} 380–500 mg kg^{-1}). The nitro group can be reduced and then eliminated, with a further deamination. The imidazole ring can be hydroxylated, and it can have the imidazole ring dehydrated to a carbon-carbon double bond, following which the imidazole ring can be cleaved (Figure 2.15).[36] The parent insecticide can have a dehydrogenation of the imidazole ring to the same carbon-carbon double bond, followed by imidazole ring cleavage.

2.8 Conclusion

As can be seen from the above discussion, there are many diverse chemistries among the historic and the current insecticides, and therefore a variety of metabolic reactions that are possible. While many metabolites of the legacy organochlorine insecticides have been identified, for the most part these metabolic pathways are not efficient and these insecticides or a few stable metabolites have bioaccumulated in the environment. This recalcitrance of the organochlorine insecticides to be metabolized by, and therefore cleared from, organisms has led to the restriction or banning of most of the compounds in the group, and the replacement by more metabolically labile insecticides. Therefore the insecticides in wide use today, *i.e.* the organophosphorus, *N*-methylcarbamate, pyrethroid and neonicotinoid classes, are readily metabolized and are not bioaccumulative. While a few important bioactivations occur among these diverse metabolic steps, most of these metabolic steps are detoxications, rendering the insecticides less toxic and more readily excreted.

References

1. H. Aizawa, *Metabolic Maps of Pesticides*, Academic Press, New York, 1982.
2. H. Aizawa, *Metabolic Maps of Pesticides*, Vol. 2, Academic Press, New York, 1989.
3. T. S. S. Dikshith, ed., *Toxicology of Pesticides in Animals*, CRC Press, Boca Raton, FL, 1991.
4. C. Klaassen, ed., Biotransformation of xenobiotics, in *Casarett & Doull's Toxicology: The Basic Science of Poisons*, McGraw-Hill, New York, 7th edn, 2008, pp. 168–304.
5. R. Smart and E. Hodgson, ed., *Molecular and Biochemical Toxicology*, Wiley, Hoboken, NJ, 4th edn, 2008.
6. R. T. Meister, ed., *Farm Chemicals Handbook 1990*, Meister Publishing Company, Willoughby, OH, 1990.
7. I. P. Kapoor, R. L. Metcalf, R. F. Nystrom and G. K. Sanga, Comparative metabolism of methoxychlor, methiochlor, and DDT in mouse, insects, and in a model ecosystem, *J. Agric. Food Chem.*, 1970, **18**, 1145–1152.
8. I. P. Kapoor, R. L. Metcalf, A. S. Hirwe, P.-Y. Lu, J. R. Coats and R. F. Nystrom, Comparative metabolism of DDT, methylchlor, and ethoxychlor in mouse, insects, and in a model ecosystem, *J. Agric. Food Chem.*, 1972, **20**, 1–6.
9. C. M. Menzie, Metabolism of pesticides, in *Special Scientific Report: Wildlife*, U.S. Bureau of Fisheries and Wildlife, Washington, DC, 1969, pp. 127–128.
10. L. J. Lawrence and J. E. Casida, Interactions of lindane, toxaphene and cyclodienes with brain-specific *t*-butylbicyclophosphorothionate receptor, *Life Sci.*, 1984, **35**, 171–178.
11. N. Kurihara, N. Hori and R. Ichinose, Cytochrome P-450 content and aldrin epoxidation to dieldrin in isolated rat hepatocytes, *Pestic. Biochem. Physiol.*, 1984, **21**, 63–73.
12. P. Sudershan and M. A. Q. Khan, Cytochrome P-450 content and aldrin epoxidation to dieldrin in isolated rat hepatocytes, *Pestic. Biochem. Physiol.*, 1984, **21**, 63.

13. M. K. Baldwin, J. Robinson and R. A. G. Carrington, Metabolism of HEOD (dieldrin) in the rat: examination of the major faecal metabolite, *Chem. Ind.*, 1970, pp. 595–597.
14. F. Korte and H. Arent, Metabolism of insecticides, IX (1) Isolation and identification of dieldrin metabolites from urine of rabbits after oral administration of dieldrin, *Life Sci.*, 1965, **4**, 2017–2026.
15. D. H. Hutson, Comparative metabolism of dieldrin in the rat (CFE) and in two strains of mouse (CF1 and LACG), *Food Cosmet. Toxicol.*, 1976, **14**, 577–591.
16. J. N. Damico, J.-Y. T. Chen, C. E. Cossello and E. O. Haenni, Structure of Klein's metabolites of aldrin and dieldrin, *J. Assoc. Off. Anal. Chem.*, 1968, **51**, 48–55.
17. V. J. Petrella, J. D. McKinney, J. P. Fox and R. E. Webb, Identification of metabolites of endrin. Metabolism in endrin susceptible and resistant strains of pine mice, *J. Agric. Food Chem.*, 1977, **25**, 393–398.
18. L. M. Cole and J. E. Casida, GABA-gated chloride channel: binding site for 4′-ethynyl-4-*n*-[2,3-$^{3}H_2$] propylbicycloorthobenzoate ([^{3}H]EBOB) in vertebrate brain and insect head, *Pestic. Biochem. Physiol.*, 1992, **44**, 1–8.
19. C. T. Bedford, D. H. Hutson and I. L. Natoff, The acute toxicity of endrin and its metabolites to rats, *Toxicol. Appl. Pharmacol.*, 1975, **33**, 115–121.
20. S. Tashiro and F. Matsumura, Metabolic routes of cis- and trans-chlordane in rats, *J. Agric. Food Chem.*, 1977, **25**, 872–880.
21. S. Tashiro and F. Matsumura, Metabolism of trans-nonachlor and related chlordane components in rat and man, *Arch. Environ. Contam. Toxicol.*, 1978, **7**, 113–127.
22. J. E. Chambers and P. E. Levi, eds., *Organophosphates: Chemistry, Fate and Effects*, Academic Press, San Diego, CA, 1992.
23. R. A. Neal, Microsomal metabolism of thiono-sulfur compounds: mechanisms and toxicological significance, in *Reviews in Biochemical Toxicology*, Vol. 2, ed. E. Hodgson, J. E. Bend and R. M. Philpot, Elsevier/North Holland, New York, 1980, pp. 131–172.
24. H. T. Appleton and T. Nakatsugawa, Paraoxon deethylation in the metabolism of parathion, *Pestic. Biochem. Physiol.*, 1972, **2**, 286–294.
25. N. M. Mallipudi and T. R. Fukuto, Penetration and metabolism of 2-isopropoxlphenyl N-methyl-N-(2-methyl-4-tert-butylplhenylsulfenyl) carbamate in the house fly and honeybee, *Pestic. Biochem. Physiol.*, 1979, **12**, 55–67.
26. A. F. Machin, M. P. Quick, H. Rogers and N. F. Janes, Isomer of hydroxydiazinon formed by metabolism in sheep, *Bull. Environ. Contam. Toxicol.*, 1972, **7**, 270–272.
27. N. F. Janes, A. F. Machin, M. P. Quick, H. Rogers, D. E. Mundy and A. J. Cross, Toxic metabolites of diazinon in sheep, *J. Agric. Food Chem.*, 1973, **21**, 121–124.
28. A. F. Machin, Isolation and possible significance of some toxic mammalian metabolites of diazinon, *J. Pestic. Sci.*, 1973, **4**, 425–430.
29. D. L. Ryan and T. R. Fukuto, The effect of isomalathion and *O,S,S*-trimethyl phosphorodithionate on the *in vivo* metabolism of malathion in rats, *Pestic. Biochem. Physiol.*, 1984, **21**, 349–357.

30. D. L. Ryan and T. R. Fukuto, The effect of impurities on the toxicokinetics of malathion in rats, *Pestic. Biochem. Physiol.*, 1985, **23**, 413–424.
31. H. W. Dorough, Metabolism of insecticidal methylcarbamates in animals, *J. Agric. Food Chem.*, 1970, **18**, 1015–1022.
32. T. H. Lin, H. H. North and R. E. Menzer, Metabolism of carbaryl (1-naphthyl *N*-methylcarbamate) in human embryonic lung cell cultures, *J. Agric. Food Chem.*, 1975, **23**, 253–256.
33. T. C. Marshall and H. W. Dorough, Bioavailability in rats of bound and conjugated plant carbamate insecticide residues, *J. Agric. Food Chem.*, 1977, **25**, 1003–1009.
34. M. Elliott, N. F. Janes, E. C. Kimmel and J. E. Casida, Metabolic fate of pyrethrin I, pyrethrin II, and allethrin administered orally to rats, *J. Agric. Food Chem.*, 1972, **20**, 300–313.
35. L. C. Gaughan, T. Unai and J. E. Casida, Permethrin metabolism in rats, *J. Agric. Food Chem.*, 1977, **25**, 9–17.
36. D. Schulz-Jander and J. E. Casida, Imidacloprid insecticide metabolism: human cytochrome P450 isozymes differ in selectivity for imidazolidine oxidation versus nitroimine reduction, *Toxicol. Lett.*, 2002, **132**, 65–70.

CHAPTER 3

DDT and Other Chlorinated Insecticides

ANDREW G SMITH

MRC Toxicology Unit, University of Leicester, Lancaster Road, Leicester, UK.
Email: ags5@leicester.ac.uk

3.1 Introduction

Treating bacterial infections and combating insect-transmitted diseases were poorly achieved until, under the pressure of wartime, the apparent 'miraculous' benefits of penicillin and DDT were discovered and developed in the 1940s. For DDT this was assisted by the availability of huge quantities of chlorine and led to the subsequent production and use of other highly chlorinated chemicals such as hexachlorocyclohexane, hexachlorobenzene, chlorocyclodienes and polychlorinated biphenyls. Like DDT, hexachlorocyclohexane (HCH) isomers and various chlorinated cyclodienes were found to have potent insecticidal properties. The employment of these insecticides had dramatic effects on insect vectors of typhus and malaria, with outstanding benefits for human health and survival worldwide, as illustrated by the use of DDT to combat typhus in Naples in early 1944.[1,2] Shortly afterwards, the agricultural industry realized that these uses could be extended to protecting crops such as cotton from insect attacks. DDT and HCH (the γ isomer, lindane being the most potent) were cheap and widely used.[3] In fact, in some areas of the world the economic advantages of their agricultural use also had knock-on health benefits for people. Although the evidence for toxicity towards humans was minimal, it was recognized early on that there might be potential effects in other animal species inadvertently exposed due to their slow metabolism and persistence, especially

Issues in Toxicology No. 12
Mammalian Toxicology of Insecticides
Edited by Timothy C. Marrs

Published by the Royal Society of Chemistry, www.rsc.org

by accumulation up the food chain. Eventually, this was publicized by the work of Rachael Carson, although at a time when the use of DDT, lindane and chlorinated insecticides was declining in preference for more degradable agents.[3] The exact consequences of the individual insecticides on wildlife are still unclear and vociferously debated, but the issue is now recognized as an important consideration.

Most of these insecticides are now not widely used and are often banned. However, because of its usefulness and apparent low toxicity compared to some of the other chlorinated chemicals, limited use of DDT in controlling malaria has been allowed by the United Nations Development Programme, especially where vectors have become resistant to later pesticides such as organophosphates and pyrethroids, and concerns have resurfaced.[4–6] This is a highly contentious issue, but it would seem pragmatic to keep effective weapons in the armoury for a future in which climate change may influence incidences and persistence of insect-borne diseases. In the last few decades there have been renewed concerns about exposure of humans to these organochlorine chemicals following indoor residual spraying of DDT and even when it is only detected at very low levels in the environment. These issues include immune-modifying activity in the young, cancer incidences (especially breast cancer) and endocrine-modulating activity. The concerns can really only be proven to be soundly based for humans using epidemiological approaches, but such a task is extremely difficult as evidence often does not fulfil the Bradford Hill criteria. Sometimes the wealth of information available from times when DDT was used at much higher levels with few apparent adverse effects, and the cost-risk benefits in the fight against malaria, are not always fully considered. On the other hand, the topic stimulates strong views.[7–12] A detailed risk analysis of using DDT in indoor residual spraying has been recently reviewed by the World Health Organization (WHO).[12a]

Despite all having insecticidal properties, DDT, lindane and the cyclodienes do not all act in a similar manner; they may be absorbed, distributed, stored and metabolized in different ways and have different additional toxic mechanisms. They can be viewed as falling into five structural classes; DDT and its analogues, hexachlorocyclohexane isomers, cyclodienes, toxaphenes and the caged structures of mirex and chlordecone. As far as the mechanism of action of the insecticides is concerned, there are two broad types, those that act on sodium channels (such as DDT) and those that act on the γ-aminobutyric acid A ($GABA_A$) receptor (such as dieldrin and lindane). Here, because of its ongoing interest, more attention will be given to DDT and less to other chlorinated insecticides that have not been used for decades (detailed descriptions of their toxicology, where available, have been summarized previously).[8]

3.1.1 General Toxicity

3.1.1.1 Symptomatology

Broadly speaking, overt poisoning in animals and people caused by different organochlorine insecticides can appear similar and the symptoms are demonstrations of neuronal hyperactivity. However, there are differences between the effects of DDT and its analogues on one hand and many of the other chlorinated hydrocarbon insecticides, reflecting their different modes of action. Tremor is characteristic of poisoning by DDT, which can start in a mild form but increase in severity with greater exposure internally to a point that can develop into convulsions, but not unless high levels are reached. In contrast, lindane and chlorinated cyclodienes (*e.g.* aldrin, dieldrin, endrin), and toxaphenes frequently produced a convulsion as the first indication of poisoning. Some people reported that they experienced no symptoms of any kind prior to the initial fit. The cyclodienes (endrin, dieldrin, and isobenzan) appear to be among the most toxic of the chlorinated insecticides to humans, perhaps more than ten times more acutely toxic than DDT.[13]

For all insecticides, the degree of stimulation of the nervous system seems to be related directly to the concentration of these chemicals in nerve tissue. In animals the effect is usually reversible after either single or multiple doses. Recovery occurs when the concentration of the chlorinated hydrocarbon insecticide in the nervous system falls below a critical level. This may often be due to a redistribution of the chemical to other tissues, especially from the blood supply to fat, rather than elimination from the body which could take much longer. However, severe convulsions may result in death before redistribution can happen.[8]

3.1.1.2 Absorption

Organochlorine insecticides can be absorbed through the skin as well as by breathing and by mouth, but the importance of dermal absorption varies markedly for the different compounds. This may be because some of them, such as methoxychlor, have only a low toxicity and a small amount absorbed by any route is of no importance. More importantly, the efficiency of skin absorption varies for the different insecticides. DDT is poorly absorbed from solutions by the skin. The absorption of solid material is so poor that it is difficult to measure either the uptake of DDT or its effect. In the Second World War soldiers and civilians were dusted with copious amounts of DDT or wore clothes impregnated with the insecticide, without apparent deleterious effects. In contrast, even finely ground dieldrin is absorbed effectively through the skin and causes toxicity. Oral absorption of all is affected by fat and fibre content of diet.

3.1.1.3 Distribution, Metabolism and Excretion

Organochlorine insecticides became infamous because of their tendency to accumulate in humans, animals, birds, and the general environment as a consequence of their lipophilicity and slow metabolism. After single or repeated doses, the tissue concentrations of most of these chemicals reach their highest in adipose tissue, with lower levels in other organs with high contents of lipids such as adrenals and brain. Although storage in fat can be partly explained by the lipophilicity of these insecticides, other factors such as structural aspects of the chemical and competition between binding sites in tissues are also important.[14] The slow rates of metabolism and excretion are also critical, partly because of the difficulty of cleaving some of the carbon-chlorine bonds *in vivo* or inhibiting cytochrome P450 catalysed oxidations at other positions. DDT and its primary metabolite DDE are stored in adipose tissue of humans, whereas methoxychlor (with no aromatic chlorines) is metabolized much more rapidly and occurs only at very low levels in fat. This difference led to the increasing use of methoxychlor as an insecticide and a decline in the use of DDT. The isomers of HCH are stored to very different degrees in a pattern that is probably due to some extent to differential metabolism. Dieldrin is stored avidly whereas its isomer endrin is stored so little that it has been detected in people only after acute exposure and not even in people who were employed in its manufacture. The unhindered *anti*-C-12 hydrogen in endrin makes this position far more susceptible to attack than any other position in either isomer.[15,16] In itself, storage of chemicals in adipose tissue could be viewed as a detoxification mechanism since the insecticides are, essentially, not available for metabolic activity.

Chlorinated hydrocarbon insecticides are metabolized by the microsomal cytochrome P-450 system to hydroxyl derivatives, perhaps with dehydrochlorination as observed for lindane, or by conversion to stable epoxides as in the case of the formation of dieldrin from endrin.[8] Other routes of metabolism involve conjugation with glutathione by substitution of aromatic chlorines or the formation of glucuronides of hydroxyl groups. Besides adipose tissue, these lipophilic chemicals and their derivatives can be localized and cause toxicity in liver, brain, kidney and adrenal tissue, all of which may be associated with different types of toxic mechanisms independent of their neuronal activities. Parent insecticides can be excreted either in the bile, or possibly through the intestinal wall, for eventual faecal excretion. Metabolites are excreted in the urine only if they are of relatively high polarity such as glucuronides. This may have previously involved resorption of conjugates from the intestinal tract and transport to the liver and kidney (enterohepatic circulation) followed by further metabolic transformations. This is the case for glutathione conjugates first excreted in bile but which may be reabsorbed and converted to the mercapturates. Mobilization of lipophilic chemicals from fat to milk is an important consideration when discussing the loss of chlorinated hydrocarbon insecticides from the body.[17] Considerable concentration of these chemicals can occur in milk, compared to some other tissues, due to the lipid content (3–5%) and the high blood flow to breast tissue. This aspect has become of increasing importance in risk assessment due to the greater exposure of neonates and young children compared with adults. With mobilization of fat due to starvation or illness chlorinated hydrocarbon insecticides can be released into the circulation, sometimes with marked toxicological effects. Even so, relationships between storage, induction of drug metabolism enzymes, and excretion of metabolites are rarely simple; studies of the influence of the metabolism of one pesticide by another and the toxicity of mixtures are still in their infancy.

3.1.1.4 Neurotoxicity

Considerable evidence suggests that the insecticides act by altering the electrophysiological and associated enzymatic properties of nerve cell membranes, causing a change in the kinetics of Na^+ and Cl^- ion flow through the membrane.[18] Disturbances of calcium transport or Ca^{2+}-ATPase activity may also be involved.[18–21] Most studies have been conducted with DDT, chlordecone and cyclodienes. DDT appears to act particularly at the nerve axon by prolonging opening of the ion gates of the sodium channel and shows many similarities with type I pyrethroids.[18]

In contrast, cyclodienes and lindane act mainly at presynaptic terminals. Lindane, toxaphene, and cyclodienes mainly act by antagonism of the γ-aminobutytric acid (GABA)-regulated chloride channel causing hyperexcitation.[22–24] In fact this is a simplified view and complex interactions with subtype GABA receptors occur, especially the $GABA_A$ receptor.[18,25–27] DDT has little effect on this receptor. Morphologically, the convulsant actions of dieldrin and lindane may be mediated by effects on the hippocampus and other limbic structures.[28,29] At lower, non-convulsant doses of chlorinated hydrocarbon insecticides the susceptibility of animals to convulsions precipitated by many other poisons or by electroshock may be increased.

Changes in the biogenic amines often parallel the toxicity of chlorinated hydrocarbon insecticides, including the phenomenon of initial illness followed by clinical recovery. Fever may be a specific result of poisoning of the temperature control centre in the brain. In humans, a high fever of sometimes late but sudden onset has been followed promptly by death. This has been observed in poisoning by lindane, dieldrin, and endrin but may also accompany convulsions in humans simply because it may be difficult to dissipate heat when it is generated by such violent activity.[8]

3.1.1.5 Hepatic Toxicity

DDT and other organochlorine insecticides cause marked changes in the liver of some rodent species, especially the mouse. These changes can progress to tumours. Historically, the propensity of these pesticides to cause liver tumours in rodents has greatly influenced their safety assessment for humans. However, the relationship of the rodent tumours to possible induction of hepatocellular carcinoma in humans is probably poor.[8,30] In mice the hepatocarcinogenicity from oral administration has been demonstrated in several strains and shows a dose–response relationship. One problem is that tumours may only develop in strains that have high spontaneous rates anyway. Increased tumour incidence (particularly lung adenomas) has also been reported in other organs of mice. At high doses DDT can also be hepatocarcinogenic to rats,[31–34] but results with other species are inconclusive.[35] Chlorinated hydrocarbon insecticides are, in general, negative in mutagenicity tests,[36] but DDT, HCH, and the cyclodiene insecticides are efficient promoters of the actions of recognized potent hepatocarcinogens *in vivo*,[37–39] although again what this means for humans is debatable. The ability of these chemicals to cause tumours in the liver and promote those initiated by other carcinogens may be associated with the induction of microsomal and other enzyme systems causing activations of genotoxic carcinogens present.

The response of the rodent liver to DDT is similar to its response to moderate dosages of HCH, chlordane, dieldrin and toxaphene,[8] and importantly phenobarbital.[40,41] The earliest changes in liver cells of rodents given DDT involve considerable hypertrophy and a huge increase in the smooth endoplasmic reticulum with whorls that may have lipid droplets as their centres. Enlargement and morphological changes of the mitochondria, increased numbers of primary lysosomes, and atrophy of the Golgi body, also occur.[42] At the early stages, the changes in liver cells in rodents are reversible if dosage is discontinued but may be slow when the inducer is persistent.[8,31,41,43]

The chlorinated hydrocarbon insecticides and phenobarbital do not, to the same extent, produce in other animals besides rodents the changes in the endoplasmic reticulum that may be associated with tumour formation.[8,44] Chlorinated hydrocarbon insecticides may be carcinogens in the mouse but not all cause tumours in rats despite similar induction of the endoplasmic reticulum. An epigenetic mechanism for the hepatic carcinogenicity of chlorinated hydrocarbon insecticides has become a common hypothesis; possibly there is a disruption in intercellular communication-perhaps leading to inhibition of exchange of growth inhibitors.[45–47] Like phenobarbital, many of the effects are probably mediated through the constitutive androstane receptor (CAR).

3.1.1.6 Endocrine Disruption and Reproductive Effects

Many of the chlorinated insecticides have endocrine-disrupting properties, perhaps by acting as weak oestrogens.[8,9,48–50] Although the atrophic effects of lindane on rat testes occur at high doses,[51] some sex-linked behavioural effects occur at doses, *e.g.* with chlordane,[52] that are compatible with those found in the general population. Other chlorinated insecticides or their metabolites may act as androgen receptor antagonists, such as *p,p'*-DDE.[53] This area is the subject of considerable investigation and though at first sight it may be separate from the issue of carcinogenicity, the same mechanisms may be highly pertinent for humans in the risk analysis of low-level environmental exposure. Some studies have implicated a chronic oxidative stress in mechanisms of actions of organochlorine pesticides.[54,55]

3.1.2 Toxicity to Humans

Although acute or chronic neurotoxicity after high exposure to the chlorinated insecticides is rare, the possibility of low-level chronic exposure continues to cause debate. Despite the greatly reduced usage in the last decades of most chlorinated insecticides, interest in this issue has been reignited by the continued use of DDT against malaria:[56–59] for instance, the neurological impairment of infants that may be exposed through their mothers' milk and at doses/body weight significantly greater than adults consuming pesticide-containing food.[60] The DDT metabolite DDE has been linked with preterm and low-birthweight babies.[61]

The evidence is weak that chlorinated hydrocarbon insecticides are associated with cancer in humans. DDT has been studied the most, especially with relationship to breast cancer,[62] but the evidence is contradictory[63–65] not only with respect to developing the disease but also to survival.[66–67] No increase in the occurrence of any tumours has been found in populations heavily exposed to chlorinated insecticides. This includes groups of workers who manufactured and formulated DDT, dieldrin, aldrin, endrin, chlordane, and heptachlor and who have been examined carefully for tumours.[8,9] However, in animals many of these insecticides are good promoters of liver and breast cancer initiated by established carcinogens[39,68] and there are reports of poorly defined human use exposures that have been linked to increased incidences.[69] Some of the areas of the world where DDT and lindane have been used in large quantities are also areas where there is a significant risk of hepatocellular carcinoma due to aflatoxin contamination of food or to carrying of the hepatitis B virus. A similarity in mode of action with phenobarbital is important as it would allow direct comparison with epidemiological studies conducted on patients administered phenobarbital chronically for therapeutic purposes. So far the evidence is very weak that phenobarbital causes liver tumours in patients.[70]

3.2 Chlorinated Insecticides Acting Primarily on Sodium Channels

Aspects of many of the DDT-type insecticides that have been used are described below but further details can be obtained from Smith.[71] The structures of DDT and some analogues are shown in Figure 3.1.

3.2.1 DDT

p,p′-DDT, an acronym for dichlorodiphenyltrichloroethane, is 1,1′-(2,2,2-trichloroethane-1,1-diyl)bis(4-chlorobenzene). It has been sold under a variety of trade names. In the past, technical DDT might contain 77.1% *p,p′*-DDT, 14.9% *o,p′*-DDT, 0.3% *p,p′*-DDD, 0.1% *o,p′*-DDD, 4.0% *p,p′*-DDE, 0.1% and *o,p′*-DDE, and was formulated in many ways. Later products probably consist of much greater proportions of *p,p′*-DDT. The impurity and metabolite DDE is dichlorodiphenyldichloroethylene 1,1′-(2,2-dichloroethene-1,1-diyl)-bis(4-chlorobenzene). Although DDT was synthesized by Zeidler in 1874, its insecticidal properties were not discovered until 1939. After testing in Switzerland, USA and the UK it was used to great effect in the 1939–1945 war for controlling typhus and malaria both in troops and in civilian populations. DDT first became available for civilian use immediately after the war had ended and it was applied not only to control malaria but for other

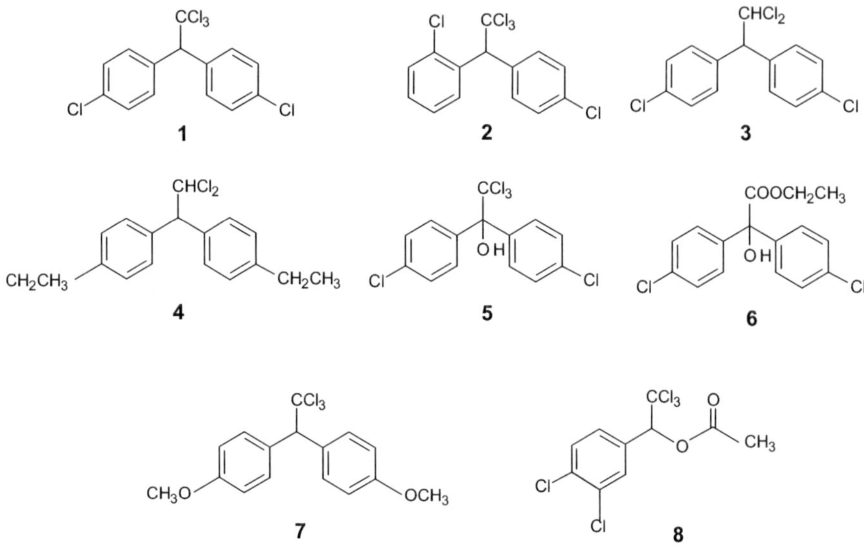

Figure 3.1 Structure of DDT and related insectides: **1**, *p,p′*-DDT; **2**, *o,p′*-DDT; **3**, TDE; **4**, ethylan; **5**, dicofol; **6**, chlorobenzilate; **7**, methoxychlor; **8**, acetofenate.

insect-transmitted diseases. The use of DDT played a major role in the eradication of malaria in the USA, many parts of Europe and elsewhere. The amount of DDT used in these capacities was relatively small.[3] Its spectacular success encouraged its use in general agriculture to protect production of economically important crops. In some countries, its use in combating days of labour lost to malaria has had huge economic benefits. However, it was soon recognized that DDT and its metabolite DDE could accumulate in animals and people and there was concern about the health repercussions for wildlife and humans. In addition, resistance of some insect populations started to be observed. The agricultural use of DDT declined although there has been a niche for it in malaria control despite attempts to have it banned worldwide.[6,72–75] Traces of DDT and DDE are detectable in many people and wildlife, although mostly too low for toxicological repercussions. Analogues of DDT, such as ethylan, dicofol, chlorobenzilate, and methoxychlor (Figure 3.1), were produced to decrease particular side effects or to make the compound effective but easily metabolized and be of much less threat to the environment. However, as with dicofol, these alternatives can contain DDT as a contaminant.[76]

3.2.1.1 Experimental Toxicity

General Symptoms. The first observed effect of DDT in rats is abnormal susceptibility to fear, with violent reaction to normally subthreshold stimuli. There is motor unrest and increased frequency of spontaneous movements.

As poisoning progresses hyperirritability develops but without the violent convulsions. A fine tremor progresses to a coarse tremor. In the later stages at very high doses, in some species, there are attacks of epileptiform, tonic-clonic convulsions with opisthotonos. Symptoms appear several hours after oral administration of the compound, and death follows after 24–72 h. The latent period after intravenous administration at about the LD_{50} levels is approximately 5 min; signs of poisoning reach a maximal level in about 30 min, and survivors are symptom-free in 18–24 h. With repeated doses of DDT the animal may be gradually debilitated, especially by malnutrition. Those metabolites that have been tested seem to be less toxic than the parent compound, except perhaps DDA which seems to be toxic to the kidney.[8] Young animals do not seem to be more susceptible than adults to the acute effects of DDT and there are no particular distinctions between sexes.[9] At chronic doses to rodents, equivalent to that reported for heavily exposed workers, very slight increases in liver hypertrophy may be detected.

Absorption and Distribution. Dissolved DDT is absorbed by all routes, although DDT powder is very weakly absorbed through the skin so that an accurate LD_{50} value for this formulation could not be determined. *o,p′*-DDT seems to be less toxic than the *p,p′*-isomer.[77]

The size of most DDT dust particles means that if inhaled they are deposited in the upper respiratory tract and probably swallowed. DDT dissolved in oils is absorbed from the gastrointestinal tract about 1.5–10 times more effectively than is undissolved DDT.[78] The initial distribution of DDT, like that of some other chlorinated chemicals, seems to occur *via* the lymph system.[79,80] Most of the DDT absorbed into the lymph is carried in lipid and thence into the plasma proteins. DDT is stored in all tissues and particularly high concentrations are found in fat. After repeated doses, storage of DDT in fat rises rapidly at first and then more gradually until a peak or plateau is reached. This may be much greater than is seen with a single fatal dose. Levels in the brain seem to be the most critical factor and gradually decline once exposure ceases. Tissues in the fetus are lower than those in corresponding tissues of the mother. Most species, including humans, store the metabolite DDE more tenaciously than they do DDT, the greater part of which is metabolized by a different pathway from that of DDA and excreted more rapidly. The result is that DDE, expressed as a percentage of total DDT-related compounds, increases in individuals after DDT intake decreases and increases in successive steps of the food chain.

Metabolism and Excretion. A simplified scheme of DDT metabolism is shown in Figure 3.2. The sequence of metabolism *in vivo* is likely to proceed *via* more than one pathway and depends on tissues and species and less is known about humans than rodents.[71,81] A major water-soluble urinary metabolite of *p,p′*-DDT was recognized in early work as 2,2,-bis(4-chlorophenyl) acetic acid (DDA).[82] There is evidence for both reductive and oxidative metabolism of the trichloroethane function.[83–86] A variety of minor

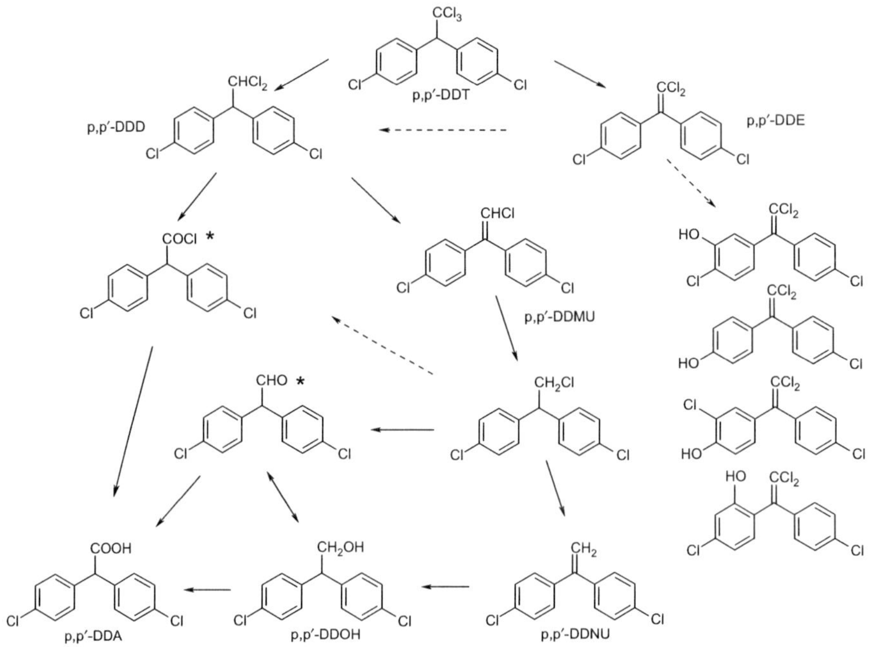

Figure 3.2 Major metabolic routes for *p,p′*-DDT in species of rodents and humans. Predomination of routes depend on species. * indicates likely reactivity with cellular constituents.

metabolites have been identified in excreta.[8,71] Following the route of metabolism of DDT by dehydrochlorination to form DDE, further metabolism *via* the acyl chloride route is possible, but most of the identified metabolites of DDE have been hydroxylated analogues such 4-hydroxy-*p*-DDE, 4-hydroxy-*m,p*-DDE, 3-hydroxy-*p,p′*-DDE and 2-hydroxy-*p,p′*-DDE. Hydroxylation of DDE to form some of these products, probably occurs by epoxidation and in some cases NIH shift rearrangements. There is evidence for glutathione conjugation of *p,p′*-DDE followed by conversion to a mercapurate, cleavage by a lyase and then methylation of the thiol.[71]

Metabolism of *o,p′*-DDT is similar to that of the *p,p′*-analogue but at a greater rate because of the ease of ring hydroxylation. Strong evidence seems to suggest that activation of *o,p′*-DDT or *o,p′*-DDD occurs by the action of cytochrome P450 particularly in the adrenal mitochondria, possibly by the intermediacy of a reactive acyl chloride.[87,88] Binding of *o,p′*-DDD in mouse lung and *p,p′*-DDD in rabbit Clara and human bronchial epithelial cells may occur by a similar mechanism.[89,90] 3-Methylsulfonyl-*p,p′*-DDE is selectively bound and toxic to the adrenal zona fasciculata of mice, even occurring in the pups of dosed mothers.[91,92] The damage may result from activation in the mitochondria by CYP11B.[93] Similar mechanisms may apply to the toxicity to the adrenals of *o,p′*-DDD, toxicity being greater because of the faster metabolism than the *p,p′*-analogue (see Smith[71]).

Most of the metabolites of DDT are excreted initially *via* the bile. Importantly, the difference between rats and birds in the excretion of DDT metabolites may be the reduced ability of the latter to metabolize DDE further.[86] As with many lipophilic chlorinated chemicals, excretion of DDT in the milk can be an important clearance route for mothers in rodents, cows and humans.[8,17,71]

Neurotoxicity and Behaviour. The major effect of DDT is on the nervous system, and this seems to occur by slowing down closing of 'gates' in axon Na^+ channels.[18] Both central and peripheral parts of the nervous system are affected, although the cerebellum and motor cortex have been viewed to be of greatest importance. Electrophysiological activity may become abnormal only a few minutes after a large dose of DDT to rats which can culminate in seizures.[94] Unlike the cyclodienes or lindane, the most characteristic feature of DDT in rats is production of tremor. There also seems to be other disturbances of thermal regulation such as coldness of the skin. Death from DDT in experimental animals is usually from respiratory arrest, often from a prolonged period of muscular activity leaving them exhausted. Large doses of DDT may directly affect the myocardium as well as producing fibrillation induced *via* the central nervous system.[8] Behavioural changes have also been shown in animals receiving low doses of DDT, attributed to an exaggeration of the stretch reflex, without other apparent symptoms.[95] At a single dose of 12.5 mg kg^{-1} of DDT to rats there is a significant increase in the acoustic startle response that can be attenuated by phenytoin.[96,97]

The effect of DDT appears to be on axonal membranes and may result in the inhibition of Na^+, K^+ and Mg^{2+} ATPases and Ca^{2+}-ATPase regulating calcium levels at the axon surface.[98] DDT is known to prolong opening of the ion gates of the sodium channel, perhaps by affecting phosphorylation of the α-subunit, and interact with the sodium channel of rat dorsal, but not ganglion, neurons in a similar manner to type I pyrethroids.[19,99] Although spinal α-adrenoceptors have been proposed as modulating DDT-induced tremor, attenuation of motor dysfunction requires blockade at receptors in regions other than solely the spinal cord.[100,101] Changes in metabolism of 5-hydroxytryptamine and noradrenaline (norepinephrine) and other messengers have been proposed as accounting for some of the neurotoxic effects of DDT.[71]

Mutation and Carcinogenesis. Most studies of the mutagenicity of DDT, *in vitro* and *in vivo*, have been inconclusive.[102] As with some other chlorinated aromatics, DDT causes inhibition of intercellular communication in cell lines, and freeze-fracture of hepatocytes from rats administered DDT showed reduction in number and size of gap junctions possibly involving expression and turnover of connexins and some phosphorylated forms.[45,46,103–106] High levels of DDT induce liver tumours in rats and mice, although studies have not always been repeatable.[31–34,56,107] DDT will stimulate the oestrogen-dependent P38 MAPK pathway with phosphorylation of p300, a general transcription factor for growth and differentiation.[108,109] Certainly DDT promotes carcinogenesis initiated by aflatoxin and other chemicals, although the significance of these findings with respect to human exposure is open to question.[39,68,110,111]

Influence on Reproduction. There is great interest in the possibility that DDT may have endocrine disruptor activity, although it was shown many years ago that most of the activity observed was due to the presence of the *o,p′*-isomer.[8] The *o,p′*-analogue was present at significant quantities in the technical product. However, the exact composition of DDT preparations used now is not clear. *o,p′*-DDT acts on the uterus, probably as a long-lasting oestrogen receptor agonist although it seems that its actions may not be identical to oestradiol.[112–114] Relative to a potent chemical such as diethylstilboestrol, *o,p′*-DDT is about 1/10 000 less active, so that at environmental exposure levels, and the fact that it is metabolized faster than *p,p′*-DDT, make it seem unlikely to be a major hazard in this respect. Levels of endogenous oestrogen will always be far greater. The DDT metabolite *p,p′*-DDE may disrupt male reproductive development by acting as an antiandrogen by binding to androgen receptors in a non-productive way.[53,115,116] DDE, like vinclozolin and flutamide, changed the expression of androgen receptor regulated genes in castrated male rats. After metabolism, DDT can activate liver and brain oestrogen receptors.[117] Exposure of rats to DDE *in utero* by lactation can show changes in anogenital distance but not testes, epididymus, seminal vesicles or ventral prostate weights and at maternal doses at or above 10 mg kg^{-1}. In Holtzman rats, effects on prostate development by *in utero* or lactational exposure to *p,p′*-DDE has been shown.[118–120] Some influence of DDT on reproduction, including multigeneration studies, has been reported,[121,122] but except for a decrease in pup viability at high doses, substantial reproductive disorders have not been observed despite hormonal charges and delay in male rat maturation.[123]

General Metabolic Responses. *p,p′*-DDT, *p,p′*-DDE, *o,p′*-DDT and various metabolites of these will cause a variety of metabolic changes to occur, principally in the liver. Foremost of these is the induction of microsomal enzymes. Multiple doses of DDT to female rats induced hepatic CYP2B and 3A proteins but not CYP1A1 or CYP2E1 and caused elevated hydroxylation at positions 16 and 6β of testosterone.[124] DDT, DDE and DDD all induced CYP2B and 3A in male rat liver, roughly to the same degree, despite marked differences in retention.[125] Inductions are probably mediated by constitutive androstane and pregnane X receptors. Changes in lipid metabolism that have been observed are either linked to microsomal proliferation and are of unknown significance, for instance in the lung.[126] Some immunotoxic effects of DDT have been ascribed to inhibition of the functional activity of macrophages.[127,128]

3.2.1.2 Toxicity to Humans

Over the last decade, the literature associated with DDT and DDE levels in humans and associated possible effects has become large and cannot be covered *in toto* here. More details can be found elsewhere.[8,9,59,62,71,129] The risks associated with indoor residual spraying have been assessed in detail by the WHO.[12a]

Levels in Fat, Blood and Milk. One of the highest reported storage levels of DDT and related compounds was that of a healthy worker whose fat contained DDT and DDE at concentrations of 648 and 483 ppm, but most have been considerably lower.[8,9] Considerable evidence suggests that with time the greatest proportion of DDT in people is as the metabolite DDE.[8] Although each person without special exposure to DDT has relatively constant serum levels of DDT and DDE, DDE values differ more than the DDT values from person to person. Levels of DDT and its metabolites in the serum of adults rose over a 12 month period following application of the pesticide to their homes in KwaZulu, South Africa. In contrast, levels fell in the age group 3–20 years, showing the complexity of any phamacokinetic interpretations.[130] In the past there has been a gradual decline in the concentrations of DDT and related compounds in human fat, serum and milk of Western countries.[8,56,65] Consumption of fish appeared to be a predictor of plasma DDE levels, but most reliable were age and serum cholesterol.[131] As with increased concern over continual DDT usage in some parts of the world, variations in serum or fat levels may be more complex. The detection of DDA in urine looks promising for assessment of occupational and environmental exposure to DDT in antimalarial campaigns.[82]

Little information is available on the secretion of DDT in the milk of women who were occupationally exposed to the compound or who were made ill by it. As with other chlorinated pollutants, women may be in negative DDT balance during lactation and this may be a significant factor in determining the lower levels of DDT found in women than men in the general population. In areas of KwaZulu, human milk levels of DDT and metabolites were significantly higher in women whose houses had been treated with DDT to interrupt malaria transmission.[132,133] Primiparous mothers had significantly more than multiparous mothers and transfer from the mother's milk to the child's blood was clearly demonstrated.[134] Other studies have suggested that estimated daily intakes of DDT metabolites by infants from mothers' milk in some Limpopo villages in South Africa may exceed WHO acceptable daily intakes.[135] DDT and DDE levels in human milk in China vary considerably, reflecting past usage and consumption of meat and fish.[136] Overall, despite the presence of very low levels of DDT in human milk and placenta, the risk to neonates in the past has been judged to be low,[60] but recently debate has resumed. Clearly, levels would have to be very high before any advice against breast-feeding could be given.

Acute Intentional and Accidental Exposures. Early studies of the exposure of human volunteers to DDT focused on oral and dermal administration of large acute doses to volunteers, particularly with respect to the storage of DDE and relative to animal studies. On the whole, few effects were observed with perhaps paraesthesia of the tongue and mouth, confusion and slight tremors at high oral doses. Very high accidental or attempted suicidal exposure led to convulsions being observed in a few cases, including children, but the involvement of other chemicals cannot be excluded; the exact doses are unknown.[8,137,138] Recovery in survivors was often within 1 or 2 days. In fact, the acute toxicity of DDT to humans seems remarkably low.[8,71]

Chronic and Environmental Exposure. Despite current concerns, the safety record for humans in the use of DDT is phenomenally good considering the huge quantities used in agriculture and applications such as delousing in the past.[2,3,102] In delousing, DDT was used in such a way that the bodies and inner clothing of thousands of people of all ages and states of health were liberally dusted. The applicators sometimes worked in a cloud of the material without noted problems. Other applicators have sprayed the interior of hundreds of millions of homes in tropical and subtropical countries under conditions of extensive dermal and respiratory exposure without the safety procedures now prescribed by WHO. Some people made or formulated DDT for many years. Although dermatitis was commonly observed among workers who used DDT solutions, the rashes were probably due to the solvent. Paraesthesia of the extremities, headache, dizziness, and tremor of the tongue and hands have been linked with DDT production but studies of other workers making or formulating DDT have shown little evidence of or adverse effects on reproductive outcome or endocrine diseases (see Smith[71] for original references). No significant overall cause of specific mortality excess among men potentially exposed at work to DDT from 1935 to 1976 was detected.[139] Similarly, a population living downstream from a defunct DDT-manufacturing plant showed no DDT-specific illnesses or ill health, but possibly some changes in clinical chemistry, despite total DDT serum levels three times the national mean.[140] The induction by DDT level of microsomal enzymes of human liver was demonstrated first in workers. DDT may be more important than DDE in this respect.[141] Although not investigated in detail, it is interesting that major health or reproductive outcomes from these exposures to DDT, which were probably much greater than has occurred subsequently, have not been associated in methodical studies. Some evidence has linked DDT exposure, as defined by serum levels, with diabetes.[59,142–146] However, the complexity of the aetiology of diabetes, as with other clinical issues, makes these associations difficult to interpret.

The most heavily exposed group of workers whose health has been investigated is those in malaria control. In Brazil, examinations failed to show consistent symptoms of DDT neurotoxicity. In Indian sprayers, serum DDT levels were 8.5 times higher than controls and visuomotor functions were slightly depressed.[147] The levels of DDT and its metabolites in the sera from applicators in malaria control in Natal were significantly higher than in the population protected by the spraying. Although serum GGT was not statistically different from controls, the mean in applicators was greater than the maximum laboratory mean level and alanine transaminase (ALT) values were significantly greater in the applicators although not deemed clinically significant and possibly associated with alcohol consumption.[148,149] Members of households that had been sprayed inside with DDT had significantly greater levels in their serum than people from non-sprayed households. Studies in Costa Rica have linked declines in neuro-behavioural functioning and an increase in neuropsychological and psychiatric symptoms with DDT exposure of retired malarial sprayers; but no clinical differences were detected.[57] A possible association between maternal blood DDE levels and preterm and low-birthweight babies has been proposed.[61] Many subsequent studies have shown mixed findings between prenatal exposure to DDE,[59] most recently in Saudi Arabia.[150] There was no association with childhood height or body mass index in Mexican boys in an area where DDT had been used extensively.[151] Neurodevelopment in Mexico farm workers' children in California was possibly associated with mothers' DDE levels.[152–154] Spanish findings suggest association between prenatal exposure to DDE and asthma development, but probably other factors are involved in such a complex issue.[155] In this same cohort, subjects with a polymorphism in a glutathione transferase P1 valine 105-allele were more at risk, apparently, to the association of DDT with adverse cognitive functioning, but it is possible the effects are indirect.[156]

Evidence for Carcinogenicity in Humans. Although there is a lot of evidence against DDT causing liver cancer in humans in Western countries, there is still the possibility of it acting as a promoter of potent carcinogens such as aflatoxin and hepatitis B.[62,129] A study in China purported to show association of serum DDT, but not DDE, with liver cancer risk. Alternatively, this could be interpreted with individual ability to metabolize DDT as a biomarker, and background incidences were comparatively high.[157] Many studies have determined DDT levels in cases of cancer or other diseases but it is difficult to judge the significance of these.[9,56,65,158] A study of deaths that occurred among men who used DDT in an antimalarial campaign in Sardinia in the late 1940s suggested that workers had a significant increased risk for liver and biliary tract cancers and multiple myeloma. However, non-exposed workers also showed elevated incidences of cancer.[69] Reports that pancreatic cancer might be associated with exposure to DDT and ethylan in a nested case-control mortality study among workers at a chemical plant could not be confirmed in wider non-occupational surveys.[158–161] In summary, the evidence is poor that DDT exposure is directly associated with pancreatic, lung, liver, prostate and other common cancers, including non-Hodgkins lymphoma.[59,62,162,163]

Plasma and tissue levels of DDT have been particularly targeted as being linked with a rising incidence of breast cancer.[164–166] However, much of the evidence has been reviewed in detail and is either not supportive or is inconclusive.[63–65,158,167,168] Elevated levels of DDT or DDE were reported in cancerous breast tissue fat compared with tissue from benign mammary disease.[169] In contrast, a number of studies found no relationship between blood levels of DDE and risk of, or progression of, breast cancer,[65,170,171] including women in North Vietnam who had generally high levels of DDT or DDE due to exposure from antimalarial use. More recently, the hypothesis has emerged that the critical issue is the exposure from a young age rather than as an adult,[172,173] in a similar way to diethylstilboestrol. Similarly, a nested case-control study reports association between exposures to DDT/DDE and testicular cancer.[174,175] Again whether this is direct or indirect is difficult to gauge. At exposure levels recorded in antimalarial campaigns, the risks extrapolated from animal data seem very low even without taking into account species and tumour site differences.

Influence on Reproductive Potential. In the last decade, the greatest issue associated with DDT use in malaria control has been the possibility of modulation of reproductive physiology, mechanistically based on the very weak oestrogenicity of DDT, especially the *o,p′*- isomer, and the *in vitro* anti-androgenicity of *p,p′*-DDE.[53,115] There is little consistent evidence that time to pregnancy or fertility is affected by prior DDT exposure.[59] Most studies have not found consistent associations between DDT and mothers' pregnancy loss or measures of fetal growth or DDT/DDE levels in children and the onset of puberty or menarche, and many authors themselves recognize the need for caution in interpretation.[59,176] Serum DDE levels in mothers at delivery were not related to length of time of subsequent lactation.[177] At present the possibility of mens' hormone levels or semen quality being affected by malarial DDT spraying, both occupationally and through residues, or environmentally, is a controversial topic even from studies within the same region of South Africa or Mexico.[178–183] Urogenital malformations and development in newborn boys have been linked, but again with results that clearly require more detailed study for consensual agreement.[184–187]

3.2.2 TDE, Ethylan, Dicofol and Chlorobenzilate

A more detailed description of these DDT analogues below can be found elsewhere.[71]

TDE is the 1,1′-dichloro-2,2-bis (4-chlorophenyl) ethane metabolite of *p,p′*-DDT (Figure 3.1) and has been used as an insecticide in its own right. It is less toxic than DDT. The *o,p′*-isomer has been used as an anticancer or anti-hyperplasia agent on the adrenal.[188–190]

Ethylan, a *p,p′*-ethyl analogue of DDT (Figure 3.1), is 1,1′-(2,2-dichloroethane-1,1-diyl)bis(4-ethylbenzene) and was introduced to control pear psylla, leaf hoppers, larvae on vegetables, and moths and carpet beetles. The oral toxicity to animals seems to be very low: more than 4 g kg^{-1} in rats and mice.[191] Minimal liver changes in rats are observed at doses of 2–5 g kg^{-1} of diet for prolonged periods.[71] Like other chemicals of this type ethylan is stored in fat but is continually metabolized. Ethylan affects corticosteroid excretion in dogs, with marked atrophy of the adrenal cortex, and weakly induces liver tumours in mice.[71] As a potential anticancer agent, ethylan was administered to men in short-term regimens with prostatic carcinoma and women with carcinoma of the breast for up to 200–300 mg kg^{-1} for 6 days.[192] Smaller doses caused diarrhoea, vomiting and nausea in some patients but there were signs that patients began to tolerate higher doses. Marked depression of plasma 17-hydroxycorticosteroid levels was observed although still within the normal range. Although an association between exposure to ethylan and development of pancreatic cancer has been reported (see DDT) the exposure data is insecure.[193]

Dicofol is 2,2,2-trichloro-1,1-bis(4-chlorophenyl)ethanol (Figure 3.1) and has been used as a miticide for citrus fruits, nuts, cotton, and beans. It seems to be metabolized to the 4,4′-dichlorobenzophenone, which is stored in fat and muscle as well as being excreted in the faeces. Nearly 50% of the administered doses is excreted in the urine within 24 h. Part may be glucuronides of 4,4′-dichlorobenzhydrol.[194,195] Some small adverse effects associated with reproduction in rats and mice have been reported.[71] Dicofol caused dosage-related increases in hepatic drug metabolizing enzymes at a potency equivalent to that of DDT and phenobarbital,[196] and may be hepatocarcinogenic in mice but not in rats.[71] The presence of *p,p′*-Cl-DDT in dicofol and conversion to DDE during analysis can be misleading for estimations of DDT/DDE presence.[76]

Chlorobenzilate is ethyl bis(4-chlorophenyl)(hydroxy)acetate (Figure 3.1). The acute oral LD_{50} to mice, rats and hamsters is about 700 mg kg^{-1}.[71,197] Symptoms in rats and mice include depressed motor activity and rapid wheezing respiration. Dogs tolerated daily oral doses of 64 mg kg^{-1} for 35 weeks and rats 500 ppm in the diet for 2 years. After daily chlorobenzilate doses of 12.8 mg kg^{-1} to dogs, 5 days per week for 35 weeks, approximately 40% of the total dose was excreted unchanged or as urinary metabolites with no significant storage in fat of dogs or rats.[71] Rat liver homogenates convert the insecticide to *p,p′*-dichlorobenzilic acid, *p,p′*-dichlorobenzophenone, *p,p′*-dichlorobenzyhydrol, and *p*-chlorobenzoic acid.[198] Chlorobenzilate induced hepatocellular carcinomas in mice and some testicular atrophy in rats.[71]

3.2.3 Methoxychlor

Methoxychlor is 1,1′-(2,2,2-trichloroethane-1,1-diyl)bis(4-methoxybenzene) (Figure 3.1), the *p,p′*-dimethoxy analogue of DDT. It is effective against many insects affecting fruits, vegetables and livestock. The acute oral toxicity of methoxychlor to rodents is much lower than other organochlorine insecticides, probably due to its rapid metabolism,[9,197] and the attractiveness of methoxychlor for use was its relatively short biological half-life and low mammalian toxicity.

3.2.3.1 Experimental Toxicity

Absorption, Distribution, Metabolism and Excretion. After oral administration of methoxychlor to mice, 93% was recovered from the excreta within 24 h

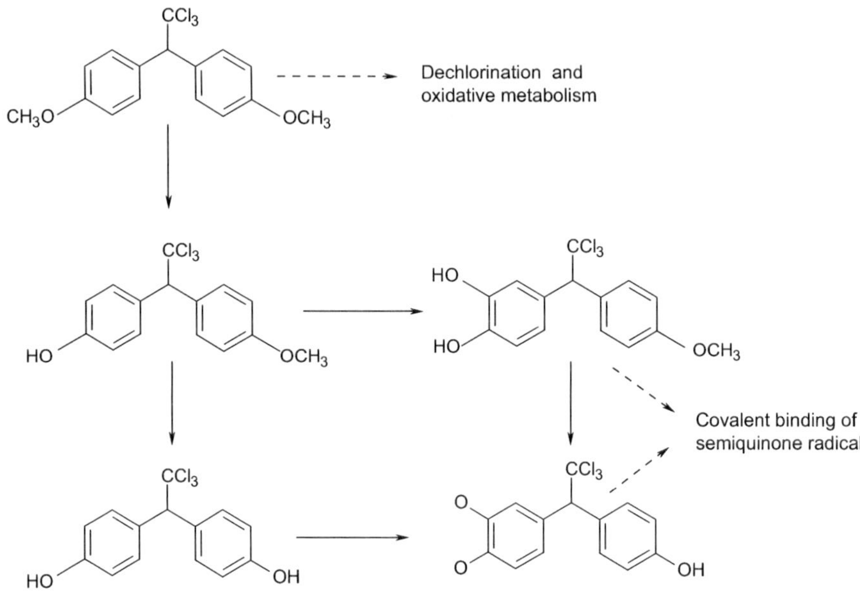

Figure 3.3 Simplified scheme of metabolism of methoxychlor in mammals.

including 2-(*p*-hydroxyphenyl)-1,1-trichloroethane, 2,2-bis(*p*-hydroxy-phenyl)-2′-(methoxyphenyl)-1,1,1-trichloroethane and 2,2-bis(*p*-hydroxyphenyl)-1,1-trichloroethane, which were eliminated largely in conjugated form.[199] Since then there have been a great deal of studies on its metabolism *in vitro* (see Figure 3.3 for a simplified scheme). In the rat demethylation seems to occur with CYP2B isoforms with some evidence for additional involvement of CYP2C6. In humans, CYP2C9 and CYP1A2 seem to be responsible for demethylation. *Ortho*-hydroxylation of the monodemethylated products is predominantly catalysed by CYP3A4. On the whole, bidemethylation and *ortho*-hydroxylation occurs less readily with human samples than with rat liver.[71,200–204] Phenolic products are probably converted to activated intermediates which bind covalently to macromolecules.[205] Microsomal metabolism of methoxychlor in rat liver has also been reported to bind to iodothyronine 5′-monodeiodinase with resulting depression of iodinase activity *in vivo*.[206] In rats, methoxychlor induced hepatic CYP2B1 and CYP3A enzymes and metabolism showing sexual dimorphism, but in regimens with multiple treatment and less efficiently than DDT.[124,207,208]

Toxicity. Mutagenicity results for methoxychlor are mixed;[209] increased occurrence of neoplasms of all sorts have been shown in rats and carcinoma of the testes in mice, but only after feeding high doses for 2 years.[210] Large doses of methoxychlor also produce chronic nephritis and hypertrophy of the kidneys, mammary glands, and uteri of swine, and cystic tubular nephropathy in rats.[211,212] The liver seems relatively insensitive to methoxychlor.

Large doses of methoxychlor may have oestrogenic effects. The potency is about 1/10 000 of diethylstilboestrol; mating and litter size of rats and mice are reduced and male and female offspring reproductive performance is decreased.[49] In studies of the effect of methoxychlor on reproductive tract development following neonatal exposure of mice, precocious vaginal opening, cornification and increased tract size, and ovarian atrophy were observed in females and reduced serum testosterone, testicular DNA content, seminal vesicles, and prostate in males.[213–216] Rats dosed with methoxychlor before and following birth had immune and reproductive changes at doses of 0, 5, 50, or 150 mg kg^{-1} day^{-1}. Primary adult effects were reproductive and 5 mg kg^{-1} day^{-1} was not a no observed adverse effect level (NOAEL).[217] Methoxychlor fed to lactating dams affected the reproductive tract of suckling females.[218] The oestrogenicity of methoxyclor is probably due to metabolites and not the parent compound. *In vitro*, pure methoxychlor itself is not oestrogenic whereas metabolites (*e.g.* bis(hydroxyphenyl)trichloroethane) are.[219–223] The oestrogenic activity of impure methoxychlor in inducing accepted parameters of uterine proliferation may be partly caused by the demethylated metabolites and impurities, probably disrupting binding of the oestrogen receptor to response elements of the *HOXA10* gene and others.[49,200,219,224–227] *In vitro* and *in vivo*, 1,1-dichloro bis(4-hydroxyphenyl)ethane is the most potent agent. Atresia of the ovarian follicles may occur possibly by the Bcl-2 and Bax-mediated pathways and oxidative damage,[228–234] together with inhibition of steroid genesis by the 2,2-bis hydroxy metabolite and ovarian epithilium proliferation.[222,235–237] The reproductive effects of methoxychlor metabolites in male rats[211,228] may be mediated, in part, by elevation of prolactin concentration and release, which in turn influences hypothalamic levels of gonadotropin-releasing hormone[238] or direct effects on cholesterol side chain cleavage.[239,240] Transgenerational imprinting effects of methoxychlor on male germline by epigenomic pathways, possibly by changes in methylation status, are of potential significant importance.[241–243] Apoptosis in testis is induced by mitochondria and FasL pathways.[244,245]

The oestrogenic effects of methoxychlor are not restricted to those on uterine or other reproductive physiology and function. Methoxychlor prevented ovariectomy-induced bone loss in the rat.[246] Both running wheel activity (oestrogen controlled) and sex behaviour in rats and hamsters were induced by 400 mg kg^{-1} day^{-1}.[247] Exposure of pregnant mice to methoxychlor may change development of male offspring[248,249] and exposure of monkeys at puberty may lead later to impaired cognitive function.[250] It may be erroneous to assume that the methoxychlor metabolites act in an identical manner to oestradiol,[49,247,251] and interaction with both oestrogen receptors α and β probably occur *in vivo*.[252,253] Methoxychlor also affects inflammatory responses.[254,255]

3.2.3.2 Toxicity to Humans

The rapid metabolism and excretion of methoxychlor suggests that at most exposure levels humans are at little risk including absorption from cows' milk.

Residues are not often found in human tissues.[71] Volunteers given methoxychlor at rates of up to 2 mg kg^{-1} day^{-1} for 8 weeks were without detectable effect on health, clinical chemistry, or the morphology of blood, bone marrow, liver, small intestine, or testis, the highest dosage administered being similar to that considered safe for occupational intake at the time.[256]

3.2.4 Acetofenate

Acetofenate is 2,2,2-trichloro-1-(3,4-dichlorophenyl)ethyl acetate (Figure 3.1). It could be considered as bearing partial resemblance to DDT and has been used in insecticides in China. However, there is little accessible information on its use, toxicity and bioaccumulation. The little available on neurotoxicity, hepatic toxicity, and immunotoxicity suggests further studies are warranted.[257–260]

3.3 Chlorinated Insecticides Acting Primarily on the $GABA_A$ Receptor

3.3.1 Lindane (γ-Hexacholorcyclohexane)

Lindane is (1*R*,2*S*,3*r*,4*R*,5*S*,6*r*)-1,2,3,4,5,6-hexachlorocyclohexane. 1,2,3,4,5,6-Hexachlorocyclohexane (HCH) has eight separable steric isomers, one of which (the α) exists in two enantiomorphic forms. Of the eight isomers, six (including the two mirror-image forms of the α isomer, plus the ß, γ, δ and ε isomers) are relatively stable, and they are the only ones commonly identified in technical HCH. The γ-isomer lindane accounts for the insecticidal properties (Figure 3.4). HCH and lindane have been used to control grasshoppers, cotton insects, rice insects, wireworms and other soil pests. Lindane has been used for protection of seeds, for treatment of poultry, livestock and household insects; HCH was little used for crops because the mixture imparted a marked flavour to food. Lindane has also been used as a scabicide and pediculocide, usually in lotions, creams and shampoos.

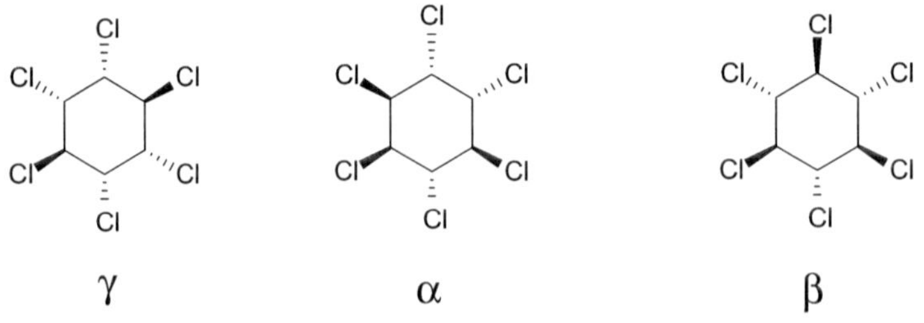

Figure 3.4 Major isomers of hexachlorocyclohexane showing gamma (lindane), α and β.

3.3.1.1 Experimental Toxicity

Acute poisoning by lindane involves increased respiratory rate, restlessness accompanied by frequency of micturition, intermittent muscular spasms of the whole body, salivation, grinding of teeth and consequent bleeding from the mouth, backward movement with loss of balance and somersaulting, retraction of the head, convulsions, gasping and biting, and collapse and death.[1] For most species, the compound has a moderate acute oral toxicity not greatly different from that of DDT but it is more readily absorbed by the skin and therefore, is more toxic by the dermal route than DDT. In rats, the acute toxicity of the isomers of hexachlorocyclohexane (HCH) decreases in the order $\gamma > \alpha > \delta > \beta$,[261] but the toxicity of repeated doses decreases in the order $\beta > \alpha > \gamma > \delta$.[262] The long-term toxicity of the different isomers is directly related to their storage and inversely related to their rate of metabolism.[263,264]

Absorption, Metabolism and Excretion. Although the α, β, γ and δ isomers of HCH are stored in the fat of rats and dogs, over 30 times more β than γ is stored at equivalent dosage levels.[263] This explains why the β isomer is more toxic when administered repeatedly even though the γ isomer is more toxic when given as a single dose. The difference in storage is explained by differences in metabolism;[8,83] α-HCH seems to accumulate more than lindane, especially in the brain.[265,266]

Multiple lindane administration induces expression of a number of drug metabolism enzymes involved in its oxidation and excretion[267–269] that may be genetically variable and be responsible in mice for strain differences in toxicity.[270,271] Isomers of HCH are metabolized by slightly different routes in mammals, and the biotransformation of lindane alone is complex.[8] Many of the products or intermediates, when given separately, are converted to other metabolites not usually detected during the metabolism of HCH isomers. Details of the metabolism of lindane and related chemicals can be found elsewhere.[8,272] Metabolism involves not only phase 1 pathways, such as oxidation by cytochrome P450, and phase 2 pathways, but conjugation of alcohol and phenol products to form glucuronides. Hydroxylations, epoxidations, *cis* and *trans*-dehydrochlorinations, isomerizations and desaturations lead to a large number of chlorinated cyclohexanols, cyclohexenols and phenols.[273–278] Under anaerobic conditions lindane is dechlorinated to chlorobenzene and benzene.[279] Conversion of lindane to other HCH isomers or to hexachlorobenzene is of little importance.[280,281] Most of the di- and trichlorophenyl mercapturates in the urine of lindane-treated rats arise by conjugation of hexachlorocyclohexenes and pentachlorocyclohexenes with glutathione, followed by dechlorination.[282,283] The alcohol and phenol metabolites are excreted as the glucuronides or sulphates but degree and types vary with species and strain, age, fibre content of diet, obesity and co-treatment with inducers of drug metabolism enzymes.[271,277,278,284,285] The metabolism of other isomers of HCH besides lindane has not been studied in as much detail as that of the γ isomer but probably occurs by variation of the routes for lindane.[8] Prenatal exposure to lindane can lead to persistence in cytochrome P450 expression in liver and brain.[286]

Neurotoxicity. The different isomers of HCH may have opposing pharmacological actions.[18] Lindane is a stimulant of the nervous system, causing violent epileptiform convulsions that are rapid in onset and generally followed by death or recovery within 24 h.[13,21,287–289] In rats, lindane may cause hypothermia and anorexia.[21,290,291] The α isomer is much weaker, and β and δ isomers are mainly depressants of the nervous system. The onset of effects are slow compared with that of lindane. Poisoning by α-HCH is characterized by tremors of the extremities and inability of the animals to make coordinated movements.

The main site of action of lindane seems to be at the synapse, with both excitatory and inhibitory effects.[292] The possible effect of inhibition of Na^+,K^+-ATPases on Ca^{2+} extrusion still requires more study.[21] These effects occur at concentrations of lindane greater than required for antagonism of the $GABA_A$ receptor complex and among different HCH isomers are not specific for the γ isomer.[293,294] The actions do not appear to be a consequence of direct inhibition.[295,296]

Lindane binds specifically to the $GABA_A$ receptor–ionophore complex, probably in the cerebellum,[297] and results in the effects of GABA and GABAergic transmission being disturbed.[22,297] The degree of binding for lindane, other HCH isomers, dieldrin, and other chlorinated insecticides to the $GABA_A$ receptor correlates with their acute toxicities[298] and abilities to induced convulsions.[18,22–24,299] Lindane also inhibits sterospecifically the GABA-induced CI-influx into rat brain membrane microsacs, whereas β-HCH has no effect.[300,301] Diazepam blocks the anorexic and hypothermic effects of lindane, and this could be taken as evidence for action at this receptor.[21] However, the protection by lindane against the convulsant properties of pentylenetetrazole, which also acts at the GABA-receptor-linked chloride channel, after the insecticide itself has disappeared, shows that the mechanism is complex.[302,303] Some work has implied that impairment of the GABA receptor–ionophore complex might not be involved in the neurotoxicity of lindane *in vivo,* but other evidence suggests differently.[304–306]

Often the symptoms of lindane poisoning are similar to those of dysfunction of the hippocampus–limbic system, and measurements of glucose uptake by brain at convulsant doses of lindane show increases in the limbic regions.[290,307] Long-term potentiation by lindane of the evoked potential elicited in the dentate gyrus after stimulation of the prepyriform cortex occurs when levels of the chemical should have greatly diminished.[308] Kindling is a sequence of changes resulting from repeated stimulation of a part of the limbic system, such as the amygdala. Progressively severe behavioural signs are observed in the rat after lindane exposure, commencing with eye-closing or chewing and climaxing in clonic motor seizure.[13,292,293,309,310] Exposure of neonates to lindane can greatly sensitize their acquisition of kindling in adulthood.[292] Studies showing that the granule cell population of the dentate gyrus becomes more excitable are comensurate with the convulsant action of lindane being explicable by interaction with the $GABA_A$ receptor.[292,311–313] Although experiments with cultured neurons from newborn rat dorsal root ganglia supported the hypothesis that lindane *in vivo* inhibits the $GABA_A$ receptor–channel complex, there may be other receptor targets.[314] Association of oxidative stress with lindane neurotoxicity has been proposed.[315]

Some biochemical changes after exposures to lindane or HCH do not seem to have direct connections with toxicity.[8] For instance, there has been interest in the turnover of inositol phospholipids after the occupation of some cell surface receptors by agonists and the involvement of Ca^{2+}.[316,317] Lindane does not act as a competitive inhibitor in enzyme systems which act on *myo*-inositol, but it does seem to inhibit phosphatidylinositol synthase although the importance of these findings *in vivo* is not clear.[318]

Mutation and Carcinogenesis. There is little evidence of genotoxic or mutagenic activity of lindane.[319,320] Rats are not susceptible to the tumorigenic effects of lindane and HCH.[8,319] On the other hand, mice seem to develop hepatic tumours in some strains after being fed lindane and the other isomers as well as technical HCH. The α isomer may be more active than the others in this respect. In mice with a dominant mutation at the agouti locus, which increases susceptibility to strain-specific spontaneous and chemically induced neoplasms, lindane acted as a tumour promoter *via* a cellular proliferation mechanism, perhaps as with other pesticides, by a nongenotoxic mechanism mediated by inhibition of gap-junction intercellular communication.[321] α-HCH, or technical HCH, acted as a promoter of rat liver tumours or foci previously initiated by carcinogens such as diethylnitrosamine.[37,39,322,323] In contrast, lindane protected rats against aflatoxin B_1-induced liver tumours.[324] As with DDT, the relevance of these findings to human exposure is questionable.

Effects on Reproduction. Chronic administration of lindane affects breeding parameters in rats.[325,326] Lindane and the β isomer are weakly oestrogenic to female rats and mice and the β isomer and lindane have adverse effects on the testes of rats and mice.[51,327–331] Seminiferous tubules become atrophied. The reduced sexual behaviour of female rats given lindane at pro-oestrus does not seem to be mediated through oestrogenic or GABAegic actions but by some other antioestrogenic mechanism.[332,333]

3.3.1.2 Toxicity to Humans

HCH was explored as a possible anthelmintic agent but it causes a burning sensation of the tongue.[334] Its use, or that of lindane, against scabies has been generally trouble-free[8] but is now greatly restricted or banned.[58] Permethrin cream rinse was found to give slightly better results against head lice than lindane with possibly fewer dermal reactions,[335] but lindane may be a better scabicide.[336] Because of the possible susceptibility of infants, pregnant women and patients with highly excoriated skin, other insecticides have been used instead.[337,338] In fact, lindane is absorbed through the skin to an appreciable degree over a period of a few days[339] and washing can aid absorption.[340,341]

Old surveys of workers have revealed a moderate number of mild and acute intoxications.[8] However, the cases often show many of the elements found in occupational poisoning by dieldrin and aldrin. No significant neurophysiological, neuromuscular or other poor health effects could be detected in a group of 60 workers producing lindane for 1–30 years,[342,343] but abnormal EEG patterns have been found in some workers exposed to HCH.[344] Toxic symptoms, including cardiac effects, were observed in workers engaged in the manufacture of technical HCH.[345]

The majority of effort to prevent poisoning by HCH and/or lindane focused on their use in vaporizing devices leading to low-level exposure. At one time, the unregulated sale of these devices led to their use in millions of homes. Many accidental poisonings were of children who ate vaporizer pellets.[8] The clinical course of fatal poisoning occurred in as little as 2 h after ingestion. Even at the peak of its use, the number of recorded deaths from HCH was very small averaging less than one a year in the US. Non-fatal cases of illness occurred when HCH or lindane was accidentally added to food or food supply. Initial symptoms included malaise, faintness, and dizziness followed by collapse and convulsions, sometimes accompanied by foaming at the mouth and biting the tongue, nausea, vomiting, severe cyanosis, facial pallor and ocular defects.[346,347] A moderate rise in temperature may be the consequence of the convulsions but the high fever that was reported in some cases may have been the direct toxic action of the pesticide, especially in children.[8] Recovery of normal health took some weeks.

The amount of HCH ingested in fatal poisonings seems to have been approximately 200 mg kg^{-1} or above, but variation in the ages of patients is a confounding factor.[8,348] Convulsions were observed at considerably lower doses.[349] A variety of chlorinated phenolic and conjugated metabolites of lindane have been reported in the urine of acutely poisoned individuals or exposed workers.[350–352]

3.3.2 Cyclodiene and Related Insecticides

The cyclodiene and related insecticides (shown in Figure 3.5 with their systematic names) have many similar modes of neurotoxic action, probably related to that of lindane, causing convulsions rather than the severe tremors usually characteristic of DDT. However, distinct differences exist in their storage and potency.[8] Even stereoisomers can differ remarkably in their potency of action or storage. Dieldrin is less toxic than endrin but is stored in humans whereas endrin is not. Of the group dieldrin is probably the most thoroughly investigated.

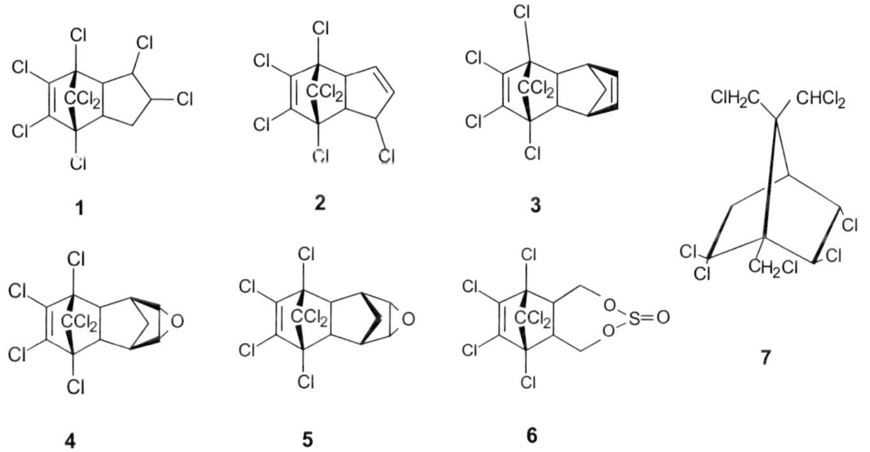

Figure 3.5 Structures of some chlorinated insecticides of the cyclodiene type: **1**, chlordane, 1,2,4,5,6,7,8,8-octachloro-2,3,3a,4,7,7a-hexahydro-1H-4,7-methanoindene; **2**, heptachlor, 1,4,5,6,7,8,8-heptachloro-3a,4,7,7a-tetrahydro-1H-4,7-methanoindene; **3**, aldrin, (1*R*,4*S*,4a*S*,5*S*,8*R*,8a*R*)-1,2,3,4,10,10-hexachloro-1,4,4a,5,8,8a-hexahydro-1,4:5,8-dimethanonaphthalene; **4**, dieldrin, (1a*R*,2*R*,2a*S*,3*S*,6*R*,6a*R*,7*S*,7a*S*)-3,4,5,6,9,9-hexachloro-1a,2,2a,3,6,6a,7,7a-octahydro-2,7:3,6-dimethanonaphtho[2,3-*b*]oxirene; **5**, endrin, (1*R*,2*S*,3*R*,6*S*,7*R*,8*S*,9*S*,11*R*)-3,4,5,6,13,13-hexachloro-10-oxapentacyclo[6.3.1.$1^{3,6}$.$0^{2,7}$.$0^{9,11}$]-tridec-4-ene; **6**, endosulfan, (5a*R*,6*R*,9*S*,9a*S*)-6,7,8,9,10,10-hexachloro-1,5,5a,6,9,9a-hexahydro-6,9-methano-2,4,3-benzodioxathiepine 3-oxide; **7**, toxaphene constituent 8-Cl Toxicant B.

Chlordane, heptachlor, aldrin, dieldrin, endrin and α-endosulfan had many domestic and agricultural formulations. Some pesticides used on crops, like aldrin, gave distinct flavours to foods. Dieldrin was also used as a residual spray in homes in the control of vectors of tropical diseases, mainly malaria. Probably aldrin and dieldrin were the most commonly used of the class, not only for agricultural purposes but also in public health. The decline in their use followed the recognition that dieldrin, especially, was accumulating in wildlife up through the food chain.

3.3.2.1 Experimental Toxicity

Absorption, Metabolism and Excretion. Although there is a wide range of potency from chlordane to endrin, all members of this group are toxic by the dermal route to an appreciable degree when compared to the oral toxicity. The exact values vary between studies, sex, and vehicle used for dosing.[8] Recovery can be rapid in survivors. With some insecticides, heptachlor and aldrin, the toxicities observed are probably due in part to their metabolism to the respective epoxides, *e.g.* heptachlor epoxide and dieldrin.[8,353] Chronic toxicity of the cyclodiene-type insecticides is a balance between detoxifying metabolism and accumulation to critical levels enough to cause convulsions or at lower accumulations to give liver changes. A dosage of about 0.35 mg kg^{-1} day^{-1} of heptachlor in the rat was found in a 2-year study to give hepatic changes whereas a dose of 0.25 mg kg^{-1} day^{-1} was a no observed effect level (NOEL).[354] Dieldrin, and perhaps some of the other chemicals, cause liver tumours after prolonged dietary exposure. Feeding or dosing of dams with most of these insecticides can cause increased mortality and decrease the number of offspring with perhaps skeletal abnormalities of the fetuses.[8] Aldrin has been reported to cause kidney damage in rats and dogs[355,356] and dieldrin to induce adrenal enlargement and functional changes.[357] Both dieldrin and chlordane can alter the immune competence in mice perhaps acting through suppression of macrophage function.[358,359] There is probably little relevance for humans at likely exposure levels.

All of these insecticides are efficiently absorbed from the intestinal tract, especially when dissolved in vegetable oil, possibly *via* the hepatic portal vein rather than *via* the lymphatic system as also occurs with DDT.[8,360] Importantly, unlike DDT, they are also toxic by absorption through the skin. Because of the retention of dieldrin in the body, unlike some of its analogues, this insecticide can also be passed on to offspring during lactation.

Metabolism can both activate and detoxify these insecticides. Aldrin is metabolized by epoxidation to both *endo*-dieldrin and the *exo* isomer (the active dieldrin) by cytochrome P450 isoforms.[361–363] In extrahepatic tissues such as seminal vesicles there is evidence for oxidation occurring by other mechanisms, including prostaglandin endoperoxide synthase.[364] As with the pairs aldrin and dieldrin, heptachlor and heptachlor epoxide, isodrin and endrin, the epoxide is probably more toxic than its parent, although this may not be apparent from the LD_{50} values due to rapid metabolism to the epoxide *in vivo*.[298,365] Metabolism of endrin to *syn*-12-hydroxyendrin and then to 12-ketoendrin, gives even greater potency. In contrast to potentiation of action, metabolism of all these chlorinated cyclodiene insecticides in the liver may lead to a variety of dechlorinated and hydroxylated products with a number of isomeric forms that can combine with cellular macromolecules and also be converted to glucuronides and sulphates for excretion. Some metabolites are stored more readily in the body organs than others. In mice *cis*- and *trans*-chlordanes are less readily stored than the metabolite oxychlordane.[366] The metabolism of dieldrin has been studied in particular detail with rats, mice, monkeys and other species. The metabolism of endosulfan by a stereoselective route in humans involves the cytochrome P450 enzymes CYP2B6 and 3A4.[367,368]

Mutagenesis and Cancer. Various studies show slight activity in mutational or chromosomal aberration assays and even DNA adducts but this seems to have little bearing on the development of tumours. With endosulfan, DNA adducts have been observed in cultured hepatic cells in parallel with CYP3A induction, although it has been concluded that the pesticide has no carcinogenic potential.[369,370] Some of the insecticides, including chlordane and dieldrin, give liver tumours in mice probably by a nongenotoxic mechanism.[8] Changes in liver histology with hypertrophy and induction of microsomal cytochrome P450 activities are observed at quite low doses, *e.g.* 10 ppm of dieldrin in the diet, but whether there are any associated increases in the rate of liver tumours is equivocal. For some of these insecticides, *e.g.* endrin, neurotoxicity occurs with lower doses than that required to accumulate the body burdens necessary for liver tumours. There is a marked formation of Mallory bodies in the liver tumours of mice fed dieldrin, possibly by hepatocyte premature ageing and polyplodization, perhaps involving an oxidative stress mechanism.[371–373] Other studies have implicated inhibition of intercellular comunication.[47]

Neurotoxicity. The neurotoxic action of dieldrin on the central nervous system typifies the neurotoxicities of the chlorinated cyclodiene-type insecticides. A major site of action appears to be at the synapse. Dieldrin binds to the picrotoxin binding site of the $GABA_A$ receptor–ionophore complex.[22,23,297–299,365] The mechanisms seem similar for endrin and endosulfan. The degree of binding broadly correlates with acute toxicity and with convulsions although other studies have been interpreted as not being completely consistent with this explanation.[374,375] The exact brain location is probably similar to that for lindane as the synaptic processes of the thalamocortical relay.[21] Parallel increases in blood flow and evoked response amplitude has been seen in rat cerebral cortex seizures initiated by dieldrin.[376] Transient hypothermia after a large dose of dieldrin has been reported and a reduction in food intake described but it is not known whether this occurs by the GABAergic pathway.[308] Endosulfan appears to increase aggressive and locomotive behaviour in rats possibly by the serotonergic system.[377–379]

Reproduction. On the whole, the experimental endocrine, reproductive and development effects of cyclodiene insecticides are minimal and have been studied *in vivo* and *in vitro*.[50,380,381]

3.3.2.2 Toxicity to Humans

Few cases have been reported for chlordane, heptachlor, isobenzan and endosulfan whereas there have been a significant numbers of deaths from aldrin, dieldrin and endrin, either accidental or suicidal.[8] Despite convulsions, accidental poisonings by dieldrin, for instance, have frequently not proved fatal. Many instances of poisoning by endrin, aldrin and dieldrin occurred by people eating contaminated sugar, rice, seed grain or bread. Sometimes this has been the consequence of contamination due to use of the insecticides near food, but in other instances were unexplained. In severe poisoning by endrin, violent epileptiform convulsions lasting a few minutes were followed by semi-consciousness or coma until the next fit. Even in moderate poisoning, especially of children, there might be no warning before the first fit. However, in a poisoning episode in Pakistan thought to be due to endrin, older patients complained of headaches, nausea and minor muscular spasms perhaps half an hour before collapsing.[382] Serious poisoning can be accompanied by hyperthermia. Acute poisonings by dieldrin and chlordane have similar, but not identical, symptoms.[8] Cerebral and pulmonary oedema has been observed following a fatal dose of dieldrin.[383] Dermal exposure can be as toxic as oral ingestion.[384] Non-fatal, non-convulsant doses of chlordane, aldrin, endrin or dieldrin have been associated with headaches, dizziness, gastrointestinal symptoms, skin irritation and weakness in the legs. Some patients went temporarily deaf. Workers making or using these pesticides have shown non-convulsant symptoms and even developed convulsions or hyperexcitability.[8] Recovery may depend on which of the insecticides is responsible. Dieldrin may have a much longer half-life in humans than, say, endrin. Recurrence of convulsions may be more likely even though the insecticide is less toxic. There is no firm evidence that exposure to these pesticides increases cancer risks despite the proposal that dieldrin may influence survival from breast cancer.[66] Heavy agricultural exposure to pesticides, particularly endrin, has been linked to an excess of deaths due to stroke and cerebrovascular diseases.[385,386]

3.3.3 Toxaphene

Extensive chlorination of camphene to 67–69% by weight gives a complex mixture of many chemicals, known as toxaphene, which has been used as an insecticide. During the chlorination procedure 2-*exo*,10-dichlorobornane is first formed by rearrangement so that many of the products are bornanes not camphenes. An example is shown in Figure 3.5. Similar products have been formed by the chlorination of pinene. Toxaphene was valued both for its power as an insecticide and because of its limited environmental persistence and rapid excretion by mammals, but investigation of its metabolism has been difficult because of the complicated mixture involved.[8] Following initial distribution throughout the body of rats, redistribution to adipose tissue and to the adrenals occurs.[387,388] Extensive metabolism by dechlorination is followed by conjugation to glucuronic acid and glutathione; levels in tissues decline rapidly.[389]

3.3.3.1 Experimental Toxicity

The acute toxicity of toxaphene is similar to that observed with lindane and the cyclodiene-type chlorinated insecticides. Animals show hyperactivity, muscle spasms, convulsions and coma but may recover quickly with metabolism of the insecticide. Toxaphene is moderately toxic to rats by mouth but substantially less toxic dermally, unlike endrin and dieldrin. The 8-chloro-B constituent seems to be the most acute orally toxic constituent.[390] Toxaphene constituents are mutagenic in some tests but not positive in the mouse dominant lethal assay.[391] As with many of these highly chlorinated chemicals, prolonged feeding causes liver and thyroid tumours[392] but probably not by a genotoxic mechanism since no adducts are observed.[393] Thyroid tumours are probably the result of proliferation driven by thyroid stimulating hormone.[394] The influence of toxaphene on reproduction is mild.[8] Biochemical studies on various ATPase functions following toxaphene poisoning in animals are difficult to put into context.[8] Exposure of cynomolgus monkeys to varying levels of toxaphene (up to 0.8 mg kg^{-1} day^{-1}) for 20 weeks showed decreased serum cholesterol levels but no other detectable changes.[395] The most important studies are those showing that the 9-chloro toxicant B is a potent ligand for the picrotoxin binding site of mouse brain synaptosomes blocking the $GABA_A$-regulated chloride ionophore,[22,23,298] thus resembling the actions of lindane and the cyclodienes.

3.3.3.2 Toxicity to Humans

Fatal and non-fatal acute poisonings by toxaphene have begun within half an hour of ingestion of the pesticide and were characterized by mental confusion, possible nausea, jerking of arms and legs and by convulsions. In many cases of survivors, recovery was rapid. Chronic exposure to chlorinated camphene and pinene mixtures similar to toxaphene gave headache, nausea and abdominal pain and weakness.[8,396] What little is documented suggests that a fatal oral dose to humans is about 10 mg kg^{-1}.[397]

3.4 Mirex and Chlordecone

Mirex is 1,1a,2,2,3,3a,4,5,5,5a,5b,6-dodecachlorooctahydro-1,3,4-metheno-1*H*-cyclobuta[*c,d*]-pentalene and chlordecone is 1,1a,3,3a,4,5,5,5a,5b,6-decachloro-octahydro-1,3,4,-metheno-2*H*-cyclobut[*c,d*] pentalen-2-one. They resemble the chlorinated cyclodiene insecticides but with a cage structure (Figure 3.6). Mirex was an insecticide for consumption by fire ants whereas chlordecone was used as a contact insecticide against leaf-cutting insects and fly larvae. They are only slowly metabolized, often concentrating many thousandfold in food chains.

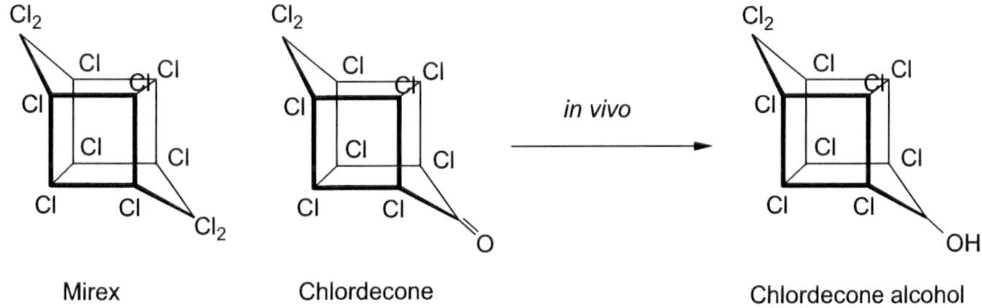

Figure 3.6 Structures of mirex and chlorodecone together with the human metabolite chlordecone alcohol.

Unlike lindane and the cyclodienes, they do not immediately cause sudden seizures. In some respects in action they resemble DDT and its analogues. These two insecticides have properties in common but also have different metabolism and toxicities because of the ketone function in chlordecone.

3.4.1 Mirex

Mirex is virtually insoluble in water and is poorly absorbed from the gastrointestinal tract with a half-life in rats of about 100 days.[398] The major dechlorination products in the rat are 2,8-dihydromirex and 5,10-dihydromirex. The latter is converted to more polar metabolites that can be excreted in the urine.[399]

3.4.1.1 *Experimental Toxicity*

The acute oral LD50 toxicity of mirex is approximately 300–3000 mg kg^{-1} for rats and 250 and 125 mg kg^{-1} for male and female hamsters respectively.[8,197,400] Rats fed diets containing mirex showed hyperexcitibility, depressed haemoglobin levels, tremors and even convulsions.[401] Photomirex (8-monohydromirex), formed in the environment, may be more toxic than the parent compound. Enlarged livers have been observed in rats, mice and rabbits fed mirex, but little in dogs,[401–404] which can persist for a year.[405] Increased biliary tree dysfunction and impaired bile acid secretion probably reflect changes in ATPase transport pumps at the cell membranes.[406–409] The proliferation of the hepatic endoplasmic reticulum in rodents is associated with induction of cytochrome P450 isoforms.[8] Mirex and chlordecone might act in a similar way to induce CYP isoforms but there are differences between these chemicals.[410,411]

Chronic administration of mirex to rodents causes hepatic adenomas and carcinomas probably by an uncertain epigenetic mechanism.[412–414] Mirex can act as a promoter of skin carcinogenesis,[415–418] but the significance of these findings for human health is questionable and a number of *in vitro* mutagenicity tests have proven negative.[8] Depression of serum triodothyronine and thyroxine levels, with reduction of colloid density in the thyroid, are possibly related to the liver effects.[404,419] Low doses of mirex (*e.g.* 5 ppm in the diet to mice) cause reduced litter size and viability,[400] and offspring of rat mothers exposed to mirex may develop cataracts;[420] this seems to be an indirect effect perhaps associated with oedema or protein insufficiency. Both mirex and photomirex damage rat testes causing hypocellularity of seminiferous tubules and decreased spermatogenesis.[404]

3.4.1.2 *Toxicity to Humans*

At its time of use there appeared to be no reports of its human toxicity although highly persistent in human and soil samples where used on ants.[8]

3.4.2 Chlordecone

3.4.2.1 Experimental Toxicity

Absorption and Metabolism. Chlordecone is the 2,2 dechlorinated 2-oxo analogue of mirex (often known by its trade name Kepone) and is slightly more soluble in water than mirex. The greater aqueous solubility means that it is easily more absorbed from the intestine and is probably more greatly distributed to the blood and tissues, other than fat. Compared to some other polyhalogenated pesticides and related chemicals, the decline of chlordecone from tissues is relatively rapid and a significant proportion of a single dose can be excreted in the milk of lactating rats.[421] In both rats and humans, a major route of excretion is in the faeces; in humans some by secretion of the pesticide through the intestinal wall.[422,423] Chlordecone induces a number of cytochrome P450-mediated oxidations in the male rat liver.[424] The presence of the ketone function in chlordecone provides a simple route for metabolism by reduction of the ketone group (Figure 3.6) to give chlordecone alcohol (chlordecol) and subsequent glucuronidation.[425,426] This is only a minor metabolism route in rats, mice and guinea-pigs but much more significant in humans and the gerbil.[427] The species differences seemed to be explained by the presence of a specific aldo-ketose reductase in humans, gerbils and pigs.[428–430] Despite the formation of chlordecone alcohol in humans, the overall rate of clearance of chlordecone is probably greater in rats and mice.

Toxicity. Morbidity of animals due to chlordecone is preceded by abnormal gait and severe tremors. The acute dermal LD_{50} value for rats is approximately 10 times that of the oral dose, reflecting the solubility of the pesticide in the intestinal tract.[8,191,431,432] Chronic administration leads to a decreased weight gain and liver enlargement as well as changes in the adrenals (hyperplasia of the zona fasiculata and zona reticularis), kidney and testes.[433,434] Hepatocellular carcinomas and adenomas occur in mice and rats after chronic administration of chlordecone at low doses in their diets.[431,435,436] Chlordecone does not seem to act as a mutagen in its hepatocarcinogenic action and is more likely acting as a epigenetic carcinogen.[8,437]

Many studies have been concerned with inhibition of mitochondrial and membrane-bound Na^+, K^+-ATPases and certain Mg^+-ATPases caused by chlordecone, perhaps associated with energy production.[8] Other studies of liver and brain implicate aspects of calcium metabolism linked to ATPases in a mechanism that does not occur with mirex.[101,438] The relationship between brain ATPases, calcium homeostasis and neurotransmitter turnover is still not understood. It was proposed that chlordecone increases free intrasynaptosomal Ca^{2+} by a non-specific leakage through the plasma membrane and by passage through voltage-sensitive Ca^{2+} channels due to membrane depolarization.[439] This is compatible with increased synaptosomal calcium levels causing depolarization of the membrane, enhancing release of neurotransmitters.

Chlordecone may also, through activation of serotonin neurons, cause decreased GABAergic activity in the striatum and thereby an increase cholinergic tone, causing tremors.[440] It is thought that chlordecone does not bind to the picrotoxin binding site of the $GABA_A$ receptor complex to the degree of lindane, cyclodienes and toxaphenes, but this has not been deeply studied.[22] In contrast, chlordecone was more potent than cyclodiene-type insecticides in releasing dopamine from striatal synaptosomes.[441]

Chlordecone potentiates CCl_4 hepatic toxicity and partial hepatectomy protects against this. Potentiation occurs at doses of chlordecone that do not increase lipid peroxidation or cytochrome P450 activity.[8,442] The findings do not seem to be explained by greater CCl_4 metabolism. Chlordecone enhances the suppression of calcium sequestration by microsomes and it is thought that chlordecone suppresses hepatocellular regeneration after necrosis caused by CCl_4[443–450] Neonatal rats are resistant.[451–455] Chlordecone also inhibits transfer of metabolites from hepatocyte to the bile canaliculus, possibly by interference of ABC transporters,[451,456] and may disturb cholesterol and steroid metabolism and homeostasis.[457,458]

Studies in rats, mice and birds showed clearly that chlordecone affects breeding performance.[8,456] Surviving rat fetuses exhibited reduced body weight, reduced ossification, oedema, undescended testicles, enlarged renal pelvis and enlarged cerebral ventricles.[459] Some of the reproductive effects are due to chlordecone acting as an oestrogen.[50] In uterotrophic assays on neonatal mice, chlordecone has marked dose-related effects on both the vagina and uterus.[215,460] As we would now expect from present knowledge of the action of oestrogens and xeno-oestrogens with the oestrogen α and β receptors, the action of chlordecone was not identical to that of oestradiol.[253] Interestingly, a stress-related protein Bip acts as a critical regulator to amplify chlordecone oestrogenic potency on the α-receptor in mouse uterus but not *o,p*′-DDT and methoxychlor.[461] Chlordecone is an antagonist of oestrogen β receptor.[462,463] In some experiments chlordecone caused decreases in spermatozoa motility and viability but does not seem to produce dominant lethal effects.[464] Chlordecone affecting autoimmune responses in mice seems different from the oestrogenic pathway of oestradiol.[465,466]

3.4.2.2 Toxicity to Humans

In an industrial incident, workers manufacturing chlordecone were chronically exposed to the pesticide.[456,467] Of those employees or former employees examined, 57% had symptoms of poisoning. Those affected had weight loss, headache, tremor especially of the upper extremities, muscle weakness, unusual eye movements, ataxia, slurred speech, skin rash and abnormalities of normal liver function.[468] It was thought that absorption probably occurred by the dermal route and some symptoms persisted.[469] Sural nerve, muscle and liver biopsies showed a variety of changes including accumulation of elongated, electron dense membranous bodies in Schwann cells below the sarcolemma and between myofibrils and proliferation of the hepatic endoplasmic reticulum.[8,456]

Of 28 chlordecone-poisoned workers examined, 8 had normal sperm counts but in only 1 was chlordecone blood level above 1 ppm.[456] Arrest of sperm maturation was observed in testicular biopsies from some patients. Neurotoxicity was associated with blood levels of chlordecone greater than 0.1–1 ppm. Chlordecone was also found in the blood of non-affected workers, relatives, workers in nearby businesses and others.[470] Plasma levels relative to adipose tissue seemed to be much higher than expected compared with some other chlorinated insecticides perhaps reflecting specific binding to albumin and lipoproteins.[471] Most of excreted chlordecone in humans is in the faeces and some as the glucuronide of chlordecol.[426] Cholestyramine treatment effected faster excretion of the chemical.[472] Subsequent work showed that expression of the aldo-keto reductase responsible for chlordecol formation varied considerably between individuals.[428,429] Recent findings showed that the incidence of prostate cancer in Antigua, where chlordecone was used for 20 years, was associated with plasma levels of the insecticide and a possibly less efficient variant of chlordecone reductase.[473]

3.5 Regulatory Aspects

In the European Union (EU) and the North American Free Trade Area (NAFTA) organochlorine insecticides are nowadays rarely used in agriculture or domestically. However, uses around the world do still occur. Lindane, the last organochlorine to be used in food production in the EU, was revoked in 2001. The remaining uses of lindane in the US for targeting lice and treating scabies as well, as some other uses, are also under pressure.[58,338] It is many years since DDT was used in food production, but because of its persistence in the soil, residues of DDT and its metabolites continue to occur in commodities. To facilitate international trade, the Joint Expert Meeting on Pesticides Residues assigned DDT a provisional tolerable daily intake in 2000.[474] The complex arguments concerning a total worldwide ban in the light of the problems in combating malaria have been well discussed elsewhere.[11,129] The ending of the use of endosulfan in the US seems likely;[475] this has been the subject of recent detailed risk and exposure assessments.[476–479] Some insecticides such as complex toxaphene mixtures and their environmental breakdown products have not been used for decades. Previous risk assessments may be out of date in the light of modern concepts such as mode of action and nonlinearity of cancer response.[480]

3.6 Summary

The use of organochlorine insecticides has massively declined worldwide over the last decades. However, their toxicology is of interest as in some countries they are still sometimes used or their levels in humans or the environment are of concern. They have been of great use in the general studies of the mechanisms of neurotoxic agents, for instance the role of different GABA receptors, but there are continual concerns over their potential to be endocrine disruptors at environmental levels or associated with residual spraying. At the time of writing, DDT is still allowed by the United Nations for combating malaria although this requires careful monitoring and assessment. Elimination of all of these pesticides from the world's arsenal may be unwise. There may still be a role for the use of some of these highly effective insecticides in the future, under strictly controlled conditions, in circumstances of world health brought about by climate change or human activities that have yet to develop.

References

1. J. R. P. Cameron, R. K. Hall, L. Rossi, S. A. Bronczyk and P. Shubik, Risks to man and animals from the use of 2,2-bis(*p*-chlorophenyl)-1,1,1-trichlorethane (DDT): With a note on the toxicity of γ-benzene hexachloride (666, Gammexane), *Br. Med. J.*, 1945, **3**, 865–871.
2. H. D. Chalke, The Naples typhus epidemic: the use of D.D.T, *Proc. R. Soc. Med.*, 1946, **39**, 165–168.
3. W. J. Hayes, Jr., Introduction, in *Handbook of Pesticide Toxicology*, ed. W. J. Hayes and E. R. Laws, 1991, Academic Press, San Diego, CA, pp. 1–37.
4. A. Mandavilli, Health agency backs use of DDT against malaria, *Nature*, 2006, **443**, 250–251.
5. A. Mandavilli, DDT returns, *Nat. Med.*, 2006, **12**, 870–871.
6. World Health Organization. *The Use of DDT in Malaria Vector Control. WHO Position Statement.* 2007 (cited 16 March 2011). http://www.who.int/ipcs/capacity_building/who_statement.pdf.
7. A. Chen, J. Zhang, L. Zhou, E. S. Gao, L. Chen, W. J. Rogan and M. S. Wolff, DDT serum concentration and menstruation among young Chinese women, *Environ. Res.*, 2005, **99**, 397–402.
8. A. G. Smith, Chlorinated hydrocarbon insecticides, in *Handbook of Pesticide Toxicology*, ed. W. J. Hayes and E. R. Laws, 1991, Academic Press, San Diego, CA, pp. 731–915.
9. A. G. Smith, DDT and its analogs, in *Handbook of Pesticide Toxicology*, ed. R. I. Krieger, Academic Press, Orlando, FL, 2nd edn, 2001, pp. 1305–1355.
10. E. D. Carter, 'God bless General Peron': DDT and the endgame of malaria eradication in Argentina in the 1940s, *J. Hist. Med. Allied Sci.*, 2009, **64**, 78–122.
11. H. van den Berg, Global status of DDT and its alternatives for use in vector control to prevent disease, *Environ. Health Perspect.*, 2009, **117**, 1656–1663.
12. R. Bate, D. B. Roberts, and R. Tren. *The United Nations' Scientific Fraud against DDT,* Outlook Series 2011. http://www.aei.org/outlook/101019.

12a. World Health Organization, IPCS, Environmental Health Criteria 241, DDT in indoor residual spraying: Human health aspects, 2011.

13. R. M. Joy, Mode of action of lindane, dieldrin and related insecticides in the central nervous system, *Neurobehav. Toxicol. Teratol.*, 1982, **4**, 813–823.
14. M. H. Bickel, The role of adipose tissue in the distribution and storage of drugs, *Prog. Drug. Res.*, 1984, **28**, 273–303.
15. C. T. Bedford and D. H. Hutson, The comparative metabolism in rodents of the isomeric insecticides dieldrin and endrin, *Chem. Ind. (London)*, 1976, 440–447.
16. J. F. Cole, L. M. Klevay and M. R. Zavon, Endrin and dieldrin: a comparison of hepatic excretion in the rat, *Toxicol. Appl. Pharmacol.*, 1970, **16**, 547–555.
17. A. A. Jensen, Chemical contaminants in human milk, *Residue Rev.*, 1983, **89**, 1–128.
18. T. Narahashi, Neurophysiological effects of insecticides, in *Handbook of Pesticide Toxicology*, 2nd edn, ed. R. Krieger, Elsevier, Amsterdam, 2010, pp. 799–816.
19. Y. Ishikawa, P. Charalambous and F. Matsumura, Modification by pyrethroids and DDT of phosphorylation activities of rat brain sodium channel, *Biochem. Pharmacol.*, 1989, **38**, 2449–2457.
20. T. Narahashi, D. B. Carter, J. Frey, K. Ginsburg, B. J. Hamilton, K. Nagata, M. L. Roy, J. H. Song and H. Tatebayashi, Sodium channels and GABA(A) receptor-channel complex as targets of environmental toxicants, *Toxicol. Lett.*, 1995, **83**, 239–245.
21. D. Woolley, L. Zimmer, D. Dodge and K. Swanson, Effects of lindane-type insecticides in mammals: unsolved problems, *Neurotoxicology*, 1985, **6**, 165–192.
22. L. J. Lawrence and J. E. Casida, Interactions of lindane, toxaphene and cyclodienes with brain-specific t-butylbicyclophosphorothionate receptor, *Life Sci.*, 1984, **35**, 171–178.
23. L. M. Cole and J. E. Casida, Polychlorocycloalkane insecticide-induced convulsions in mice in relation to disruption of the GABA-regulated chloride ionophore, *Life Sci.*, 1986, **39**, 1855–1862.
24. K. Nagata and T. Narahashi, Differential effects of hexachlorocyclohexane isomers on the GABA receptor-chloride channel complex in rat dorsal root ganglion neurons, *Brain Res.*, 1995, **704**, 85–91.
25. J. Huang and J. E. Casida, Role of cerebellar granule cell-specific GABA(A) receptor subtype in the differential sensitivity of [^{3}H]ethynylbicycloorthobenzoate binding to GABA mimetics, *Neurosci. Lett.*, 1997, **225**, 85–88.
26. G. S. Ratra, S. G. Kamita and J. E. Casida, Role of human GABA(A) receptor beta3 subunit in insecticide toxicity, *Toxicol. Appl. Pharmacol.*, 2001, **172**, 233–240.
27. G. S. Ratra, B. E. Erkkila, D. S. Weiss and J. E. Casida, Unique insecticide specificity of human homomeric ρ1 $GABA_C$ receptor, *Toxicol. Lett.*, 2002, **129**, 47–53.
28. K. Swanson and D. Woolley, Neurotoxic effects of dieldrin, *Toxicol. Appl. Pharmacol.*, 1978, **45**, 339.

29. M. E. Gilbert, Repeated exposure to lindane leads to behavioral sensitization and facilitates electrical kindling, *Neurotoxicol. Teratol.*, 1995, **17**, 131–141.
30. M. P. Holsapple, H. C. Pitot, S. M. Cohen, A. R. Boobis, J. E. Klaunig, T. Pastoor, V. L. Dellarco and Y. P. Dragan, Mode of action in relevance of rodent liver tumors to human cancer risk, *Toxicol. Sci.*, 2006, **89**, 51–56.
31. O. G. Fitzhugh and A. A. Nelson, The chronic oral toxicity of DDT (2,2-bis-*p*-chlorophenyl-1,1,1-trichloroethane). *J. Pharmacol. Exp. Ther.*, 1947, **89**, 18–30.
32. L. Rossi, M. Ravera, G. Repetti and L. Santi, Long-term administration of DDT or phenobarbital-Na in Wistar rats, *Int. J. Cancer*, 1977, **19**, 179–185.
33. J. R. Cabral, R. K. Hall, L. Rossi, S. A. Bronczyk and P. Shubik, Effects of long-term intake of DDT on rats, *Tumori*, 1982, **68**, 11–17.
34. T. Harada, R. Ohtsuka, M. Takeda, T. Yoshida, A. Enomoto, S. Kojima, N. Tomiyama, N. Nakashima and M. Ozaki, Hepatocarcinogenesis by DDT in rats, *J. Toxicol. Pathol.*, 2006, **19**, 155–167.
35. IARC and World Health Organization, Some organochlorine pesticides., in *IARC Monogr. Eval. Carcinog. Risk Chem. Man*, International Agency for Research on Cancer, Lyon, 1974.
36. C. Wildemauwe, J. F. Lontie, L. Schoofs and N. van Larebeke, The mutagenicity in procaryotes of insecticides, acaricides, and nematicides, *Residue Rev.*, 1983, **89**, 129–178.
37. M. A. Pereira, S. L. Herren, A. L. Britt and M. M. Khoury, Sex difference in enhancement of GGTase-positive foci by hexachlorobenzene and lindane in rat liver, *Cancer Lett.*, 1982, **15**, 95–101.
38. G. M. Williams and S. Numoto, Promotion of mouse liver neoplasms by the organochlorine pesticides chlordane and heptachlor in comparison to dichlorodiphenyltrichloroethane, *Carcinogenesis*, 1984, **5**, 1689–1696.
39. R. Schulte-Hermann, Tumor promotion in the liver, *Arch. Toxicol.*, 1985, **57**, 147–158.
40. A. I. Walker, E. Thorpe and D. E. Stevenson, The toxicology of dieldrin (HEOD). I. Long-term oral toxicity studies in mice, *Food Cosmet. Toxicol.*, 1973, **11**, 415–432.
41. A. S. Wright, D. Potter, M. F. Wooder, C. Donninger and R. D. Greenland, The effects of dieldrin on the subcellular structure and function of mammalian liver cells, *Food Cosmet. Toxicol.*, 1972, **10**, 311–332.
42. D. Obuchowska and A. Pawlowska-Tochman, [The effect of gamma-isomer HCH (Lindane) on the ultrastructure of the liver cell (author's transl)], *Ann. Univ. Mariae Curie Skłodowska [Med]*, 1973, **28**, 63–66.
43. W. Kunz, G. Schaude, W. Schmid and M. Siess, Liver hypertrophy caused by foreign agents, *Naunyn-Schmiedebergs Arch. Pharmakol. Exp. Pathol.*, 1966, **254**, 470–488.
44. D. E. Stevenson, E. F. Walborg, Jr., D. Warner North, Jr., R. L. Sielken, C. E. Ross, A. S. Wright, Y. Xu, L. M. Kamendulis and J. E. Klaunig,

Monograph: Reassessment of human cancer risk of aldrin/dieldrin, *Toxicol. Lett.*, 1999, **109**, 123–186.
45. G. Tsushimoto, C. C. Chang, J. E. Trosko and F. Matsumura, Cytotoxic, mutagenic, and cell-cell communication inhibitory properties of DDT, lindane, and chlordane on Chinese hamster cells, *in vitro*, *Arch. Environ. Contam. Toxicol.*, 1983, **12**, 721–729.
46. L. Warngard, H. Hemming, S. Flodstrom, S. K. Duddy and G. E. Kass, Mechanistic studies on the DDT-induced inhibition of intercellular communication, *Carcinogenesis*, 1989, **10**, 471–476.
47. L. Zhong-Xiang, T. Kavanagh, J. E. Trosko and C. C. Chang, Inhibition of gap junctional intercellular communication in human teratocarcinoma cells by organochlorine presticides, *Toxicol. Appl. Pharmacol.*, 1986, **83**, 10–19.
48. A. M. Soto, C. Sonnenschein, K. L. Chung, M. F. Fernandez, N. Olea and F. O. Serrano, The E-SCREEN assay as a tool to identify estrogens: an update on estrogenic environmental pollutants, *Environ. Health Perspect.*, 1995, **103 Suppl 7**, 113–122.
49. A. M. Cummings, Methoxychlor as a model for environmental estrogens, *Crit. Rev. Toxicol.*, 1997, **27**, 367–379.
50. R. J. Kavlock, Pesticides as endocrine-disrupting chemicals, in *Handbook of Pesticide Toxicology*, 2nd edn, ed. R. Krieger, Elsevier, Amsterdam, 2001, pp. 727–746.
51. R. B. Raizada, P. Misra, P. Saxena, K. K. Datta and T. S. Dikshith, Weak estrogenic activity of lindane in rats, *J. Toxicol. Environ. Health*, 1980, **6**, 483–492.
52. R. A. Cassidy, C. V. Vorhees, D. J. Minnema and L. Hastings, The effects of chlordane exposure during pre- and postnatal periods at environmentally relevant levels on sex steroid-mediated behaviors and functions in the rat, *Toxicol. Appl. Pharmacol.*, 1994, **126**, 326–337.
53. W. R. Kelce, C. R. Stone, S. C. Laws, L. E. Gray, J. A. Kemppainen and E. M. Wilson, Persistent DDT metabolite *p,p′*-DDE is a potent androgen receptor antagonist, *Nature*, 1995, **375**, 581–585.
54. M. Bagchi, E. Hassoun, P. Akubue, D. Bagchi and S. J. Stohs, Comparative effects of endrin on hepatic lipid peroxidation and DNA damage, and nitric oxide production by peritoneal macrophages from C57BL/6J and DBA/2 mice, *Comp. Biochem.Physiol. C, Comp. Pharmacol. Toxicol.*, 1993, **105**, 525–529.
55. E. A. Hassoun and S. J. Stohs, Comparative teratological studies on TCDD endrin and lindane in C57BL/6J and DBA/2J mice, *Comp. Biochem.Physiol. C, Comp. Pharmacol. Toxicol.*, 1996, **113**, 393–398.
56. V. Turusov, V. Rakitsky and L. Tomatis, Dichlorodiphenyltrichloroethane (DDT): ubiquity, persistence, and risks, *Environ. Health Perspect.*, 2002, **110**, 125–128.
57. B. Van Wendel de Joode, C. Wesseling, H. Kromhout, P. Monge, M. Garcia and D. Mergler, Chronic nervous-system effects of long-term occupational exposure to DDT, *Lancet*, 2001, **357**, 1014–1016.

58. B. Weinhold, Last call for Lindane, *Environ. Health Perspect.*, 2001, **109**, A254.
59. B. Eskenazi, J. Chevrier, L. G. Rosas, H. A. Anderson, M. S. Bornman, H. Bouwman, A. Chen, B. A. Cohn, C. de Jager, D. S. Henshel, F. Leipzig, J. S. Leipzig, E. C. Lorenz, S. M. Snedeker and D. Stapleton, The Pine River statement: human health consequences of DDT use, *Environ. Health Perspect.*, 2009, **117**, 1359–1367.
60. L. Hardell, G. Lindstrom and B. Van Bavel, Is DDT exposure during fetal period and breast feeding associated with neurological impairment?, *Environ. Res.*, 2002, **88**, 141–144.
61. M. P. Longnecker, M. A. Klebanoff, H. Zhou and J. W. Brock, Association between maternal serum concentration of the DDT metabolite DDE and preterm and small-for-gestational-age babies at birth, *Lancet*, 2001, **358**, 110–114.
62. J. Beard, DDT and human health, *Sci. Total Environ.*, 2006, **355**, 78–89.
63. M. S. Wolff, G. S. Berkowitz, S. Brower, R. Senie, I. J. Bleiweiss, P. Tartter, B. Pace, N. Roy, S. Wallenstein and A. Weston, Organochlorine exposures and breast cancer risk in New York City women, *Environ. Res.*, 2000, **84**, 151–161.
64. M. S. Wolff, A. Zeleniuch-Jacquotte, N. Dubin and P. Toniolo, Risk of breast cancer and organochlorine exposure, *Cancer Epidemiol. Biomarkers Prev.*, 2000, **9**, 271–277.
65. S. M. Snedeker, Pesticides and breast cancer risk: a review of DDT, DDE, and dieldrin, *Environ. Health Perspect.*, 2001, **109 Suppl 1**, 35–47.
66. A. P. Hoyer, T. Jorgensen, J. W. Brock and P. Grandjean, Organochlorine exposure and breast cancer survival, *J. Clin. Epidemiol.*, 2000, **53**, 323–330.
67. A. Demers, P. Ayotte, J. Brisson, S. Dodin, J. Robert and E. Dewailly, Risk and aggressiveness of breast cancer in relation to plasma organochlorine concentrations, *Cancer Epidemiol. Biomarkers Prev.*, 2000, **9**, 161–166.
68. J. D. Scribner and N. K. Mottet, DDT acceleration of mammary gland tumors induced in the male Sprague-Dawley rat by 2-acetamidophenanthrene, *Carcinogenesis*, 1981, **2**, 1235–1239.
69. P. Cocco, A. Blair, P. Congia, G. Saba, C. Flore, M. R. Ecca and C. Palmas, Proportional mortality of dichloro-diphenyl-trichloroethane (DDT) workers: a preliminary report, *Arch. Environ. Health*, 1997, **52**, 299–303.
70. J. H. Olsen, G. Schulgen, J. D. Boice, Jr., J. Whysner, L. B. Travis, G. M. Williams, F. B. Johnson and J. O. McGee, Antiepileptic treatment and risk for hepatobiliary cancer and malignant lymphoma, *Cancer Res.*, 1995, **55**, 294–297.
71. A. G. Smith, Toxicology of DDT and some analogues, in *Handbook of Pesticide Toxicology*, 2nd edn, ed. R. Krieger, Elsevier, Amsterdam, 2010, pp. 1975–2032.
72. G. L. Ross, The DDT question, *Lancet*, 2000, **356**, 1189.

73. A. Attaran and R. Maharaj, Ethical debate: doctoring malaria, badly: the global campaign to ban DDT, *BMJ*, 2000, **321**, 1403–1405.
74. A. Attaran, D. R. Roberts, C. F. Curtis and W. L. Kilama, Balancing risks on the backs of the poor, *Nat. Med.*, 2000, **6**, 729–731.
75. C. F. Curtis and J. D. Lines, Should DDT be banned by international treaty, *Parisitol. Today*, 1999, **16**, 119–121.
76. X. Qiu, T. Zhu, B. Yao, J. Hu and S. Hu, Contribution of dicofol to the current DDT pollution in China, *Environ. Sci. Technol.*, 2005, **39**, 4385–4390.
77. W. E. Dale, M. F. Copeland, G. W. Pearce and J. W. Miles, Concentration of *o,p'*-DDT in rat brain at various intervals after dosing, *Arch. Int. Pharmacody. Ther.*, 1966, **162**, 40–43.
78. K. J. Palin, C. G. Wilson, S. S. Davis and A. J. Phillips, The effect of oils on the lymphatic absorption of DDT, *J. Pharm. Pharmacol.*, 1982, **34**, 707–710.
79. D. E. Pocock and A. Vost, DDT absorption and chylomicron transport in rat, *Lipids*, 1974, **9**, 374–381.
80. S. M. Sieber, The lymphocytic absorption of *p,p'*-DDT and some structurally-related compounds in the rat, *Pharmacology*, 1976, **14**, 443–454.
81. O. Tebourbi, M. R. Driss, M. Sakly and K. B. Rhouma, Metabolism of DDT in different tissues of young rats, *Environ. Sci. Health, B*, 2006, **41**, 167–176.
82. Z. Chen, F. Maartens, H. Vega, S. Kunene, J. Gumede and R. I. Krieger, 2,2-bis(4-chlorophenyl)acetic acid (DDA), a water-soluble urine biomarker of DDT metabolism in humans, *Int. J. Toxicol.*, 2009, **28**, 528–533.
83. M. Kujawa, R. Macholz and R. Knoll, [The enzymatic degradation of DDT. 5. Direct transformation of DDD (dichlorodiphenyldichloroethane) into an aldehyde], *Nahrung*, 1985, **29**, 517–522.
84. B. Gold and G. Brunk, Metabolism of 1,1,1-trichloro-2,2-bis(*p*-chlorophenyl)ethane (DDT), 1,1-dichloro-2,2-bis(*p*-chlorophenyl)ethane, and 1-chloro-2,2-bis(*p*-chlorophenyl)ethene in the hamster, *Cancer Res.*, 1983, **43**, 2644–2647.
85. B. Gold and G. Brunk, A mechanistic study of the metabolism of 1,1-dichloro-2,2-bis(*p*-chlorophenyl)ethane (DDD) to 2,2-bis(*p*-chlorophenyl)acetic acid (DDA), *Biochem. Pharmacol.*, 1984, **33**, 979–982.
86. S. C. Fawcett, L. J. King, P. J. Bunyan and P. I. Stanley, The metabolism of 14C-DDT, 14C-DDD, 14C-DDE and 14C-DDMU in rats and Japanese quail, *Xenobiotica*, 1987, **17**, 525–538.
87. F. Martz and J. A. Straw, Metabolism and covalent binding of 1-(*o*-chlorophenyl)-1-(*p*-chlorophenyl)-2,2-dichloroethane (*o,p,'*-DDD). Correlation between adrenocorticolytic activity and metabolic activation by adrenocortical mitochondria, *Drug Metab. Dispos.*, 1980, **8**, 127–130.
88. R. C. Pohland and R. E. Counsell, In vitro and in vivo metabolism of a radioiodinated analog of 1-(2-chlorophenyl)-1-(4-chlorophenyl)-2,-dichloroethane, *Drug Metab. Dispos.*, 1985, **13**, 113–115.

89. B. O. Lund, H. Ghantous, A. Bergman and I. Brandt, Covalent binding of four DDD isomers in the mouse lung: lack of structure specificity, *Pharmacol. Toxicol.*, 1989, **65**, 282–286.
90. W. K. Nichols, C. M. Terry, N. S. Cutler, M. L. Appleton, P. K. Jesthi and G. S. Yost, Oxidation at C-1 controls the cytotoxicity of 1,1-dichloro-2,2-bis(*p*-chlorophenyl)ethane by rabbit and human lung cells, *Drug Metab. Dispos.*, 1995, **23**, 595–599.
91. C. J. Jonsson, B. O. Lund, A. Bergman and I. Brandt, Adrenocortical toxicity of 3-methylsulphonyl-DDE; 3: Studies in fetal and suckling mice, *Reprod. Toxicol.*, 1992, **6**, 233–240.
92. B. O. Lund, A. Bergman and I. Brandt, Metabolic activation and toxicity of a DDT-metabolite, 3-methylsulphonyl-DDE, in the adrenal zona fasciculata in mice, *Chem. Biol. Interact.*, 1988, **65**, 25–40.
93. J. Jonsson, H. Rodriguezmartinez and I. Brandt, Transplacental toxicity of 3-methylsulfonyl-DDE in the developing adrenal cortex in mice, *Reprod. Toxicol.*, 1995, **9**, 257–264.
94. R. M. Joy, Electrical correlates of preconvulsive and convulsive doses of chlorinated hydrocarbon insecticides in the CNS, *Neuropharmacology*, 1973, **12**, 63–76.
95. M. Khairy and Changes in behaviour associated with a nervous system poison (DDT), *Q. J. Exper. Physiol.*, 1959, **11**, 91–94.
96. K. Saitoh, S. Shaw and H. A. Tilson, Noradrenergic influence on the prepulse inhibition of acoustic startle, *Toxicol. Lett.*, 1986, **34**, 209–216.
97. H. A. Tilson, P. M. Hudson and J. S. Hong, 5,5-Diphenylhydantoin antagonizes neurochemical and behavioral effects of *p,p′*-DDT but not of chlordecone, *J. Neurochem.*, 1986, **47**, 1870–1878.
98. S. M. Ghiasuddin and F. Matsumura, DDT inhibition of Ca-ATPase of the peripheral-nerves of the American lobster, *Pest. Biochem. Physiol.*, 1979, **10**, 151–161.
99. J. H. Song, K. Nagata, H. Tatebayashi and T. Narahashi, Interactions of tetramethrin, fenvalerate and DDT at the sodium channel in rat dorsal root ganglion neurons, *Brain Res.*, 1996, **708**, 29–37.
100. D. W. Herr, R. B. Mailman and H. A. Tilson, Blockade of only spinal alpha 1 adrenoceptors is insufficient to attenuate DDT-induced alterations in motor function, *Toxicol. Appl. Pharmacol.*, 1989, **101**, 11–26.
101. D. W. Herr, J. A. Gallus and H. A. Tilson, Pharmacological modification of tremor and enhanced acoustic startle by chlordecone and *p,p′*- DDT, *Psycopharmacology*, 1987, **91**, 320–325.
102. F. Coulston, Reconsideration of the dilemma of DDT for the establishment of an acceptable daily intake, *Regul. Toxicol. Pharmacol.*, 1985, **5**, 332–383.
103. S. Flodstrom, H. Hemming, L. Warngard and U. G. Ahlborg, Promotion of altered hepatic foci development in rat liver, cytochrome P450 enzyme induction and inhibition of cell-cell communication by DDT and some structurally related organohalogen pesticides, *Carcinogenesis*, 1990, **11**, 1413–1417.

104. S. Sugie, H. Mori and M. Takahashi, Effect of *in vivo* exposure to the liver tumor promoters phenobarbital or DDT on the gap junctions of rat hepatocytes: a quantitative freeze-fracture analysis, *Carcinogenesis*, 1987, **8**, 45–51.
105. C. Tateno, S. Ito, M. Tanaka, M. Oyamada and A. Yoshitake, Effect of DDT on hepatic gap junctional intercellular communication in rats, *Carcinogenesis*, 1994, **15**, 517–521.
106. R. J. Ruch, W. J. Bonney, K. Sigler, X. Guan, D. Matesic, L. D. Schafer, E. Dupont and J. E. Trosko, Loss of gap junctions from DDT-treated rat liver epithelial cells, *Carcinogenesis*, 1994, **15**, 301–306.
107. National Cancer Institute, *Bioassay of DDT, TDE, and p,p′-DDE for Possible Carcinogenicity*. Carcinogen Program, Division of Cancer Cause and Prevention, U.S. Government Printing Office, Washington, D.C., 1978.
108. D. E. Frigo, Y. Tang, B. S. Beckman, A. B. Scandurro, J. Alam, M. E. Burow and J. A. McLachlan, Mechanism of AP-1-mediated gene expression by select organochlorines through the p38 MAPK pathway, *Carcinogenesis*, 2004, **25**, 249–261.
109. M. R. Bratton, D. E. Frigo, K. A. Vigh-Conrad, D. Fan, S. Wadsworth, J. A. McLachlan and M. E. Burow, Organochlorine-mediated potentiation of the general coactivator p300 through p38 mitogen-activated protein kinase, *Carcinogenesis*, 2009, **30**, 106–113.
110. D. T. Williams, G. L. LeBel and E. Junkins, A comparison of organochlorine residues in human adipose tissue autopsy samples from two Ontario municipalities, *J. Toxicol. Environ. Health*, 1984, **13**, 19–29.
111. C. Peraino, Fry, R. J. M, Staffeldt, E. Christopher and J. P., Comparative enhancing effects of phenobarbital, amobarbital, diphenylhydantoin, and dichlorodiphenyltrichloroethane on 2-acetylaminofluorene-induced hepatic tumorigenesis in the rat, *Cancer Res.*, 1975, **35**, 2884–2890.
112. P. Galand, N. Mairesse and C. Degraef, J. Rooryck, *o,p′*-DDT (1,1,1-trichloro-2(*p*-chlorophenyl) 2-(*o*-chlorophenyl) ethane is a purely estrogenic agonist in the rat uterus *in vivo* and *in vitro, Biochem. Pharmacol.*, 1987, **36**, 397–400.
113. A. K. Robison, W. A. Schmidt and G. M. Stancel, Estrogenic activity of DDT: estrogen-receptor profiles and the responses of individual uterine cell types following *o,p′*-DDT administration, *J. Toxicol. Environ. Health*, 1985, **16**, 493–508.
114. A. K. Robison, D. A. Sirbasku and G. M. Stancel, DDT supports the growth of an estrogen-responsive tumor, *Toxicol. Lett.*, 1985, **27**, 109–113.
115. W. R. Kelce, C. R. Lambright, L. E. Gray, Jr. and K. P. Roberts, Vinclozolin and *p,p′*-DDE alter androgen-dependent gene expression: *in vivo* confirmation of an androgen receptor-mediated mechanism, *Toxicol. Appl. Pharmacol.*, 1997, **142**, 192–200.
116. B. J. Danzo, Environmental xenobiotics may disrupt normal endocrine function by interfering with the binding of physiological ligands to steroid

receptors and binding proteins, *Environ. Health Perspect.*, 1997, **105**, 294–301.

117. P. Mussi, P. Ciana, M. Raviscioni, R. Villa, S. Regondi, E. Agradi, A. Maggi and D. Di Lorenzo, Activation of brain estrogen receptors in mice lactating from mothers, *Brain Res. Bull.*, 2005, **65**, 241–247.
118. L. You, M. Casanova, S. Archibeque-Engle, M. Sar, L. Q. Fan and H. A. Heck, Impaired male sexual development in perinatal Sprague-Dawley and Long-Evans hooded rats exposed in utero and lactationally to *p,p′*-DDE, *Toxicol. Sci.*, 1998, **45**, 162–173.
119. J. C. O'Connor, S. R. Frame, L. G. Davis and J. C. Cook, Detection of the environmental antiandrogen p,p-DDE in CD and long-evans rats using a tier I screening battery and a Hershberger assay, *Toxicol. Sci.*, 1999, **51**, 44–53.
120. I. K. Loeffler and R. E. Peterson, Interactive effects of TCDD and *p,p′*-DDE on male reproductive tract development in in utero and lactationally exposed rats, *Toxicol. Appl. Pharmacol.*, 1999, **154**, 28–39.
121. A. Ottoboni, Effect of DDT on reproduction in the rat, *Toxicol. Appl. Pharmacol.*, 1969, **14**, 74–81.
122. A. Ottoboni, G. D. Bissell and A. C. Hexter, Effects of DDT on reproduction in multiple generations of beagle dogs, *Arch. Environ. Contam. Toxicol.*, 1977, **6**, 83–101.
123. H. Hojo, H. Aoyama, K. L. Takahashi, N. Shimizu, M. Araki, Y. Takizawa, K. Sakasai, M. Kuwahara, M. Saka and S. Teramoto, Two-generation reproduction toxicity study in rats with 1,1,1-trichloro-2,2-bis(4-chlorophenyl)ethane (*p,p′*-DDT), *Congenital Anomalies*, 2006, **46**, 105–114.
124. H. C. Li, S. S. Dehal and D. Kupfer, Induction of the hepatic CYP2B and CYP3A enzymes by the proestrogenic pesticide methoxychlor and by DDT in the rat. Effects on methoxychlor metabolism, *J. Biochem. Toxicol.*, 1995, **10**, 51–61.
125. R. W. Nims, R. A. Lubet, S. D. Fox, C. R. Jones, P. E. Thomas, A. B. Reddy and T. A. Kocarek, Comparative pharmacodynamics of CYP2B induction by DDT, DDE, and DDD in male rat liver and cultured rat hepatocytes, *J. Toxicol. Environ. Health, A*, 1998, **53**, 455–477.
126. S. Narayan, H. M. Dani and U. K. Misra, Lung subcellular fractions and surfactant lipid metabolism of rats exposed with DDT or endosulfan intratracheally, *Environ. Sci., Health, B*, 1990, **25**, 259–272.
127. M. A. Nunez G, I. Estrada and E. S. Calderon-Aranda, DDT inhibits the functional activation of murine macrophages and decreases resistance to infection by *Mycobacterium microti, Toxicology*, 2002, **174**, 201–210.
128. R. Dutta, A. M. Mondal, V. Arora, T. C. Nag and N. Das, Immunomodulatory effect of DDT (bis[4-chlorophenyl]-1,1,1-trichloroethane) on complement system and macrophages, *Toxicology*, 2008, **252**, 78–85.
129. W. J. Rogan and A. Chen, Health risks and benefits of bis(4-chlorophenyl)-1,1,1-trichloroethane (DDT), *Lancet*, 2005, **366**, 763–773.

130. H. Bouwman, P. J. Becker and C. H. Schutte, Malaria control and longitudinal changes in levels of DDT and its metabolites in human serum from KwaZulu, *Bull. World Health Organ.*, 1994, **72**, 921–930.
131. F. Laden, L. M. Neas, D. Spiegelman, S. E. Hankinson, W. C. Willett, K. Ireland, M. S. Wolff and D. J. Hunter, Predictors of plasma concentrations of DDE and PCBs in a group of U.S. women, *Environ. Health Perspect.*, 1999, **107**, 75–81.
132. H. Bouwman, R. M. Cooppan, A. J. Reinecke and P. J. Becker, Levels of DDT and metabolites in breast milk from Kwa-Zulu mothers after DDT application for malaria control, *Bull. World Health Organ.*, 1990, **68**, 761–768.
133. H. Bouwman, A. J. Reinecke, R. M. Cooppan and P. J. Becker, Factors affecting levels of DDT and metabolites in human breast milk from Kwazulu, *J. Toxicol. Environ. Health*, 1990, **31**, 93–115.
134. H. Bouwman, P. J. Becker, R. M. Cooppan and A. J. Reinecke, Transfer of DDT used in malaria control to infants via breast milk, *Bull. World Health Organ.*, 1992, **70**, 241–250.
135. J. O. Okonkwo, T. N. Mutshatshi, B. Botha and N. Agyei, DDT, DDE and DDD in human milk from South Africa, *Bull. Environ. Contam. Toxicol.*, 2008, **81**, 348–354.
136. M. H. Wong, A. O. Leung, J. K. Chan and M. P. Choi, A review on the usage of POP pesticides in China, with emphasis on DDT loadings in human milk, *Chemosphere*, 2005, **60**, 740–752.
137. R. E. Cunningham and F. S. Hill, Convulsions and deafness following ingestion of DDT, *Pediatrics*, 1952, **9**, 745–747.
138. H. C. Hsich, D.D.T. intoxication of a family of Southern Taiwan, *Arch. Ind. Health*, 1954, **10**, 344–346.
139. O. Wong, W. Brocker, H. V. Davis and G. S. Nagle, Mortality of workers potentially exposed to organic and inorganic brominated chemicals, DBCP, TRIS, PBB, and DDT, *Br. J. Indust. Med.*, 1984, **41**, 15–24.
140. K. Kreiss, M. M. Zack, R. D. Kimbrough, L. L. Needham, A. L. Smrek and B. T. Jones, Cross-sectional study of a community with exceptional exposure to DDT, *JAMA*, 1981, **245**, 1926–1930.
141. A. Poland, D. Smith, R. Kuntzman, M. Jacobson and A. H. Conney, Effect of intensive occupational exposure to DDT on phenylbutazone and cortisol metabolism in human subjects, *Clin. Pharmacol. Ther.*, 1970, **11**, 724–732.
142. L. Rylander, A. Rignell-Hydbom and L. Hagmar, A cross-sectional study of the association between persistent organochlorine pollutants and diabetes, *Environ. Health*, 2005, **4**, 28.
143. D. H. Lee, I. K. Lee, K. Song, M. Steffes, W. Toscano, B. A. Baker and D. R. Jacobs, Jr., A strong dose-response relation between serum concentrations of persistent organic pollutants and diabetes: results from the National Health and Examination Survey 1999–2002, *Diabetes Care*, 2006, **29**, 1638–1644.

144. C. J. Everett, I. L. Frithsen, V. A. Diaz, R. J. Koopman, W. M. Simpson, Jr. and A. G. Mainous, 3rd, Association of a polychlorinated dibenzo-*p*-dioxin, a polychlorinated biphenyl, and DDT with diabetes in the 1999–2002 National Health and Nutrition Examination Survey, *Environ. Res.*, 2007, **103**, 413–418.
145. S. Cox, A. S. Niskar, K. M. Narayan and M. Marcus, Prevalence of self-reported diabetes and exposure to organochlorine pesticides among Mexican Americans: Hispanic health and nutrition examination survey, 1982–1984, *Environ. Health Perspect.*, 2007, **115**, 1747–1752.
146. M. Turyk, H. Anderson, L. Knobeloch, P. Imm and V. Persky, Organochlorine exposure and incidence of diabetes in a cohort of Great Lakes sport fish consumers, *Environ. Health Perspect.*, 2009, **117**, 1076–1082.
147. U. K. Misra, D. Nag and C. R. Murti, A study of cognitive functions in DDT sprayers, *Indust. Health*, 1984, **22**, 199–206.
148. H. Bouwman, R. M. Cooppan, P. J. Becker and S. Ngxongo, Malaria control and levels of DDT in serum of two populations in Kwazulu, *J. Toxicol. Environ. Health*, 1991, **33**, 141–155.
149. H. Bouwman, R. M. Cooppan, M. J. Botha and P. J. Becker, Serum levels of DDT and liver function of malaria control personnel, *S. Afr. Med. J.*, 1991, **79**, 326–329.
150. I. Al-Saleh, I. Al-Doush, A. Alshabbeen and A. Rabbah, Prenatal exposure to DDT and its effect on reproductive outcome measures, *Toxicol. Sci. (Toxicologist Supplement)*, 2011, **120**, Abstract no. 1208.
151. L. A. Cupul-Uicab, M. Hernandez-Avila, E. A. Terrazas-Medina, M. L. Pennell and M. P. Longnecker, Prenatal exposure to the major DDT metabolite 1,1-dichloro-2,2-bis(p-chlorophenyl)ethylene (DDE) and growth in boys from Mexico, *Environ. Res.*, 2010, **110**, 595–603.
152. L. Torres-Sanchez, S. J. Rothenberg, L. Schnaas, M. E. Cebrian, E. Osorio, M. d. C. Hernandez, R. M. Garcia-Hernandez, C. del Rio-Garcia, M. S. Wolff and L. Lopez-Carrillo, In utero *p,p'*-DDE exposure and infant neurodevelopment: A perinatal cohort in Mexico, *Environ. Health Perspect.*, 2007, **115**, 435–439.
153. B. Eskenazi, A. R. Marks, A. Bradman, L. Fenster, C. Johnson, D. B. Barr and N. P. Jewell, In utero exposure to dichlorodiphenyltrichloroethane (DDT) and dichlorodiphenyldichloroethylene (DDE) and neurodevelopment among young Mexican American children, *Pediatrics*, 2006, **118**, 233–241.
154. B. Eskenazi, L. G. Rosas, A. R. Marks, A. Bradman, K. Harley, N. Holland, C. Johnson, L. Fenster and D. B. Barr, Pesticide toxicity and the developing brain, *Basic Clin. Pharmacol. Toxicol.*, 2008, **102**, 228–236.
155. J. Sunyer, M. Torrent, L. Munoz-Ortiz, N. Ribas-Fito, D. Carrizo, J. Grimalt, J. M. Anto and P. Cullinan, Prenatal dichlorodiphenyldichloroethylene (DDE) and asthma in children, *Environ. Health Perspect.*, 2005, **113**, 1787–1790.
156. E. Morales, J. Sunyer, F. Castro-Giner, X. Estivill, J. Julvez, N. Ribas-Fito, M. Torrent, J. O. Grimalt and R. de Cid, Influence of glutathione

S-transferase polymorphisms on cognitive functioning effects induced by *p,p'*-DDT among preschoolers, *Environ. Health Perspect.*, 2008, **116**, 1581–1585.
157. K. A. McGlynn, C. C. Abnet, M. Zhang, X. D. Sun, J. H. Fan, T. R. O'Brien, W. Q. Wei, B. A. Ortiz-Conde, S. M. Dawsey, J. P. Weber, P. R. Taylor, H. Katki, S. D. Mark and Y. L. Qiao, Serum concentrations of 1,1,1-Trichloro-2,2-bis(p-chlorophenyl)ethane (DDT) and 1,1-Dichloro-2,2-bis(p-chlorophenyl)ethylene (DDE) and risk of primary liver cancer, *J. Natl. Cancer Inst.*, 2006, **98**, 1005–1010.
158. P. Cocco, N. Kazerouni and S. H. Zahm, Cancer mortality and environmental exposure to DDE in the United States, *Environ. Health Perspect.*, 2000, **108**, 1–4.
159. N. Malats, F. X. Real and M. Porta, DDT and pancreatic cancer, *J. Natl. Cancer Inst.*, 1993, **85**, 328–329.
160. D. H. Garabrant, J. Held, B. Langholz, J. M. Peters and T. M. Mack, DDT and related compounds and risk of pancreatic cancer. [see comment], *J. Natl. Cancer Inst.*, 1992, **84**, 764–771.
161. D. T. Garabrant, J. Held and D. Homa, DDT and pancreatic cancer, *J. Natl. Cancer Inst.*, 1993, **85**, 328–329.
162. P. Cocco, D. Fadda, B. Billai, M. D'Atri, M. Melis and A. Blair, Cancer mortality among men occupationally exposed to dichlorodiphenyltrichloroethane, *Cancer Res.*, 2005, **65**, 9588–9594.
163. M. P. Purdue, J. A. Hoppin, A. Blair, M. Dosemeci and M. C. R. Alavanja, Occupational exposure to organochlorine insecticides and cancer incidence in the Agricultural Health Study, *Int. J. Cancer*, 2007, **120**, 642–649.
164. M. S. Wolff, P. G. Toniolo, E. W. Lee, M. Rivera and N. Dubin, Blood levels of organochlorine residues and risk of breast cancer. [see comment], *J. Natl. Cancer Inst.*, 1993, **85**, 648–652.
165. E. Dewailly, P. Ayotte, J. Brisson and S. Dodin, Breast cancer and organochlorines, *Lancet*, 1994, **344**, 1707–1708.
166. E. Dewailly, P. Ayotte and S. Dodin, Could the rising levels of estrogen receptor in breast cancer be due to estrogenic pollutants?, *J. Natl. Cancer Inst.*, 1997, **89**, 888–889.
167. T. Key and G. Reeves, Organochlorines in the environment and breast cancer, *Br. Med. J.*, 1994, **308**, 1520–1521.
168. U. G. Ahlborg, L. Lipworth, L. Titus-Ernstoff, C. C. Hsieh, A. Hanberg, J. Baron, D. Trichopoulos and H. O. Adami, Organochlorine compounds in relation to breast cancer, endometrial cancer, and endometriosis: an assessment of the biological and epidemiological evidence, *Crit. Rev. Toxicol.*, 1995, **25**, 463–531.
169. S. Guttes, K. Failing, K. Neumann, J. Kleinstein, S. Georgii and H. Brunn, Chlororganic pesticides and polychlorinated biphenyls in breast tissue of women with benign and malignant breast disease, *Arch. Environ. Contam. Toxicol.*, 1998, **35**, 140–147.
170. A. Schecter, P. Toniolo, L. C. Dai, L. T. Thuy and M. S. Wolff, Blood levels of DDT and breast cancer risk among women living in the north of Vietnam, *Arch. Environ. Contam. Toxicol.*, 1997, **33**, 453–456.

171. D. J. Hunter, S. E. Hankinson, F. Laden, G. A. Colditz, J. E. Manson, W. C. Willett, F. E. Speizer and M. S. Wolff, Plasma organochlorine levels and the risk of breast cancer, *N. Engl. J. Med.*, 1997, **337**, 1253–1258.
172. B. A. Cohn, M. S. Wolff, P. M. Cirillo and R. I. Scholtz, DDT and breast cancer in young women: New data on the significance of age at exposure, *Environ. Health Perspect.*, 2007, **115**, 1406–1414.
173. B. A. Cohn, Developmental and environmental origins of breast cancer: DDT as a case study, *Reprod. Toxicol.*, 2010.
174. M. P. Purdue, L. S. Engel, H. Langseth, L. L. Needham, A. Andersen, D. B. Barr, A. Blair, N. Rothman and K. A. McGlynn, Prediagnostic serum concentrations of organochlorine compounds and risk of testicular germ cell tumors, *Environ. Health Perspect.*, 2009, **117**, 1514–1519.
175. B. A. Cohn, P. M. Cirillo and R. E. Christianson, Prenatal DDT exposure and testicular cancer: a nested case-control study, *Arch. Environ. Occup. Health*, 2010, **65**, 127–134.
176. S. A. Korrick, C. Chen, A. I. Damokosh, J. Ni, X. Liu, S. I. Cho, L. Altshul, L. Ryan and X. Xu, Association of DDT with spontaneous abortion: a case-control study, *Ann. Epidemiol.*, 2001, **11**, 491–496.
177. L. A. Cupul-Uicab, B. C. Gladen, M. Hernandez-Avila, J. P. Weber and M. P. Longnecker, DDE, a degradation product of DDT, and duration of lactation in a highly exposed area of Mexico, *Environ. Health Perspect.*, 2008, **116**, 179–183.
178. R. Hauser, Z. Y. Chen, L. Pothier, L. Ryan and L. Altshul, The relationship between human semen parameters and environmental exposure to polychlorinated biphenyls and p,p′-DDE, *Environ. Health Perspect.*, 2003, **111**, 1505–1511.
179. N. H. Aneck-Hahn, G. W. Schulenburg, M. S. Bornman, P. Farias and C. De Jager, Impaired semen quality associated with environmental DDT exposure in young men living in a malaria area in the Limpopo Province, South Africa, *J. Androl.*, 2007, **28**, 423–434.
180. M. A. Dalvie, J. E. Myers, M. Lou Thompson, S. Dyer, T. G. Robins, S. Omar, J. Riebow, J. Molekwa, P. Kruger and R. Millar, The hormonal effects of long-term DDT exposure on malaria vector-control workers in Limpopo Province, South Africa, *Environ. Res.*, 2004, **96**, 9–19.
181. M. A. Dalvie, J. E. Myers, M. L. Thompson, T. G. Robins, S. Dyer, J. Riebow, J. Molekwa, M. Jeebhay, R. Millar and P. Kruger, The long-term effects of DDT exposure on semen, fertility, and sexual function of malaria vector-control workers in Limpopo Province, South Africa, *Environ. Res.*, 2004, **96**, 1–8.
182. C. de Jager, P. Farias, A. Barraza-Villarreal, M. H. Avila, P. Ayotte, E. Dewailly, C. Dombrowski, F. Rousseau, V. D. Sanchez and J. L. Bailey, Reduced seminal parameters associated with environmental DDT exposure and *p,p′*-DDE concentrations in men in Chiapas, Mexico: A cross-sectional study, *J. Androl.*, 2006, **27**, 16–27.
183. C. de Jager, N. H. Aneck-Hahn, M. S. Bornman, P. Farias, G. Leter, P. Eleuteri, M. Rescia and M. Spano, Sperm chromatin integrity in DDT-

exposed young men living in a malaria area in the Limpopo Province, South Africa, *Hum. Reprod.*, 2009, **24**, 2429–2438.
184. M. P. Longnecker, M. A. Klebanoff, J. W. Brock, H. Zhou, K. A. Gray, L. L. Needham and A. J. Wilcox, Maternal serum level of 1,1-dichloro-2,2-bis(*p*-chlorophenyl)ethylene and risk of cryptorchidism, hypospadias, and polythelia among male offspring, *Am. J. Epidemiol.*, 2002, **155**, 313–322.
185. M. J. Lopez-Espinosa, E. Lopez-Navarrete, A. Rivas, M. F. Fernandez, M. Nogueras, C. Campoy, F. Olea-Serrano, P. Lardelli and N. Olea, Organochlorine pesticide exposure in children living in southern Spain, *Environ. Res.*, 2008, **106**, 1–6.
186. R. Bornman, C. de Jager, Z. Worku, P. Farias and S. Reif, DDT and urogenital malformations in newborn boys in a malarial area, *Br. J. Urol. Int.*, 2010, **106**, 405–411.
187. N. Lubick, Examining DDT's urogenital effects, *Environ. Health Perspect.*, 2010, **118**, A18.
188. F. Daffara, S. De Francia, G. Reimondo, B. Zaggia, E. Aroasio, F. Porpiglia, M. Volante, A. Termine, F. Di Carlo, L. Dogliotti, A. Angeli, A. Berruti and M. Terzolo, Prospective evaluation of mitotane toxicity in adrenocortical cancer patients treated adjuvantly, *Endocr. Relat. Cancer*, 2008, **15**, 1043–1053.
189. P. Igaz, Z. Tombol, P. M. Szabo, I. Liko and K. Racz, Steroid biosynthesis inhibitors in the therapy of hypercortisolism: theory and practice, *Curr. Med. Chem.*, 2008, **15**, 2734–2747.
190. M. Terzolo and A. Berruti, Adjunctive treatment of adrenocortical carcinoma, *Curr. Opin. Endocrinol. Diabetes Obes.*, 2008, **15**, 221–226.
191. T. B. Gaines, Acute toxicity of pesticides, *Toxicol. Appl. Pharmacol.*, 1969, **14**, 515–534.
192. I. Taliaferro and L. Leone, Inhibitory effect of perthane (2,2-bis-[para-ethyl-phenyl]-1,1-dichloroethane) on adrenocortical function in human subjects, *N. Engl. J. Med.*, 1957, **257**, 855–860.
193. D. H. Garabrant, J. Held, B. Langholz, J. M. Peters and T. M. Mack, DDT and related compounds and risk of pancreatic cancer, *J. Natl Cancer Inst.*, 1992, **84**, 764–771.
194. K. Tabata, T. Miyata and T. Saito, Water-soluble metabolites of dicofol in mouse urine, *Appl. Entomol. Zool.*, 1979, **14**, 490–493.
195. M. A. Brown and J. E. Casida, Metabolism of a dicofol impurity alpha-chloro-DDT, but not dicofol or dechlorodicofol, to DDE in mice and a liver microsomal system, *Xenobiotica*, 1987, **17**, 1169–1174.
196. B. A. Narloch, M. P. Lawton, D. E. Moody, B. D. Hammock and L. R. Shull, The effects of dicofol on induction of hepatic microsomal metabolism in rats, *Pest. Biochem. Physiol.*, 1987, **28**, 362–370.
197. J. R. P. Cabral, F. Raitano, T. Mollner, S. Bronczyk and P. Shubik, Acute toxicity of pesticides in hamsters, *Toxicol. Appl. Pharmacol.*, 1979, **48**, A192.
198. C. O. Knowles and S. Ahmad, Comparative metabolism of chlorobenzilate, chloropropylate, and bromopropylate acaricides by rat hepatic enzymes, *Canad. J. Physiol. Pharmacol.*, 1971, **49**, 590–597.

199. I. P. Kapoor, R. L. Metcalf, R. F. Nystrom and G. K. Sangha, Comparative metabolism of methoxychlor, methiochlor, and DDT in mouse, insects, and in a model ecosystem, *J. Agric. Food Chem.*, 1970, **18**, 1145–1152.
200. D. Kupfer and W. H. Bulger, Metabolic activation of pesticides with proestrogenic activity, *Fed. Proc.*, 1987, **46**, 1864–1869.
201. D. Kishimoto and N. Kurihara, Effect of cytochrome P450 antibodies on the oxidative demethylation of methoxychlor catalysed by rat liver microsomal cytochrome P450 enzymes: Isozyme specificity and enantiotopic selectivity, *Pest. Biochem. Physiol.*, 1996, **51**, 44–52.
202. D. M. Stresser and D. Kupfer, Human cytochrome P450-catalyzed conversion of the proestrogenic pesticide methoxychlor into an estrogen. Role of CYP2C19 and CYP1A2 in *O*-demethylation, *Drug Metab. Dispos.*, 1998, **26**, 868–874.
203. D. M. Stresser and D. Kupfer, Prosubstrates of CYP3A4, the major human hepatic cytochrome P450: transformation into substrates by other P450 isoforms, *Biochem. Pharmacol.*, 1998, **55**, 1861–1871.
204. Y. Hu, K. Krausz, H. V. Gelboin and D. Kupfer, CYP2C subfamily, primarily CYP2C9, catalyses the enantioselective demethylation of the endocrine disruptor pesticide methoxychlor in human liver microsomes: use of inhibitory monoclonal antibodies in P450 identification, *Xenobiotica*, 2004, **34**, 117–132.
205. W. H. Bulger and D. Kupfer, Studies on the formation of methoxychlor-protein adduct in rat and human liver microsomes. Is demethylation of methoxychlor essential for cytochrome P450 catalyzed covalent binding?, *Biochem. Pharmacol.*, 1990, **40**, 937–945.
206. L. X. Zhou, S. S. Dehal, D. Kupfer, S. Morrell, B. A. McKenzie, Jr. E. D. Eccleston and J. L. Holtzman, Cytochrome P450 catalyzed covalent binding of methoxychlor to rat hepatic, microsomal iodothyronine 5′-monodeiodinase, type I: does exposure to methoxychlor disrupt thyroid hormone metabolism?, *Arch. Biochem. Biophys.*, 1995, **322**, 390–394.
207. K. Ohyama, S. Maki, K. Sato and Y. Kato, *In vitro* metabolism of [^{14}C]methoxychlor in rat, mouse, Japanese quail and rainbow trout in precision-cut liver slices, *Xenobiotica*, 2004, **34**, 741–754.
208. K. Ohyama, S. Maki, K. Sato and Y. Kato, Comparative *in vitro* metabolism of methoxychlor in male and female rats, *Xenobiotica*, 2005, **35**, 683–695.
209. T. J. Oberly, K. C. Michaelis, M. A. Rexroat, B. J. Bewsey and M. L. Garriott, A comparison of the CHO/HGPRT+ and the L5178Y/TK+/- mutation assays using suspension treatment and soft agar cloning: results for 10 chemicals, *Cell Biol. Toxicol.*, 1993, **9**, 243–257.
210. M. D. Reuber, Interstitial cell carcinomas of the testis in Balb/C male mice ingesting methoxychlor, *J. Cancer Res. Clin. Oncol.*, 1979, **93**, 173–179.
211. W. W. Tullner, Cystic tubular nephropathy and decreases in testicular weight in rats following oral methoxychlor treatment, *J. Pharmacol. Exper. Ther.*, 1962, **138**.

212. A. S. Tegeris, F. L. Earl, H. E. Smalley Jr. and J. M. Curtis, Methoxychlor toxicity. Comparative studies in the dog and the swine, *Arch. Environ. Health*, 1966, **13**, 776–787.
213. P. S. Cooke and V. P. Eroschenko, Inhibitory effects of technical grade methoxychlor on development of neonatal male mouse reproductive organs, *Biol. Reprod.*, 1990, **42**, 585–596.
214. V. P. Eroschenko, Ultrastructure of vagina and uterus in young mice after methoxychlor exposure, *Reprod. Toxicol.*, 1991, **5**, 427–435.
215. V. P. Eroschenko and F. Osman, Scanning electron microscopic changes in vaginal epithelium of suckling neonatal mice in response to estradiol or insecticide chlordecone (Kepone) passage in milk, *Toxicology*, 1986, **38**, 175–185.
216. V. P. Eroshenko, A. A. Abuel-Atta and M. S. Grober, Neonatal exposures to technical methoxychlor alters ovaries in adult mice., *Reprod. Toxicol.*, 1995, **9**, 379.
217. R. E. Chapin, M. W. Harris, B. J. Davis, S. M. Ward, R. E. Wilson, M. A. Mauney, A. C. Lockhart, R. J. Smialowicz, V. C. Moser, L. T. Burka and B. J. Collins, The effects of perinatal/juvenile methoxychlor exposure on adult rat nervous, immune, and reproductive system function, *Fundam. Appl. Toxicol.*, 1997, **40**, 138–157.
218. R. J. Appel and V. P. Eroschenko, Passage of methoxychlor in milk and reproductive organs of nursing female mice; 1. Light and scanning electron microscopic observations, *Reprod. Toxicol.*, 1992, **6**, 223–231.
219. W. H. Bulger, R. M. Muccitelli and D. Kupfer, Interactions of methoxychlor, methoxychlor base-soluble contaminant, and 2,2-bis(p-hydroxyphenyl)-1,1,1-trichloroethane with rat uterine estrogen receptor, *J. Toxicol. Environ. Health*, 1978, **4**, 881–893.
220. W. H. Bulger, R. M. Muccitelli and D. Kupfer, Studies on the *in vivo* and *in vitro* estrogenic activities of methoxychlor and its metabolites. Role of hepatic mono-oxygenase in methoxychlor activation, *Biochem. Pharmacol.*, 1978, **27**, 2417–2423.
221. E. P. Murono and R. C. Derk, The reported active metabolite of methoxychlor, 2,2-bis(*p*-hydroxyphenyl)-1,1,1-trichloroethane, inhibits testosterone formation by cultured Leydig cells from neonatal rats, *Reprod. Toxicol.*, 2005, **20**, 503–513.
222. K. P. Miller, R. K. Gupta and J. A. Flaws, Methoxychlor metabolites may cause ovarian toxicity through estrogen, *Toxicol. Sci.*, 2006, **93**, 180–188.
223. Z. R. Craig, T. C. Leslie, K. P. Hatfield, R. K. Gupta and J. A. Flaws, Mono-hydroxy methoxychlor alters levels of key sex steroids and steroidogenic enzymes in cultured mouse antral follicles, *Toxicol. Appl. Pharmacol.*, 2010, **249**, 107–113.
224. W. H. Bulger, V. J. Feil and D. Kupfer, Role of hepatic monooxygenases in generating estrogenic metabolites from methoxychlor and from its identified contaminants, *Mol. Pharmacol.*, 1985, **27**, 115–124.

225. A. M. Cummings and J. L. Metcalf, Mechanisms of the stimulation of rat uterine peroxidase activity by methoxychlor, *Reprod. Toxicol.*, 1994, **8**, 477–486.
226. J. L. Metcalf, S. C. Laws and A. M. Cummings, Methoxychlor mimics the action of 17 beta-estradiol on induction of uterine epidermal growth factor receptors in immature female rats, *Reprod. Toxicol.*, 1996, **10**, 393–399.
227. X. Fei, H. Chung and H. S. Taylor, Methoxychlor disrupts uterine Hoxa10 gene expression, *Endocrinology*, 2005, **146**, 3445–3451.
228. H. S. Bal, Effect of methoxychlor on reproductive systems of the rat, *Proc. Soc. Exp. Biol. Med.*, 1984, **176**, 187–196.
229. R. K. Gupta, G. Aberdeen, J. K. Babus, E. D. Albrecht and J. A. Flaws, Methoxychlor and its metabolites inhibit growth and induce atresia of baboon antral follicles, *Toxicol. Pathol.*, 2007, **35**, 649–656.
230. A. E. Armenti, A. M. Zama, L. Passantino and M. Uzumcu, Developmental methoxychlor exposure affects multiple reproductive parameters and ovarian folliculogenesis and gene expression in adult rats, *Toxicol. Appl. Pharmacol.*, 2008, **233**, 286–296.
231. K. P. Miller, R. K. Gupta, C. R. Greenfeld, J. K. Babus and J. A. Flaws, Methoxychlor directly affects ovarian antral follicle growth and atresia, *Toxicol. Sci.*, 2005, **88**, 213–221.
232. R. K. Gupta, R. A. Schuh, G. Fiskum and J. A. Flaws, Methoxychlor causes mitochondrial dysfunction and oxidative damage in the mouse ovary, *Toxicol. Appl. Pharmacol.*, 2006, **216**, 436–445.
233. R. A. Schuh, T. Kristian, R. K. Gupta, J. A. Flaws and G. Fiskum, Methoxychlor inhibits brain mitochondrial respiration and increases hydrogen peroxide production and CREB phosphorylation, *Toxicol. Sci.*, 2005, **88**, 495–504.
234. D. A. Symonds, I. Merchenthaler and J. A. Flaws, Methoxychlor and estradiol induce oxidative stress DNA damage in the mouse ovarian surface epithelium, *Toxicol. Sci.*, 2008, **105**, 182–187.
235. R. Zachow and M. Uzumcu, The methoxychlor metabolite, 2,2-bis-(*p*-hydroxyphenyl)-1,1,1 inhibits steroidgenesis in rat ovarian granule cells *in vitro*, *Reprod. Toxicol.*, 2006, **22**, 659–665.
236. C. Borgeest, D. Symonds, L. P. Mayer, P. B. Hoyer and J. A. Flaws, Methoxychlor may cause ovarian follicular atresia and proliferation of the ovarian epithelium in the mouse, *Toxicol. Sci.*, 2002, **68**, 473–478.
237. D. A. Symonds, D. Tomic, K. P. Miller and J. A. Flaws, Methoxychlor induces proliferation of the mouse ovarian surface, *Toxicol. Sci.*, 2005, **83**, 355–362.
238. J. M. Goldman, R. L. Cooper, G. L. Rehnberg, J. F. Hein, W. K. McElroy and L. E. Gray, Jr., Effects of low subchronic doses of methoxychlor on the rat hypothalamic-pituitary reproductive axis, *Toxicol. Appl. Pharmacol.*, 1986, **86**, 474–483.
239. E. P. Murono and R. C. Derk, The effects of the reported active metabolite of methoxychlor, 2,2-bis(*p*-hydroxyphenyl)-1,1,1-trichloroethane,

on testosterone formation by cultured Leydig cells from young adult rats, *Reprod. Toxicol.*, 2004, **19**, 135–146.

240. E. P. Murono, R. C. Derk and Y. Akgul, In vivo exposure of young adult male rats to methoxychlor reduces serum testosterone levels and ex vivo Leydig cell testosterone formation and cholesterol side-chain cleavage activity, *Reprod. Toxicol.*, 2006, **21**, 148–153.
241. M. D. Anway, A. S. Cupp, M. Uzumcu and M. K. Skinner, Epigenetic transgenerational actions of endocrine disruptors and male fertility, *Science*, 2005, **308**, 1466–1469.
242. A. M. Zama and M. Uzumcu, Fetal and neonatal exposure to the endocrine disruptor methoxychlor causes epigenetic alterations in adult ovarian genes, *Endocrinology*, 2009, **150**, 4681–4691.
243. C. Stouder and A. Paoloni-Giacobino, Specific transgenerational imprinting effects of the endocrine disruptor methoxychlor on male gametes, *Reproduction*, 2011, **141**, 207–216.
244. S. Vaithinathan, B. Saradha and P. P. Mathur, Methoxychlor-induced alteration in the levels of HSP70 and clusterin is accompanied with oxidative stress in adult rat testis, *J. Biochem. Mol. Toxicol.*, 2009, **23**, 29–35.
245. S. Vaithinathan, B. Saradha and P. P. Mathur, Methoxychlor induces apoptosis via mitochondria- and FasL-mediated pathways in adult rat testis, *Chem. Biol. Interact.*, 2010, **185**, 110–118.
246. J. A. Dodge, A. L. Glasebrook, D. E. Magee, D. L. Phillips, M. Sato, L. L. Short and H. U. Bryant, Environmental estrogens: effects on cholesterol lowering and bone in the ovariectomized rat, *J. Steroid Biochem. Mol. Biol.*, 1996, **59**, 155–161.
247. L. E. Gray, Jr., J. S. Ostby, J. M. Ferrell, E. R. Sigmon and J. M. Goldman, Methoxychlor induces estrogen-like alterations of behavior and the reproductive tract in the female rat and hamster: effects on sex behavior, running wheel activity, and uterine morphology, *Toxicol. Appl. Pharmacol.*, 1988, **96**, 525–540.
248. F. S. vom Saal, S. C. Nagel, P. Palanza, M. Boechler, S. Parmigiani and W. V. Welshons, Estrogenic pesticides: binding relative to estradiol in MCF-7 cells and effects of exposure during fetal life on subsequent territorial behaviour in male mice, *Toxicol. Lett.*, 1995, **77**, 343–350.
249. S. Y. Amstislavsky, E. A. Kizilova, A. N. Golubitsa, A. A. Vasilkova and V. P. Eroschenko, Preimplantation exposures of murine embryos to estradiol or methoxychlor change postnatal development, *Reprod. Toxicol.*, 2004, **18**, 103–108.
250. M. S. Golub, S. L. Germann and C. E. Hogrefe, Endocrine disruption and cognitive function in adolescent female rhesus monkeys, *Neurotoxicol. Teratol.*, 2004, **26**, 799–809.
251. D. L. Hall, L. A. Payne, J. M. Putnam and Y. M. Huet-Hudson, Effect of methoxychlor on implantation and embryo development in the mouse, *Reprod. Toxicol.*, 1997, **11**, 703–708.
252. K. W. Gaido, S. C. Maness, D. P. McDonnell, S. S. Dehal, D. Kupfer and S. Safe, Interaction of methoxychlor and related compounds with

estrogen receptor alpha and beta, and androgen receptor: structure-activity studies, *Mol. Pharmacol.*, 2000, **58**, 852–858.
253. E. Swedenborg, I. Pongratz and J. A. Gustafsson, Endocrine disruptors targeting ERbeta function, *Int. J. Androl.*, 2010, **33**, 288–297.
254. K. L. White, Jr., D. R. Germolec, C. D. Booker, D. M. Hernendez, J. A. McCay, K. B. Delclos, R. R. Newbold, C. Weis and T. L. Guo, Dietary methoxychlor exposure modulates splenic natural killer cell activity, antibody-forming cell response and phenotypic marker expression in F0 and F1 generations of Sprague Dawley rats, *Toxicology*, 2005, **207**, 271–281.
255. J. Y. Kim, K. N. Oh, E. H. Han, D. H. Kim, T. C. Jeong, E. S. Lee and H. G. Jeong, Methoxychlor-induced inducible nitric oxide synthase and proinflammatory cytokines expression in macrophages via NF-kappaB, ERK, and p38 mitogen-activated protein kinases, *Biochem. Biophys. Res. Commun.*, 2005, **333**, 1234–1240.
256. A. A. Stein, D. M. Serrone and F. Coulston, Safety evaluation of methoxychlor in volunteers, *Toxicol. Appl. Pharmacol.*, 1965, **7**, 499.
257. R. Fytizas and A. Ioannou, Toxicity distribution and elimination of plifenate in albino rats, *Ann. Inst. Phytopathol. Benaki*, 1982, **13**, 168–177.
258. H. Wu, C. Cui, B. Huang and H. Wang, Toxicity study on acetofenate, *Jinan Daxue Xuebao, Ziran Kexue Yu Yixueban*, 1991, **12**, 34–37.
259. C. Xu, M. Zhao, W. Liu, S. Chen and J. Gan, Enantioselectivity in zebrafish embryo toxicity of the insecticide acetofenate, *Chem. Res. Toxicol.*, 2008, **21**, 1050–1055.
260. M. Zhao, Y. Zhang, C. Wang, Z. Fu, W. Liu and J. Gan, Induction of macrophage apoptosis by an organochlorine insecticide acetofenate, *Chem. Res. Toxicol.*, 2009, **22**, 504–510.
261. G. Woodard and E. C. Hagan, Toxicological studies on the isomers and mixtures of isomers of benzene hexachloride, *Fed. Proc.*, 1947, **6**, 386.
262. O. G. Fitzhugh, A. A. Nelson and J. P. Frawley, The chronic toxicities of technical benzene hexachloride and its alpha, beta, and gamma isomers, *J. Pharmacol. Exper. Ther.*, 1950, **100**, 59–66.
263. B. Davidow and J. P. Frawley, Tissue distribution, accumulation, and elimination of the isomers of benzene hexachloride, *Proc. Soc. Exp. Biol.*, 1951, **76**, 780–783.
264. R. M. Macholz, D. W. Bleyl, H. Klepel, R. Knoll, M. Kujawa, H. J. Lewerenz, D. Muller and R. Plass, Comparison of the distribution and toxicity of alpha-, beta-, and gamma-hexachlorocyclohexane (HCH) following 30 days of administration to rats, *Nahrung*, 1986, **30**, 701–708.
265. K. Stein, J. Portig, H. Fuhrmann, W. Koransky and G. Noack, Steric factors in the pharmacokinetics of lindane and alpha-hexachlorocyclohexane in rats, *Xenobiotica*, 1980, **10**, 65–77.
266. D. Eichler, W. Heupt and W. Paul, Comparative study on the distribution of α- and γ-hexachlorocyclohexane in the rat with particular reference to the problem of isomerization, *Xenobiotica*, 1983, **13**, 639–647.

267. P. Kraus, B. Gross and H. D. Kloft, The elevation of rat liver glutathione-*S*-transferase activity by alpha-hexachlorocyclohexane, *Biochem. Pharmacol.*, 1981, **30**, 355–361.
268. A. Kumar and P. P. Dwivedi, Relative induction of molecular forms of cytochrome P-450 in gamma-hexachlorocyclohexane exposed rat liver microsomes, *Arch. Toxicol.*, 1988, **62**, 479–481.
269. G. L. Wolff and R. L. Suber, Hepatic glutathione *S*-transferase activity in mice: effects of Avy/-genotype, obesity, lindane treatment, and sex, *Proc. Soc. Exp. Biol. Med.*, 1986, **181**, 535–541.
270. J. R. Robinson, J. S. Felton, R. C. Levitt, S. S. Thorgeirsson and D. W. Nebert, Relationship between 'aromatic hydrocarbon responsiveness' and the survival times in mice treated with various drugs and environmental compounds, *Mol. Pharmacol.*, 1975, **11**, 850–865.
271. P. T. Liu and D. P. Morgan, Comparative toxicity and biotransformation of lindane in C57BL/6 and DBA/2 mice, *Life Sci.*, 1986, **39**, 1237–1244.
272. R. M. Macholz and M. Kujawa, Recent state of lindane metabolism. Part III, *Residue Rev.*, 1985, **94**, 119–149.
273. K. Tanaka, N. Kurihgara and M. Nakajima, Pathways of chlorophenol formation in oxidative biodegradation of BHC, *Agric. Biol. Chem.*, 1977, **41**, 723–725.
274. K. Tanaka, N. Kurihara and M. Nakajima, Oxidative metabolism of tetrachlorocyclohexenes, pentachlorocyclohexenes, and hexachlorocyclohexenes with microsomes from rat liver and house fly abdomen., *Pestic. Biochem. Physiol.*, 1979, **10**, 79–95.
275. K. Stein, J. Protig and W. Koransky, Oxidative transformation of hexachlorocyclohexane in rats and with rat liver microsomes, *Naunyn-Schmiederberg's Arch. Pharmacol.*, 1977, **298**, 115–128.
276. J. F. Fitzloff and J. C. Pan, Epoxidation of the lindane metabolite, beta-PCCH, by human- and rat-liver microsomes, *Xenobiotica*, 1984, **14**, 599–604.
277. R. W. Chadwick, J. J. Freal, G. W. Sovocool, C. C. Bryden and M. F. Copeland, The identification of three previously unreported lindane metabolites from mammals, *Chemosphere*, 1978, **7**, 633–640.
278. R. W. Chadwick, M. F. Copeland, G. L. Wolff, A. G. Stead, M. L. Mole and D. A. Whitehouse, Saturation of lindane metabolism in chronically treated (YS x VY) F1 hybrid mice, *J. Toxicol. Environ. Health*, 1987, **20**, 411–434.
279. M. T. Baker, R. M. Nelson and R. A. Van Dyke, The formation of chlorobenzene and benzene by the reductive metabolism of lindane in rat liver microsomes, *Arch. Biochem. Biophys.*, 1985, **236**, 506–514.
280. R. W. Chadwick and M. F. Copeland, Investigation of HCB as a metabolite from female rats treated daily for 6 days with lindane, *J. Anal. Toxicol.*, 1985, **9**, 262–266.
281. M. F. Copeland and R. W. Chadwick, Bioisomerization of lindane in rats, *J. Environ. Pathol. Toxicol.*, 1979, **2**, 737–749.

282. N. Kurihara, K. Tanaka and M. Nakajima, Mercapturic acid formation from lindane in rats., *Pestic. Biochem. Physiol.*, 1979, **10**, 137–150.
283. J. Portig, P. Kraus, K. Stein, W. Koransky, G. Noack, B. Gross and S. Sodomann, Glutathione conjugate formation from hexachlorocyclohexane and pentachlorocyclohexene by rat liver, *in vitro, Xenobiotica*, 1979, **9**, 353–378.
284. R. W. Chadwick, M. F. Copeland, G. L. Wolff, N. Cook, D. A. Whitehouse and M. L. Mole, Effects of age and obesity on the metabolism of lindane by black *a*/*a*, yellow A^{vy}/a and pseudoagouti A^{vy}/a phenotypes of (YS x VY)F_1 hybrid mice, *J. Toxicol. Environ. Health*, 1986, **16**, 771–796.
285. R. W. Chadwick, M. F. Copeland, M. L. Mole, S. Nesnow and N. Cook, Comparative effect of pretreatment with phenobarbital, Aroclor 1254 and β–naphthoflavone on the metabolism of lindane, *Pest. Biochem. Physiol.*, 1981, **15**, 120–136.
286. A. Johri, S. Yadav, A. Dhawan and D. Parmar, Responsiveness of cerebral and hepatic cytochrome P450s in rat offspring prenatally exposed to lindane, *Toxicol. Appl. Pharmacol.*, 2008, **231**, 10–16.
287. B. P. McNamara and S. Krop, Observations on the pharmacology of the isomers of hexachlorocyclohexane, *J. Pharmacol. Exp. Ther.*, 1948, **92**, 140–146.
288. H. Coper, H. Herken and I. Klempau, On the pharmacology and toxicology of chlorinated cyclohexane, *Nauryn-Schmiedebergs Arch. Exp. Pathol. Pharmakol.*, 1951, **212**, 463–479.
289. K. van Asperen, Interaction of the isomers of benzene hexachloride in mice, *Arch. Int. Pharmacodyn. Ther.*, 1954, **99**, 368–377.
290. L. Camon, E. Martinez, F. Artigas, C. Sola and E. Rodriguez-Farre, The effect of non-convulsant doses of lindane on temperature and body weight, *Toxicology*, 1988, **49**, 389–394.
291. M. Aldegunde Villar, I. Martin Gargueiro, I. Miguez Besada and M. P. Fernandez Otero, Study of the mechanism of the hypothermic action of γ-hexachlorocyclohexane, *Acta Cient. Compostelana*, 1981, **18**, 145–154.
292. T. E. Albertson, R. M. Joy and L. G. Stark, Facilitation of kindling in adult rats following neonatal exposure to lindane, *Dev. Brain Res.*, 1985, **17**, 263–266.
293. R. M. Joy and V. W. Burns, Exposure to lindane and two other hexachlorocyclohexane isomers increases free intracellular calcium levels in neurohybridoma cells, *Neurotoxicology*, 1988, **9**, 637–643.
294. S. C. Bondy and L. Halsall, Lindane-induced modulation of calcium levels within synaptosomes, *Neurotoxicology*, 1988, **9**, 645–652.
295. M. Kamijima and J. E. Casida, Regional modification of [^{3}H]ethynylbicycloorthobenzoate binding in mouse brain GABA(A) receptor by endosulfan, fipronil, and avermectin B(1a), *Toxicol. Appl. Pharmacol.*, 2000, **163**, 188–194.
296. S. Magour, H. Maser and I. Steffen, Effect of lindane on synaptosomal Na^+/K^+-ATPase in relation to its subcellular distribution in the brain, *Acta Pharmacol. Toxicol.*, 1984, **54**, 299–303.

297. F. Matsumura and S. M. Ghiasuddin, Evidence for similarities between cyclodiene type insecticides and picrotoxin in their action mechanisms, *J. Environ. Sci. Health Part B*, 1983, **B18**, 1–14.
298. J. E. Casida and L. J. Lawrence, Structure-activity correlations for interactions of bicyclophosphorus esters and some polychlorocycloalkane and pyrethroid insecticides with the brain-specific t-butylbicyclophosphorothionate receptor, *Pestic. Biochem. Physiol.*, 1985, **61**, 123–132.
299. I. M. Abalis, M. E. Eldefrawi and A. T. Eldefrawi, High-affinity sterospecific binding of cyclodiene and γ–aminobutyric acid receptors of rat brain, *J. Toxicol. Environ. Health*, 1985, **24**, 95–102.
300. I. M. Abalis, M. E. Eldefrawi and A. T. Eldefrawi, Effects of insecticides on GABA-induced chloride influx into rat brain microsacs, *J. Toxicol. Environ. Health*, 1986, **18**, 13–23.
301. J. R. Bloomquist and D. M. Soderlund, Neurotoxic insecticides inhibit GABA-dependent chloride uptake by mouse brain vesicles, *Biochem. Biophys. Res. Commun.*, 1985, **133**, 37–43.
302. H. W. Vohland, J. Portig and K. Stein, Neuropharmacological effects of isomers of hexachlorocyclohexane. 1. Protection against pentylenetetrazol-induced convulsions, *Toxicol. Appl. Pharmacol.*, 1981, **57**, 425–438.
303. B. E. Fishman and G. Gianutsos, Differential effects of gamma-hexachlorocyclohexane (lindane) on pharmacologically-induced seizures, *Arch. Toxicol.*, 1987, **59**, 397–401.
304. F. Cattabeni, M. C. Pastorello and M. Eli, Convulsions induced by lindane and the involvement of the GABAergic system, *Arch. Toxicol. Suppl*, 1983, **6**, 244–249.
305. J. R. Bloomquist, P. M. Adams and D. M. Soderlund, Inhibition of gamma-aminobutyric acid-stimulated chloride flux in mouse brain vesicles by polychlorocycloalkane and pyrethroid insecticides, *Neurotoxicology*, 1986, **7**, 11–20.
306. C. Sunol, J. M. Tusell, E. Gelpi and E. Rodriguez-Farre, GABAergic modulation of lindane (gamma-hexachlorocyclohexane)-induced seizures, *Toxicol. Appl. Pharmacol.*, 1989, **100**, 1–8.
307. D. Woolley, L. Zimmer, Z. Hasson and K. Swanson, Do some insecticides and heavy metals produce long-term potentiation in the limbic system?, in *Cellular and Molecular Toxicology*, ed. T. Narahashi, Raven Press, New York, 1984, pp. 45–69.
308. D. E. Woolley, Introduction to Symposium—Application of neurophysiological techniques to toxicological problems—an overview, *Fund. Appl. Toxicol.*, 1985, **5**, 1–8.
309. R. M. Joy, The effects of neurotoxicants on kindling and kindled seizures, *Fund. Appl. Toxicol.*, 1985, **5**, 41–65.
310. L. G. Stark, T. E. Albertson and R. M. Joy, Effects of hexachlorocyclohexane isomers on the acquisition of kindled seizures, *Neurobehav. Toxicol. Teratol.*, 1986, **8**, 487–491.

311. R. M. Joy and T. E. Albertson, Interactions of lindane with synaptically mediated inhibition and facilitation in the dentate gyrus, *Neurotoxicology*, 1987, **8**, 529–542.
312. R. M. Joy and T. E. Albertson, Factors responsible for increased excitability of dentate gyrus granule cells during exposure to lindane, *Neurotoxicology*, 1987, **8**, 517–527.
313. R. M. Joy and T. E. Albertson, Convulsant-induced changes in perforant path-dentate gyrus excitability in urethane anesthetized rats, *J. Pharmacol. Exp. Ther.*, 1988, **246**, 887–895.
314. N. Ogata, S. M. Vogel, T. Narahashi and T. Lindane, but not deltamethrin blocks a component of FABA-activated chloride channels, *FASEB J.*, 1988, **2**, 2895–2900.
315. D. Mladenovic, D. Djuric, N. Petronijevic, T. Radosavljevic, N. Radonjic, D. Matic, D. Hrncic, A. Rasic-Markovic, D. Vucevic, D. Dekanski and O. Stanojlovic, The correlation between lipid peroxidation in different brain regions and the severity of lindane-induced seizures in rats, *Mol. Cell Biochem.*, 2010, **333**, 243–250.
316. A. Holian, M. A. Marchiarullo and D. F. Stickle, gamma-Hexachlorocyclohexane activation of alveolar macrophage phosphatidylinositol cycle, calcium mobilization of O_2-production, *FEBS Lett.*, 1984, **176**, 151–154.
317. L. G. Stark, R. Y. Chuang and R. M. Joy, Biochemical markers of exposure to proconvulsant and anticonvulsant chlorinated hydrocarbons, *Epilepsia*, 1987, **8**, 44.
318. G. S. Parries and M. Hokin-Neaverson, Inhibition of phosphatidylinostitol synthase and other membrane-associated enzymes by steroisomers of hexachlorocyclohexane, *J. Biol. Chem.*, 1985, **260**, 2687–2693.
319. IPCS, *Lindane*. 1992, International Programme on Chemical Safety, World Health Organization: Geneva.
320. F. Iverson, J. J. Ryan, R. Lizotte and S. L. Hierlihy, *In vivo* and *in vitro* binding of alpha- and gamma-hexachlorocyclohexane to mouse liver macromolecules, *Toxicol. Lett.*, 1984, **20**, 331–335.
321. X. Guan and R. J. Ruch, Gap junction endocytosis and lysosomal degradation of connexin43-P2 in WB-F344 rat liver epithelial cells treated with DDT and lindane, *Carcinogenesis*, 1996, **17**, 1791–1798.
322. K. M. Munir, K. V. Rao and S. V. Bhide, Effect of hexachlorocyclohexane on diethylnitrosamine-induced hepatocarcinogenesis in rat and its failure to promote skin tumors on dimethylbenz[a]-anthracene initiation in mouse, *Carcinogenesis*, 1984, **5**, 479–481.
323. C. Schroter, W. Parzefall, H. Schroter and R. Schulte-Hermann, Dose-response studies on the effects of alpha-, beta-, and gamma-hexachlorocyclohexane on putative preneoplastic foci, monooxygenases, and growth in rat liver, *Cancer Res.*, 1987, **47**, 80–88.
324. S. Angsubhakorn, N. Bhamarapravati, A. Pradermwong, N. Im-Emgamol and S. Sahaphong, Minimal dose and time protection by lindane (gamma-isomer of 1,2,3,4,5,6 hexachlorocyclohexane) against liver tumors induced by aflatoxin B1, *Int. J. Cancer*, 1989, **43**, 531–534.

325. T. K. Trifonova, In. N. Gladenko and V. D. Shulyak, Effect of gamma-BHC and Sevin on reproduction, *Veterinariya (Kiev)*, 1970, **47**, 91–93.
326. S. Y. Nashteyn and D. L. Leybovich, Low doses of DDT, γ-HCCH and mixtures of these: Effect on sexual function and embryogenesis in rats, *Gig. Sanit*, 1971, **36**, 19–22.
327. F. L. Van Velsen, L. H. Danse, F. X. Van Leeuwen, J. A. Dormans and M. J. Van Logten, The subchronic oral toxicity of the beta-isomer of hexachlorocyclohexane in rats, *Fundam. Appl. Toxicol.*, 1986, **6**, 697–712.
328. S. K. Nigam, B. C. Lakkad, A. B. Karnik, K. N. Thakore, D. K. Bhatt, K. Aravindra Babu and S. K. Kashyap, Effect of hexachlorocyclohexane feeding on testicular tissue of pure inbred Swiss mice, *Bull. Environ. Contam. Toxicol.*, 1979, **23**, 431–437.
329. T. Shivanandappa and M. K. Krishnakumari, Hexachlorocyclohexane-induced testicular dysfunction in rats, *Acta. Pharmacol. Toxicol.*, 1983, **52**.
330. A. R. Chowdhury, H. Venkatakrishna-Bhatt, A. K. Gautam and A. K. Testicular changes of rats under lindane treatment, *Bull. Environ. Contam. Toxicol.*, 1987, **38**, 154–156.
331. Q. Huang and X. Huang, The effect of benzene hexachloride on mouse sperm, *Zhejiang Yike Daxue Xuebao*, 1987, **16**, 9–12.
332. R. L. Cooper, R. W. Chadwick, G. L. Rehnberg, J. M. Goldman, K. C. Booth, J. F. Hein and W. K. McElroy, Effect of lindane on hormonal control of reproductive function in the female rat, *Toxicol. Appl. Pharmacol.*, 1989, **99**, 384–394.
333. L. Uphouse and J. Williams, Diestrous treatment with lindane disrupts the female rat reproductive cycle, *Toxicol. Lett.*, 1989, **48**, 21–28.
334. J. Klosa, On the toxicology of hexachlorocyclohexane, *Pharmazie*, 1950, **5**, 615–616.
335. J. G. Bowerman, M. P. Gomez, R. D. Austin and D. E. Wold, Comparative study of permethrin 1% creme rinse and lindane shampoo for the treatment of head lice, *Pediatr. Infect. Dis. J.*, 1987, **6**, 252–255.
336. B. Shacter, Treatment of scabies and pediculosis with lindane preparations: an evaluation, *J. Am. Acad. Dermatol.*, 1981, **5**, 517–527.
337. V. J. Brown, Life after lindane in California, *Environ. Health Perspect.*, 2008, **116**, A128.
338. E. H. Humphreys, S. Janssen, A. Heil, P. Hiatt, G. Solomon and M. D. Miller, Outcomes of the California ban on pharmaceutical lindane: clinical and ecologic impacts, *Environ. Health Perspect.*, 2008, **116**, 297–302.
339. R. J. Feldmann and H. I. Maibach, Pesticide percutaneous penetration in man, *J. Invest. Dermatol.*, 1970, **54**, 435–436.
340. M. Lange, K. Nitzsche and A. Zesch, Percutaneous absorption of lindane in healthy volunteers and scabies patients. Dependency of penetration kinetics in serum upon frequency of application, time and mode of washing, *Arch. Dermatol. Res.*, 1981, **271**, 387–399.
341. K. Nitsche, M. Lange, E. Bauer and A. Zesch, Quantitative distribution of locally applied lindane in human skin and subcutaneous fat *in vitro*.

Dependence of penetration on the applied concentration, skin state, duration of action and nature and time of washing, *Derm. Beruf. Umwelt.*, 1984, **32**, 161–165.

342. K. Baumann, K. Behling, H.-L. Brassow and K. Stapel, Occupational exposure to hexachlorocyclohexane. III. Neurophysiological findings and neuromuscular function in chronically exposed workers, *Int. Arch. Occup. Environ. Health*, 1981, **48**, 165–172.
343. H. L. Brassow, K. Baumann and G. Lehnert, Occupational exposure to hexachlorocyclohexane. II. Health conditions of chronically exposed workers, *Int. Arch. Occup. Environ. Health.*, 1981, **48**, 81–87.
344. D. Muller, H. Klepel, R. Macholz and R. Knoll, Electroneurographic and electroencephalographic finding in pateints exposed to hexachlorocyclohexane, *Psychiatr. Neurol. Med. Psychol.*, 1981, **33**, 468–472.
345. S. K. Kashyap, Health surveillance and biological monitoring of pesticide formulators in India, *Toxicol. Lett.*, 1986, **33**, 107–114.
346. S. B. Khare, A. G. Rizvi, O. P. Shukla, R. R. Singh, O. Perkash, V. D. Misra, J. P. Gupta and P. K. Sethi, Epidemic outbreak of neuro-ocular manifestations due to chronic BHC poisoning, *J. Assoc. Physicians India*, 1977, **25**, 215–222.
347. C. Bambov, M. Chomakov and N. Dimitrova, Group intoxication with lindane, *Suvrem. Med.*, 1966, **17**, 477–481.
348. U. Jaeger, A. Podczeck, A. Haubenstock, K. Pirich, A. Donner and K. Hruby, Acute oral poisoning with lindane-solvent mixtures, *Vet. Hum. Toxicol.*, 1984, **26**, 11–14.
349. H. E. Klein-Natrop, H. Roder and H. Kadner, Note on timely scabies treatment., *Dtsch. Gesundheitswes*, 1970, **25**, 2082–2084.
350. H. G. Starr, Jr. and N. J. Clifford, Acute lindane intoxication: a case study, *Arch. Environ. Health*, 1972, **25**, 374–375.
351. J. Angerer, R. Maass and R. Heinrich, Occupational exposure to hexachlorocyclohexane. VI. Metabolism of gamma-hexachlorocyclohexane in man, *Int. Arch. Occup. Environ. Health*, 1983, **52**, 59–67.
352. L. Drummond, E. M. Gillanders and H. K. Wilson, Plasma gamma-hexachlorocyclohexane concentrations in forestry workers exposed to lindane, *Br. J. Ind. Med.*, 1988, **45**, 493–497.
353. F. Sperling, H. K. Ewenike and T. Farber, Changes in LD_{50} of parathion and hepatachlor following turpentine pretreatment, *Environ. Res.*, 1972, **5**, 164–171.
354. M. Eisler, Heptachlor: Toxicology and safety evaluation, *Ind. Med. Surg.*, 1968, **37**, 667–673.
355. J. F. Treon and F. P. Cleveland, Toxicity of certain chlorinated hydrocarbon insecticides for laboratory animals with special reference to aldrin and dieldrin, *J. Agric. Food Chem.*, 1955, **3**, 402–408.
356. M. D. Reuber, Significance of acute and chronic renal disease in Osborne-Mendel rats ingesting dieldrin or aldrin, *Clin. Toxicol.*, 1980, **17**, 159–170.
357. T. S. Foster, Effect of some pesticides on the adrenal glands in the rat, *J. Biochem. (Tokyo)*, 1968, **46**, 115–1120.

358. L. D. Loose, Macrophage induction of T-suppressor cells in pesticide-exposed and protozoan-infected mice, *Environ. Health Perspect.*, 1982, **43**, 89–97.
359. J. B. Barnett, D. Holcomb, J. H. Menna and L. S. Soderberg, The effect of prenatal chlordane exposure on specific anti-influenza cell-mediated immunity, *Toxicol. Lett.*, 1985, **25**, 229–238.
360. W. F. Mueller, M. J. Iatropoulos, K. Rozman, F. Korte and F. Coulston, Comparative kinetic, metabolic, and histopathologic effects of chlorinated hydrocarbon pesticides in rhesus monkeys., *Toxicol. Appl. Pharmacol.*, 1978, **45**, 283–284.
361. S. L. Newman and P. S. Guzelian, Identification of the cyanopregnenolone-inducible form of hepatic cytochrome P-450 as a catalyst of aldrin epoxidation, *Biochem. Pharmacol.*, 1983, **32**, 1529–1531.
362. T. Wolff, H. Greim, M. T. Huang, G. T. Miwa and A. Y. Lu, Aldrin epoxidation catalyzed by purified rat-liver cytochromes P-450 and P-448. High selectivity for cytochrome P-450, *Eur. J. Biochem.*, 1980, **111**, 545–551.
363. T. Wolff and F. P. Guengerich, Rat liver cytochrome P-450 isozymes as catalysts of aldrin epoxidation in reconstituted monooxygenase systems and microsomes, *Biochem. Pharmacol.*, 1987, **36**, 2581–2588.
364. B. Lang, K. Frei and P. Maier, Prostaglandin synthase dependent aldrin epoxidation in hepatic and extrahepatic tissues of rats, *Biochem. Pharmacol.*, 1986, **35**, 3643–3645.
365. M. E. Eldefrawi, S. M. Sherby, I. M. Abalis and A. T. Eldefrawi, Interactions of pyrethroid and cyclodiene insecticides with nicotinic acetylcholine and GABA receptors, *Neurotoxicology*, 1985, **6**, 47–62.
366. F. Hirasawa and Y. Takizawa, Accumulation and declination of chlordane congeners in mice, *Toxicol. Lett.*, 1989, **47**, 109–117.
367. R. C. Casabar, A. D. Wallace, E. Hodgson and R. L. Rose, Metabolism of endosulfan-alpha by human liver microsomes and its utility as a simultaneous *in vitro* probe for CYP2B6 and CYP3A4, *Drug Metab. Dispos.*, 2006, **34**, 1779–1785.
368. H. K. Lee, J. K. Moon, C. H. Chang, H. Choi, H. W. Park, B. S. Park, H. S. Lee, E. C. Hwang, Y. D. Lee, K. H. Liu and J. H. Kim, Stereoselective metabolism of endosulfan by human liver microsomes and human cytochrome P450 isoforms, *Drug Metab. Dispos.*, 2006, **34**, 1090–1095.
369. P. Dubois, A. Pfohl-Leszkowicz, I. De Waziers and P. Kremers, Selective induction of the CYP3A family by endosulfan and DNA-adduct formation in different hepatic and hepatoma cells, *Environ. Toxicol. Pharmacol.*, 1996, **1**, 249–256.
370. R. Hack, E. Ebert and K. H. Leist, Chronic toxicity and carcinogenicity studies with the insecticide endosulfan in rats and mice, *Food Chem. Toxicol.*, 1995, **33**, 941–950.
371. E. F. Meierhenry, B. H. Ruebner, M. E. Gershwin, L. S. Hsieh and S. W. French, Dieldrin-induced Mallory bodies in hepatic tumors of mice of different strains, *Hepatology*, 1983, **3**, 90–95.

372. B. van Ravenswaay, H. J. M. Toussaint and R. L. Schmitt, Dieldrin-induced changes in isozyme composition in the livers of CF-1 mice, *Int. J. Cancer*, 1988, **41**, 305–308.
373. J. E. Klaunig, Y. Xu, S. Bachowski, C. A. Ketcham, J. S. Isenberg, K. L. Kolaja, T. K. Baker, E. F. Walborg, Jr. and D. E. Stevenson, Oxidative stress in nongenotoxic carcinogenesis, *Toxicol. Lett.*, 1995, **83**.
374. K. Nagata and T. Narahashi, Dual action of the cyclodiene insecticide dieldrin on the gamma-aminobutyric acid receptor-chloride channel complex of rat dorsal root ganglion neurons, *J. Pharmacol. Exp. Ther.*, 1994, **269**, 164–171.
375. R. M. Joy and T. E. Albertson, Lindane and limbic system excitability, *Neurotoxicology*, 1985, **6**, 193–214.
376. D. E. Ray, T. Lister and R. M. Joy, The action of dieldrin on regional brain blood flow in the rat., *Pestic. Biochem. Physiol.*, 1986, **25**, 205–210.
377. A. K. Agrawal, M. Anand, N. F. Zaidi and P. K. Seth, Involvement of serotonergic receptors in endosulfan neurotoxicity, *Biochem. Pharmacol.*, 1983, **32**, 3591–3593.
378. M. Anand, S. Mehrotra, K. Gopal, R. N. Sur and S. V. Chandra, Role of neurotransmitters in endosulfan-induced aggressive behaviour in normal and lesioned rats, *Toxicol. Lett.*, 1985, **24**, 79–84.
379. P. K. Seth, N. F. Saidi, A. K. Agrawal and M. Anand, Neurotoxicity of endosulfan in young and adult rats, *Neurotoxicology*, 1986, **7**, 623–635.
380. P. A. Fowler, D. R. Abramovich, N. E. Haites, P. Cash, N. P. Groome, A. Al-Qahtani, T. J. Murray and R. G. Lea, Human fetal testis Leydig cell disruption by exposure to the pesticide dieldrin at low concentrations, *Hum. Reprod.*, 2007, **22**, 2919–2927.
381. M. H. Silva and D. Gammon, An assessment of the developmental, reproductive, and neurotoxicity of endosulfan, *Birth Defects Res. B. Dev. Reprod. Toxicol.*, 2009, **86**, 1–28.
382. D. L. Rowley, M. A. Rab, W. Hardjotanojo, J. Liddle, V. W. Burse, M. Saleem, D. Sokal, H. Falk and S. L. Head, Convulsions caused by endrin poisoning in Pakistan, *Pediatrics*, 1987, **79**, 928–934.
383. A. Steentoft, A case of fatal dieldrin poisoning, *Med. Sci. Law*, 1979, **19**, 268–269.
384. V. J. Derbes, J. H. Dent, W. W. Forrest and M. F. Johnson, Fatal clhordane poisoning, *J. Am. Med. Assoc.*, 1955, **158**, 1367–1369.
385. M. E. Loevinsohn, Insecticide use and increased mortality in rural Central Luzon, Philippines, *Lancet*, 1987, **1**, 1359–1362.
386. S. Shindell and S. Ulrich, Mortality of workers employed in the manufacture of chlordane: an update, *J. Occup. Med.*, 1986, **28**, 497–501.
387. Y. H. Wen and H. M. Chan, A pharmacokinetic model for predicting absorption, elimination, and tissue burden of toxaphene in rats, *Toxicol. Appl. Pharmacol.*, 2000, **168**, 235–243.
388. A. Mohammed, E. Hallberg, J. Rydstrom and P. Slanina, Toxaphene: accumulation in the adrenal cortex and effect on ACTH-stimulated corticosteroid synthesis in the rat, *Toxicol. Lett.*, 1985, **24**, 137–143.

389. G. A. Pollock and W. W. Kilgore, Excretion and storage of [^{14}C]toxaphene fractions, *J. Toxicol. Environ. Health*, 1980, **a6**, 127–140.
390. W. V. Turner, J. L. Engel and J. E. Casida, Toxaphene components and related compounds: preparation and toxicity of some hepta-, octa-, and nonachlorobornanes, hexa- and heptachlorobornenes, and a hexachlorobornadiene, *J. Agric. Food. Chem.*, 1977, **25**, 1394–1401.
391. J. I. Goodman, D. J. Brusick, W. M. Busey, S. M. Cohen, J. C. Lamb and T. B. Starr, Reevaluation of the cancer potency factor of toxaphene: Recommendations from a peer review panel, *Toxicol. Sci.*, 2000, **55**, 3–16.
392. NCI, *Bioassay of Toxaphene for Possible Carcinogenesis*. U. S. Govenment Printing Office, Washington, D. C., 1979.
393. C. C. Hedli, R. Snyder, F. K. Kinoshita and M. Steinberg, Investigation of hepatic cytochrome P-450 enzyme induction and DNA adduct formation in male CD/1 mice following oral administration of toxaphene, *J. Appl. Toxicol.*, 1998, **18**, 173–178.
394. R. S. Waritz, M. Steinberg, F. K. Kinoshita, C. M. Kelly and W. R. Richter, Thyroid function and thyroid tumors in toxaphene-treated rats, *Regul. Toxicol. Pharmacol.*, 1996, **24**, 184–192.
395. D. L. Arnold, F. Bryce, C. Baccanale, S. Hayward, J. R. Tanner, E. MacLellan, T. Dearden and S. Fernie, Toxicological consequences of toxaphene ingestion by cynomolgus (*Macaca fascicularis*) monkeys. Part 1: Pre-mating phase, *Food Chem. Toxicol.*, 2001, **39**, 467–476.
396. V. P. Bezuglyi and N. D. Mukhtarova, [Clinical aspects of polychlorpinene poisoning], *Gig. Tr. Prof. Zabol.*, 1969, **13**, 53–55.
397. L. C. McGee, H. L. Reed and J. P. Fleming, Accidental poisoning by toxaphene. Review of toxicology and case reports., *J. Am. Med. Assoc.*, 1952, **149**, 1124–1126.
398. H. M. Mehendale, L. Fishbein, M. Fields and H. B. Matthews, Fate of mirex-^{14}C in the rat and plants, *Bull. Environ. Contam. Toxicol.*, 1972, **8**, 200–207.
399. J. D. Yarbrough, J. M. Grimley, P. I. Karl, J. E. Chambers, R. S. Case and E. G. Alley, Tissue disposition, metabolism, and excretion of *cis*- and *trans*-5,10-dihydrogen mirex, *Drug Metab. Dispos.*, 1983, **11**, 611–614.
400. T. B. Gaines and R. D. Kimbrough, Oral toxicity of mirex in adult and suckling rats. With notes on the ultrastructure of liver changes, *Arch. Environ. Health*, 1970, **21**, 7–14.
401. P. S. Larson, J. L. Egle, Jr., G. R. Hennigar and J. F. Borzelleca, Acute and subchronic toxicity of mirex in the rat, dog, and rabbit, *Toxicol. Appl. Pharmacol.*, 1979, **49**, 271–277.
402. R. J. Warren, R. L. Kirkpatrick and R. W. Young, Barbiturate-induced sleeping times, liver weights, and reproduction of cottontail rabbits after mirex ingestion, *Bull. Environ. Contam. Toxicol.*, 1978, **19**, 223–228.
403. E. P. Pittz, D. Rourke, R. Abraham and F. Coulston, Alterations in hepatic microsomal proteins of mice administered mirex orally, *Bull. Environ. Contam. Toxicol.*, 1979, **21**, 344–351.

404. J. D. Yarbrough, J. E. Chambers, J. M. Grimley, E. G. Alley, M. M. Fang, J. T. Morrow, B. C. Ward and J. D. Conroy, Comparative study of 8-monohydromirex and mirex toxicity in male rats, *Toxicol. Appl. Pharmacol.*, 1981, **58**, 105–117.
405. I. Chu, D. C. Villeneuve, B. L. MacDonald, V. E. Secours and V. E. Valli, Reversibility of the toxicological changes induced by photomirex and mirex, *Toxicology*, 1981, **21**, 235–250.
406. L. R. Curtis and H. M. Mehendale, Hepatobiliary dysfunction and inhibition of adenosine triphosphatase activity in bile canaliculi-enriched fractions following *in vivo* mirex, photomirex, and chlordecone exposures, *Toxicol. Appl. Pharmacol.*, 1981, **61**, 429–440.
407. L. R. Curtis and D. Hoyt, Impaired biliary excretion of taurocholate associated with increased biliary tree permeability in mirex- or chlordecone-pretreated rats, *J. Pharmacol. Exp. Ther.*, 1984, **231**, 495–501.
408. H. M. Mehendale, Mirex-induced impairment of hepatobiliary function. Suppressed biliary excretion of imipramine and sulfobromophthalein, *Drug Metab. Dispos.*, 1977, **5**, 56–62.
409. S. Teo and M. Vore, Mirex inhibits bile acid secretory function *in vivo* and in the isolated perfused rat liver, *Toxicol. Appl. Pharmacol.*, 1991, **109**, 161–170.
410. R. E. Ebel, Hepatic microsomal *p*-nitroanisole *O*-demethylase. Effects of chlordecone or mirex induction in male and female rats, *Biochem. Pharmacol.*, 1984, **33**, 559–564.
411. L. S. Crouch and R. E. Ebel, Influence of chlordecone and mirex exposure on benzo[a]pyrene metabolism of rat liver micrososmes, *Xenobiotica*, 1987, **17**, 25–34.
412. R. Abraham, K. F. Benitz and R. Mankes, Ploidy patterns in hepatic tumors induced by Mirex, *Exp. Mol. Pathol.*, 1983, **38**, 271–282.
413. J. R. Innes, B. M. Ulland, M. G. Valerio, L. Petrucelli, L. Fishbein, E. R. Hart, A. J. Pallotta, R. R. Bates, H. L. Falk, J. J. Gart, M. Klein, I. Mitchell and J. Peters, Bioassay of pesticides and industrial chemicals for tumorigenicity in mice: a preliminary note, *J. Natl. Cancer Inst.*, 1969, **42**, 1101–1114.
414. B. M. Ulland, N. P. Page, R. A. Squire, E. K. Weisburger and R. L. Cypher, A carcinogenicity assay of Mirex in Charles River CD rats, *J. Natl. Cancer Inst.*, 1977, **58**, 133–140.
415. T. W. Kim, K. L. Porter, J. F. Foley, R. R. Maronpot and R. C. Smart, Evidence that mirex promotes a unique population of epidermal cells that cannot be distinguished by their mutant Ha-ras genotype, *Mol. Carcinogenesis*, 1997, **20**, 115–124.
416. G. J. Moser, S. A. Meyer and R. C. Smart, The chlorinated pesticide mirex is a novel nonphorbol ester-type tumor promoter in mouse skin, *Cancer Res.*, 1992, **52**, 631–636.
417. G. J. Moser, C. L. Robinette and R. C. Smart, Characterization of skin tumor promotion by mirex: Structure-activity relationships, sexual dimorphism and presence of Ha-ras mutation, *Carcinogenesis*, 1993, **14**, 1155–1160.

418. S. A. Meyer, G. J. Moser, N. A. Monteiro-Riviere and R. C. Smart, Minimal role of enhanced cell proliferation in skin tumor promotion by mirex: a nonphorbol ester-type promoter, *Environ. Health Perspect.*, 1993, **101**, 265–269.
419. A. Singh, M. K. Bhatnagar, D. C. Villeneuve and V. E. Valli, Ultrastructure of the thyroid glands of rats fed photomirex: a 48-week recovery study, *J. Environ. Pathol. Toxicol. Oncol.*, 1985, **6**, 115–126.
420. J. M. Rogers and C. T. Grabowski, Postnatal mirex cataractogenesis in rats: lens cation balance, growth and histology, *Exp. Eye Res.*, 1984, **39**, 563–573.
421. R. J. Kavlock, N. Chernoff, E. Rogers and D. Whitehouse, Comparative tissue distribution of mirex and chlordecone in fetal and neonatal rats, *Pestic. Biochem. Physiol.*, 1980, **14**, 227–235.
422. J. J. Boylan, J. L. Egle and P. S. Guzelian, Cholestyramine: use as a new therapeutic approach for chlordecone (Kepone) poisoning, *Science*, 1978, **199**, 893–895.
423. J. L. J. Egle, S. B. Fernandez, P. S. Guzelian and J. F. Borzelleca, Distribution and excretion of chlordecone (Kepone) in the rat, *Drug. Metab. Dispos.*, 1978, **6**, 91–95.
424. J. E. Chambers and C. A. Trevathan, Effect of mirex, dechlorinated mirex derivatives and chlordecone on microsomal mixed-function oxidase activity and other hepatic parameters, *Toxicol. Lett.*, 1983, **16**, 109–115.
425. R. V. Blanke, M. W. Fariss, P. S. Guzelian, A. R. Paterson and D. E. Smith, Identification of a reduced form of chlordecone (Kepone) in human stool, *Bull. Environ. Contam. Toxicol.*, 1978, **20**, 782–785.
426. M. W. Fariss, R. V. Blanke, J. J. Saady and P. S. Guzelian, Demonstration of major metabolic pathways for chlordecone (Kepone) in humans, *Drug Metab. Dispos.*, 1980, **8**, 434–438.
427. T. E. Houston, L. C. Mutter, R. V. Blanke and P. S. Guzelian, Chlordecone alcohol formation in the Mongolian gerbil (*Meriones unguiculatus*): a model for human metabolism of chlordecone (Kepone), *Fundam. Appl. Toxicol.*, 1981, **1**, 293–298.
428. D. T. Molowa, A. G. Shayne and P. S. Guzelian, Purification and characterization of chlordecone reductase from human liver, *J. Biol. Chem.*, 1986, **261**, 12624–12627.
429. D. T. Molowa, S. A. Wrighton, R. V. Blanke and P. S. Guzelian, Characterization of a unique aldo-keto reductase responsible for the reduction of chlordecone in the liver of the gerbil and man, *J. Toxicol. Environ. Health*, 1986, **17**, 375–384.
430. P. J. Soine, R. V. Blanke and C. C. Schwartz, Chlordecone metabolism in the pig, *Toxicol. Lett.*, 1983, **17**, 35–41.
431. P. S. Larson, Jr. J. L. Egle, G. R. Hennigar, R. W. Lane and J. F. Borzelleca, Acute, subchronic, and chronic toxicity of chlordecone, *Toxicol. Appl. Pharmacol.*, 1979, **48**, 29–41.
432. J. L. Egle, Jr., P. S. Guzelian and J. F. Borzelleca, Time course of the acute toxic effects of sublethal doses of chlordecone (Kepone), *Toxicol. Appl. Pharmacol.*, 1979, **48**, 533–536.

433. J. S. Klingensmith and H. M. Mehendale, Chlordecone-induced fat depletion in the male rat, *J. Toxicol. Environ. Health*, 1982, **10**, 121–129.
434. J. M. Baggett, A. Thureson-Klein and R. L. Klein, Effects of chlordecone on the adrenal medulla of the rat, *Toxicol. Appl. Pharmacol.*, 1980, **52**, 313–322.
435. National Cancer Institute, *Report on Carcinogenesis Bioassay of Technical Grade Chlordecone (Kepone)*. Carcinogen Program, Divisio of Cancer Cause and Prevention, U. S. Government Printing Office, Washington, D. C., 1976.
436. M. D. Reuber, Carcinomas of the liver in Osborne-Mendel rats ingesting DDT, *Tumori*, 1978, **64**, 571–577.
437. A. E. Sirica, C. S. Wilkerson, L. L. Wu, R. Fitzgerald, R. V. Blanke and P. S. Guzelian, Evaluation of chlordecone in a two-stage model of hepatocarcinogenesis: a significant sex difference in the hepatocellular carcinoma incidence, *Carcinogenesis*, 1989, **10**, 1047–1054.
438. B. Hoskins and I. K. Ho, Chlordecone-induced alterations in content and subcellular distribution of calcium in mouse brain, *J. Toxicol. Environ. Health*, 1982, **9**, 535–544.
439. H. Komulainen and S. C. Bondy, Modulation of levels of free calcium within synaptosomes by organochlorine insecticides, *J. Pharmacol. Exp. Ther.*, 1987, **241**, 575–581.
440. O. Gandolfi, D. L. Cheney, J. S. Hong and E. Costa, On the neurotoxicity of chlordecone: a role for gamma-aminobutyric acid and serotonin, *Brain Res.*, 1984, **303**, 117–123.
441. M. L. Kirby, R. L. Barlow and J. R. Bloomquist, Selective effects of cyclodiene insecticides on dopamine release in mammalian synaptosomes, *Toxicol. Appl. Pharmacol.*, 2002, **181**, 89–92.
442. H. M. Mehendale, Mechanism of the lethal interaction of chlordecone and CCl_4 at non-toxic doses, *Toxicol. Lett.*, 1989, **49**, 215–241.
443. J. M. Hegarty, E. A. Glende, Jr. and R. O. Recknagel, Potentiation by chlordecone of the defect in hepatic microsomal calcium sequestration induced by carbon tetrachloride, *J. Biochem. Toxicol.*, 1986, **1**, 73–78.
444. M. G. Soni and H. M. Mehendale, Hepatic failure leads to lethality of chlordecone-amplified hepatotoxicity of carbon tetrachloride, *Fundam. Appl. Toxicol.*, 1993, **21**, 442–450.
445. M. G. Soni and H. M. Mehendale, Role of tissue repair in toxicologic interactions among hepatotoxic organics, *Environ. Health Perspect.*, 1998, **106**, 1307–1317.
446. P. R. S. Kodavanti, U. P. Kodavanti, O. M. Faroon and H. M. Mehendale, Pivotal role of hepatocellular regeneration in the ultimate hepatotoxicity of CCl4 in chlordecone-, mirex-, or phenobarbital-pretreated rats, *Toxicol. Pathol.*, 1992, **20**, 556–569.
447. P. R. S. Kodavanti, V. C. Rao and H. M. Mehendale, Loss of calcium homeostasis leads to progressive phase of chlordecone-potentiated carbon tetrachloride hepatotoxicity, *Toxicol. Appl. Pharmacol.*, 1993, **122**, 77–87.

448. S. Chaudhury and H. M. Mehendale, Amplification of CCl4 toxicity by chlordecone: Destruction of rat hepatic microsomal cytochrome P-450 subpopulation, *J. Toxicol. Environ. Health*, 1991, **32**, 277–294.
449. H. M. Carpenter, O. R. Hedstrom, L. K. Siddens, J. R. Duimstra, Z. W. Cai, K. A. Fisher and L. R. Curtis, Ultrastructural, protein, and lipid changes in liver associated with chlordecone treatment of mice, *Fundam. Appl. Toxicol.*, 1996, **34**, 157–164.
450. L. G. Rochelle and L. R. Curtis, Distribution of chlordecone to liver plasma membranes and recovery from hepatobiliary dysfunction in rats, *Toxicology*, 1994, **86**, 123–134.
451. D. J. Gilroy, H. M. Carpenter and L. R. Curtis, Chlordecone pretreatment alters [^{14}C]chlordecone and [^{14}C]cholesterol transport kinetics in the perfused rat liver, *Fundam. Appl. Toxicol.*, 1994, **22**, 286–292.
452. Z. Cai and H. M. Mehendale, Resiliency to amplification of carbon tetrachloride hepatotoxicity by chlordecone during postnatal development in rats, *Pediat. Res.*, 1993, **33**, 225–232.
453. A. Dalu, A. Warbritton, T. J. Bucci and H. M. Mehendale, Age-related susceptibility to chlordecone-potentiated carbon tetrachloride hepatotoxicity and lethality is due to hepatic quiescence, *Pediat. Res.*, 1995, **38**, 140–148.
454. A. Dalu, P. S. Rao and H. M. Mehendale, Colchicine antimitosis abolishes resiliency of postnatally developing rats to chlordecone-amplified carbon tetrachloride hepatotoxicity and lethality, *Environ. Health Perspect.*, 1998, **106**, 597–606.
455. A. Dalu and H. M. Mehendale, Efficient tissue repair underlies the resiliency of postnatally developing rats to chlordecone + CCl_4 hepatotoxicity, *Toxicology*, 1996, **111**, 29–42.
456. P. S. Guzelian, Comparative toxicology of chlordecone (Kepone) in humans and experimental animals, *Ann. Rev. Pharmacol. Toxicol.*, 1982, **22**, 89–113.
457. J. Lee, R. C. Scheri, Y. Zhang and L. R. Curtis, Chlordecone, a mixed pregnane X receptor (PXR) and estrogen receptor alpha (ERalpha) agonist, alters cholesterol homeostasis and lipoprotein metabolism in C57BL/6 mice, *Toxicol. Appl. Pharmacol.*, 2008, **233**, 193–202.
458. J. Lee, R. C. Scheri and L. R. Curtis, Chlordecone altered hepatic disposition of [14C]cholesterol and plasma cholesterol distribution but not SR-BI or ABCG8 proteins in livers of C57BL/6 mice, *Toxicol. Appl. Pharmacol.*, 2008, **229**, 265–272.
459. N. Chernoff and E. H. Rogers, Fetal toxicity of Kepone in rats and mice, *Toxicol. Appl. Pharmacol.*, 1976, **38**, 189–194.
460. L. Uphouse and K. Eckols, Serotonin receptors in striatum after chlordecone treatment of adult female rats, *Neurotoxicology*, 1986, **7**, 25–32.
461. S. Ray, F. Xu, P. Li, N. S. Sanchez, H. Wang and S. K. Das, Increased level of cellular Bip critically determines estrogenic potency for a xenoestrogen Kepone in the mouse uterus, *Endocrinology*, 2007, **148**, 4774–4785.

462. G. G. Kuiper, J. G. Lemmen, B. Carlsson, J. C. Corton, S. H. Safe, P. T. van der Saag, B. van der Burg and J. A. Gustafsson, Interaction of estrogenic chemicals and phytoestrogens with estrogen receptor beta, *Endocrinology*, 1998, **139**, 4252–4263.
463. G. Lemaire, W. Mnif, P. Mauvais, P. Balaguer and R. Rahmani, Activation of alpha- and beta-estrogen receptors by persistent pesticides in reporter cell lines, *Life Sci.*, 2006, **79**, 1160–1169.
464. R. E. Linder, T. M. Scotti, W. K. McElroy, J. W. Laskey, L. F. Strader and K. Powell, Spermotoxicity and tissue accumulation of chlordecone (Kepone) in male rats, *J. Toxicol. Environ. Health*, 1983, **12**, 183–192.
465. F. Wang, S. M. Roberts, E. J. Butfiloski, L. Morel and E. S. Sobel, Acceleration of autoimmunity by organochlorine pesticides: a comparison of splenic B-cell effects of chlordecone and estradiol in (NZBxNZW)F1 mice, *Toxicol. Sci.*, 2007, **99**, 141–152.
466. F. Wang, S. M. Roberts, E. J. Butfiloski and E. S. Sobel, Diminished prolactin from chlordecone treatment in ovariectomized (NZBxNZW)F(1) mice, *Int. Immunopharmacol.*, 2007, **7**, 1808–1812.
467. P. S. Guzelian, The clinical toxicology of Chlordecone as an example of toxicological risk assessment for man, *Toxicol. Lett.*, 1992, **64–65**, 589–596.
468. J. R. Taylor, J. B. Selhorst, S. A. Houff and A. J. Martinez, Chlordecone intoxication in man. I. Clinical observations, *Neurology*, 1978, **28**, 626–630.
469. J. R. Taylor, Neurological manifestations in humans exposed to chlordecone: follow-up results, *Neurotoxicology*, 1985, **6**, 231–236.
470. S. B. Cannon, J. M. Veazey, Jr., R. S. Jackson, V. W. Burse, C. Hayes, W. E. Straub, P. J. Landrigan and J. A. Liddle, Epidemic kepone poisoning in chemical workers, *Am. J. Epidemiol.*, 1978, **107**, 529–537.
471. P. J. Soine, R. V. Blanke, P. S. Guzelian and C. C. Schwartz, Preferential binding of chlordecone to the protein and high density lipoprotein fractions of plasma from humans and other species, *J. Toxicol. Environ. Health*, 1982, **9**, 107–118.
472. W. J. Cohn, J. J. Boylan, R. V. Blanke, M. W. Fariss, J. R. Howell and P. S. Guzelian, Treatment of chlordecone (Kepone) toxicity with cholestyramine. Results of a controlled clinical trial, *N. Engl. J. Med.*, 1978, **298**, 243–248.
473. L. Multigner, J. R. Ndong, A. Giusti, M. Romana, H. Delacroix-Maillard, S. Cordier, B. Jegou, J. P. Thome and P. Blanchet, Chlordecone exposure and risk of prostate cancer, *J. Clin. Oncol.*, 2010, **28**, 3457–3462.
474. FAO/WHO, *Joint Meeting of the FAO and WHO Panels of Experts on Pesticides Residues in Food-2000 Evaluations. Part II Toxicological*. World Health Organization, Geneva, 2001.
475. N. Lubick, Environment. Endosulfan's exit: U. S. EPA pesticide review leads to a ban, *Science*, 2010, **328**, 1466.
476. M. H. Silva and S. L. Beauvais, Human health risk assessment of endosulfan. I: Toxicology and hazard identification, *Regul. Toxicol. Pharmacol.*, 2010, **56**, 4–17.

477. M. H. Silva and W. C. Carr, Jr., Human health risk assessment of endosulfan: II. Dietary exposure assessment, *Regul. Toxicol. Pharmacol.*, 2010, **56**, 18–27.
478. S. L. Beauvais, M. H. Silva and S. Powell, Human health risk assessment of endosulfan. Part III: Occupational handler exposure and risk, *Regul. Toxicol. Pharmacol.*, 2010, **56**, 28–37.
479. S. L. Beauvais, M. H. Silva and S. Powell, Human health risk assessment of endosulfan. Part IV: Occupational reentry and public non-dietary exposure and risk, *Regul. Toxicol. Pharmacol.*, 2010, **56**, 38–50.
480. J. C. Lamb, B. H. Neal and J. I. Goodman, Risk assessment of toxaphene and its breakdown products: time for a change?, *Crit. Rev. Toxicol.*, 2008, **38**, 805–815.

CHAPTER 4

Toxicity of Organophosphates and Carbamates

RAMESH C GUPTA[1] AND DEJAN MILATOVIC[2]

[1] Murray State University, Breathitt Veterinary Center, Toxicology Department, Hopkinsville, KY, USA. Email: ramesh1.gupta@murraystate.edu; [2] Vanderbilt University, Department of Pediatrics, Nashville, TN, USA

4.1 Introduction

Since the early 1950s, organophosphate (OP) and carbamate (CM) insecticides have gained wide popularity around the world. This has been partly due to their lack of residue persistence in the environment and in mammalian systems, and also due to the development of lesser resistance in insects compared to the organochlorine insecticides. OPs and CMs, commonly referred to as anti-cholinesterase (anti-ChE) agents, constitute a large class of synthetic compounds that are primarily used to protect crops, gardens, homes and offices from insects as well as to protect humans and animals from external parasites. These compounds are heavily used to protect public health from diseases like malaria, West Nile disease, Lyme disease, and others by controlling vectors, such as mosquitoes, ticks, *etc*. These chemicals are also involved in intentional poisonings in humans, and malicious poisonings in animals and wildlife. Therefore, when insecticides are mishandled or misused, overexposure can have deadly health effects.

When used properly, insecticides offer a variety of benefits to society, as they increase crop yield, preserve produce, reduce diseases in production animals, and may also be used in combating insect-vector-mediated human diseases.

Issues in Toxicology No. 12
Mammalian Toxicology of Insecticides
Edited by Timothy C. Marrs

Published by the Royal Society of Chemistry, www.rsc.org

The anti-ChE insecticides lack species selectivity and being extremely toxic chemicals, they pose a serious threat to the environment as well as to the health of humans, domestic animals, wildlife, and aquatic species. Overexposure of individuals can occur at the sites of production or transportation and at end-user level, and often result in poisoning.[1–6] Acute poisoning with these insecticides most often results from suicide attempts or accidental ingestion, but occupational exposure may also occur. Concern is even greater for special populations such as children and elderly people.[7,8] These insecticides have been shown to adversely affect various body organs and systems, including the central nervous system (CNS), skeletal muscles, cardiovascular, respiratory, ocular, reproductive, endocrine, dermal, and immune systems.[5,8] Of course, the nervous system and skeletal muscles are the primary target organs for the toxicity of anti-ChE insecticides. This chapter describes mammalian toxicity of AChE-inhibiting OPs and CMs.

4.2 Brief Historical Perspective

Tetraethyl pyrophosphate was the first OP compound synthesized in France in 1854 by Phillipe de Clermont. In 1932, Lange and Kruger in Germany synthesized diethyl and dimethyl phosphorofluoridate and observed that inhalation of vapours of either compound caused dimness of vision and a choking sensation. These findings led Gerhard Schrader to the exploration of the OP class of compounds, while he was engaged in the development of insecticides for I.G. Farbenindustrie. Parathion was one of the earliest OP pesticides synthesized by the Schrader group, and is still commonly used throughout the world. Since the 1950s, thousands of OPs have been synthesized worldwide for the quest of insecticidal activity with species selective toxicity, *i.e.* highly toxic to insects and least toxic to non-target mammalian, avian and aquatic species. The advent of malathion fulfilled some of these expectations since it has been the most popular insecticide worldwide for more than half a century. Today, more than 200 OP insecticides are on the market.

During the mid-19th century, the first carbamate (CM) compound, physostigmine (eserine), was extracted from the Calabar beans (ordeal poison) of a perennial plant *Physostigma venenosum*, commonly found in tropical West Africa. The Calabar beans were used for witchcraft and physostigmine was used for the treatment of glaucoma. More than 50 years later, neostigmine (an aromatic ester of carbamic acid) was synthesized and used in the treatment of myasthenia gravis. During the 1960s and 1970s, dozens of CMs (esters of carbamic acid) were synthesized for pesticidal use. Carbaryl was the first CM to be used as an insecticide. To mimic the structure of acetylcholine (ACh), aldicarb was synthesized. Aldicarb has the maximal potential for mammalian toxicity and is commonly marketed globally under the trade name Temick. Presently, more than two dozen CM insecticides are on the market. Together, OPs and CMs are involved in thousands of formulations, where they are present singly or in combinations.

4.3 Brief Epidemiology of OP and CM Poisonings

Epidemiological studies suggest that each year approximately 1 million people are poisoned with anti-ChE pesticides, and several hundred thousand of them die. Pesticide poisoning incidents often occur from accidental, intentional or illegal use, or occupational exposure, and the number of people involved in each incident can vary from a few to several thousand.[9–13] Massive poisonings with anti-ChE insecticides are cited in many reports. In 1976 in Pakistan, out of 7500 spray men, 2800 were poisoned and 5 died from isomalathion that was produced during storage of formulated malathion. In 1984, an estimated 400 000 people were exposed to a toxic gas, methyl isocyanate (a precursor used in the production of CMs), that leaked from the Union Carbide plant in Bhopal, India. As a result of this catastrophic incident, more than 15 000 people died. In 1985, an unprecedented outbreak of aldicarb poisoning occurred in which approximately 2000 California residents became sick due to consumption of contaminated melons. Many such poisoning incidences have occurred in the past with many other OPs and CMs, and may occur in the future from accidental or inadvertent use.

Acute pesticide poisoning is a large-scale problem worldwide, especially in developing countries due to poor regulations, lack of surveillance systems, inadequate enforcement, lack of training, and inadequate access to information systems.[7,14,15] These papers describe the creation of a standard case definition and classification scheme (*i.e.* probable, possible, and unlikely/unknown cases) to facilitate the identification and diagnosis of all cases of acute pesticide poisoning, especially at the field level, in rural clinics, and in primary health-care systems. Its use is intended to be applicable worldwide in order to contribute to identification of the scope of existing problems and promote actions for improved poisoning management and prevention. In the United States, there are several state and federal pesticide exposure reporting and information systems. Surveillance systems are set up for pesticide exposure and illness reporting. These systems primarily use standardized classification systems[16] and U.S. EPA toxicity categories (acute toxicity category I for the most toxic pesticide products and category IV for the least toxic) for recording the reported incidents. For further details on epidemiological studies of OP and CM poisonings at national and international levels, readers are referred to recent publications elsewhere.[6,13]

4.4 Types of OPs and CMs

Currently, there are more than 200 OPs and two dozen CMs commercially available for a variety of purposes. Basic structures of OP and CM compounds are shown in Figures 4.1 and 4.2. OPs are derivatives of phosphoric, phosphonic,or phosphinic acid.

There are at least 13 types of OPs, which are listed in Table 4.1[18] Despite differences in chemical structures, all OPs share one thing in common in that

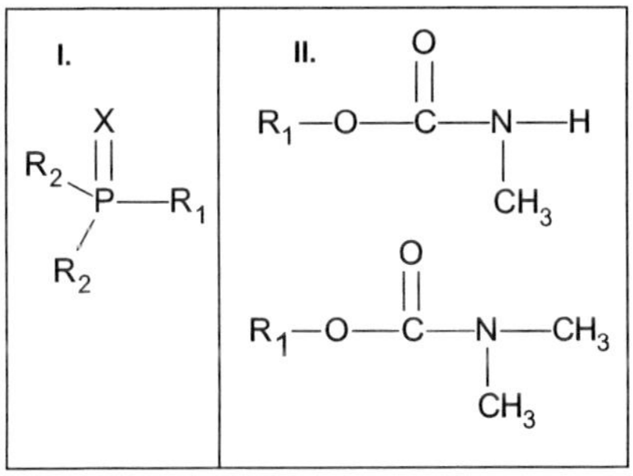

Figure 4.1 General structure for organophosphorus (I) and carbamate (II) insecticides. (Adapted from Timchalk *et al.*[17]).

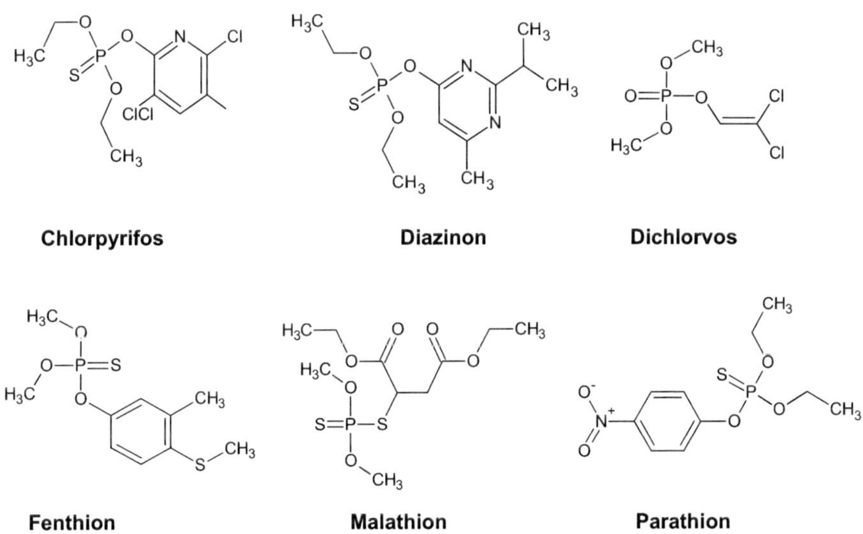

Figure 4.2 Chemical structures of some commonly used OP insecticides.

they all have a pentavalent phosphorus atom and a characteristic phosphoryl bond (P=O) or thiophosphoryl bond (P=S). Essentially, OPs are esters of phosphoric acid with varying combinations of oxygen, carbon, sulfur, and/or nitrogen attached. Usually, OP compounds have two alkyl substituents and an additional substituents group (the leaving group), which is more labile to hydrolysis than the alkyl group[18]. Of course, the chemistry of these compounds is much more complex. The OPs that are derivatives of phosphoric or phosphonic acid possess anticholinesterase activity, unlike those that are derivatives of phosphinic acid. Basically, some OPs (such as chlorfenvinphos, dichlorvos, monocrotophos and trichlorfon) are direct

Table 4.1 Types of organophosphorus compounds.

Type of OP	*Chemical structure*	*Example*
Phosphates	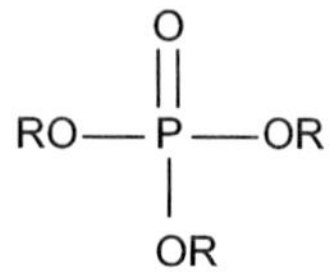	Chlorfenvinphos Dichlorvos Monocrotophos
Phosphonates	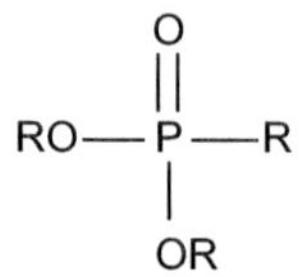	Trichlorfon
Phosphinates	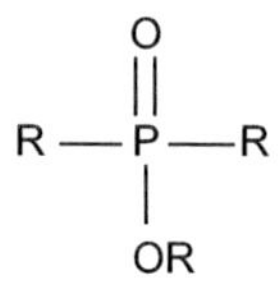	Glufosinate
Phosphorothioates (S=)	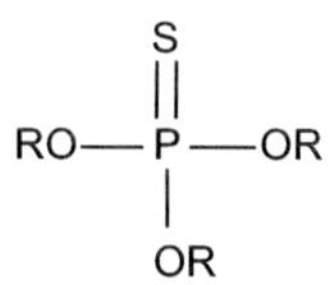	Bromophos Diazinon Fenthion Parathion Pirimiphos-methyl
Phosphonothioates (S=)	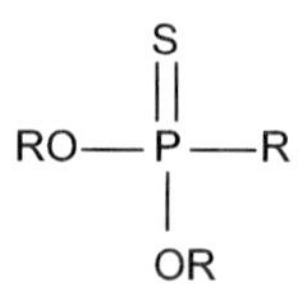	EPN Leptophos
Phosphorothioates (S-substituted)	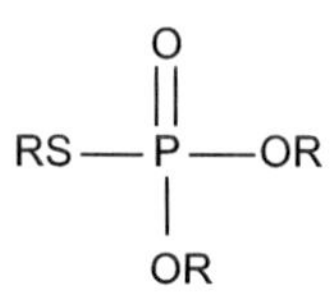	Demeton-*S*-methyl Echothiophate
Phosphonothioates (S-substituted)	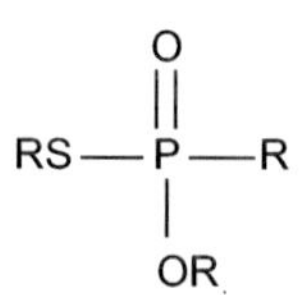	VX
Phosphorodithioates	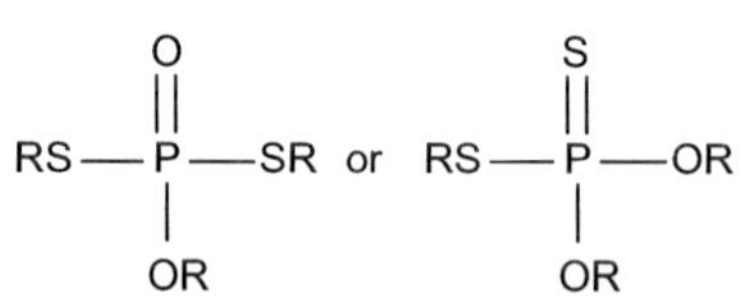	Azinphos-ethyl Azinphos-methyl Dimethoate Disulfoton Malathion Methidathion

Table 4.1 (*Continued*)

Type of OP	*Chemical structure*	*Example*
Phosphorotrithioates	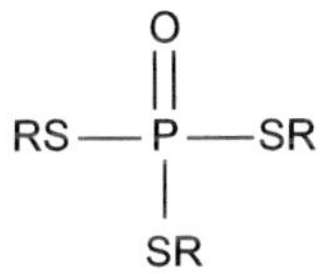	DEF (tribufos)
Phosphoramidates	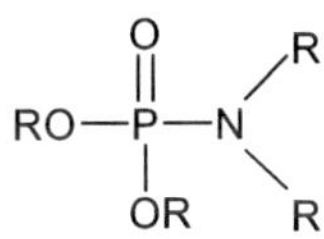	Fenamiphos
Phosphoramidothioates	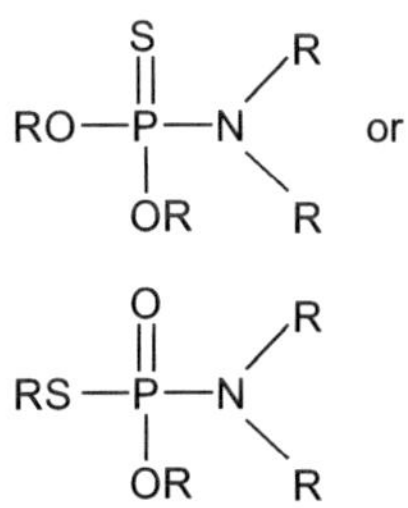	Methamidophos Isofenphos
Phosphorofluoridates	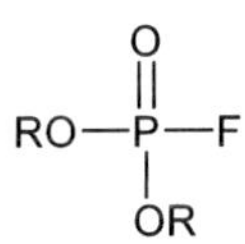	Diisopropyl phosphorofluoridate (DFP)
Phosphonofluoridates	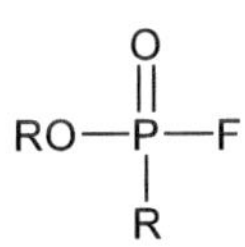	Cyclosarin Sarin Soman

Source: Adapted from Marrs.[18]

AChE inhibitors, while those of phosphorothioates (P=S) type (such as bromophos, diazinon, fenthion, malathion, parathion and many others) possess minimal anticholinesterase (anti-AChE) activity and require desulfuration to the analogous oxon before acquiring anti-AChE activity. Also, OPs which are used as defoliants (*S,S,S*-tributyl phosphorotrithioate and *S,S,S*-tributyl phosphorotrithioite) and herbicides (glyphosate and glufosinate) are of very low acute mammalian toxicity, but can induce delayed polyneuropathy.

In general, the carbamate (CM) compounds are esters of carbamic acid, and their structures are not as complex as that of OPs. Chemical structures of some CM insecticides are shown in Figure 4.3. For the details of other carbamates, readers are referred to Gupta[5] and Satoh and Gupta.[6]

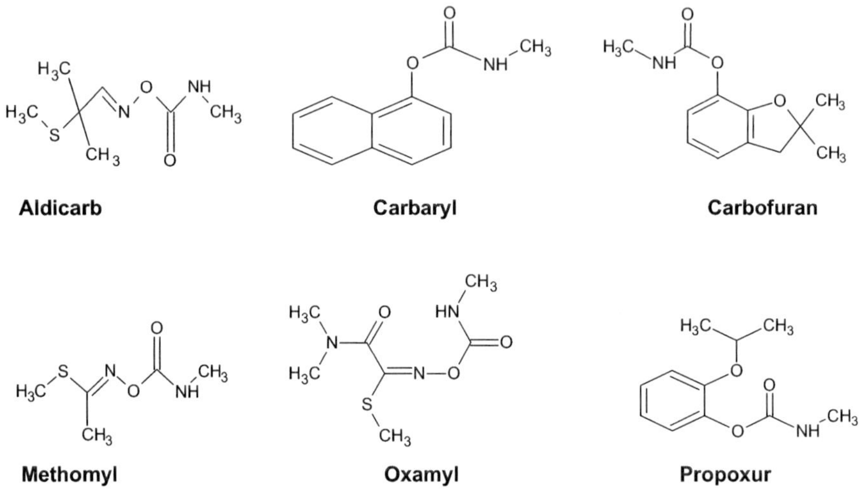

Figure 4.3 Chemical structures of some commonly used CM insecticides.

4.5 Toxicity

4.5.1 Acute Toxicity

4.5.1.1 Symptoms of Acute Toxicity

Signs and symptoms of acute poisoning with anti-ChE agents are well described.[19,20] Symptoms of acute poisoning with OPs and CMs are similar and result from overstimulation of both muscarinic and nicotinic ACh receptors due to accumulation of ACh because of AChE inactivation. Since the cholinergic system is widely distributed within both the central and peripheral nervous systems, chemicals that inhibit AChE are known to produce a broad range of well-characterized symptoms of anti-ChEs.[17,21] The muscarinic symptoms include hypersalivation, excessive tracheobronchial secretions, gastrointestinal cramps, broncho- and laryngospasms, lacrimation, dacryorrhea, nausea, excessive sweating, urinary incontinence, diarrhoea, miosis, and bradycardia. The major nicotinic symptoms include muscle fasciculations, tremors, muscle weakness, flaccid paralysis, blurred vision, vomiting, and paralysis of respiratory muscles. Exposure to a high dose of an OP or CM insecticide also leads to the occurrence of symptoms of CNS origin and may include restlessness, tremors, convulsions, partial or generalized seizures, mental disturbance, incoordination, cyanosis, and coma. Eventually, death ensues due to cardiac arrest and respiratory failure. The respiratory failure is of both central and peripheral origin.[22] Onset of symptoms usually occur within less than an hour and death occurs within a span of few hours to 24 h.

The surviving patient, past the day of poisoning, may suffer from personality changes, mood swings, aggressive events and psychotic episodes including schizoid reactions, paranoid delusions, and exacerbations of pre-existing psychiatric problems. In addition, there can be poor sleep because of nightmares and hallucinations, and deficits in memory and attention.[23]

At non-lethal doses, the signs of toxicity caused by each OP or CM can vary due to multiple factors.[24–26] It needs to be emphasized that each OP or CM reacts not only with AChE but also with additional targets. This can be explained in multiple ways. First, that in spite of similar levels of AChE inhibition, different OPs can exert different degrees of toxicity. For example, when AChE in rat brain is inhibited to a level of 80%, the rat shows signs of toxicity. But the severity of signs can be greater with chlorpyrifos than with parathion,[27] because chlorpyrifos may have additional mechanism(s) involved in toxicity. Secondly, variability in symptoms can be due to brain regional heterogeneity of cholinergic and non-cholinergic sites for OPs and CMs interaction.[25] For example, the hypothalamus binds two to five times more OP than the striatum, despite the fact that the striatum has severalfold greater AChE activity than the hypothalamus. This suggests that in the hypothalamus, targets other than cholinesterase bind OP.[28] In essence, most OPs and CMs react with AChE, but there are still many other proteins that each of these insecticides can react with, such as BuChE, CarbE, neuropathy target esterase, acylpeptide hydrolase, arylformamidase, fatty acid amide hydrolase, phospholipase, proteases, albumin, cannabinoid CB1 receptor, and many other enzymes, receptors and proteins.[26] Lockridge *et al.*[29] further substantiated this fact in the AChE knockout mouse, which has no AChE activity in any tissue. These authors not only explained the basis of survivability of the mouse without AChE due to the presence of BuChE, but demonstrated the existence of non-AChE protein targets. A more complete understanding of the proteins modified by exposure to OPs/CMs will aid in understanding why some people cannot tolerate doses of these insecticides that are harmless to the majority.

4.5.1.2 Mechanism of Action in Acute Toxicity

Cholinergic Mechanisms. There is general agreement that acute clinical signs of OPs and CMs are primarily associated with AChE inhibition at synapses in the brain and at neuromuscular junctions in skeletal muscles.[30] Normally, AChE is responsible for the hydrolysis of the neurotransmitter ACh and termination of its biological activity in a microsecond (Figure 4.4). Both OPs and CMs react covalently with a serine residue in AChE, in a similar manner to ACh. But OPs and CMs inhibit enzyme activity by phosphorylation and carbamylation, respectively, and differ quantitatively in rates of dephosphorylation and decarbamylation of inhibited AChE. It is important to mention that the AChE enzyme can 'age' with OPs and not with CMs.[31,32] Following AChE inhibition, free ACh accumulates at the nerve endings of all cholinergic nerves and causes overstimulation of electrical activity.[33,34] Inhibition of AChE (>70%) leads to a toxic-level accumulation of ACh at central and peripheral sites. A comparison of the AChE inhibition dynamics for

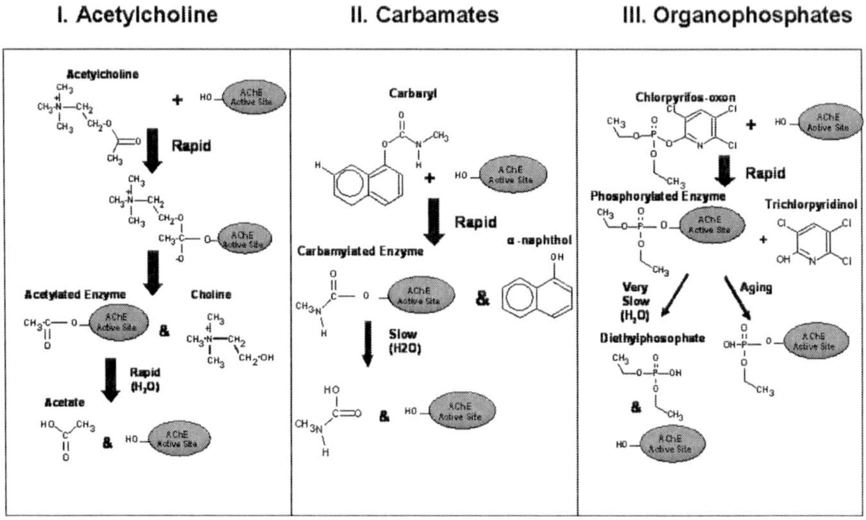

Figure 4.4 Interaction of acetylcholine (I), the carbamate carbaryl (II), and the organophosphate chlorpyrifos-oxon (III) with the active site of acetylcholinesterase (AChE). The general rate of bound AChE hydrolysis is ACh > carbaryl > chlorpyrifos-oxon (Adapted from Timchalk *et al.*[17]).

the interaction of ACh, of a CM (carbaryl), or of an OP (chlorpyrifos-oxon) with AChE is shown in Figure 4.4.

The cholinesterases (ChEs) are serine hydrolases that catalyse the breakdown of ACh through an acyl transfer, where water is the acceptor molecule to which the substrate acyl moiety is transferred. A serine oxygen of the active site gorge in ChEs carries out a nucleophilic attack on the electrophilic carbon of the carbonyl group of ACh, resulting in an acetylated enzyme intermediate and the release of choline (Ch). Deacetylation occurs when an attacking water molecule (hydroxyl ion) acts as a more effective nucleophile, thereby releasing acetate.[35] The molecular interactions between OPs and AChE (Figure 4.5) have been studied in much more detail than those between CMs and AChE.

The rates of hydrolysis and reactivation of AChE following carbamylation and phosphorylation of the active site appear to be drastically slower than for the hydrolysis of the acetylated enzyme. The turnover time for ACh is of the order of 150 μs, whereas the carbamylated enzyme $t_{\frac{1}{2}}$ for hydrolysis is substantially slower ($\sim$15–30 min). The phosphorylated enzyme is highly stable ($t_{\frac{1}{2}} \sim$ days), and further dealkylation of the phosphorylation group produces an 'aged' AChE that is irreversibly inhibited.[17,31,32,39,40] Details of ChEs, interaction of OPs and CMs, and reactivation/regeneration of ChEs are described elsewhere.[17,35,40] Generally, clinical signs of acute poisoning resolve within days to weeks. Some symptoms of a neuropsychological nature appear to persist for months or even longer.[7]

Evidence suggests that some OPs and CMs directly interact with muscarinic and nicotinic ACh receptors (mAChRs and nAChRs). Physostigmine and related ChE inhibitors displace the mAChR agonist [^{3}H]-oxotremorine-M from

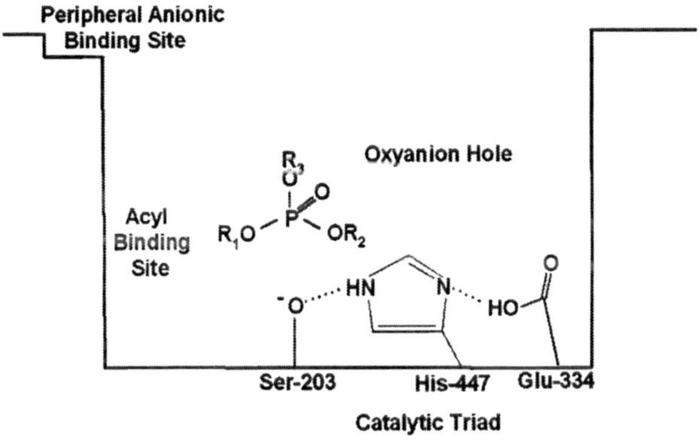

Figure 4.5 Schematic drawing of the active site gorge of AChE, with the entry of an OP molecule. R_1 and R_2 on the OP are usually identical alkyl chains, whereas R_3 is the leaving group. The catalytic triad consists of Ser203, His447 and Glu334. The acyl binding site is likely important in positioning the inhibitor for the nucleophilic attack from Ser203 (Ordentlich *et al.*[36]), whereas the oxyanion hole may polarize the P=O bond, thereby facilitating the nucleophilic attack (Ordentlich *et al.*[37]). Binding of ligand to the peripheral anionic site can lead to inhibition or activation. Additionally, the peripheral anionic site plays an important role in the stereoselectivity of AChE toward methylphosphonates (Ordentlich *et al.*[38]). (Adapted from Sultatos.[35]).

its receptors in rat cortex and brainstem.[41,42] Additional evidence suggests that some of these ChE inhibitors are mAChR agonists. By now, the agonistic, antagonistic, potentiating and inhibitory effects of physostigmine and related ChE inhibitors on nAChRs have also been described.[43] Smulders *et al.*[43] reported the potencies of some CMs for binding with nAChRs and inhibition of AChE activity in the brain. Based on nicotinic receptor ($\alpha4\beta4$ nAChRs) binding study, potency of different CMs followed the order fenoxycarb > EPTC (*S*-ethyl *N*, *N*-dipropylthiocarbamate) > carbaryl > bendiocarb > propoxur > aldicarb, and based on inhibition of AChE, CMs followed the order bendiocarb > propoxur, aldicarb > carbaryl > EPTC > fenoxycarb. These authors concluded that the CMs which have more potent interaction with nAChRs are the less potent inhibitors of AChE. The findings also suggested that nAChRs are additional non-AChE targets for the CM pesticides, and contribute to the toxicity of some CMs.

Non-Cholinergic Mechanisms. It needs to be pointed out that not all OPs and CMs exert poisoning by the same mechanism.[24,25] In addition to the well-understood cholinergic mechanisms, these insecticides are demonstrated to induce a myriad of toxic effects through multiple non-cholinergic mechanisms.[44–47] Involvement of non-cholinergic mechanisms in OP/CM-induced seizures was supported by evidence that glutamate release leads to activation of *N*-methyl-D-aspartate (NMDA) receptors.[48] In addition, the adinosinergic, γ-aminobutergic (GABAergic), and monoaminergic systems may also be involved in the seizures and lethality associated with OP- or CM- induced poisonings.[20] Recent studies provide evidence that neuroinflammatory consequences of exposure to anti-ChE may contribute to seizures and pathology associated with these agents (Milatovic and Gupta, unpublished).

The persistence of excitotoxicity for more than an hour can lead to neuronal degeneration in cortex, amygdala, hippocampus and some other brain regions.[44,45] These are the brain areas primarily involved in initiation and propagation of convulsions and seizures. The early morphological alterations include dendritic swelling of pyramidal neurons in the CA1 sector of the hippocampus.[49] The AChEI- induced neuronal cell death is a consequence of a series of extra- and intracellular events leading to the intracellular accumulation of Ca^{+2} ions and the generation of free radicals.[50–53] Excessive production of free radicals causes oxidative and nitrosative stress to which the brain is especially vulnerable. Lipids are readily attacked by free radicals, resulting in the formation of a number of peroxidation products, such as F_2-isoprostanes, isofurans, and F_4-neuroprostanes, which are formed non-enzymatically. F_4-neuroprostanes are specific markers of oxidative damage to the neurons (Figure 4.6). These

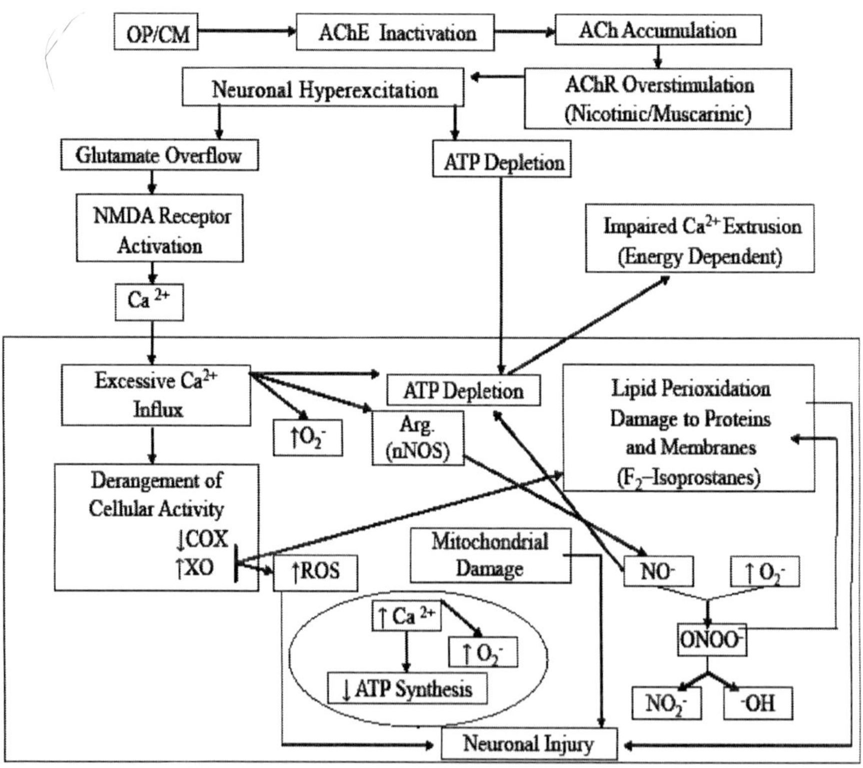

Figure 4.6 A schematic diagram showing possible mechanisms involved in OP/CM-induced neuronal injury/death by excessive generation of reactive oxygen species (ROS) and reactive nitrogen species (RNS) leading to lipid peroxidation.

events, in addition to many more described elsewhere, cause irreversible destruction of cellular components, including the proteins, DNA, and especially mitochondria.[25,44,45,55–58] For further details on the involvement of oxidative stress in anti-ChE toxicity, readers are referred to some recent publications elsewhere.[59–61]

4.5.2 Intermediate Syndrome

OP insecticide-induced intermediate syndrome (IMS) was reported for the first time in human patients in Sri Lanka in 1987.[62] The observations were made in 10 patients who presented 24–96 h after acute cholinergic crisis from exposure to methamidophos, fenthion, dimethoate and monocrotophos. These patients had acute muscle paralysis and some required ventilator support. Since then IMS has been diagnosed in many countries, including South Africa (1989), Turkey (1990), Belgium (1992), the United States (1992), Venezuela (1998), France (2000) and elsewhere. To date, more than a dozen OPs (bromophos, chlorpyrifos, diazinon, dicrotophos, dimethoate, disulfoton, fenthion, malathion, tributyl phosphorotrithioite, methamidophos, monocrotophos, omethoate, parathion, phosmet and trichlorfon) are known to cause IMS. In the case of OPs, IMS appears to be due to insufficient oxime therapy and lack of oxygenation or ventilation. In 2005, the carbamate insecticide carbofuran was also demonstrated to cause IMS in patients accidentally or intentionally exposed to large doses of this insecticide.[63]

IMS is usually observed in individuals who have ingested a massive dose of an OP insecticide either accidentally or intentionally in a suicide attempt. IMS is clearly a separate clinical entity from acute toxicity and delayed neuropathy. Clinically, IMS occurs 24–96 h after exposure and is characterized by acute respiratory paresis and muscular weakness, primarily in the facial, neck and proximal limb muscles. In addition, it is often accompanied by generalized weakness, cranial nerve palsies, depressed deep tendon reflexes, ptosis and diplopia. These symptoms may last for several days or weeks, depending on the OP involved. It is important to mention that despite severe AChE inhibition, muscle fasciculations and muscarinic receptor- associated hypersecretory activities are absent.

Although the exact mechanism involved in pathogenesis of IMS is unclear, studies suggest that a decrease of AChE and nAChR mRNA expression occurs after oral poisoning with disulfoton in rats. Involvement of oxidative stress is also suggested.[64] Based on electromyographic (EMG) findings from OP-poisoned patients and experimental studies on laboratory animals, it has been found that the defect in IMS is at the neuromuscular endplate and postsynaptic level, but the effects of neural and central components in producing muscular weakness have not been ruled out. It seems clear that some anti-ChE insecticides are greatly distributed to muscles and have a higher affinity for nAChR. Currently, very little is known about the type of damage at the motor endplate or about risk factors contributing to its development. Jayawardane *et al.*[65] described a large cohort of OP-poisoned patients and delineated the electrophysiological abnormalities that occur at the neuromuscular junction in conjunction with the clinical entity of IMS. There is no specific treatment, and therapy relies upon atropine sulfate and 2-PAM (pralidoxime). The atropine and 2-PAM therapy should be continued for a long period, even if efficacy of these drugs on the development of IMS appears to be limited. Pawar *et al.*[66] observed improved neck flexor function and reduced morbidity and mortality in patients receiving high-dose pralidoxime infusions. Recently, Aaron[67] suggested that a large randomized trial of OP-poisoned patients receiving no oximes, low-dose oximes or high-dose oximes would be useful in establishing appropriate treatment in relation to reducing muscle weakness and mortality. In such a trial, studying additional outcomes such as recording the repetitive nerve stimulus and nerve conduction, particularly phrenic nerve conduction velocity, would answer the question of whether IMS is a clearly defined entity, and whether oximes truly make a difference to patients' outcome. For further details on IMS, readers are referred to Gupta[68] and De Bleecker.[69]

4.5.3 Chronic Toxicity

Certain acetylcholinesterase inhibitors can also cause another type of toxicity known as OP-induced delayed polyneuropathy (OPIDP). Signs and symptoms of this neurodegenerative disorder include tingling of the hands and feet, followed by sensory loss, progressive muscle weakness and flaccidity of the distal skeletal muscles of the lower and upper extremities, and ataxia.[70–73] Symptoms of OPIDP usually appear 2–3 weeks after a single-dose exposure to certain OPs. This delay in symptom development depends on both the kinetic characteristics of the compound and the dose. After a high dose, it can be as short as 10 days, whereas the onset after poisoning by an OP with prolonged kinetics or at a relatively low dose may be up to 5 weeks. Notably, both cholinergic and IMS signs have subsided before the onset of OPIDP.

Extensive studies carried out in the past 40 years have identified an OPIDP target enzyme, neuropathy target esterase (NTE).[72,74–78] It is a large polypeptide of 1327 amino acids, a membrane-bond esterase with a molecular weight of 155 kDa, and its physiological role has not yet been clarified.[79,80] As an integral membrane protein in vertebrate neurons, NTE is present in the endoplasmic reticulum of neurons, but is absent from glia.[81,82] Two forms of NTE are distributed differently in brain (90% in microsomes and 10% activity in cytosol) and sciatic nerve (45% activity in microsomes and 55% in cytosole). Several OPs, depending on their structure, can inhibit NTE, as do some non-OPs, such as carbamates and sulfonyl fluorides. However, the initial biochemical event in the development of OPIDP involves inhibition of NTE by phosphorylation. Phopshorylation of NTE is a rapid event and degree of inhibition varies according to the OP and the source of NTE. Inhibition of peripheral nerve NTE is required to develop OPIDP, and no clinical deficits appear if only brain NTE is inhibited. For neuropathy to occur, a second reaction called 'ageing' of the enzyme must take place and this involves cleavage of the lateral side chain from the phosphorylated NTE. This reaction is called ageing because it is a slow progressive process and the product is no longer responsive to nucleophilic reactivating agents such as oximes. Some OPs, as well as non-OP inhibitors such as carbamates and sulfonyl fluorides, also covalently react with NTE but cannot undergo the ageing reaction. As a consequence, these inhibitors do not cause OPIDP, and when given to experimental animals before a neuropathic OP, may protect from OPIDP when they occupy at least 30% of the NTE active site. In addition, several studies have also suggested that severity of the OPIDP does not appear to correlate with the amount of aged NTE.[83] Therefore other mechanisms may also be involved in pathogenesis of OPIDP.

Recent studies suggest that NTE is also involved in intracellular membrane trafficking, and regulation of calcium and cell-signalling pathways between neurons and glia.[84–86] Genetic studies with progressive spastic paraplegia patients found an association on the disease with mutations of NTE gene encoding the catalytic domain of the enzyme and thus raised the possibility that NTE pathway disturbance may contribute to other motor neuron diseases, including amyotrophic lateral sclerosis.[87]

The first and largest OPIDP epidemics occurred in the southern United States during accidental ingestion of tri-*ortho*-cresyl phosphate (TOCP). Additional outbreaks of this neurodegenerative disorder causing paralysis of tens of thousands of individuals have occurred in Morocco, Fiji and India due to the consumption of cooking oil contaminated with lubricating oil containing TOCP.[88,89] In addition to TOCP, OPIDP has been also caused by certain OP pesticides such as leptophos, dichlorvos, fenhtion, trichloronat, trichlorfon, tributyl phosphorotrithioite, methamidofos and chlorpyrifos.[72,90] Clinical, electrophysiological and histopathological details of these cases have been discussed by Lotti and Moretto.[72] However, toxicological data regarding the amount of OP doses, impurities and isomers in OP formulations with possible additional neurotoxic potential are not detailed.

Another potential neurotoxic consequence of OP exposure is chronic OP-induced neuropsychiatric disorder (COPIND)[91,92] with symptoms of anxiety and depression. Humans exposed to low levels of OP agents in industrial or agricultural settings have reported difficulty in concentrating and memory impairment long after such exposure has ended.[93–96] In addition, dystonic reactions, schizophrenia, cog-wheel rigidity, choreoathetosis and electro-encephalographical changes have been reported with high-dose exposures. These extrapyramidal symptoms are thought to be due to the inhibition of the AChE in the human extrapyramidal area. Psychosis, delirium, aggression, hallucination and depression may also be seen during recovery from the cholinergic syndrome. High levels of anxiety in commercial sprayers of insecticides and other types of delayed neurobehavioural effects are seen among people exposed to low doses of OP compounds for prolonged periods.[97] It has been observed that clinical features of psychological syndromes occurring after chronic exposure to OP compounds have a great similarity to chronic fatigue syndrome.[98]

4.6 Development of Tolerance

The phenomenon of tolerance development to the toxicity of AChE-inhibiting OPs was noted more than half a century ago. During tolerance development, the clinical effects are often diminished more than would be expected from the degree of AChE inhibition, in spite of repeated exposure of an OP. Tolerance to OPs (demeton, DFP, dichlorvos, disulfoton, paraoxon, parathion-methyl, schradan and others) has been observed in different species (such as mice, rats and guinea-pigs) following different routes of administration.

Tolerance to OP toxicity can develop in several ways. Brodeur and DuBois[99] proposed that tolerance to chronic treatment with some OPs may be due to inactivation of the cholinergic receptors. This was later shown to be one of the most important mechanisms for a cell or an individual to obtain tolerance toward AChE-inhibiting OPs. Both mAChRs and nAChRs are involved in the development of tolerance to OP toxicity. Costa *et al.*[100] reported a decrease in both the numbers/density (B_{max}) and the affinity (K_D) of mAChRs to the ligand in ileum and brain striatum, but most authors have found only a decrease in the B_{max} of mAChRs in the brain, and the decrease was found to be similar in different brain regions. It is noteworthy that the down-regulation of mAChRs (G-protein-linked receptors) is common during tolerance development and occurs by means of receptor internalization and degradation. There is a correlation between AChE inhibition and a reduction in mAChRs binding, as the decrease in mAChRs depends on both the time of AChE inhibition and degree of inhibition. In general, the degree of AChE inhibition is greater than that of reduction in receptor binding.

Like mAChRs, significant reductions in B_{max}, without change in K_D, of nAChRs were also found in the brain of tolerant rats treated with disulfoton,[101] and in skeletal muscles of rats treated with DFP.[102] Fonnum and Sterri[103] also described the involvement of muscarinic and nicotinic autoreceptors in tolerance development to OP toxicity. There is no evidence for any change in choline acetyltransferase activity or ACh synthesis in brains of rats tolerant to DFP or paraoxon (desulfurated metabolite of parathion). In AChE knockout mice, Mesulam *et al.*[104] demonstrated that tolerance development is obtained by a reduction in ACh receptors and not by regulating ACh synthesis.

For some OPs, the spontaneous reactivation of inhibited AChE may explain part of the tolerance. However, it can also occur due to the presence of other enzymes that can bind or inactivate the inhibitor and thereby make it less readily available for AChE inhibition. Some examples for binding to the OPs are serine-containing enzymes, such as carboxylesterases (CarbEs), and butyrylcholinesterase (BuChE). Although the functional role of CarbEs and BuChE is unknown, binding to and inhibition of these enzymes reduces the free concentration of OPs. Stimulation of *in vivo* synthesis of AChE and BuChE was observed in skeletal muscles, but not in brain, of rats receiving DFP (0.5 mg^{-1} kg^{-1} day^{-1}subcutaneously) for 2 weeks.[102,105] The observed recovery of AChE and to some extent of BuChE in muscles was a result of *de novo* synthesis. This was confirmed by our finding of *in vivo* protein synthesis using L-[1-^{14}C]valine incorporation in different brain regions and skeletal muscles during toxicity and tolerance. Following a reduction in *in vivo* protein synthesis during the initial 5 days of treatment (toxicity phase), protein synthesis recovered after 14 days treatment (tolerance phase) in muscles but not in brain.[106] Also, modifications in the affinity of AChE for ACh in paraoxon-tolerant rats was reported by Milatovic and Dettbarn,[107] who observed a 20–25% increase in affinity (decrease in K_m) for ACh by both the brain and the diaphragm AChE. In general, the spontaneous reactivation of dimethoxy-phosphorylated ChEs is much more rapid than that of the corresponding diethoxy or phosphonyl compounds. Accordingly, repetitive exposure at adequate time intervals to a dimethoxy OP at a sublethal dose may be highly tolerable due to the recovery of AChE, as was observed for dichlorvos.[108]

Along with the recovery of AChE activity and the loss of cholinergic toxic symptoms, there was recovery of CarbE activity in the liver, which suggested the renewed availability of non-specific binding sites for DFP and other OPs, and may have contributed to the tolerance development.[105] Also, the possibility exists for CarbE's spontaneous reactivation due to histidine, which is located in the active site of CarbE, but not present in AChE and BuChE. Different levels of plasma CarbEs in different species partly offers an explanation for variability in tolerance level to OPs. It is important to mention that rat, guinea-pig and primates including humans are species with high, medium and minimal/absent concentrations of plasma CarbE, respectively.

Tolerance can also be achieved through more rapid metabolism of the OP compounds by OP-hydrolysing enzymes such as aryldialkylphosphatase/paraoxonase (PON) and DFPase/somanase,[103] and binding to proteins such as albumin.[109] PON and DFPase are enzymes that can hydrolyse OPs with generally low affinity (K_m 0.1–10 mM). Based on experimental studies, it can be concluded that animals can tolerate accumulative doses of an OP several times the acute LD_{50} value when non-lethal doses are given daily for a period of 2 weeks or longer, and tolerance development occurs through multiple mechanisms. To date, no tolerance development is reported for any CM insecticides. For further details on tolerance development readers are referred to Fonnum and Sterri.[103]

4.7 Interaction of Anticholinesterase Insecticides

It is highly unlikely that humans or animals are exposed to a single pesticide. Instead, they are more often exposed to multiple pesticides at low doses. OP and CM insecticides are often used in combination, with the objective of achieving synergistic interaction and to control a wide range of insects including those that are hard to kill (resistant). In many cases, these insecticides are also used simultaneously as ectoparasiticides and anthelminthics in veterinary medicine. Under most circumstances, exposure to a single ChE-inhibiting insecticide will be at the subtoxic level, but simultaneous exposure to more than one can sometimes lead to devastating health effects because of an additive or potentiating interaction.[110] Ito *et al.*[111] stated that the exposure of agricultural workers and the general population to pesticides is of major concern, and a possible summation or synergistic effects could occur at some levels.

OP and CM insecticides are often used simultaneously to control pests in both agricultural and domestic environments. Many interaction studies have dealt with two or more OPs, but interactions between OP and CM compounds in mammalian species have rarely been studied. The toxicities of both classes of compounds have been attributed to their ability to inhibit the activity of AChE, which catalyses the hydrolysis of the neurotransmitter ACh at nerve synapses and neuromuscular junctions. More than half a century ago, the landmark publication of Frawley *et al.*[112] demonstrated for the first time marked potentiation in mammalian toxicity from the simultaneous administration of two anti-ChE insecticides, malathion and EPN (*O*-ethyl-*O*-*P*-nitrophenyl phosphonothionate). Potentiation appears to result from an increased persistence rather than an increased concentration of malaoxon in the tissues. Since then, the phenomenon of potentiation has been confirmed in several studies using different anti-ChE insecticides and various laboratory animal species. DuBois[113] investigated 43 pairs of OPs for interaction and found four pairs (malathion + EPN, malathion + Dipterex (trichlorfon), Dipterex + Guthion (azinphos-methyl), and malathion + Co-Ral (coumaphos)) to have potentiation interaction. Cohen and Murphy[114] investigated the mechanisms involved in the potentiation interaction between malathion and each of the other three OP insecticides (EPN, abate and parathion) in mice. EPN and abate inhibited CarbEs to the same degree in liver, lung, kidney and plasma, but EPN potentiated malathion's toxicity at twice the extent of abate. However, both compounds potentiated malaoxon's toxicity to the same degree. This suggested that abate inhibited the oxidation of malathion to malaoxon. It was evident from this and a few other investigations that CarbE inhibition alone is not sufficient to predict the relative capacities of various compounds to potentiate the toxicity of malathion and related OP compounds. Timchalk *et al.*[115] studied a pharmacokinetic-based interaction between chlorpyrifos and diazinon, and discovered that anything more than an additive interaction is unlikely at environmentally relevant concentrations. However, some studies have observed potentiation at higher doses of OPs.

Compared to the interaction between OPs, only a few studies investigated the interaction between OPs and CMs.[116–119] These investigators demonstrated the interaction of toxicity in mice of a CM insecticide, 2-*sec*-butylphenyl *N*-methylcarbamate (BPMC) by P=S type OPs (cyanophos, fenitrothion, and malathion) through metabolic blockage and not through CarbE inhibition. Miyaoka *et al.*[120] reported synergism in mice treated with fenitrothion and BPMC. It was demonstrated that competitive inhibition of BPMC metabolism by fenitrothion appears to play a role in inhibition of BPMC detoxification, resulting in potentiation of its toxicity. It is interesting to note that no such potentiation has been found with P=O-type OP insecticides such as dichlorvos.

With regard to OP and CM interactions, CarbEs appear to play two major roles. First, CarbEs provide partial protection against OPs and CMs. This hypothesis was deduced from the findings that low-level exposure to either OPs or CMs caused marked inhibition of CarbE, particularly in the liver and serum, without significantly affecting AChE activity. It implies that CarbE has a greater affinity than AChE for these inhibitors and thus only a reduced amount of free inhibitor concentration is left to interact with AChE. Secondly, in the case of pre-exposure to tetraethylpyrophosphoramide (*iso*-OMPA) or any other CarbE inhibitor, these non-specific binding sites are blocked. This raises the free concentration of the OP or CM compound, which would consequently be available to inhibit AChE to a great extent. As a result potentiation of OP or CM toxicity occurs.

Several technical grade OP insecticides are known to contain some OP impurities that potentiate the toxicity of the parent compounds. For example, technical malathion contains three major impurities, malaoxon, isomalathion, and *O,S,S*-trimethyl phosphorodithioate (OSS-Me). All three impurities are known to inhibit CarbEs in liver and serum, which could also account for their ability to potentiate malathion toxicity. Technical phenthoate contains two impurities (*O,O,S*-trimethyl phosphorothioate, OOS-Me and OSS-Me). These impurities caused a marked inhibition of CarbE and greatly diminished the amount of phenthoate acid formed. It needs to be mentioned that unlike malathion and phenthoate, during certain storage conditions for acephate or fenitrothion their impurities are formed, but they do not potentiate their toxicity.

In essence, it is obvious that exposure of humans and animals, as well as the environment, to multiple insecticides is inevitable. In such scenarios, toxic interactions between ChE-inhibiting insecticides are expected to occur in mammalian systems. Therefore, predicting and avoiding such interactions are the major challenges in this area of research.

4.8 Biomarkers and Biomonitoring

Biomarkers for biomonitoring of pesticide exposures are typically divided into three broad categories: biomarkers of exposure, effect or susceptibility.[121]

4.8.1 Biomarkers of Exposure

Biomarkers of exposures include measurements of pesticides, pesticide metabolites, and modified cells or their molecules (*e.g.* DNA and protein adducts) in biological tissues/fluids. Among OPs, the most common metabolite measured is 3,5,6-trichloropyridinol, which is a metabolite of chlorpyrifos. Specific metabolites of malathion, malathion dicarboxylic acid and α and β isomers of malathion monocarboxylic acid have also been measured. Other less frequently measured specific OP metabolites include 2-isopropyl-4-methyl-6-hydroxypyrimidine, a metabolite of diazinon, and 4-nitrophenol, a metabolite of parathion-methyl and parathion, EPN, and some non-OP pesticide chemicals.[122] Presently, six dialkyl phosphate (DAP) metabolites of OP pesticides are the most commonly measured metabolites for biomonitoring. These DAP metabolites include dimethylphosphate, diethylphosphate, dimethylthiophosphate, diethylthiophosphate, dimethyldithiophosphate and diethyldithiophosphate. DAP concentrations provide non-specific information about exposure to a class of OPs, rather than a single OP compound. Such information is certainly useful when determining exposure prevalence to most members of a class of compounds; however, it may not accurately reflect the toxicity associated with the exposure.

A CM insecticide, carbaryl exposure is usually estimated based upon urinary measurements of 1-naphthol. However, 1-naphthol, as well as 2-naphthol, is a metabolite also of naphthalene, a ubiquitous polyaromatic hydrocarbon. Thus, the measurement of 1-naphthol does not distinguish these two pesticides. Measurement of other less abundant metabolites of carbaryl, such as 4-hydroxycarbarylglucuronide, usually helps to circumvent this problem. Other CMs that have been measured in urine include benomyl, carbofuran, carbosulfan, aldicarb and pirimicarb. In addition, several CMs have been measured in serum, plasma or blood. Methomyl was measured in the whole blood of a pilot who died during aerial application of the pesticide. In general, the CMs are particularly unstable in blood, so sometimes their metabolites must be measured as well. For instance, carbaryl is hydrolysed to its major metabolite 1-naphthol; propoxur is metabolized to 2-isopropoxyphenol; and carbofuran is metabolized to 3-ketocarbofuran and 3-hydroxycarbofuran, and these metabolites can be successfully quantified in serum or plasma. In recent studies, OP and CM pesticides and their metabolites have also been measured in saliva and the data were used for pharmacokinetics and dosimetry.[17,123] Recently, liquid chromatography-mass spectrometry has been employed to identify new biomarkers of OP and CM pesticide exposure by detecting their adducts on serine of butyrylcholinesterase and tyrosine of albumin.[124–127] All these biological measurements are directly related to the dose of a pesticide and are a function of pesticide exposure.

RBC-AChE inhibition has also been identified by the scientific and regulatory communities as a sensitive biomarker of exposure to OPs and CMs because it serves as a sensitive surrogate endpoint for the inhibition of brain ChE. AChE inhibition, especially in blood, is still considered one of the most sensitive and reliable biomarkers of exposure to these insecticides. It is important to mention that human blood-AChE inhibition by OPs is no greater than in laboratory animals, suggesting equal sensitivity of AChE inactivation.[128] In recent years egasyn-β-glucuronidase has also gained significant attention for being a most sensitive biomarker of an acute OP exposure.[129,130]

Recently, Barr and Buckley[122] stated that biomarkers of exposure from samples of human tissue, fluids and excreta offer qualitative and quantitative evidence of pesticide exposure. These measurements are particularly useful in exposure research because they can highlight population-based exposure trends and improve estimates of pesticide exposure and dose.

4.8.2 Biomarkers of Effect

Biomarkers of effect(s) include measurements of biochemical, physiological or behavioural alterations that result as a consequence of pesticide exposure. Some examples of biomarkers of effect include biological measurements of endogenous and inflammatory responses, measurements of DNA, protein, cell, tissue and organ damage/modification, and observations of tumours or cancer cell clusters. These biological measurements reflect exposure and biological effect, but are often difficult to ascribe to a specific pesticide exposure event.

4.8.3 Biomarkers of Susceptibility

These are measurements of an individual's inherent ability to respond to pesticide exposures. These measurements include observations of molecular properties and functions, such as genetic polymorphisms and enzyme activities, which can affect the pharmacokinetics (rates of absorption, distribution, metabolism and elimination) of pesticides, along with an individual's biochemical disposition towards disease progression or repair. It should be noted that both OPs and CMs are readily absorbed, widely distributed and extensively metabolized before being excreted in the urine and/or bile. Biomarkers of susceptibility are affected by a suite of exogenous sources, and may therefore be difficult to link to a specific pesticide exposure event. It should be taken into account that a broad range of the population (workers involved in manufacturing, applicators, field workers, and the general population) can be exposed to pesticides either *via* domestic use or by consumption of food or water containing low-level residues.

Biomonitoring data are useful for a variety of applications, from exposure assessment to risk assessment. Developing methods for measurement of pesticide biomarkers and understanding the resultant data are complex processes that require a great deal of time and intellectual input. Recently, Barr and Buckley[122] specified that the biomarkers of exposure should be, at a minimum, sensitive, specific, valid, biologically relevant and easy to collect (*i.e.* practical) in order to be useful as a surveillance tool and for improving quantitative estimates of exposure and dose. Furthermore, a meaningful biomonitoring strategy is based on sound pharmacokinetic principles that enable quantification of chemical specific dosimetry and biological response, which then can be used to inform risk in humans who are exposed to these pesticides.[123] Further, pharmacokinetic studies provide important data on the amount of toxicant delivered to a target site as well as species-, age-, gender-specific or dose-dependent differences in biological response in both animals (relevant to toxicology) and humans (biomonitoring/risk assessment). It is important to mention that some pesticide or metabolite measurements may be more meaningful in one matrix as opposed to another. These data all come into consideration when determining the appropriate uses for biomonitoring data.

4.9 Treatment

4.9.1 Supportive Measures

In acute OP or CM poisoning cases, supportive measures should be directed toward the maintenance of respiratory and cardiovascular systems, such as maintenance of oxygenation by ventilation, cardiac rhythm, and blood pressure. As early as possible, ingested pesticide(s) should be removed by gastric aspiration followed by lavage with protection of the airway because they are mostly dissolved in aromatic hydrocarbons. Gastric lavage can be effective until 4 h after ingestion, as OPs and CMs are rapidly absorbed from the gastrointestinal tract. In a series of clinical trials conducted in Sri Lanka, Eddleston *et al.*[131,132] found no differences in mortality between patients treated with or without charcoal. It was concluded that the routine use of multiple-dose activated charcoal in poisonings with OP or CM pesticides cannot be recommended, and suggested that further studies of early charcoal administration might be useful.

4.9.2 Antidotal Therapy

OPs and CMs are highly toxic chemicals and present a serious threat to human health. The high mortality ascribed to these pesticides can be due to delayed diagnosis and improper treatment. Development of effective antidotes against OPs and CMs has been a big challenge since the inception of anti-ChE agents, and medical treatment has not been changed for years.

4.9.2.1 Atropine

Presently, humans acutely poisoned with OPs are treated with a combination of an antimuscarinic agent (atropine), an AChE reactivator such as one of the pyridinium oximes (pralidoxime, trimedoxime, methoxime, obidoxime, asoxime chloride) and diazepam. In most OP or CM poisoning cases, atropine is considered a life-saving agent. Atropine acts by blocking the effects of excess concentrations of ACh at the muscarinic cholinergic synapsis following OP- or CM-induced inhibition of AChE. Atropine can relieve the symptoms, such as salivation, sweating, lacrimation, rhinorrhea, nausea, vomiting and diarrhoea. In addition, atropine can help control bradycardia and in dilating the bronchi and abolishing bronchorrhea.[19] In experimental studies, it has been demonstrated that atropine can also partly block convulsions because in OP or CM poisoning other transmitter systems (GABA and glutamate) become involved in cholinergic overstimulation in the brain.[133] This suggests that the effects of atropine are far more complex than muscarinic receptor blockade.

4.9.2.2 Glycopyrolate

Glycopyrolate is used as an alternative to atropine as it offers similar outcomes using continuous infusion. Ampoules of 7.5 mg of glycopyrolate are added to

200 mL of saline and this infusion is then titrated to the desired effects of dry mucous membranes and heart rate above 60 beats/min. This regimen offers an obvious advantage when there is a concern for respiratory infections due to excessive and difficult to control secretions.[134]

4.9.2.3 Oximes

Pyridinium oximes have been used in OP-poisoned patients for more than half a century. Oximes can be divided into two groups, the monopyridinium and bispyridinium oximes. Currently, the only used monopyridinium oxime is pralidoxime (2-PAM), and the most significant bispyridinium oximes are trimedoxime (B-4), obidoxime (LüH-6, Toxogonin), methoxime and HI-6. Oximes reactivate phosphorylated cholinesterases by displacing the phosphoryl moiety from the enzyme by virtue of their high affinity for the enzyme and their powerful nucleophilicity.[19,135] Various studies suggest that the ability to reverse AChE inhibition with oximes varies with the insecticide involved. For example, AChE inhibited by diethyl OP insecticides, such as parathion and quinalphos, seems to be effectively reactivated by oximes, but AChE inhibited by dimethyl OP, such as dimethoate, monocrotophos and oxydemeton-methyl, apparently responds poorly. AChE inhibited by *S*-alkyl-linked OPs such as profenofos is not reactivated by oximes at all. This difference is probably caused by variations in the rate of ageing on inhibited AChE induced by different OP pesticides. It is important to mention that oximes permeate weakly to the CNS, but they are very efficient at the neuromuscular junctions, where atropine is considered to be less effective. It is well established that the combination of atropine and an oxime affords virtually complete peripheral protection. In a recent report, Stefanovic *et al.*[136] suggested that the addition of sodium bicarbonate to atropine sulfate improves the protective effects of standard antidotes (atropine and oxime) against OPs. It should be noted that pralidoxime is of limited value and may actually be hazardous in poisonings by the ChE-inhibiting CMs.

4.9.2.4 Diazepam

In OP-poisoned patients, benzodiazepines such as diazepam offer beneficial effects by reducing anxiety and restlessness, increasing muscle relaxation, arresting seizures and convulsions, controlling apprehension and agitation and possibly reducing morbidity and mortality when used in conjunction with atropine and an oxime.[19] Diazepam should be given to patients poisoned with OPs whenever convulsions or muscle fasciculations are present. The mechanism by which diazepam, like other benzodiazepines, may act is by altering binding of γ-amino butyric acid (GABA) at the $GABA_A$ receptor in an allosteric fashion, but these drugs do not directly activate the receptors.[137–139] The recommended dose of diazepam in cases of OP poisoning is 5–10 mg intravenously (IV) in the absence of convulsions and 10–20 mg IV in cases with convulsions, and can be repeated if necessary.[133,140] It has been hypothesized that benzodiazepines further potentiate a widely described depression of these centres by OPs and CMs[141]. Special care is needed while using these antidotes in children and pregnant women. For further details on antidotes against anti-ChE poisonings, readers are referred to Marrs and Vale[22], Sellström and Marrs[139] and Jokanovic.[19]

4.10 Conclusions and Future Directions

OPs and CMs constitute a large number of chemicals that are used in agriculture primarily as insecticides and in veterinary medicine as parasiticides. These chemicals exert a broad range of toxic effects, varying from mild effects such as salivation and tremors to those as serious as convulsions, seizures, paralysis, and death. Basically, OPs and CMs are neurotoxicants, but directly or indirectly they affect several vital organs, as these chemicals produce a variety of toxicological effects on the CNS, peripheral nervous system, cardiovascular, pulmonary, ocular, neurobehavioural, immunological, reproductive, placental, cutaneous, and other body systems. In addition, these insecticides cause neurodegeneration, oxidative stress, endocrine disruption and many other complications. In general, OPs produce more serious and lingering health effects than CMs. For example, some of the complex syndromes like OPIDP and IMS have devastating effects, have yet to be thoroughly characterized mechanistically and need to be well defined. It is expected that newer compounds of both OP and CM classes will be developed with greater selective toxicity. Also, newer antidotes need to be developed that can be effective in patients with OPIDP or IMS, or against CMs.

Acknowledgements

The authors would like to thank Mrs Robin B. Doss and Mrs Kristie M. Rohde for their technical assistance in preparation of this chapter.

References

1. R. J. Zwiener and G. M. Ginsburg, Organophosphate and carbamate poisoning in infants and children, *Pediatrics*, 1988, **81**, 121–126.
2. L. Karalliedde and N. Senanayake, Organophosphorus insecticide poisoning, *Br. J. Anaesth.*, 1989, **63**, 736–750.
3. M. O'Malley, Clinical evaluation of pesticide exposure and poisoning, *Lancet*, 1997, **349**, 1161–1166.
4. M. Eddleston, Patterns and problems of deliberate self-poisoning in the developing world, *Q. J. Med.*, 2002, **93**, 715–731.
5. R. C. Gupta, in *Toxicology of Organophosphate and Carbamate Compounds*, ed. R. C. Gupta, Academic Press/Elsevier, Amsterdam, 2006, pp. 3–24.

6. T. Satoh and R. C. Gupta, *Anticholinesterase Pesticides: Metabolism, Neurotoxicity, and Epidemiology*, John Wiley & Sons, Hoboken, NJ, 2010.
7. A. M. Fan, Epidemiology of anticholinesterase pesticide poisoning in the United States, in *Anticholinesterase Pesticides: Metabolism, Neurotoxicity, and Epidemiology*, ed. T. Satoh and R. C. Gupta, John Wiley & Sons, Hoboken, NJ, 2010, pp. 541–566.
8. R. C. Gupta, J. K. Malik and D. Milatovic, Organophosphate and carbamate pesticides, in *Reproductive and Developmental Toxicology*, ed. R. C. Gupta, Academic Press/Elsevier, Amsterdam, 2011, pp. 471–486.
9. L. R. Goldman, O. F. Smith, R. R. Neutra, *et al.*, Pesticide food poisoning from contaminated watermelons in California, Arch. Environ. Health, 1990, **45**, 229–232.
10. T. Satoh, Global epidemiology of organophosphate and carbamate poisonings, in *Toxicology of Organophosphate and Carbamate Compounds*, ed. R. C. Gupta, Academic Press/Elsevier, Amsterdam, 2006, pp. 89–100.
11. D. R. Varma and S. Mulay, The Bhopal accident and methyl isocyanate toxicity, in *Toxicology of Organophosphate and Carbamate Compounds*, ed. R. C. Gupta, Academic Press/Elsevier, Amsterdam, 2006, pp. 79–88.
12. A. Dewan, A. B. Patel, R. Pal, U. Jani, V. C. Singel and M. D. Panchal, Mass ethion poisoning with high mortality, *Clin. Toxicol.*, 2008, **46**, 85–88.
13. C. Colosio, F. Vellere and A. Moretto, Epidemiological studies of anticholinesterase pesticide poisoning.: Global impact, in *Anticholinesterase Pesticides: Metabolism, Neurotoxicity, and Epidemiology*, ed. T. Satoh, R. C. Gupta, John Wiley & Sons, Hoboken, NJ, 2010, pp. 343–355.
14. J. G. Thundiyil, J. Stober, N. Besbelli and J. Pronczuk, Acute pesticide poisoning: A proposed classification tool, *Bull. World Health Org.*, 2008, **86**, 205–209.
15. K. Savolainen, Regulatory aspects of anticholinesterase pesticides, in *Anticholinesterase Pesticides: Metabolism, Neurotoxicity, and Epidemiology*, ed. T. Satoh and R. C. Gupta, John Wiley & Sons, Hoboken, 2010, pp. 569–579.
16. G. M. Calvert, W. T. Sanderson, M. Barnett, J. M. Blondell and L. N. Mehler, Surveillance of pesticide-related illness and injury in humans, in *Handbook of Pesticide Toxicology*, ed. R. Krieger, Academic Press, San Diego, 2nd edn, 2001, pp. 603–641.
17. C. Timchalk, Physiologically based pharmacokinetic modeling of organophosphorus and carbamate pesticides, in *Toxicology of Organophosphate and Carbamate Compounds*, ed. R. C. Gupta, Academic Press, Amsterdam/Elsevier, 2006, pp. 103–125.
18. T. C. Marrs, Organophosphate poisoning, *Pharmacol. Ther.*, 1993, **58**, 51–66.
19. M. Jokanovic, Medical treatment of poisoning by organophosphates and carbamates, in *Anticholinesterase Pesticides: Metabolism, Neurotoxicity, and Epidemiology*, ed. T. Satoh and R. C. Gupta, John Wiley & Sons, Hoboken, NJ, 2010, pp. 583–597.

20. A. Dekundy and R. M. Kaminski, Central mechanisms of seizures and lethality following anticholinesterase pesticide exposure, in *Anticholinesterase Pesticides: Metabolism, Neurotoxicity, and Epidemiology*, ed. T. Satoh and R. C. Gupta, John Wiley & Sons, Hoboken, NJ, 2010, pp. 149–164.
21. K. Savolainen, Understanding the toxic actions of organophosphates, in *Handbook of Pesticide Toxicology*, ed. R. I. Krieger. Academic Press, San Diego, 2nd edn, 2001, pp. 1013–1014.
22. T. C. Marrs and J. A. Vale. Management of organophosphorus pesticide poisoning, in *Toxicology of Organophosphate and Carbamate Compounds*, ed. R. C. Gupta, Academic Press/Elsevier, Amsterdam, 2006, pp. 715–733.
23. A. G. Karczmar, Anticholinesterases and war gases, in *Exploring the Vertebrate Central Cholinergic Nervous System*. ed. A. G. Karczmar, Springer, New York, 2007, pp. 237–310.
24. C. N. Pope, Organophosphorus pesticides: Do they all have the same mechanism of toxicity?, *J. Toxicol. Environ. Health. B*, 1999, **2**, 161–181.
25. R. C. Gupta, Brain regional heterogeneity and toxicological mechanism of organophosphates and carbamates., *Toxicol. Mechan. Methods*, 2004, **14**, 103–143.
26. O. Lockridge and L. M. Schopfer, Biomarkers of organophosphorus pesticide poisoning, in *Toxicology of Organophosphate and Carbamate Compounds*, ed. R. C. Gupta, Academic Press/Elsevier, Amsterdam, 2006, pp. 703–711.
27. J. Chaudhuri, T. K. Chakraborti, S. Chanda and C. N. Pope, Differential modulation of organophosphate-sensitive muscarinic receptors in rat brain by parathion and chlorpyrifos, *J. Biochem. Toxicol.*, 1993, **8**, 207–216.
28. P. J. Little, J. A. Scimeca and B. R. Martin, Distribution of [^{3}H]diisopropylfluorophosphate, [^{3}H]soman, [^{3}H]sarin and their metabolites in mouse brain, *Drug Metab. Dispos.*, 1988, **16**, 515–520.
29. O. Lockridge, E. G. Duysen, T. Voelker, C. M. Thompson and L. M. Schopfer, Life without acetylcholinesterase: the implications of cholinesterase inhibitor toxicity in AChE knockout mice, *Environ. Toxicol. Pharmacol.*, 2005, **19**, 463–469.
30. B. E. Mileson, J. E. Chambers, W. L. Chen, *et al.*, Common mechanism of toxicity: A case study of organophosphorus pesticides, Toxicol. Sci., 1998, **41**, 8–20.
31. B. Clothier, M. K. Johnson and E. Reiner, Interaction of some trialkyl phosphorothiolates with acetylcholinesterase. Characterization of inhibition, aging and reactivation, *Biochim. Biophys. Acta*, 1981, **660**, 306–316.
32. T. R. Fukoto, Mechanism of action or organophosphorus and carbamate insecticides, *Environ. Health Perspect.*, 1990, **87**, 245–254.
33. S. Karanth, J. Liu, N. Mirajkar and C. N. Pope, Effects of acute chlorpyrifos exposure on *in vivo* acetylcholine accumulation in rat striatum, *Toxicol. Appl. Pharmacol.*, 2006, **216**, 150–156.

34. A. Ray, J. Liu, S. Karanth, Y. Gao, S. Brimijoin and C. N. Pope, Cholinesterase inhibition and acetylcholine accumulation following intracerebral administration of paraoxon in rats, *Toxicol. Appl. Pharmacol.*, 2009, **236**, 341–347.
35. L. G. Sultatos, Interaction of organophosphorus and carbamate compounds with cholinesterases, in *Toxicology of Organophosphate and Carbamate Compounds*, ed. R. C. Gupta, Academic Press/Elsevier, Amsterdam, 2006, pp. 209–218.
36. A. Ordentlich, D. Barak, C. Kronman, N. Ariel, Y. Segall, B. Velan and A. Shafferman, The architecture of human acetylcholinesterase active center probed by interactions with selected organophosphate inhibitors, *J. Biol. Chem.*, 1996, **271**, 11953–11962.
37. A. Ordentlich, D. Barak, C. Kronman, N. Ariel, Y. Segall, B. Velan and A. Shafferman, Functional characteristics of the oxyanion hole in human acetylcholinesterase, *J. Biol. Chem.*, 1998, **273**, 19509–19517.
38. A. Ordentlich, D. Barak, G. Sod-Moriah, D. Kaplan, D. Mizrahi, Y. Segall, C. Kronman, Y. Karton, A. Lazar, D. Marcus, B. Velan and A. Shafferman, Stereoselectivity toward VX is determined by interactions with residues of the acyl pocket as well as of the peripheral anionic site of AChE, *Biochemistry*, 2004, **43**, 11255–11265.
39. M. A. Sogorb and E. Vilanova, Enzymes involved in the detoxification of organophosphorus, carbamate and pyrethroid insecticides through hydrolysis, *Toxicol. Lett.*, 2002, **128**, 215–228.
40. Z. Radic and P. Taylor, Structure and function of cholinesterases, in *Toxicology of Organophosphate and Carbamate Compounds*, ed. R. C. Gupta, Academic Press/Elsevier, Amsterdam/Elsevier, 2006, pp. 161–186.
41. I. Van den Beukel, F. A. Dijcks, P. Vanderheijden, G. Vaucquelin and M. Oortgiesen, Differential muscarinic receptor binding of acetylcholinesterase inhibitors in rat brain, human brain and Chinese hamster ovary cells expressing human receptors, *J. Pharmacol. Exp. Therap.*, 1997, **281**, 1113–1119.
42. B. Lockhart, M. Closier, K. Howard, C. Steward and P. Lestage, Differential inhibition of [^{3}H]-oxotremorine-M and [^{3}H]-quinuclidinyl benzilate binding to muscarinic receptors in rat brain membranes with acetylcholinesterase inhibitors, *Naunyn-Schmiedeberg's Arch. Pharmacol.*, 2001, **363**, 429–438.
43. C. J. G. M. Smulders, T. J. H. Bueters, R. G. D. M. Van Kleef and H. P. M. Vijverberg, Selective effects of carbamate pesticides on rat neuronal nicotinic acetylcholine receptors and rat brain acetylcholinesterase, *Toxicol. Appl. Pharmacol.*, 2003, **193**, 139–146.
44. R. C. Gupta, S. Milatovic, W.-D. Dettbarn, M. Aschner and D. Milatovic, Neuronal oxidative injury and dendritic damage induced by carbofuran: Protection by memantine, *Toxicol. Appl. Pharmacol.*, 2007, **219**, 97–105.
45. S. Zaja-Milatovic, R. C. Gupta, M. Aschner and D. Milatovic, Protection of DFP-induced oxidative damage and neurodegeneration by

antioxidants and NMDA receptor antagonist, *Toxicol. Appl. Pharmacol.*, 2009, **240**, 124–131.
46. D. Milatovic, M. Aschner, R. C. Gupta, S. Zaja-Milatovic and G. Barnes, Involvement of oxidative stress in anticholinesterase pesticide toxicity, in *Anticholinesterase Pesticides: Metabolism, Neurotoxicity, and Epidemiology*, ed. T. Satoh and R. C. Gupta, John Wiley & Sons, Hoboken, NJ, 2010, pp. 135–147.
47. R. C. Gupta and D. Milatovic, Oxidative injury and neurodegeneration by OPs: Protection by NMDA receptor antagonists and antioxidants, in *The Neurochemical Consequences of Organophosphate Poisoning in the CNS*, ed. B. A. Weissman and L. Raveh, Transworld Network, Trivendrum, 2010, pp. 19–39.
48. Y. Solberg and M. Belkin, The role of excitotoxicity in organophosphorus nerve agents central poisoning, *Trends Pharmacol.*, 1997, **18**, 183–185.
49. P. Carpentier, M. Lambrinis and G. Blanchet, Early dendritic changes in hippocampal pyramidal neurons (field CA1) of rats subjected to acute soman intoxication—a light microscopic study, *Brain Res.*, 1991, **541**, 293–299.
50. D. Milatovic, Z. Radic, M. Zivin and W-D. Dettbarn, Atypical effect of some spin trapping agents: Reversible inhibition of acetylcholinesterase, *Free Rad. Biol. Med.*, 2000, **28**, 597–603.
51. D. Milatovic, M. Zivin and W-D. Dettbarn, The spin trapping agent phenyl-*N-tert*-butylnitron (PBN) prevents excitotoxicity in skeletal muscle, *Neurosci. Lett.*, 2000, **278**, 25–28.
52. R. C. Gupta, D. Milatovic and W-D. Dettbarn, Depletion of energy metabolites following acetylcholinesterase inhibitor-induced status epilepticus: prevention by antioxidants, *Neurotoxicology*, 2001, **22**, 271–282.
53. R. C. Gupta, D. Milatovic and W-D. Dettbarn, Nitric oxide modulates high-energy phosphates in brain regions of rats intoxicated with diisopropylphosphorofluoridate or carbofuran: Prevention by *N-tert*-butyl-α-phenylnitrone or vitamin E, *Arch. Toxicol.*, 2001, **75**, 346–356.
54. R. C. Gupta and J. T. Goad, Biochemical and microscopic changes in brain by acute carbofuran toxicity, *Toxicol. Sci.*, 1998, **42**, 156.
55. S. J. Garcia, F. J. Seidler, T. L. Crumpton and T. A. Slotkin, Does the developmental neurotoxicity of chlorpyrifos involve glial targets? Macromolecule synthesis, adenylyl cyclase signaling, nuclear transcription factors, and formation of reactive oxygen in C6 glioma cells, *Brain Res.*, 2001, **891**, 54–68.
56. S. H. Lee and I. A. Blair, Oxidative DNA damage and cardiovascular disease, *Trends Cardiovasc. Med.*, 2001, **11**, 148–155.
57. D. Qiao, F. J. Seidler and T. A. Slotkin, Developmental neurotoxicity of chlorpyrifos modeled *in vitro*: Comparative effects of metabolites and other cholinesterase inhibitors on DNA synthesis in PC12 and C6 cells, *Environ. Health Perspect.*, 2001, **109**, 909–913.

58. E. Maran, M. Fernandez-Franzon, G. Font and M. J. Ruiz, Effects of aldicarb and propoxur on cytotoxicity and lipid peroxidation in CHO-K1 cells, *Food Chem. Toxicol.*, 2010, **48**, 1592–1596.
59. A. Ranjbar, H. Solhi, F. J. Mashayekhi, A. Susanabdi, A. Rezaie and M. Abdollahi, Oxidative stress in acute human poisoning with organophosphorus insecticides: a case control study, *Environ. Toxicol. Pharmacol.*, 2005, **20**, 88–91.
60. D. Milatovic, R. C. Gupta and M. Aschner, Anticholinesterase toxicity and oxidative stress, *Scientific World J.*, 2006, **6**, 295–310.
61. A. Lukaszewicz-Hussain, Role of oxidative stress in organophosphate insecticide toxicity—short review, *Pest. Biochem. Physiol.*, 2010, **98**, 145–150.
62. N. Senanayake and L. Karalliedde, Neurotoxic effects of organophosphorus insecticides: An intermediate syndrome, *N. Engl. J. Med.*, 1987, **316**, 761–763.
63. N. Paul and T. J. Mannathukkaran, Intermediate syndrome following carbamate poisoning, *Clin. Toxicol.*, 2005, **43**(7), 867–868.
64. M. Dandapani, A. Zachariah, M. R. Kavitha, L. Jeyaseelan and A. Oommen, Oxidative damage in intermediate syndrome of acute organophosphorous poisoning, *Indian J. Med. Res.*, 2003, **117**, 253–259.
65. P. Jayawardane, A. H. Dawson, V. Weerasinghe, L. Karalliedde, N. A. Buckley and N. Senanayake, The spectrum of intermediate syndrome following acute organophosphate poisoning: a prospective cohort study from Sri Lanka, *PLoS Med.*, 2008, **5**(7), e147. doi: 10.1371/journal.pmed.0050147.
66. K. S. Pawar, R. R. Bhoite, C. P. Pillay, S. C. Chavan and D. S. Mashikare, Continuous pralidoxime infusion versus repeated bolus injection to treat organophosphorus pesticide poisoning, *Lancet.*, 2006, **368**, 2136–2141.
67. C. K. Aaron, Organophosphate poisoning-induced intermediate syndrome: can electrophysiological changes help predict outcome, *PLos Med.* 2008, **5**(7), e154:doi: 10.1371/journal.pmed.0050154.
68. R. C. Gupta, Organophosphate poisoning, Intermediate syndrome, in *Encyclopedia of Toxicology*, ed. P. Wexler, Elsevier, San Diego, 2005, pp. 306–308.
69. J. De Bleecker, Intermediate syndrome in organophosphate poisoning, in *Toxicology of Organophosphate and Carbamate Compounds*, ed. R. C. Gupta, Academic Press/Elsevier, Amsterdam, 2006, pp. 371–380.
70. M. Lotti, The pathogenesis of organophosphate neuropathy, *Crit. Rev. Toxicol.*, 1992, **21**, 465–487.
71. M. Ehrich and B. S. Jortner, Organophosphorous-induced delayed neuropathy, in *Handbook of Pesticide Toxicology*, ed. R. I. Krieger. Academic Press, San Diego, 2nd edn, 2001, pp. 987–1012.
72. M. Lotti and A. Moretto, Organophosphate-induced delayed polyneuropathy, *Toxicol. Rev.*, 2005, **24**, 37–49.

73. S. Little, H. Lewellyn and S. F. J. Clarke, Sensory ataxia with chronic organophosphate pesticide exposure, *Hum. Exp. Toxicol.*, 2010, **29**, 679–688.
74. M. K. Johnson, The target for initiation of delayed neurotoxicity by organophosphorus esters: biochemical studies and toxicological applications, *Rev. Biochem. Toxicol.*, 1982, **4**, 141–212.
75. M. K. Johnson and P. Glynn, Neuropathy target esterase, in *Handbook of Pesticide Toxicology*, ed. R. Krieger, Academic Press, San Diego, 2nd edn, 2001, pp. 953–965.
76. D. Milatovic and M. K. Johnson, Reactivation of phosphorodiamidated acetylcholinesterase and neuropathy target esterase by treatment of inhibited enzyme with potassium fluoride, *Chem. Biol. Interact.*, 1993, **87**, 425–430.
77. D. Milatovic, A. Moretto, K. A. Osman and M. Lotti, Phenyl valerate esterase other than neuropathy target esterase and the promotion of organophosphate polyneuropathy, *Chem. Res. Toxicol.*, 1997, **10**, 1045–1048.
78. M. Jokanović and M. Kosanović, Neurotoxic effects in patients poisoned with organophosphorus pesticides, *Environ. Toxicol. Pharmacol.*, 2010, **29**, 195–201.
79. P. Glynn, A mechanism for organophosphate-induced delayed neuropathy, *Toxicol. Lett.*, 2006, **162**, 94–97.
80. P. Glynn, Axonal degeneration and neuropathy target esterase, *Arh. Hig. Rada Toksikol.*, 2007, **58**, 355–358.
81. P. Glynn, Neuropathy target esterase, *Biochem. J.*, 1999, **344**, 625–631.
82. Y. Ly, D. Dinsdale and P. Glynn, Protein domains, catalytic activity and subcellular distribution of neuropathy target esterase in mammalian cells, *J. Biol. Chem.*, 2003, **278**, 8820–8825.
83. M. Lotti, A. Moretto, M. Bertolazzi, F. Fioroni, M. Peraica and M. L. Scapellato, Neuropathy target esterase and the mechanisms of toxicity in organophosphate polyneuropathy, in *Mechanisms of Toxicity and Their Relevance in Industrial Toxicology*, SGEditoriali, Padova 1993, pp. 1–11.
84. J. Atkins and P. Glynn, Membrane association of and critical residues in the catalytic domain of human neuropathy target esterase, *J. Biol. Chem.*, 2000, **275**, 24477–24483.
85. M. van Tienhoven, J. Atkins, Z. Li and P. Glynn, Human neuropathy target esterase catalyzes hydrolysis of membrane lipids, *J. Biol. Chem.*, 2002, **277**, 20942–20948.
86. G. L. Emerick, R. G. Peccinini and G. H. de Oliveira, Organophosphorus-induced delayed neuropathy: A simple and efficient therapeutic strategy, *Toxicol. Lett.*, 2010, **192**, 238–244.
87. S. Rainier, M. Bui, E. Mark, D. Thomas, D. Tokarz, L. Ming, C. Delaney, R. J. Richardson, J. W. Albers, N. Matsunami, J. Stevens, H. Coon, M. Leppert and J. K. Fink, Neuropathy target esterase gene mutations cause motor neuron disease, *Am. J. Human Genet.*, 2008, **82**, 780–785.

88. N. Inoue, K. Fujishiro, K. Mori and M. Matsuoka, Triorthocresyl phosphate poisoning—a review of human cases, *J. UOEH*, 1988, **10**, 433–442.
89. P. A. Chang and Y. J. Wu, Motor neuron diseases and neurotoxic substances: a possible link?, *Chem. Biol. Interact.*, 2009, **180**, 127–130.
90. M. Jokanović, M. Kosanović, D. Brkić and P. Vukomanović, Organophosphate induced delayed polyneuropathy in man: an overview, *Clin. Neurol. Neurosurg.*, 2011, **113**(1), 7–10.
91. G. A. Jamal, Neurological syndromes of organophosphorus compounds, *Adverse Drug React. Toxicol. Rev.*, 1997, **16**, 133–170.
92. G. A. Jamal and P. O. O. Julu, Low level exposures to organophosphorus esters may cause neurotoxicity, *Toxicology*, 2002, **181**, 23–33.
93. S. Gershon and F. B. Shaw, Psychiatric sequelae of chronic exposure to organophosphate insecticides, *Lancet*, 1961, **1**, 1371–1374.
94. D. R. Metcalf and J. H. Holmes, EEG, psychological and neurological alterations in human with organophosphorus exposure, *Ann. N. Y. Acad. Sci.*, 1969, **160**, 357–365.
95. D. E. Ray, Chronic effects of low level exposure to anticholinesterases—a mechanistic review, *Toxicol. Lett.*, 1998, **102–103**, 527–533.
96. F. Sánchez-Santed, F. Cañadas, P. Flores, M. López-Grancha and D. Cardona, Long-term functional neurotoxicity of paraoxon and chlorpyrifos: Behavioural and pharmacological evidence, *Neurotoxicol. Teratol.*, 2004, **2**(26), 305–17.
97. H. S. Levin, R. L. Rodnitzky and D. L. Mick, Anxiety associated with exposure to organophosphate compounds, *Arch. Gen. Psychiatr.*, 1976, **33**, 225–228.
98. P. D. Behan, Chronic fatigue syndrome as a delayed reaction to low dose organophosphate exposure, *J. Nutr. Environ. Med.*, 1996, **6**, 341–350.
99. J. Brodeur and K. P. DuBois, Studies on the mechanism of acquired tolerance by rats to *O, O*-diethyl-S-2-(ethylthio)ethyl phosphorodithioate (Di-Syston), *Arch. Int. Pharmacodyn. Ther.*, 1964, **149**, 560–570.
100. L. G. Costa, B. W. Schwab, H. Hand and S. D. Murphy, Reduced [^{3}H]quinuclidinyl benzilate binding tomuscarinic receptors in disulfoton-tolerant mice, *Toxicol. Appl. Pharmacol.*, 1981, **60**, 441–450.
101. L. G. Costa and S. D. Murphy, [^{3}H]-nicotine binding in rat brain: alterations after chronic acetylcholinesterase inhibition, *J. Pharmacol. Exp. Ther.*, 1983, **226**, 392–397.
102. R. C. Gupta, G. T. Patterson and W-D. Dettbarn, Mechanisms of toxicity and tolerance to diisopropylphosphorofluoridate at the neuromuscular junction of the rat, *Toxicol. Appl. Pharmacol.*, 1986, **84**, 541–550.
103. F. Fonnum and S. H. Sterri, Tolerance development to toxicity of cholinesterase inhibitors, in *Toxicology of Organophosphate and Carbamate Compounds*, ed. R. C. Gupta, Academic Press/Elsevier, Amsterdam, 2006, pp. 257–267.
104. M. M. Mesulam, A. Guillozet, P. Shaw, A. Levey, E. G. Duysen and O. Lockridge, Acetylcholinesterase knockouts establish central cholinergic

pathways and can use butyrylcholinesterase to hydrolyze acetylcholine, *Neuroscience*, 2002, **110**, 627–639.
105. R. C. Gupta, G. T. Patterson and W-D. Dettbarn, Mechanisms involved in the development of tolerance to DFP toxicity, *Fundam. Appl. Toxicol.*, 1985, **5**, S17–28.
106. R. C. Gupta and W-D. Dettbarn, Role of uptake of [^{14}C]valine into protein in the development of tolerance to diisopropylphosphorofluoridate (DFP) toxicity, *Toxicol. Appl. Pharmacol.*, 1986, **84**, 551–560.
107. D. Milatovic and W-D. Dettbarn, Modification of acetylcholinesterase during adaptation to chronic, subacute paraoxon application in rat, *Toxicol. Appl. Pharmacol.*, 1996, **136**, 20–28.
108. S. H. Sterri, Factors modifying the toxicity of organophosphorus compounds including dichlorvos, *Acta Pharmacol. Toxicol.*, 1981, **49(Suppl. 5)**, 67–71.
109. M. A. Sogorb and E. Vilanova, Detoxication of anticholinesterase pesticides, in *Anticholinesterase Pesticides: Metabolism, Neurotoxicity, and Epidemiology*, ed. T. Satoh and R. C. Gupta, John Wiley & Sons, Hoboken, NJ, 2010, pp. 121–132.
110. R. C. Gupta and D. Milatovic, Anticholinesterase pesticides interactions, in *Anticholinesterase Pesticides: Metabolism, Neurotoxicity, and Epidemiology*, ed. T. Satoh and R. C. Gupta, John Wiley & Sons, Hoboken, NJ, 2010, pp. 315–327.
111. N. Itoh, R. Hasegawa, K. Imaida, Y. Kurata, A. Hasegawa and T. Shirai, Effect of ingestion of 20 pesticides in combination at acceptable daily intake levels on rat liver carcinogenesis, *Food Chem. Toxicol.*, 1995, **33**, 159–163.
112. J. P. Frawley, H. N. Fuyat, E. C. Hagan, J. R. Blake and O. G. Fitzhugh, Marked potentiation in mammalian toxicity from simultaneous administration of two anticholinesterase compounds, *J. Pharmacol. Exp. Ther.*, 1957, **121**, 96–106.
113. K. P. DuBois, Potentiation of the toxicity of organophosphorus compounds, *Adv. Pestic. Count. Res.*, 1961, **4**, 117–151.
114. S. D. Cohen and S. D. Murphy, Carboxylesterase inhibition as an indicator of malathion potentiation in mice, *J. Pharmacol. Exp. Ther.*, 1971, **176**, 733–776.
115. C. Timchalk, T. S. Poet, M. N. Hinman, A. L. Busby and A. A. Kousba, Pharmacokinetic and pharmacodynamic interaction for a binary mixture of chlorpyrifos and diazinon in the rat, *Toxicol. Appl. Pharmacol.*, 2005, **205**, 31–42.
116. R. C. Gupta and W. L. Kadel, Concerted role of carboxylesterases in the potentiation of carbofuran toxicity by *iso*-OMPA, *J. Toxicol. Environ. Health*, 1989, **26**, 447–457.
117. H. Takahashi, T. Miyaoka, S. Tsuda and Y. Shirasu, Combined toxicity of malathion and carbamate insecticides in mice and rats, *J. Pest. Sci.*, 1983, **8**, 41–45.

118. H. Takahashi, T. Miyaoka, S. Tsuda and Y. Shirasu, Potentiated toxicity of 2-*sec*-butylphenyl methylcarbamate (BPMC) by *O,O*-dimethyl-*O* -(3-methyl-4-nitrophenyl) phosphorothioate (fenitrothion) in mice. Relationship between acute toxicity and metabolism of BPMC, *Fundam. Appl. Toxicol.*, 1984, **4**, 718–723.
119. S. Tsuda, T. Miyaoka, M. Iwasaki and Y. Shirasu, Pharmacokinetic analysis of increased toxicity of 2-*sec*-butylphenyl methylcarbamate (BPMC) by fenitrothion pretreatment in mice, *Fundam. Appl. Toxicol.*, 1984, **4**, 724–730.
120. T. Miyaoka, H. Takahashi, S. Tsuda and Y. Shirasu, Potentiation of acute toxicity of 2-*sec*-butylphenyl methylcarbamate (BPMC) by fenitrothion in mice, *Fundam. Appl. Toxicol.*, 1984, **4**, 802–807.
121. D. B. Barr, J. R. Barr, W. J. Driskell, R. H. Hill, Jr., D. L. Ashley, L. L. Needham, S. L. Head and E. J. Sampson., Strategies for biological monitoring of exposure for contemporary-use pesticides, *Toxicol. Ind. Health*, 1999, **15**, 168–179.
122. D. B. Barr and B. Buckley. In vivo biomarkers and biomonitoring in reproductive and developmental toxicity, in *Reproductive and Developmental Toxicology*, ed. R. C. Gupta, Academic Press/Elsevier, Amsterdam, 2011, pp. 253–265.
123. C. Timchalk, Biomonitoring of pesticides: Pharmacokinetics of organophosphorus and carbamate insecticides, in *Anticholinesterase Pesticides: Metabolism, Neurotoxicity, and Epidemiology*, ed. T. Satoh and R. C. Gupta, John Wiley & Sons, Hoboken, NJ, 2010, pp. 267–287.
124. J. Sun and B. C. Lynn, Development of a MALDI-TOF-MS method to identify and quantify butyrylcholinesterase inhibition resulting from exposure to organophosphate and carbamate pesticides, *J. Am. Soc. Mass Spectrom.*, 2007, **18**, 698–706.
125. O. Tacal and O. Lockridge, Methamidophos, dichlorvos, *O*-methoate and diazinon pesticides used in Turkey make a covalent bond with butyrylcholinesterase detected by mass spectrometry, *J. Appl. Toxicol.*, 2009, **30**, 469–475.
126. B. Li, I. Ricordel, L. M. Schopfer, F. Baud, B. Megarbane, P. Mason and O. Lockridge, Dichlorvos, chlorpyrifos oxon and aldicarb adducts of butyrylcholinesterase, detected by mass spectrometry in human plasma following deliberate overdose, *J. Appl. Toxicol.*, 2010, **30**, 559–565.
127. B. Li, I. Ricordel, L. M. Schopfer, F. Baud, B. Megarbane, F. Nachon, P. Mason and O. Lockridge, Detection of adduct on tyrosine 411 of albumin in humans poisoned by dichlorvos, *Toxicol. Sci.*, 2010, **116**, 23–31.
128. J. A. MacGregor, L. M. Plunkett, S. H. Youngren, A. Manley, J. B. Plunkett and T. B. Starr, Humans appear no more sensitive than laboratory animals to the inhibition of red blood cell cholinesterase by dichlorvos., *Regul. Toxicol. Pharmacol.*, 2005, **43**, 150–167.
129. Y. Fujikawa, T. Satoh, A. Suganuma, S. Suzuki, Y. Niikura, S. Yui and Y. Tamura, Extremely sensitive biomarker of acute organophosphorus insecticide exposure, *Hum. Exp. Toxicol.*, 2005, **24**, 1–4.

130. J. Ueyama, T. Satoh, T. Kondo, K. Takagi, E. Shibata, M. Goto, A. Kimata, I. Saito, T. Hasegawa, S. Wakusawa and M. Kamijima, β-glucuronidase activity is a sensitive biomarker to assess low-level organophosphorus insecticide biomarker, *Toxicol. Lett.*, 2010, **193**, 115–119.
131. M. Eddleston, E. Juszczak and N. A. Buckley, Randomised controlled trial of routine single or multiple dose superactivated charcoal for self-poisoning in a region with high mortality, *Clin. Toxicol.*, 2005, **43**, 442–443.
132. M. Eddleston, E. Juszczak, N. A. Buckley, *et al.*, Multiple dose activated charcoal in acute self-poisoning: a randomised controlled trial, Lancet, 2008, **371**, 579–587.
133. B. Antonijevic and M. P. Stojiljkovic, Unequal efficacy of pyridinium oximes in acute organophosphorus poisoning, *Clin. Med. Res.*, 2007, **5**, 71–78.
134. P. G. Bardin and S. F. Van Eden, Organophosphate poisoning: Grading the severity and comparing treatment between atropine and glycopyrolate, *Crit. Care Med.*, 1990, **18**, 956–960.
135. J. Kassa and K. Kuca, Exposure to organophosphorus compounds—oximes, neuroprotection and cognitive functions, in *The Neurochemical Consequences of Organophosphate Poisoning in the CNS*, ed. B. A. Weissman and L. Raveh, Transworld Network, Trivendrum, 2010, pp. 77–91.
136. D. Stefanovic, B. Antonijevic, D. Bokonjic, M. P. Stojiljkovic, Z. A. Milovanovic and M. Nedeljkovic, Effect of sodium carbonate in rats acutely poisoned with dichlorvos, *Basic Clin. Pharmacol. Toxicol.*, 2006, **98**, 173–180.
137. Å. Sellström, Anticonvulsants in anticholinesterase poisoning, in *Clinical and Experimental Toxicology of Organophosphates and Carbamates*, ed. B. Ballantyne and T. C. Marrs. Butterworth-Heinemann, Oxford, 1992, pp. 578–586.
138. T. C. Marrs, Diazepam, in *Antidotes for Poisoning by Organophosphorus Pesticides*, International Programme on Clinical Safety, World Health Organization, Geneva, 2004.
139. Å. Sellström and T. C. Marrs, The use of benzodiazepines in organophosphorus nerve agent intoxication, in *Chemical Warfare Agents Toxicology and Treatment*, ed. T. C. Marrs, R. L. Maynard and F. R. Sidell, Wiley, Chichester, 2007, pp. 343–354.
140. M. K. Johnson and J. A. Vale, Clinical management of acute organophosphate poisoning: An overview, in *Clinical and Experimental Toxicology of Organophosphates and Carbamates*, ed. B. Ballantyne and T. C. Marrs, Butterworth-Heinemann, Oxford, 1992, pp. 528–535.
141. N. B. Munro, A. P. Watson, K. R. Ambrose and G. D. Griffin, Treating exposure to chemical warfare agents: Implications for health care providers and community emergency planning, *Environ. Health Perspect.*, 1990, **89**, 205–215.

CHAPTER 5

Comparative Metabolism and Toxicology of Pyrethroids in Mammals

DEREK W GAMMON,* APPAVU CHANDRASEKARAN* AND SHAABAN F ELNAGGAR

FMC Corporation, Ewing, NJ 08628, USA. *Email: Derek.Gammon@fmc.com and Appavu.Chandrasekaran@fmc.com

5.1 Introduction

Early studies to understand the mode of action (MOA) of pyrethroids used nerves from cockroach, squid, and crayfish and they showed that the voltage-gated-sodium channel (VGSC) was an important target site.[1–5] Studies using the electrode-implanted cockroach identified sensory axons, which contain a tetrodotoxin (TTX)-sensitive VGSC, to be sites of action for allethrin, the first synthetic analogue of the pyrethrins.[6–8] After the discovery of a large number of further analogues in the 1970s,[9] interest in understanding the pyrethroid MOA increased dramatically.

In 1980, two syndromes were reported in the rat dosed intravenously with a series of pyrethroids.[10] These syndromes, termed T (tremors) and CS (choreoathetosis/salivation), were found in most cases to classify non-cyano and cyano pyrethroids, respectively. Studies in the mouse[11] and cockroach[12] largely confirmed these classifications. In the latter case, neurophysiological differences were also found between non-cyano (Type I) and cyano-containing (Type II) pyrethroids. Type I pyrethroids, unlike Type II, caused repetitive firing in nerve axons following stimulation.[12] A limited number of pyrethroids, such as fenpropathrin, appeared to show a mixture of Type I/II effects. In the last decade, using

Issues in Toxicology No. 12
Mammalian Toxicology of Insecticides
Edited by Timothy C. Marrs

Published by the Royal Society of Chemistry, www.rsc.org

newer techniques, it has been possible to study pyrethroid effects on isolated ion channel subtypes, either grown in culture or expressed in *Xenopus* oocytes. In this chapter, effects on several sodium ion channel subtypes are reviewed, along with data showing pyrethroid effects on other types of ion channels.

Experimental data have demonstrated that brain concentrations of parent pyrethroids correlate well with acute toxicity and that metabolites generally have little effect on neurotoxicity. It is apparent that clearance of pyrethroids from blood will influence their brain levels, since brain concentrations of pyrethroids have been shown to closely parallel those of plasma. Therefore, the metabolic disposition characteristics of pyrethroids, such as absorption and clearance rates, play a critical role in the toxicity of these insecticides in mammals.

Pyrethroid-induced clinical signs have also been given increasing attention. This has partly resulted from US legislation (the Food Quality Protection Act 1996 (FQPA)) requiring classes of pesticide to be considered cumulatively as well as from an aggregate exposure perspective in risk assessment.[13] In practice, this means that exposure to all chemicals sharing a common MOA should be determined collectively (cumulative) and by all routes of exposure, such as diet, drinking water and home and garden uses (aggregate), combined. Special attention is also paid under FQPA guidelines to the potential for increased susceptibility of the fetus or juvenile to a chemical class of pesticides with a common MOA. Because the effects of concern are generally associated with acute toxicity, this review concentrates mainly on the pharmacokinetics and pharmacodynamics of pyrethroid action in mammals.

Previous reviews of pyrethroid toxicology have considered neurotoxicity,[14,15] developmental neurotoxicity,[16] neurobehavioral toxicity[17] and MOA.[18,19] In the present review, the available data describing the biotransformation and excretion of 13 pyrethroids in the rat are described. The MOA and neurotoxicity data are then reviewed and summarized.

5.2 Metabolic Chemistry

It is now well recognized that neurotoxicity of pyrethroids in mammals is not only directly related to their potencies, but it is very much influenced by metabolism. This, in a broad sense, includes absorption, distribution, biotransformation and elimination. Metabolites of pyrethroids, especially hydrolytic products, are considered not to contribute to neurotoxicity. Recently regulatory agencies such as the US Environmental Protection Agency (EPA) have proposed using physiologically based pharmacokinetic (PBPK) models to assess internal exposure to pyrethroids. Indeed, the PBPK model is a powerful method for determining relative toxicities of pyrethroids since it takes into account all aspects of metabolism including tissue/organ concentrations and elimination profiles and relates them to toxicity endpoints. Deltamethrin has been used to develop a PBPK model called the 'Delta-model' since it contains a single enantiomer. The PBPK model can be used for all pyrethroids as long as there are sensitive analytical methods available for all isomers/enantiomers present in each insecticide.

Available metabolic data in mammals for 13 major pyrethroids are summarized below and listed in Table 5.1. Previous reviews have been considered here[20–24] as well as some unpublished data. Since pharmacokinetic data for all pyrethroids are not available in humans, the focus of this review is on the data obtained in rats. However, it should be recognized that although the metabolism of pyrethroids appears to be similar in rats and humans, newly obtained data indicate that there may be some species differences such as the individual enzymes involved and the rates of metabolism. It should also be noted that the reported values may not be directly used in PBPK models without additional studies, since there may be differences in experimental methods such as the type and volume of vehicles. Both the type of vehicle and volume of vehicle used in the dosing are known to affect the pharmacokinetics including C_{max} and the bioavailability of pyrethroids.[20–24]

Pyrethroids in general have similar chemical structures, esters containing a chrysanthemic acid or its analogues and pyretholone alcohol, phenoxybenzyl alcohol or their analogues. Pyrethroids are lipophilic in nature, with log P values ranging from 4.5 to 7.0. However, due to rapid metabolism by hydrolytic and/or oxidative processes, pyrethroids are quickly eliminated from the body and little significant accumulation occurs in tissues. Also, subtle changes in the chemical structures of individual pyrethroids can lead to significant differences in their disposition. For example, in terms of biotransformation, steric hindrance of the ester bond by a secondary cyclic alcohol or an α-cyano group is known to limit the hydrolytic process while favouring oxidative pathway(s). The nature of isomer has profound effects on hydrolysis, *e.g.* 1*R*-*trans*-isomers are hydrolysed much more rapidly than 1*R*-*cis* isomers contained in synthetic pyrethroids. Only one of the four possible diastereomers of fenvalerate undergoes an unusual esterification process with cholesterol, leading to the formation of a more lipophilic compound that tends to accumulate in tissues, causing granulomatous changes in rats.

Excretion data indicate both renal and faecal routes are major routes of elimination for pyrethroids. As a salient feature of pyrethroids, biliary excretion accounts for about one-third of the dose, which is then excreted in the faeces. The *cis* isomers of sterically hindered analogues appear to be more prevalently excreted *via* the faeces as metabolized intact esters. The bioavailability after a single oral dose is generally high and ranges from 40% to 60%. The blood elimination half-lives were estimated to be from about 8 to 30 h. Tissue concentrations after a single oral dose ranged from 1 to 5% within 168 h. Brain concentrations reflected plasma levels for the majority of the pyrethroids.

As mentioned earlier, toxicity of pyrethroids has been shown to be closely associated with internal exposures. Blood and tissue exposures are dependent on absorption and clearance rates, biotransformation pathways, type of metabolites formed, and route of elimination. Excretion and kinetic data, where available, are summarized in this section. Since metabolism and disposition are influenced by the configuration of the molecule, chemical structures are also included in the summaries. For quick and easy comparison/reference purpose, key chemical and pharmacokinetic parameters of the major pyrethroids are also provided in tabular form (Table 5.1). For example, a quick

Table 5.1 Comparative evaluation of common metabolic parameters of representative pyrethroids.

		Chrysanthemic acid moiety		Alcohol moiety			Excretion		Blood kinetics		
Compound name	*Log K_{ow}*	*Substitution*	*Stereo-chemistry*	*Order*	*Substitution*	*Phase 1 biotrans-formation mode*	*% ^{14}C-Urine*	*% ^{14}C-Faeces*	t_{max} *(h)*	$t_{1/2}$ *(h)*	*% Oral Bioavailability*
Pyrethrin I	5.9	Dimethylvinyl	*t*, 1*R*	2°ry	*S*-alicyclic	Oxidation of intact ester	30	41	NA	NA	NA
S-Bioallethrin	4.96	Dimethylvinyl	*c*, *t*, 1*R*	2°ry	*S*-alicyclic	Oxidation of intact ester	47–51	27–29	NA	NA	NA
Resmethrin	>4.7	Dimethylvinyl	*c*, *t*, 1*R*	1°ry	Benzyl-furyl-methyl	Hydrolysis by carbox-yesterases, some oxidation	36–41	33–64	NA	NA	NA
Permethrin	6.1	Dichlorovinyl-CPC	*c*, *t*, 1*RS*	1°ry	Phenoxybenzyl	*t*, Carboxyesterases *c*, Hydrolysis/oxidation	*t*, 70–71 *c*, 37–39	*t*, 8–13 *c*, 31	3.5	4.9	~61
Cypermethrin	6.6	Dichlorovinyl-CPC	c, *t*, 1*RS*	2°ry	R*S*-α-*cyano*-phenoxybenzyl	*c*, *t*, Mostly hydrolysis with some oxidation	*t*, 71–74 *c*, 50–61	*t*, 23–28 *c*, 31–40	3	NA	63–73
Bifenthrin	>6	Chlorotrifluoro-methylvinyl-CPC(Z)	*c*, *(Z)-RS*	1°ry	Methylbiphenyl system	Mainly oxidation	13–20	73–83	4–6	11–12	34–56
Tefluthrin	6.5	Chlorotrifluoro-methylvinyl-CPC (Z)	*RS*	1°ry	Tetrafluoro-methyl-benzyl	Exclusively oxidation	25–30	63–69	NA	NA	NA
Fenvalerate/ Esfenvalerate	6.22	Phenyl-valeroyl-	*RS*	2°ry	R*S*-α-*cyano*-phenoxybenzyl	Hydrolysis/oxidation	29–39/ 24–33	59–72/ 66–71	3	NA	NA
Cyfluthrin	6	Dichlorovinyl-CPC	*c*, *t*, 1*RS*	2°ry	R*S*-α-*cyano*-phenoxyfluorobenzyl	*c,t*, Mostly hydrolysis with some oxidation	55–70	25–35	2	9–12	~80
Deltamethrin	4.6	DibromoVinyl-CPC	*c*, 1*R*	2°ry	*S*-α-*cyano*-phenoxybenzyl	Oxidation and hydrolysis	31–56	36–59	1	13.3	58.4
Fenpropathrin	6	2,2,3,3-DimethylCPC	*c*, *t*, 1*RS*	2°ry	*RS*-α-*cyano*-phenoxybenzyl	Hydrolysis/oxidation	34–44	62–63	6	NA	~57
Cyhalothrin	6.8	Dhlorotrifluoro-methylvinyl-CPC(Z)	*c,(Z)-RS*	2°ry	*RS*-α-*cyano*-phenoxyfluorobenzyl	Mostly hydrolysis and some oxidation	20–40	40–65	2.7	7.6	67
Ethofenprox/ Etofenprox	7.05	Ether: ethoxyphenyl t-butyl	NA	1°ry	Phenoxybenzyl	Exclusively oxidation	8–11	86–88	2–7	NA	51

c, *cis*; CPC, cyclopropanecarboxylate; NA, not available; *t*, *trans*.

review of these data indicates that maximum plasma concentrations for permethrin and cypermethrin are considerably higher than the values observed for other pyrethroids. Although it is recognized that the differences in plasma concentrations may just be due to differences in dosages used in those studies, the overall implication of such differences in these values on toxicity cannot be ruled out. Pyrethroid derivatives of 1*R*-*trans*-chrysanthemic derivatives were easier to eliminate due to ease of hydrolysis by carboxyesterases, and therefore were relatively less toxic than 1*R*-*cis*-chrysanthemic acid pyrethroid derivatives.

A survey of 13 pyrethroids with chemical properties, biotransformation, and metabolic pathways in mammals is summarized below and listed in Table 5.1.

5.2.1 Pyrethrin I[25,26]

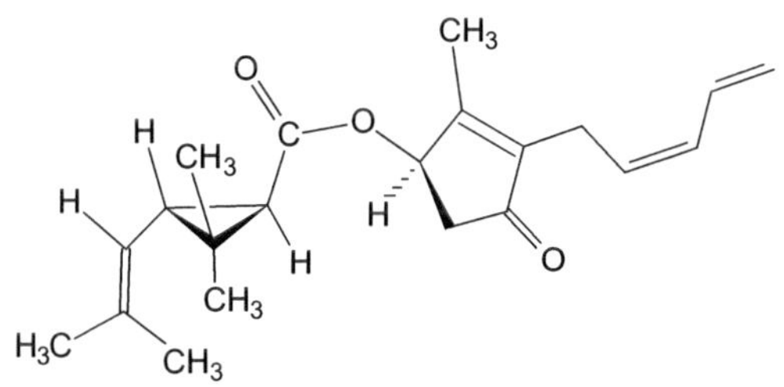

Structural formula of pyrethrin I.

Name: Pyrethrin I: (*Z*)-(*S*)-2-methyl-4-oxo-3-(penta-2,4-dienyl)cyclopent-2-enyl (1*R*)-*trans*-2,2-dimethyl-3-(2-methylprop-1-enyl)cyclopropanecarboxylate, CAS # 121-21-1
Test system: Male Sprague-Daley rat; LD_{50}: 260–450 mg kg^{-1}
Dose: 3 mg kg^{-1} ♂
Route: Gavage in 0.1 mL DMSO
Test duration: 100 h
% of dose in urine: 30% (Figure 5.1)
% of dose in faeces: 41% (Figure 5.1)

5.2.1.1 Biotransformation

Although natural pyrethrins are esters, their metabolism proceeds almost exclusively *via* oxidative reactions rather than hydrolysis. The majority of pyrethrin I metabolites are those of the intact ester. Steric hindrance of the secondary alcohol, cyclopentenyl group, likely blocks the hydrolytic degradation route by denying carboxyesterases accessibility to the ester function. Alternatively, metabolic oxidation would be the main biodegradation route of pyrethrins. The primary most accessible and enzymatically labile targets for oxidative enzymes are those of the terminal *trans*-vinylic methyl group followed by the *cis*-2,4-pentadienyl ethylene bonds of the pyrethronyl alcohol tail.

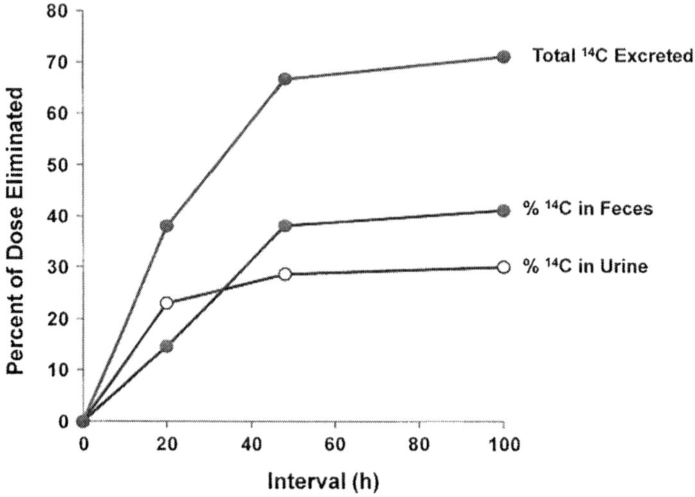

Figure 5.1 Excretion of radioactivity in male rats following oral administration of carbon-14 pyrethrin I at 3 mg kg^{-1} in DMSO.

5.2.1.2 Major Metabolites of Pyrethrin I

- Phase 1 metabolites (Figure 5.2):
  1. *trans*-OH-Methyl-pyrethrin I-ester
  2. *trans*-Pyrethrin I-dicarboxylic acid chrysanthemate ester (dicarboxylate intact)
  3. Pyrethrin I,*trans*-carboxy-*cis*-4′,5′-dihydroxypent-2′-enyl-chrysanthemate ester
  4. Pyrethrin I,*trans*-carboxy-*cis*-2′,5′-dihydroxypent-3′-enyl-chrysanthemate ester
- Phase 2 metabolites:
  - Glucuronides of the metabolic oxidation products of the intact ester
- MOA: Type I

5.2.2 *S*-Bioallethrin/Allethrin[25,26]

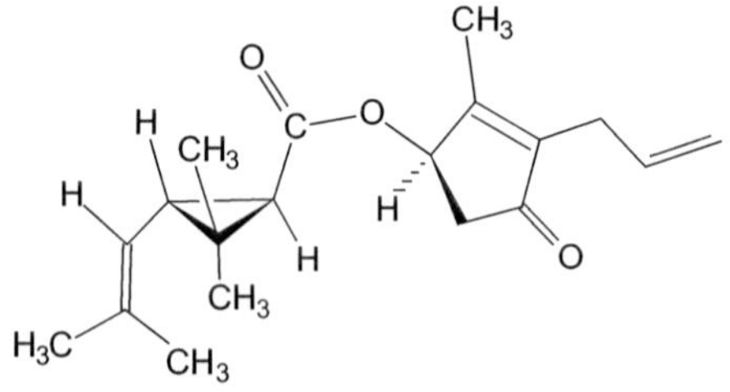

Structural formula of *S*-bioallethrin.

Name: Allethrin: *(RS)*-3-allyl-2-methyl-4-oxocyclopent-2-enyl *(*1*R*, 3*R*; 1*R*, 3*S)*-2,2-dimethyl-3-(2-methylprop-1-enyl)cyclopropane carboxylate, CAS # 584-79-2
S-Bioallethrin: *(S)*-3-allyl-2-methyl-4-oxocyclopent-2-enyl *(*1*R*, 3*R)*-2,2-dimethyl-3-(2-methylprop-1-enyl)cyclopropane carboxylate, CAS # 28434-00-6
Test system: Male Sprague Daley rat; *cis*/*trans* LD_{50}: 700 mg kg^{-1}
Male: 8–10 weeks old
Dose: 1–5 mg kg^{-1} of rat body weight
Route: Gavage in 0.1 mL DMSO
Test duration: 72 h

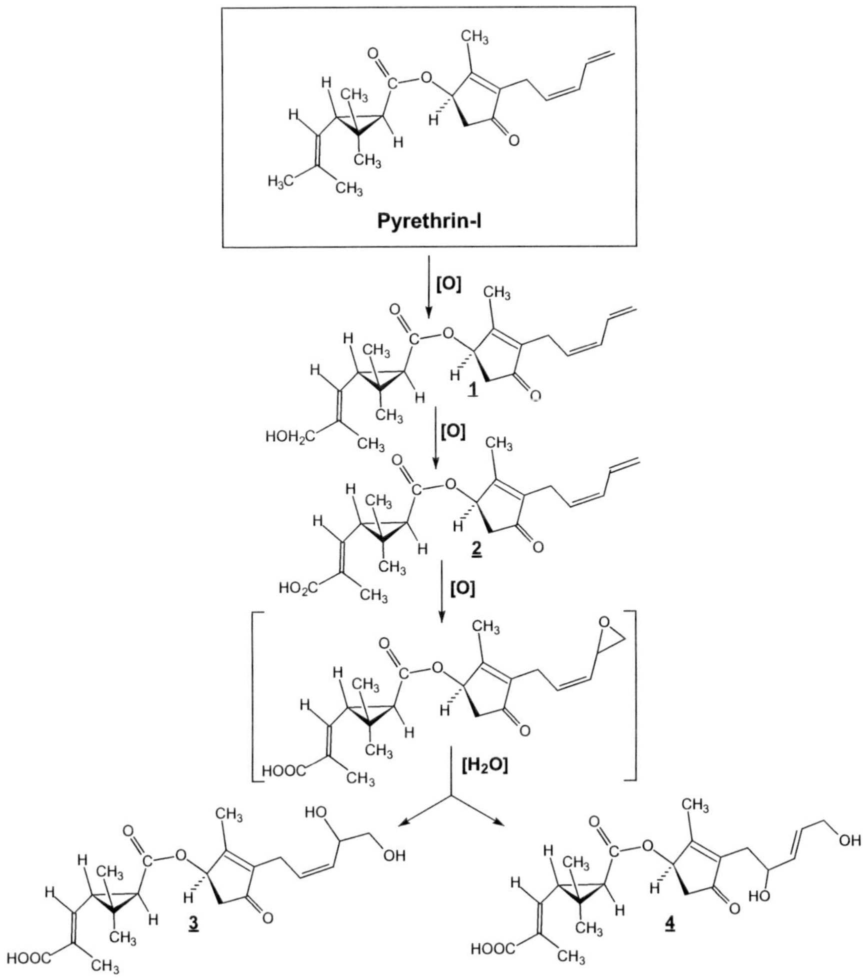

Figure 5.2 Proposed phase 1 metabolic pathways of pyrethrin I in the rat.

% of dose in urine: 47–51%
% of dose in faeces: 27–29%

5.2.2.1 *Biotransformation*

Similar to natural pyrethrins, the majority of metabolites are those of the intact ester. Hydrolysis metabolites are less prevalent. The steric hindrance of the cyclopentenyl ring protects the ester function, denying access to the carboxyesterases. This results in more oxidative metabolic products of the terminal vinylic ethylene and the *trans*-chrysanthemic acid allylic methyl group.

5.2.2.2 *Major Metabolites of Bioallethrin*

- Phase 1 metabolites (Figure 5.3):

    1. *trans*-Methyl-OH-bioallethrin
    2. *trans*-Bioallethrin-methyl-aldehyde
    3. Allethrolin carboxylic acid
    4. Allethrolone
    5. Chrysanthemic-dicarboxylic acid
    6. Allethrolone-2,5-diol-ester of chrysanthemic dicarboxylate ester
    7. *trans*-OH-Methyl-bioallethrin-dicarboxylic acid ester

- MOA: Type I

5.2.3 Resmethrin[27–30]

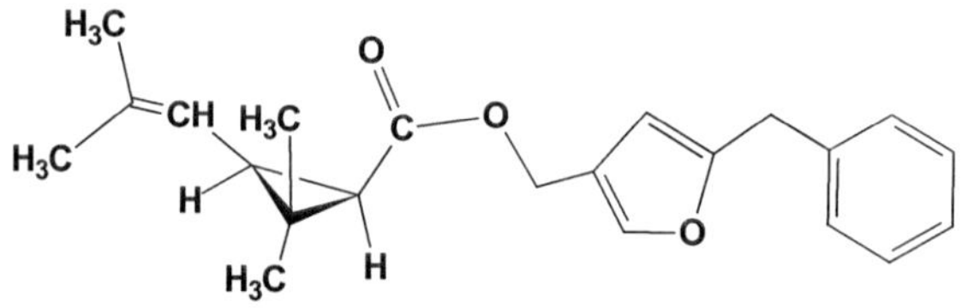

Structural formula of resmethrin.

Name: Resmethrin: 5-benzyl-3-furylmethyl (1*RS*,3*RS*;1*RS*,3*SR*)-2,2-dimethyl-3-(2-methylprop-1-enyl)cyclopropanecarboxylate. CAS # 10453-86-8
Test system: Rat, SD (Sprague-Dawley); LD_{50}: >2500 mg kg^{-1}
Male: 160–180 g (~7 weeks old)
Dose: (0.79–1.32) ~1 mg kg^{-1}
Route: Gavage, in DMSO, single doses, also 500 mg kg^{-1} was used in Sorpol emulsion
Test duration: 20 days
% of dose in urine: 41% ♂, 36% ♀ (Figure 5.4)
% of dose in faeces: 33% ♂, 64% ♀ (Figure 5.4)

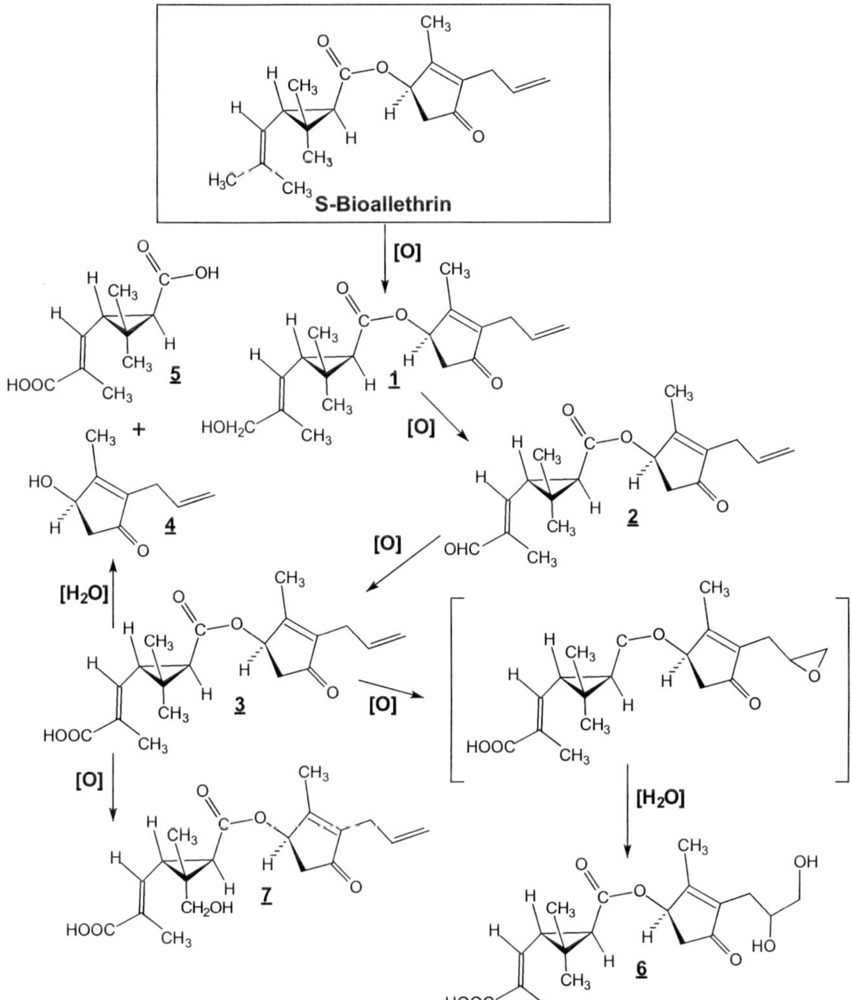

Figure 5.3 Proposed phase 1 metabolic pathways of *S*-bioallethrin in the rat.

Significant residue in tissues: Liver ♂ 0.140 ppm, fat ♂ 1.5 ppm, kidney ♂ 0.210 ppm

Brain tissue conc: ♂ 0.010 ppm (1 mg kg^{-1} dose), 11 ppm (500 mg kg^{-1} dose)

5.2.3.1 Biotransformation

Unlike natural pyrethrins, resmethrin is an ester of *cis*- and *trans*-chrysanthemic acid and a primary alcohol. The alcohol moiety is aromatic and lacks chirality. As a result the ester function for the *trans*-isomer is sterically

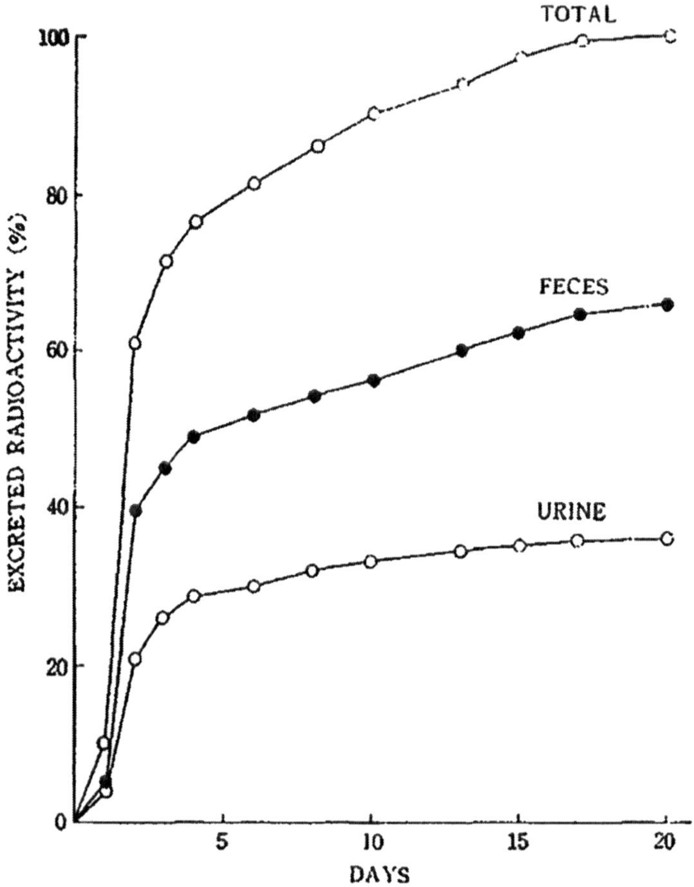

Figure 5.4 Cumulative excretion of radioactivity after oral administration of 500 mg kg^{-1} of carbon-14 resmethrin to rats.

unhindered compared to the *cis* isomer, unlike pyrethrin I and allethrin. The *trans*-isomer metabolism proceeds almost exclusively *via* carboxyesterases hydrolysis rather than oxidative metabolism, while the *cis* isomer proceeds *via* both carboxyesterases and oxidative metabolism. *trans*-Resmethrin is readily hydrolysable about 48-fold more rapidly than *cis*-resmethrin. As a result, the slower hydrolysis rate of *cis*-resmethrin allows enough time for oxidative metabolism to take place at the vinylic methyl of the intact *cis*-resmethrin ester. This is a minor biodegradation route. The slower rate of degradation of 1-*R*-*cis*-chrysanthemic acid ester, resulting from added steric hindrance and proximity of the dimethylvinyl group to the ester function, leads to higher toxicity of 1-*R*-*cis*-chrysanthemic acid ester, relative to the natural 1-*R*-*trans*-chrysanthemic acid where only less than 4% of the dose is metabolized *via* oxidation, and up to 29% of *cis*-resmethrin was metabolized *via* oxidase activity.

5.2.3.2 Major Metabolites of Resmethrin

- Phase 1 metabolites (Figure 5.5):
  1. a, *trans-* and *cis*-Chrysanthemic acid; b, Benzyl-furylmethanol
  2. Chrysanthemic-dicarboxylic acid-resmethrin
  3. Chrysanthemic-dicarboxylic acid
  4. Benzylfuroic acid
  5. 4′-OH-Benzylfuroic acid
  6. α-(4-Carboxy-2-furyl)-benzyl alcohol
  7. Benzoylfuroic acid
  8. α-OH-Benzylfuroic acid
- Phase 2 metabolites:
  - Glucuronide and sulfate conjugates of phase 1 metabolites
- MOA: Type I

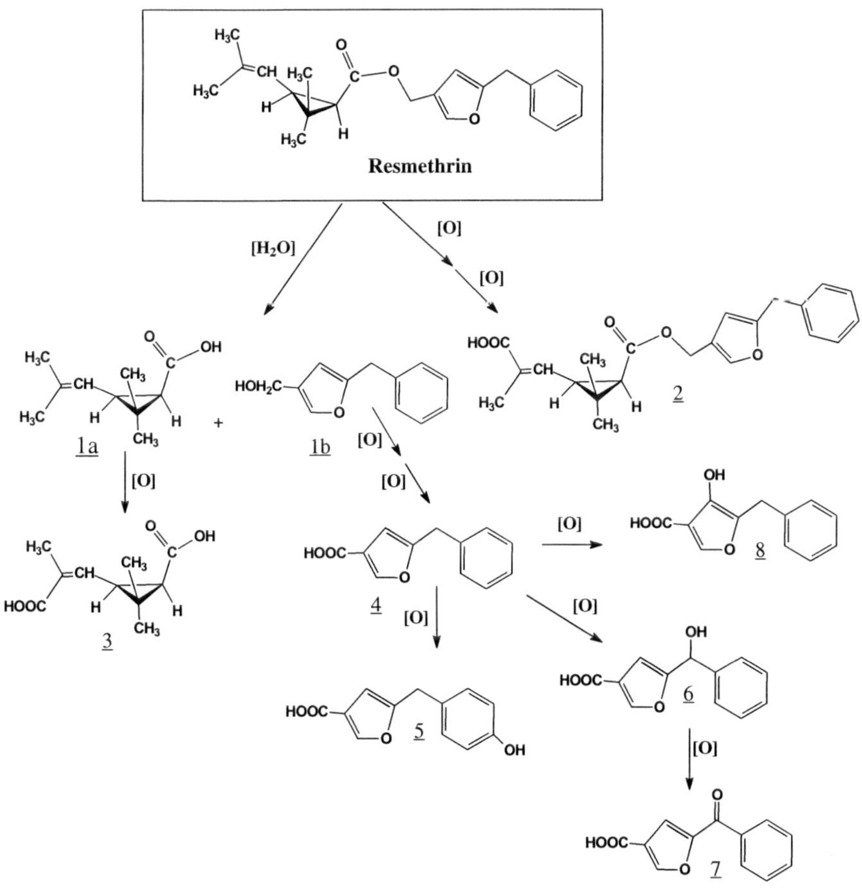

Figure 5.5 Proposed phase 1 metabolic pathways of resmethrin in the rat.

5.2.4 Permethrin[31–33]

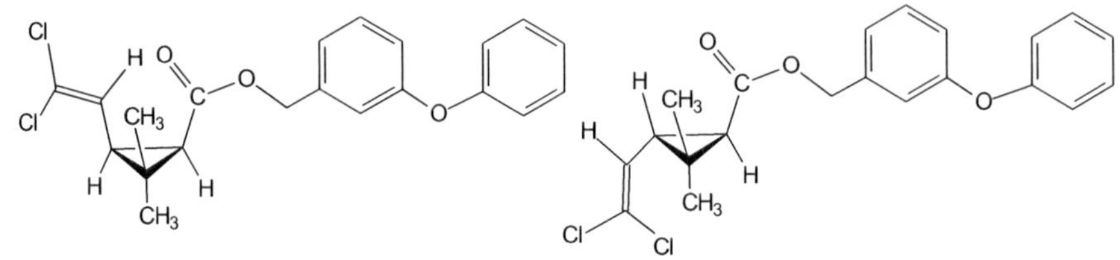

Structural formulas of *cis*-permethrin (left) and *trans*-permethrin (right).

Name: Permethrin: 3-phenoxybenzyl (1*RS*,3*RS*;1*RS*,3*SR*)-3-(2,2-dichlorovinyl)-2,2-dimethylcyclopropanecarboxylate. CAS # 52645-53-1.
Test system: Rat, Sprague-Dawley rats; LD_{50}: 1200 mg kg^{-1}; *trans* isomer LD_{50}: >2000 mg kg^{-1}
Male: 160–200 g each, ~8–9 weeks old
Dose: 1.6–4.8 mg kg^{-1}
Route: Gavage in 0.5 mL DMSO
Test duration: 4–12 days
% of dose in urine: 1*R*, *cis* 37–39% ♂; *trans* 70–71% ♂
% of dose in faeces: 1*R*, *cis* 31% ♂; 39–48% ♀; *trans* 8–13% ♂
% of dose in 1-carbon metabolism pool (CO_2): 0.5%
% unmetabolized: 1.3–7.3% of dose in faeces
Significant residue in tissues: Liver <25–55 ppb; fat <25–618 ppb
Brain tissue conc: <25 ppb ♂ after 12 days
Blood kinetics:[32] ♂ SD rat, *RS cis/trans* 460 mg kg^{-1}:
C_{max}, 49.46 μg mL^{-1} (plasma), $t_{1/2\alpha}$ 4.85 h; $t_{1/2\beta}$ 412.37 h; t_{max}, 3.52 h; $AUC_{0-\infty}$: 965 h mg L^{-1};
Bioavailability: RS 60.69% (*cis/trans*) for ♂

5.2.4.1 *Biotransformation*

Unlike natural pyrethrins, permethrin is an ester of *cis*-*trans*-3-(2,2-dichlorovinyl)-2,2-dimethylcyclopropane carboxylic acid with a primary alcohol, *m*-phenoxybenzyl alcohol. The alcohol moiety is aromatic in nature and lacks chirality. Similar to resmethrin, where the ester function is relatively sterically unhindered, permethrin metabolism proceeds almost exclusively *via* carboxyesterases, rather than oxidative metabolism routes, and eliminated as a hydrolytic product of the parent ester. Since permethrin is a mixture of *trans* and *cis* isomers, *trans*-permethrin is hydrolysable more rapidly than *cis*-permethrin. As a result, up to 71% of the *trans* isomer is eliminated in urine as hydrolytic products. The slower hydrolytic metabolism rate of the *cis*-1*R* isomer has clearly shown longer residence in the system than *trans*-1*R* isomer, therefore exhibiting higher toxicity.[33]

5.2.4.2 *Major Metabolites of Permethrin*

- Phase 1 metabolites (*c*-,*t*-Cl_2CA) = (*cis*-,*trans*-3-(2,2-dichlorovinyl)-2,2-dimethylcyclopropane carboxylate) (Figure 5.6):

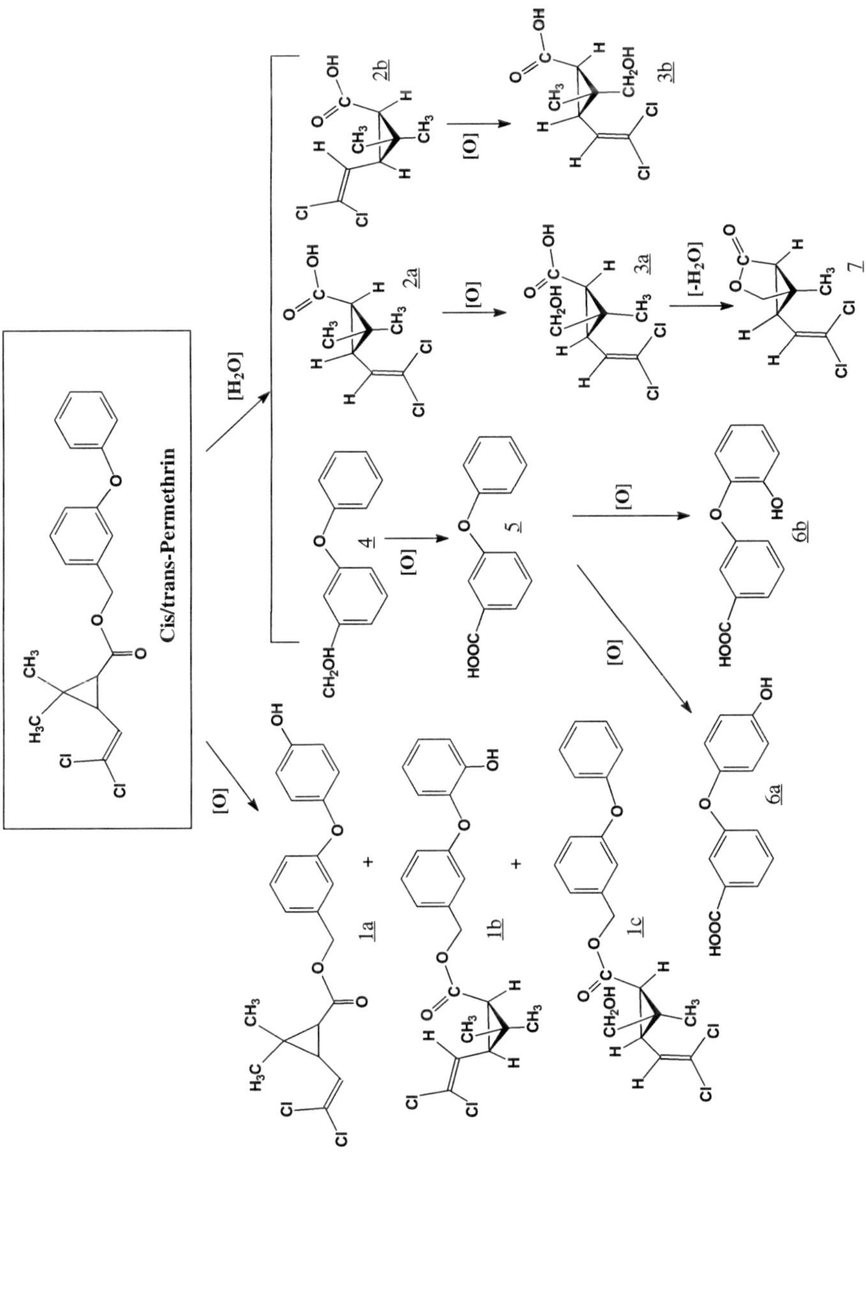

Figure 5.6 Proposed phase 1 metabolic pathways of permethrin in the rat.

1. OH-Permethrin
2. *c*-,*t*-Cl_2CA
3. *c*-,*t*-OH-Methyl-Cl_2CA
4. *m*-Phenoxybenzyl alcohol
5. *m*-Phenoxybenzoic acid
6. 4′-OH-*m*-Phenoxybenzoic acid
7. *c*-OH-methyl-Cl_2CA lactone

- Phase 2 metabolites:
  - Glucuronide, glycine and sulfate conjugates of hydrolysis and oxidation products
- MOA: Type I

5.2.5 Cypermethrins[34–37]

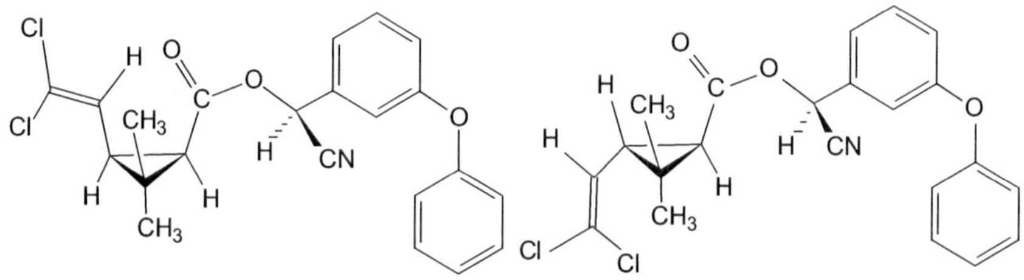

Structural formulas of *cis*-cypermethrin (left) and *trans*-cypermethrin (right).

Name: Cypermethrin: (*RS*)-α-cyano-3-phenoxybenzyl (1*RS*, 3*RS*, 1*RS*, 3*SR*)-3-(2, 2-dichlorovinyl)-2,2-dimethylcyclopropane carboxylate. CAS # 52315-07-8 (*cis*-, *trans*-3-(2,2-dichlorovinyl)-2,2-methylcyclopropane carboxylate = *c*-,*t*-Cl_2CA)
Test system: Rat, Wistar; *cis* LD_{50}: 160–300 mg kg^{-1}; *trans* LD_{50}: > 2000 mg kg^{-1}
Male: 12 weeks old
Female: 12 weeks old
Dose: 1.7–2.5 mg kg^{-1}. Rat weights ♂ 320 g, ♀ 210 g
Route: Gavage in 0.5 mL corn oil
Test duration: 8 days
% of dose in urine: *cis* 61% ♂, 50% ♀; *trans* 71% ♂, 74% ♀ (Figure 5.7)
% of dose in faeces: *cis* 31% ♂, 40% ♀; *trans* 28% ♂, 23% ♀ (Figure 5.7)
% unmetabolized: 2% of dose
% biliary excretion: ~1–1.6%
% of dose in tissues: 2.5–3.3% of applied dose
Significant residue tissues: Liver 0.048% of administered dose, fat 0.930 ppm
Brain tissue conc.: 0.002 ppm ♂, 0.004 ppm ♀ after 8 days
Blood kinetics: α-Cypermethrin, *cis* 10 mg kg^{-1} ♂/rapeseed oil:
C_{max}, 123.9 ± 8.92 ng mL^{-1} (plasma), $t_{1/2}$ (selected tissues only, fat 12d); t_{max}, 3 h (300 g ♂ Wistar rat); $AUC_{0-\infty}$: 671.9 ± 42.1 h ng mL^{-1}
Bioavailability: 63% (*cis*) to 73% (*trans*) for ♂; 52% (*cis*) to 76% (*trans*) for ♀ (sum of urine and bile)

5.2.5.1 Biotransformation

Although cypermethrin is an analogue of permethrin where a cyano group was introduced at the permethrin benzylic primary alcohol carbon bridge, this modification resulted in two critical alterations. First, the alteration converted the primary alcohol moiety to a secondary alcohol, thereby decreasing carboxyesterases access to the ester function. Second, the alteration produced a chiral centre at the benzylic α-carbon. Similar to natural pyrethrolone alcohol, the *S*-α-carbon-product is the highly bioactive enantiomer, while the *R*-α-carbon-product is relatively less active. The *trans*-1*R*-α-*S* isomer exhibits insecticidal activity. However, the *cis*-1*R*-α-*S* isomer contributes up to 85% of the total insecticidal activity of the total eight cypermethrin isomers, the remaining seven isomers contributing only about 15% of the total activity. Biotransformation studies have clearly shown that *cis*-1*R*- and *cis*-1*S*-isomers are metabolized and eliminated at a slower rate than the *trans*-1*R* and *trans*-1*S* isomers. It was noticed that metabolism of *cis* isomers proceeds mostly *via* oxidative metabolism routes (Figure 5.8), and elimination occurs as oxidation products of the parent ester.

5.2.5.2 Major Metabolites of Cypermethrin

- Phase 1 metabolites (Figure 5.8):

  1. 4′-OH-Cypermethrin
  2. *c*-,*t*-Cl_2CA
  3. *c*-,*t*-OH-Methyl-Cl_2CA
  4. *cis*-OH-Cl_2CA-lactone
  5. *m*-Phenoxybenzoic acid
  6. (a) 4′-OH-*m*-Phenoxybenzoic acid, (b) 2′-OH-*m*-phenoxybenzoic acid

- Phase 2 metabolites:

  1. *c*-,*t*-Cl_2CA-glucuronide
  2. *c*-,*t*-OH-Methyl-Cl_2CA-glucuronide
  3. *m*-Phenoxybenzoic acid-glucuronide
  4. 4′-OH-*m*-Phenoxybenzoic acid-glucuronide
  5. 4′-OH-*m*-Phenoxybenzoic acid-sulfate

- MOA: Type II

5.2.6 Bifenthrin[38–42]

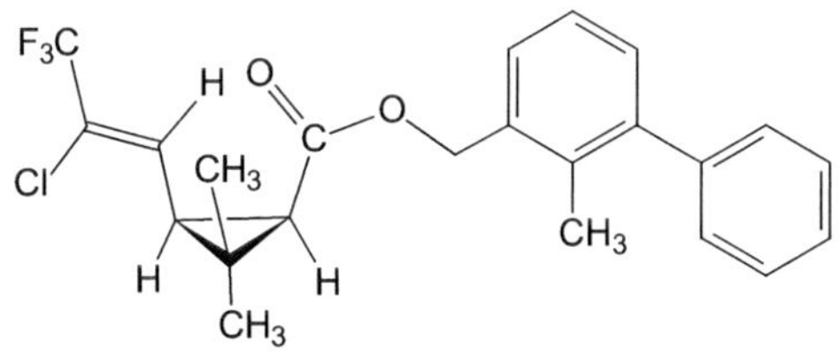

Structural formula of *cis*-(*Z*)-bifenthrin.

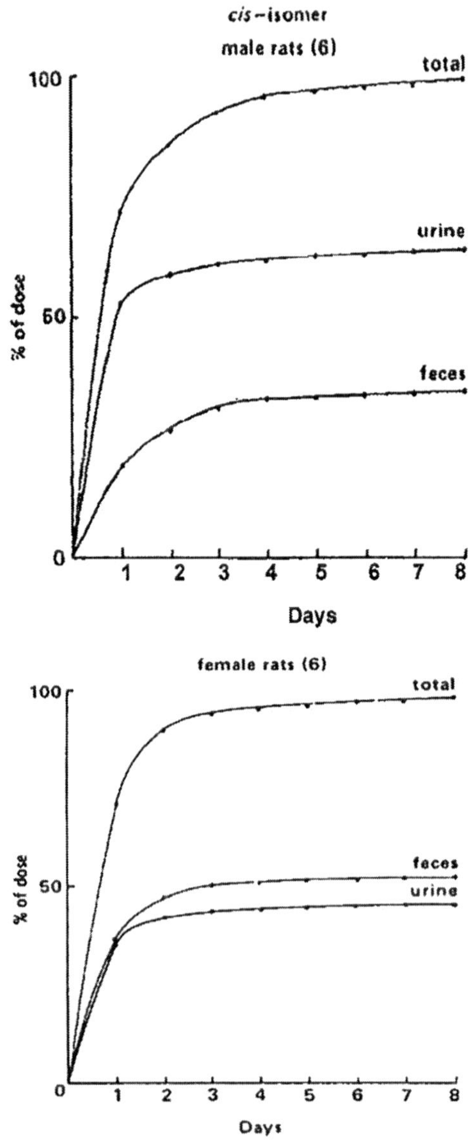

Figure 5.7 Excretion of radioactivity in male and female rats following oral administration of *cis*-cypermethrin.

Name: Bifenthrin: (2-Methylbiphenyl-3-yl)methyl(Z)(1*RS*,3*RS*)-3-*cis*-(2-chloro-3,3,3-trifluoroprop-1-enyl)- 2,2-dimethylcyclopropane-1-carboxylate. CAS 82657-04-3

Test system: Rat, SD; LD_{50}: 53.4–55.5 mg kg^{-1}

Male: 7–8 weeks old

Female: 8 weeks old

Dose: 6.4 mg kg^{-1} ♂, 4.1 mg kg^{-1} ♀

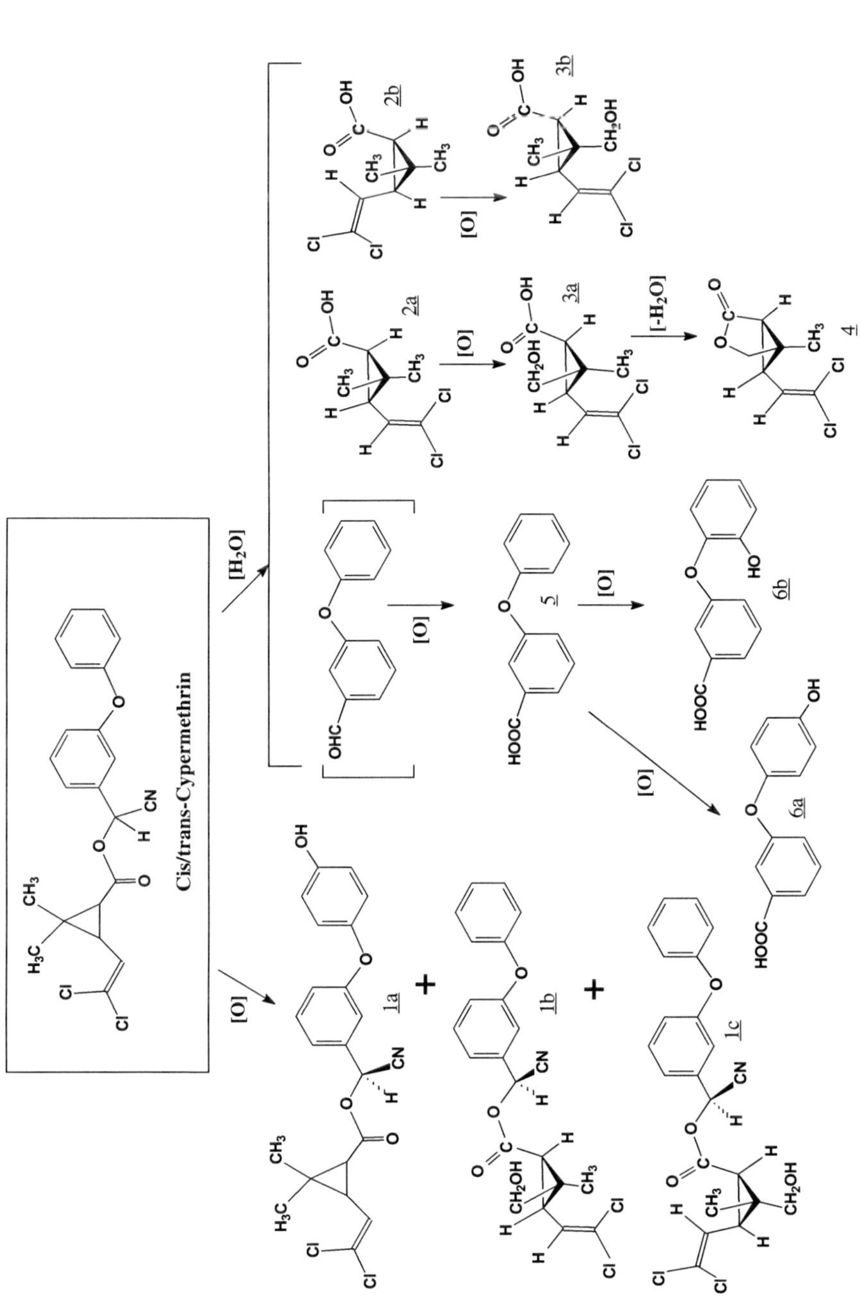

Figure 5.8 Proposed phase 1 metabolic pathways of cypermethrin in the rat.

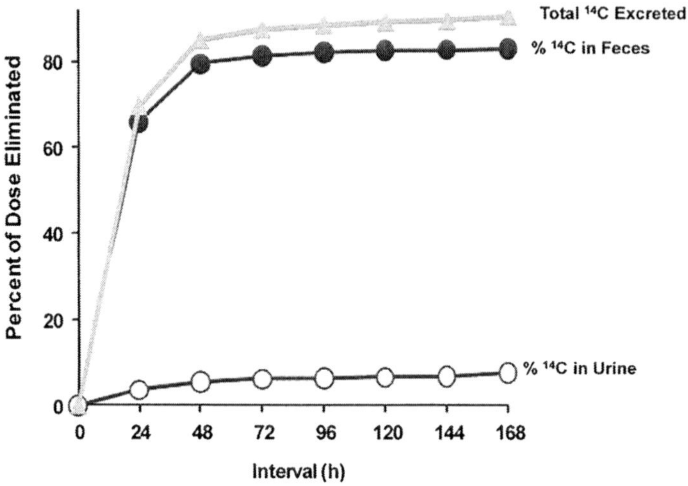

Figure 5.9 Excretion of radioactivity in rats following oral administration of carbon-14 bifenthrin.

Route: Gavage, in corn oil, single doses
Test duration: 7 days
% of dose in urine: 13.4% ♂; 19.7% ♀ (Figure 5.9)
% of dose in faeces: 82.8% ♂; 72.9% ♀ (Figure 5.9)
% of dose in tissue: 3.8% and 3.9% ♂ and ♀ respectively
% unmetabolized: 44.2% and 26.4% ♂ and ♀ respectively
% biliary excretion: ♀@ 2.5 mg kg^{-1}, 31.9%; ♂@ 5 mg kg^{-1} 18.2%
Significant residue in tissues: Liver ♂ 0.120 ppm, ♀ 0.144 ppm; fat ♂ 0.640 ppm, ♀ 2.053 ppm
Brain tissue conc.: ♂ 0.007 ppm, ♀ 0.010 ppm;
Blood kinetics: @ 4 mg kg^{-1} of BW for ♂ and ♀:
C_{max}, @ 4–6 h, plasma ^{14}C-Conc.1.885 μg L^{-1} (Figure 5.10); 29.4% of ^{14}C is BP alcohol, 16.8% of ^{14}C is BP acid, and 39.5% bifenthrin; t_{max}, 4–6 h for ^{14}C
Bioavailability: 34.4–35.6% of orally applied dose for ♂ @ 2.5 mg kg^{-1}, and 49.8–55.5% of orally applied dose for ♀

5.2.6.1 *Biotransformation*

There are two structural modifications in bifenthrin that are different from earlier pyrethroids. First, the substituted chrysanthemic acid is exclusively the *cis* isomer where one of the dimethylvinyl groups is replaced with chloro-*(Z)* and the second with a trifluoromethyl group. This acid is esterified with a 2-methyl,3-phenylphenyl methanol. Both the *Z*-acid system and the biphenyl alcohol system present a semi-rigid molecule with enough steric hindrance that denies access to carboxyesterases, but provides ample targets to oxidase enzymes. As a result more than 80% of bifenthrin metabolism proceeds *via* oxidative biodegradation rather than hydrolysis. This allows for a longer-acting

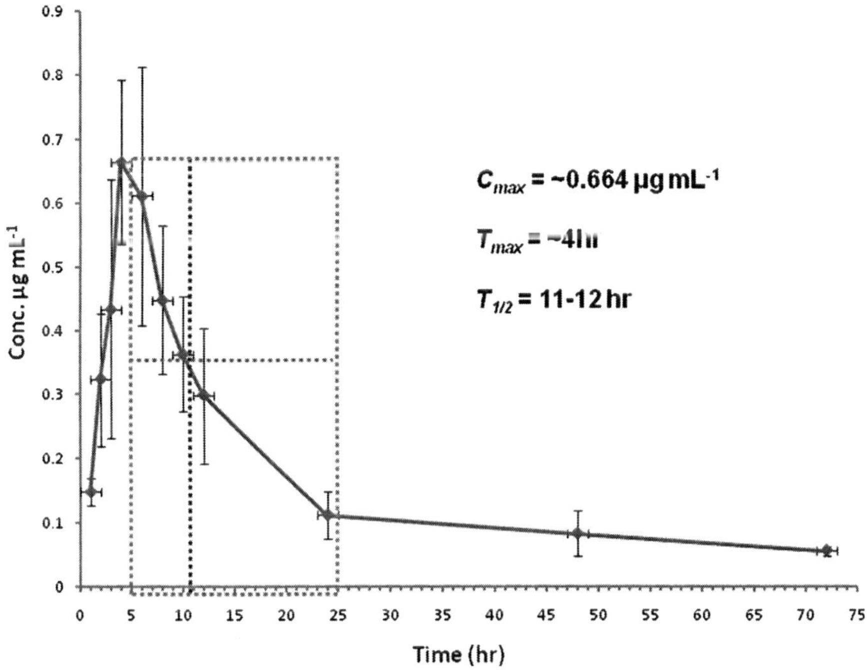

Figure 5.10 Concentrations of radioactivity in blood of rats following oral administration of carbon-14 bifenthrin.

pyrethroid than earlier analogues. Nevertheless, around 60% of the bifenthrin dose is bioavailable and undergoes exclusively oxidative biotransformation and elimination.

5.2.6.2 Major Metabolites of Bifenthrin

- Phase 1 metabolites (Figure 5.11):
  1. 4′-OH-Bifenthrin
  2. OH-Methylbifenthrin
  3. 3′- and 4′-Monomethylcatechol-bifenthrin
  4. (a) 4′-OH-Hydroxymethyl-bifenthrin, (b) 3′-OH-hydroxymethyl-bifenthrin
  5. TFP acid
  6. BP alcohol
  7. OH-Methyl-TFP acid
  8. 4′-OH-BP acid
  9. Monomethyl-catechol-BP acid
- Phase 2 metabolites:
  - Glucuronide and sulfate conjugates of phase 1 metabolites
- MOA: not Type I, perhaps Type II

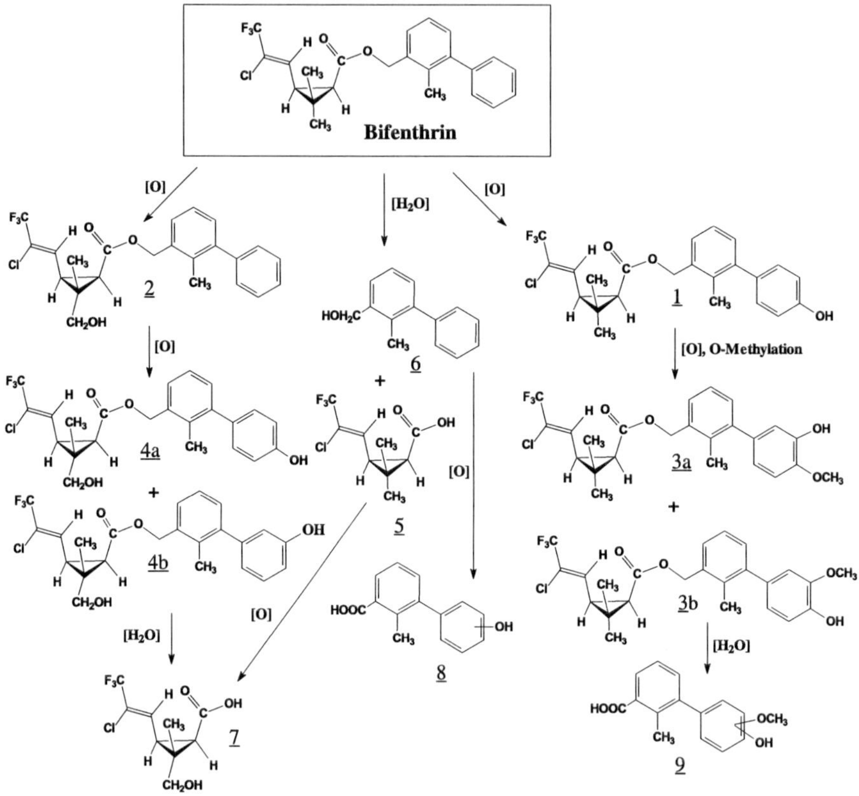

Figure 5.11 Proposed phase 1 metabolic pathways of bifenthrin in the rat.

5.2.7 Tefluthrin[43–48]

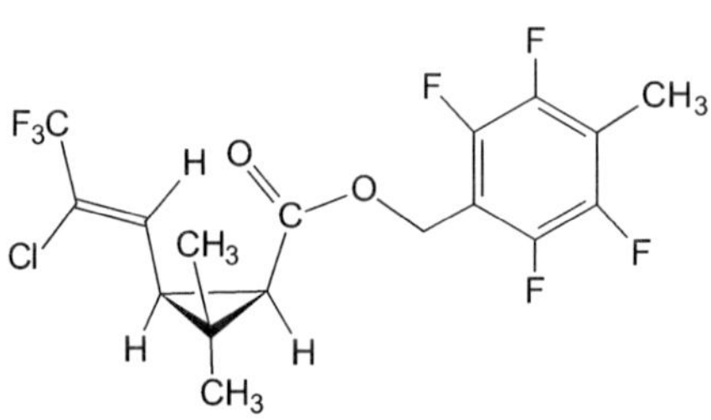

Structural formula of *cis*-(*Z*)-tefluthrin.

Name: Tefluthrin: 2,3,5,6-Tetrafluoro-4-methylbenzyl (*Z*)-(1*RS*, 3*RS*)-3-(2-chloro-3,3,3-trifluoroprop-1-enyl]- 2,2-dimethylcyclopropanecarboxylate. CAS # 79538-32-2

Test system: Rat, SD; LD_{50}: ♀ 21.8 mg kg^{-1}, ♂ 34.6 mg kg^{-1}

Dose: 1 mg kg^{-1} and 10 mg kg^{-1}

Route: Gavage, in corn oil, single doses

Test duration: 7 days

% of dose in urine: 25–30%
% of dose in faeces: 63–69%
% of dose in tissues: 2.6–10.6% of applied dose

5.2.7.1 Biotransformation

Tefluthrin is an ester containing a tetrafluoro-substituted benzyl alcohol and a trifluoro-substituted cyclopropyl acid. Metabolism of tefluthrin proceeds *via* both oxidation and hydrolysis. Fatty acid esters of hydroxytefluthrin have been identified in the fat of rats treated with tefluthrin.

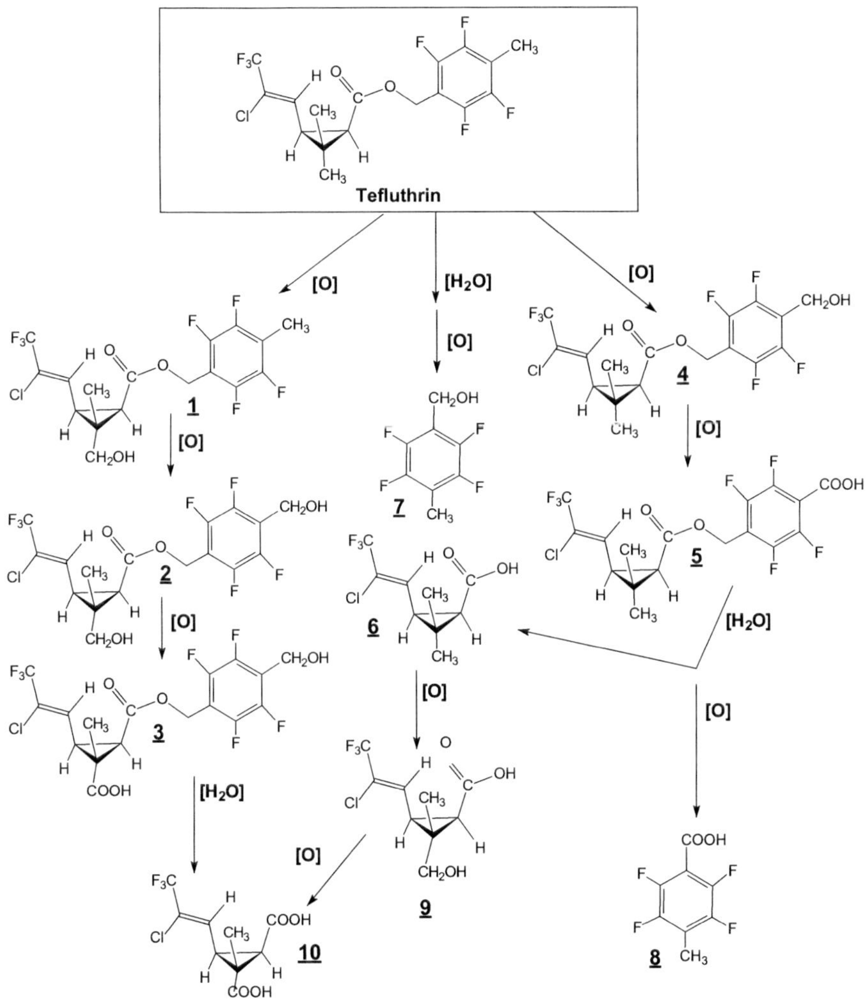

Figure 5.12 Proposed phase 1 metabolic pathways of tefluthrin in rats.

5.2.7.2 Major Metabolites of Tefluthrin

- Phase 1 metabolites (Figure 5.12):
    1. OH-Methyltefluthrin
    2. OH-Methyl-TF-benzyl alcohol
    3. Tefluthrin-benzoylcarboxylic acid
    4. TF-Benzyl alcohol
    5. TF-Benzoic acid
    6. TFP acid
    7. TFT-Methyl alcohol
    8. TFT-Carbxoylic acid
    9. OH-Methyl-TFDC acid
    10. TFD di-acid
- Phase 2 metabolites:
    - Glucuronides of the OH-methyl- and methyl-TF-benzyl alcohol and Phase 1 metabolites
    - Glucuronides of *cis*- and *trans*-dicarboxylic acids, in addition to palmetic and oleic acid esters of the hydroxymethyl metabolites
    - No *RS* selective conjugation reactions have been described
- MOA: Type I

5.2.8 Fenvalerate/Esfenvalerate[49–52]

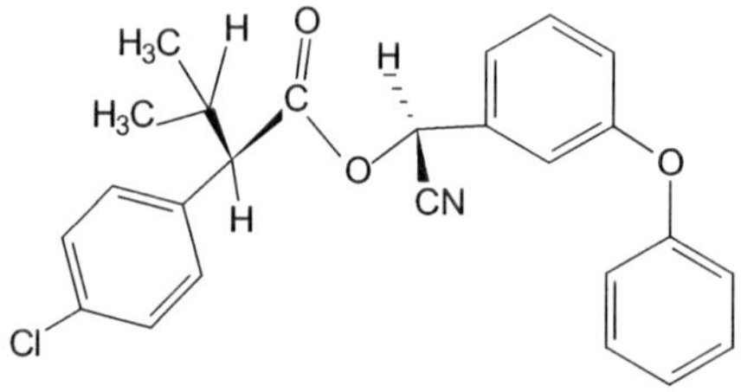

Structural formula of esfenvalerate.

Name: Fenvalerate: (*RS*)-α-Cyano-3-phenoxybenzyl (*RS*)-2-(4-chlorophenyl)-3-methylbutyrate. CAS # 51630-58-1
Esfenvalerate: (*S*)-α-Cyano-3-phenoxybenzyl (*S*)-2-(4-chlorophenyl)-3-methylbutyrate. CAS # 66230-04-4
Test system: Rat, SD, LD_{50}: 451 mg kg^{-1} fenvalerate; 458 mg kg^{-1} esfenvalerate
Dose: Fenvalerate: 10 mg kg^{-1}; esfenvalerate: 1.7 or 2.5 mg kg^{-1}
Route: Gavage, in corn oil or 10% Tween
Test duration: 6, 14 days
% of dose in urine: 29–35% ♂, 33–39 ♀ fenvalerate (24–27 ♂, 32–33 ♀% esfenvalerate)
% of dose in faeces: 61–72% ♂, 59–67 ♀ fenvalerate (71% ♂, 66/67% ♀ esfenvalerate) (59–79% of faecal ^{14}C is parent fenvalerate)

% of applied dose in tissues: fenvalerate 4–7%, esfenvalerate 1–2% of dose
Significant residue in tissues (after 8.5 mg kg^{-1} dose fenvalerate): Liver 0.01 ppm, fat 2 ppm
Brain tissue conc.: 0.01 ppm after 1.7 mg/kg of esfenvalerate
Blood kinetics: C_{max}, plasma = 0.5 μg mL^{-1} (after 1.7 mg/kg dose of esfenvalerate); t_{max}, 3 h

5.2.8.1 *Biotransformation*

Fenvalerate is a mixture of four optical isomers because of the presence of two asymmetric carbons. Esfenvalerate is one of the these isomers, derived from α-cyano-3-phenoxybenzyl alcohol and (*S*)-2-(4-chlorophenyl)isovaleric acid. Metabolic reactions of fenvalerate and esfenvalerate are similar and include hydroxylation at the 4′ position of the alcohol moiety and 2 and 3 positions of the acid moiety, hydrolysis of the ester function, release of the CN group as SCN ion and conjugation of the hydrolysis products. One of the fenvalerate isomers also forms a more lipophilic cholesterol ester. This metabolite tends to accumulate in certain tissues, leading to granulomatous changes.

5.2.8.2 *Major Metabolites of Fenvalerate/Esfenvalerate*

- Phase 1 metabolites (Figure 5.13):
  1. 4′-Hydroxyesfenvalerate
  2. PB acid
  3. 4′-OH-PB acid
  4. 2-(4-chlorophenyl) isovaleric acid (CPIA)
  5. 2-(3-hydroxy-4-chlorophenyl)isovaleric acid
  6. 2-(3-hydroxy-4-chlorophenyl)isovaleric acid lactone
- Phase 2 metabolites:
  - Glucuronide and sulfate conjugates of hydrolysis and oxidation products (and cholesteryl [2*R*]-2-(4-chlorophenyl)isovalerate—fenvalerate only).
- MOA: Type II

5.2.9 Cyfluthrin[53]

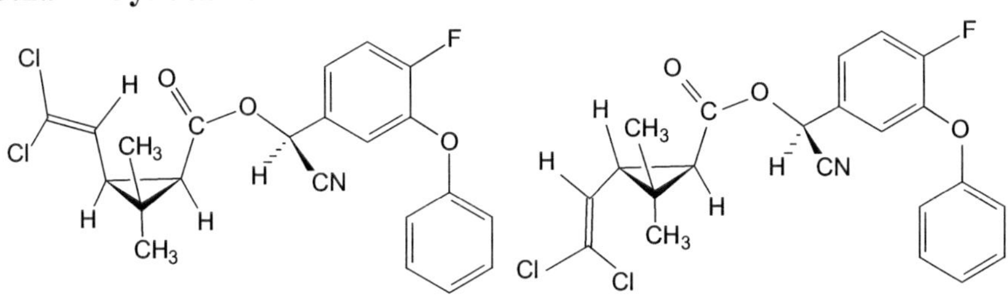

Structural formulas of *cis*-cyfluthrin (left) and *trans*-cyfluthrin (right).

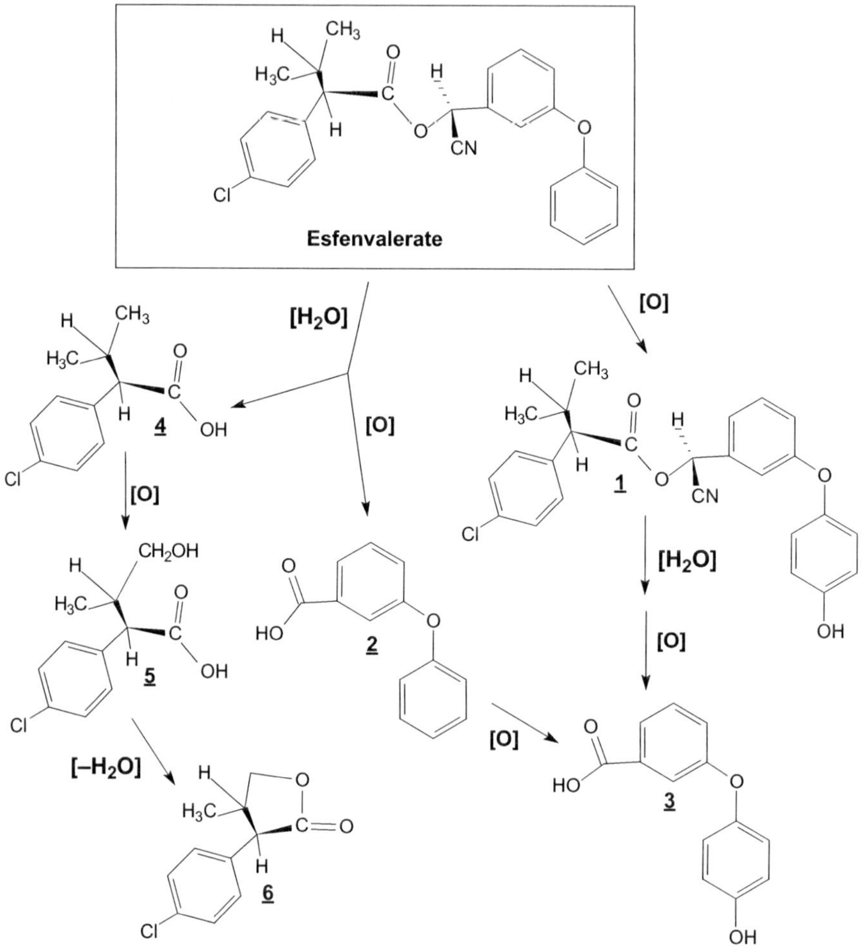

Figure 5.13 Proposed phase 1 metabolism of esfenvalerate in rats.

Name: Cyfluthrin: (*RS*)-α-cyano-4-fluoro-3-phenoxybenzyl-(1*RS*, 3*RS*; 1*RS*, 3*SR*)-3-(2,2-dichlorovinyl)-2,2-dimethylcyclopropanecarboxylate. CAS # 68359-37-5
Test system: Rat, Wistar, LD_{50}: 900 mg kg^{-1}
Dose: 0.5 mg kg^{-1} and 10 mg kg^{-1}
Route: Gavage, in cremophor in water or PEG 400
Test duration: 14 days
% of dose in urine: 55–70%
% of dose in faeces: 25–35%
% biliary excretion: 34% of dose
% Dose in tissues (after 0.5 mg kg^{-1} dose): <2
Blood kinetics: t_{max}, 2 h; $t_{1/2}$ 12 h male, 9 h female
Bioavailability: ~80% absorption of orally applied dose

5.2.9.1 Biotransformation

Cyfluthrin consists of four diastereomeric pairs forming eight isomers. The major metabolic routes of cyfluthrin are hydrolysis of the ester linkage and hydroxylation at the 4′ position of the alcohol moiety. The acid moiety in cyfluthrin is the same as that in permethrin and cypermethrin, therefore the acid moiety is expected to undergo the same metabolic reactions. The hydrolysis products are eliminated after further oxidization and/or conjugation with glucuronic acid, sulfuric acid and glycine.

5.2.9.2 Major Metabolites of Cyfluthrin

- Phase 1 metabolites (Figure 5.14):
  1. 4′-OH-Cyfluthrin
  2. 4-Fluoro-3-phenoxybenzoic acid (FPBA)
  3. 4′-OH-4-Fluoro-3-phenoxybenzoic acid (FPBA)
  4. 3-(2,2-Dichloroethenyl)-2,2-dimethylcyclopropanecarboxylic acid (DCCA)
- Phase 2 metabolites:
  - Glucuronide, glycine and sulfate conjugates of hydrolysis and oxidation products
- MOA: Type II

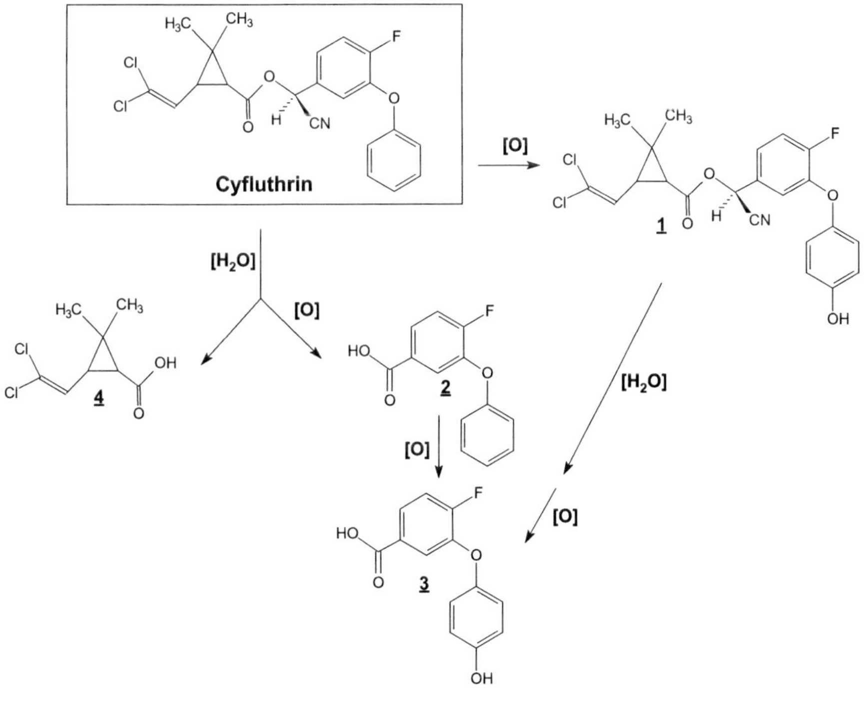

Figure 5.14 Proposed major phase 1 metabolic pathways of cyfluthrin in rats.

5.2.10 Deltamethrin[54–57]

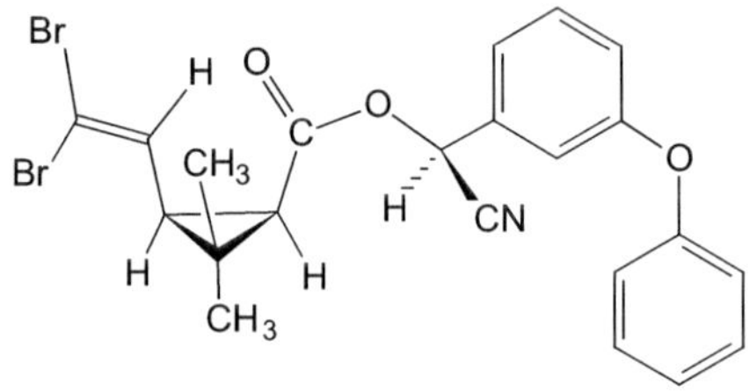

Structural formula of deltamethrin.

Name: Deltamethrin; *S*-cyano-3-phenoxybenzyl-*cis*-(1*R*,3*R*)-3-(2,2-dibromovinyl)-2,2-dimethylcyclopropanecarboxylate. CAS # 52918-63-5
Test system: Rat, SD, LD_{50}: 30 –> 5000 mg kg^{-1}
Dose: 0.64 mg kg^{-1} and 1.60 or 2.0 mg kg^{-1}
Route: Gavage, in PEG
% of dose in urine: 31–56%
% of dose in faeces: 36–59% (17–46% of faecal ^{14}C is parent)
% of dose unmetabolized: 13–20%
% biliary excretion: 27% of dose
% of applied dose in tissues: <2%
Significant residue in tissues (after 2.0 mg kg^{-1} dose): Skin 0.34 ppm, fat 0.80 ppm
Brain tissue conc.: 0.04 ppm
Blood kinetics:
C_{max}, plasma = 0.53 μg mL^{-1} (after 2 mg kg^{-1} dose)
t_{max}, 1.0 h; $t_{1/2}$ 13.3 h; $AUC_{0-\infty}$: 2.7 μg h mL^{-1}
Bioavailability: 18% (58.4% absorption) of orally applied dose.

5.2.10.1 Biotransformation

Deltamethrin is one of eight possible stereoisomers of α-cyano-phenoxybenzyl ester pyrethroid. Hydroxylation at the 2′, 4′ and 5 positions of the alcohol moiety and at the *trans* methyl group relative to the carbonyl function of the acid moiety, hydrolysis of the ester linkage, and oxidation and conjugation of the hydrolysis products constitute major metabolic pathways for deltamethrin.

5.2.10.2 Major Metabolites of Deltamethrin

- Phase 1 metabolites (Figure 5.15):

  1. 4′-OH-Deltamethrin
  2. 3-Phenoxybenzoic acid (PB acid)
  3. 4′-OH-PB acid
  4. (2,2-Dibromovinyl)-2,2-dimethylpropanecaboxylic acid (Br_2CA)
  5. *trans*-OH-Br_2CA

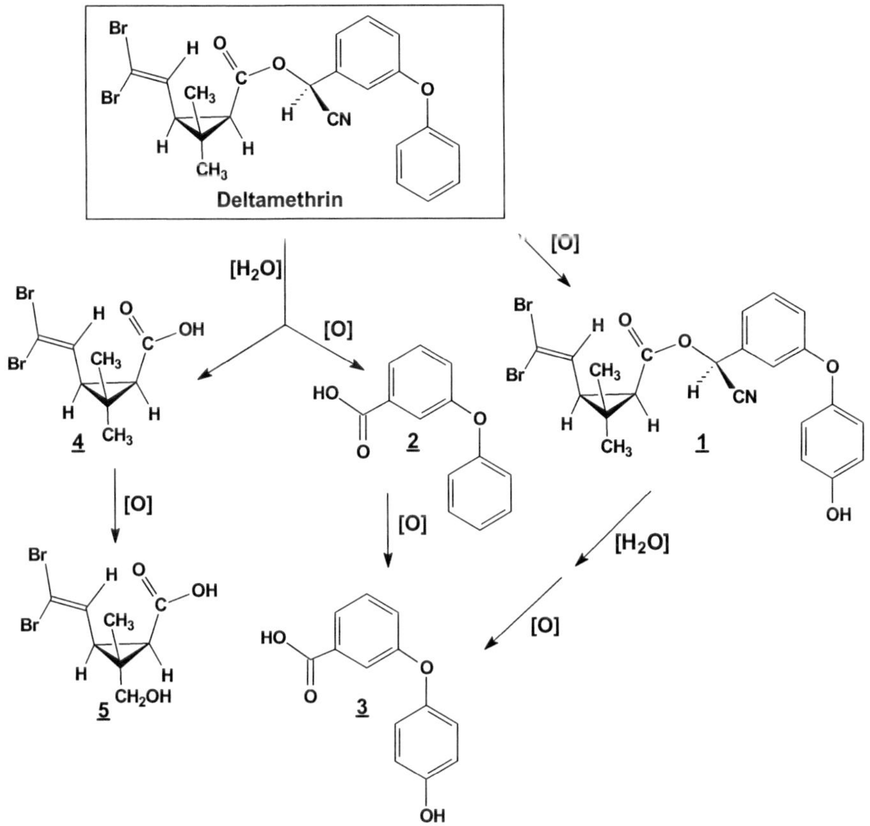

Figure 5.15 Proposed phase1 metabolic pathways of deltamethrin in rats.

- Phase 2 metabolites:
  - Glucuronide, glycine and sulfate conjugates of hydrolysis and oxidation products
- MOA: Type II

5.2.11 Fenpropathrin[58–59]

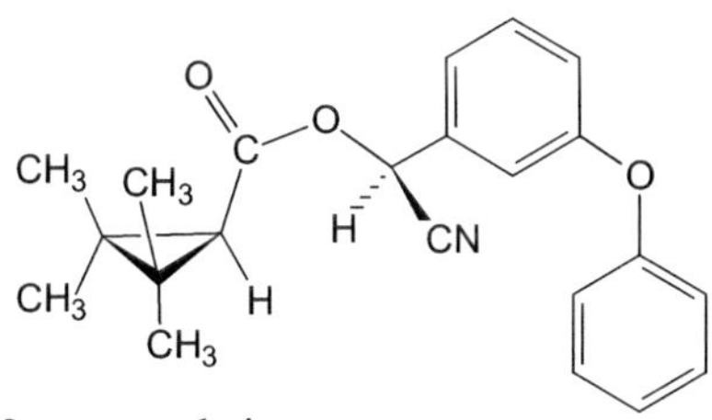

Structural formula of fenpropathrin.

Name: Fenpropathrin: (*RS*)-α-Cyano-3-phenoxybenzyl 2,2,3,3-tetramethylcyclopropanecarboxylate. CAS # 64257-84-7

Test system: Rat, SD, LD_{50}: 70.6 ♂, 66.7 ♀ mg kg^{-1}
Dose: 1.5 mg kg^{-1}, 2.4 mg kg^{-1} and 26.8 mg kg^{-1}
Route: Gavage, in corn oil
Test duration: 8 days
% of dose in urine: 33.7–43.8 ♂, 42.7–43.8 ♀%
% of dose in faeces: 61.4–62.7 ♂, 54.1–58.1 ♀% (13–34% of faecal ^{14}C is parent)
% of dose unmetabolized: 13–34%
% of dose in tissues (after 1.5 mg kg^{-1} dose): <1.5% of applied dose
Significant residue in tissues (after 2.4 mg kg^{-1} dose): Liver 0.02 ppm, fat 0.1
Brain tissue conc.: 0.02 ppm
Blood kinetics:
C_{max}, plasma = 14 ng mL^{-1} (after 2.4 mg kg^{-1} dose); t_{max}: 6 h
Bioavailability: ~57% absorption of orally applied dose.

5.2.11.1 Biotransformation

Fenpropathrin, an α-cyano-3-phenoxybenzyl ester, is a racemic mixture of two isomers (*R* and *S*) due to the asymmetric nature of the benzyl carbon. The major metabolic reactions of this pyrethroid are cleavage of the ester linkage, hydroxylation at 4′-position of the alcohol moiety, and oxidation at the methyl groups of the acid moiety and conjugation of the resultant carboxylic acids with sulfate, glucuronic acid and glycine. *In vitro* and *in vivo* studies with cytochrome P450 inhibitors suggest that the ester cleavage of fenpropathrin is a result of oxidative rather than hydrolytic reactions.

5.2.11.2 Major Metabolites of Fenpropathrin

- Phase 1 metabolites (Figure 5.16):

  1. 4′-OH-Fenpropathrin
  2. Hydroxymethylfenpropathrin
  3. Dihydroxyfenpropathrin
  4. Tetramethylcyclopropanecarboxyl acid (TMPA)
  5. TMPA-CH_2OH (*cis*)
  6. TMPA-CH_2OH (*trans*)
  7. TMPA-CH_2OH-lactone
  8. TMPA-COOH (*trans*)
  9. Carboxymethylfenpropathrin
  10. PB acid
  11. 4′-OH-PB acid

- Phase 2 metabolites:

  - Glucuronide, sulfate and glycine conjugates of hydrolysis and oxidation products

- MOA: Type I/II hybrid

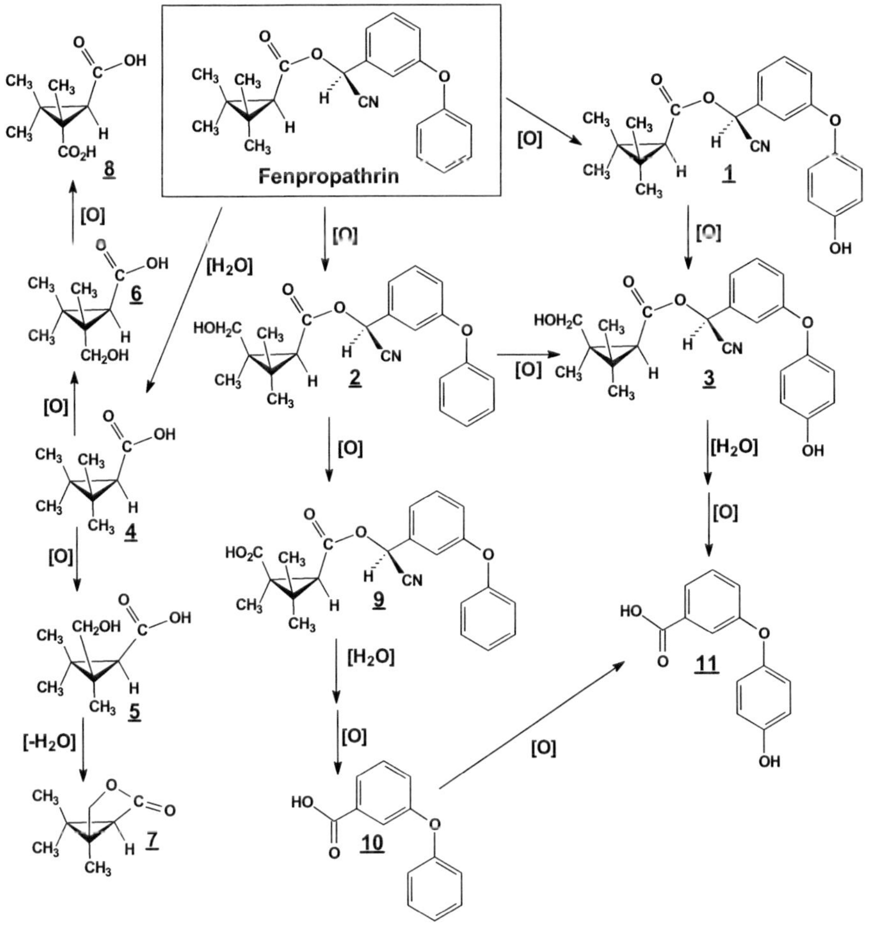

Figure 5.16 Proposed phase 1 metabolic pathways of fenpropathrin in rats.

5.2.12 Cyhalothrin[60]

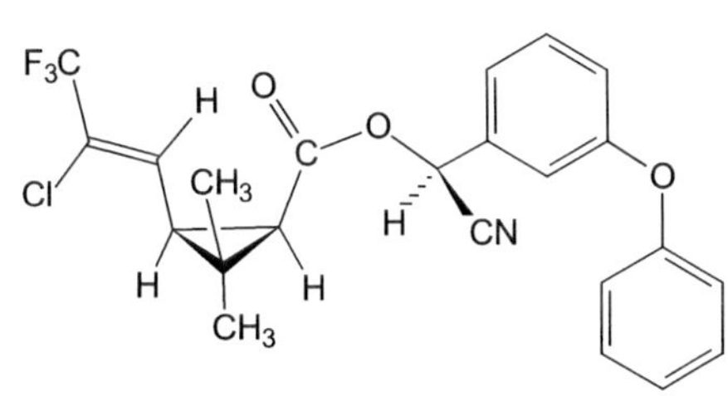

Structural formula of cyhalothrin.

Name: Cyhalothrin: (*RS*)-α-Cyano-3-phenoxybenzyl (*Z*) (1*RS*, 3*RS*)-3-(2-chloro-3,3,3-trifluoroprop-1-enyl)-2,2-dimethylcyclopropanecarboxylate. CAS # 68085-85-8

Test system: Rat, SD, LD_{50}: 166 ♂, 114 ♀ mg kg^{-1}

Dose: 1, 3, 20 and 25 mg kg^{-1}
Route: Gavage, in corn oil
Test duration: 7 days
% of dose in urine: 20–40%
% of dose in faeces: 40–65% (80% of faecal ^{14}C is parent)
% of dose unmetabolized: 30%
% biliary excretion: 4.2–12.9% of dose
% of applied dose in tissues: 2–3%
Significant residue in tissues (after 1 mg kg^{-1} dose): Liver 2.5 ppm, fat 10 ppm
Brain tissue conc.: 24–25 ppm after 20 mg kg^{-1}
Blood Kinetics:
C_{max}, plasma = 15.65 μg mL^{-1} (after 1 mg kg^{-1} dose); t_{max}, 2.69 h; $t_{1/2}$ 7.55 h ♂
Bioavailability: ~67% of orally applied dose

5.2.12.1 Biotransformation

Cyhalothrin is a mixture of four isomeric pairs and λ-cyhalothrin is composed of two of the more active isomers of these esters. Both undergo similar metabolism. Major metabolic reactions of cyhalothrin are ester hydrolysis and hydroxylation at the 4′ position of the alcohol moiety. The hydrolytic products are excreted as conjugates following further oxidation.

5.2.12.2 Major Metabolites of Cyhalothrin

- Phase 1 metabolites (Figure 5.17):
    1. 4′-OH-Cyhalothrin
    2. Cyclopropane carboxylic acid
    3. 3-Phenoxybenzoic acid
    4. 4′-OH-Phenoxybenzoicacid
- Phase 2 metabolites:
    - Glucuronide and sulfate conjugates of hydrolysis and oxidation products
- MOA: Type II

5.2.13 Ethofenprox[61]

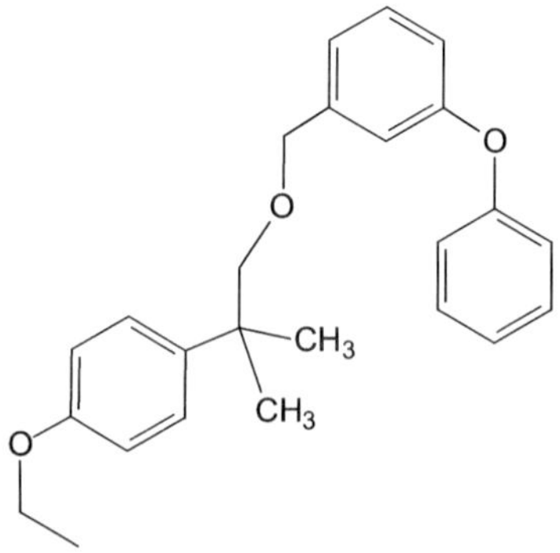

Structural formula of ethofenprox.

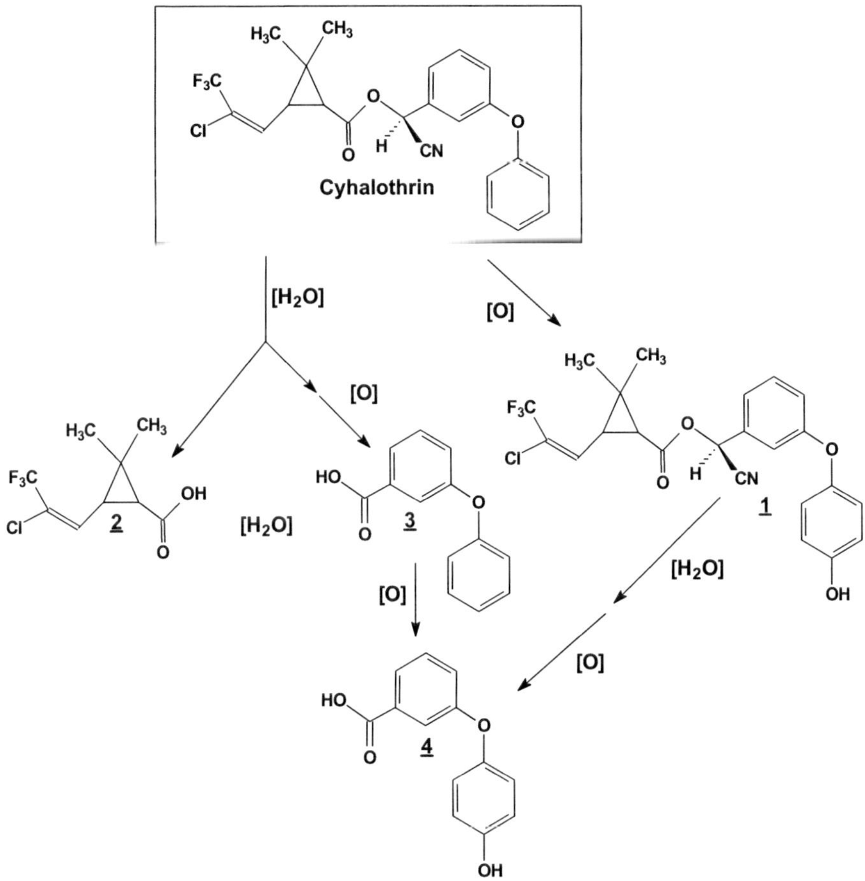

Figure 5.17 Proposed phase 1 metabolic pathways of cyhalothrin in rats.

Name: Ethofenprox/etofenprox: 2-(4-ethoxyphenyl)-2-methylpropyl 3-phenoxybenzyl ether. CAS # 80844-07-1
Test system: Rat, SD, LD_{50}: >42880 mg kg^{-1} (♂ and ♀)
Dose: 30 mg kg^{-1} and 180 mg kg^{-1}
Route: Gavage, in PEG
Test duration: 5 days, 7 days
% of dose in urine: 10.8% ♂, 8.0% ♀
% of dose in faeces: 88% ♂, 86.4% ♀ (6.6% ♂, 14.0% ♀ of faecal ^{14}C is parent)
% of dose unmetabolized: 5–13%
% of dose biliary excreted: 15–30%
% of applied dose in tissues: 3.4% ♂, 3.55% ♀
Significant residue in tissues (after 30 mg kg^{-1} dose): Liver 0.34 ppm, fat 16.6 ppm
Brain tissue conc.: 0.002 ppm ♂–0.004 ppm ♀

Blood kinetics: C_{max}, plasma = 5 μg mL^{-1} (after 30 mg kg^{-1} dose), 16 μg mL^{-1} after 180 mg kg^{-1} dose; t_{max}, 2–7 h
Bioavailability: ~14–51% absorption of orally applied dose

5.2.13.1 *Biotransformation*

Ethofenprox is a structurally unusual pyrethroid in that it contains an ether linkage instead of the traditional ester group. The major biotransformation routes of ethofenprox include *O*-deethylation of the ethoxyphenyl moiety, hydroxylation of the phenoxymethyl moiety and ether bond cleavage by α-hydroxylation. The cleavage products undergo further oxidation followed by conjugation to form sulfates and glucuronides.

5.2.13.2 *Major Metabolites of Ethofenprox*

- Phase 1 metabolites (Figure 5.18):
  1. Desethylethofenprox
  2. 4′-Hydroxyethofenprox
  3. α-CO (2-(4-ethoxyphenyl)-2-methyl propyl-3-phenoxybenzoate)
  4. Ethoxyphenyl-2-methylpropanol
  5. Phenyl-2-methylpropanol
  6. 3-Phenoxybenzoic acid (mPBA)
  7. 4′-OH-3-Phenoxybenzoic acid
- Phase 2 metabolites:
  - Glucuronide and sulfate conjugates of oxidation products
- MOA: Type I

5.3 Mode of Action

The absorption of pyrethroids in mammals through the skin is very low, but following oral administration using corn oil as vehicle it appears to be consistently 40–60%. Oxidases and esterases, primarily in the liver, metabolize pyrethroids at varying rates. Because metabolites are less toxic than the parents, the most rapidly cleared pyrethroids have the lowest toxicity, *i.e. trans* acids without halogen substituents, such as pyrethrin I. Peak blood plasma concentrations tend to correlate with clinical signs of neurotoxicity after acute oral administration. Such signs dissipate within hours after a single gavage dosing, correlating with the reduction in blood plasma pyrethroid concentration. Thus, the principal effects can be considered acute rather than chronic in nature. The nervous system is exposed to pyrethroids through the blood and a variety of target sites in nerve and muscle has been described. In general, voltage-gated ion channels are the predominant site(s) but transmitter-gated ion channels may also play a role in some clinical signs. The target sites are reviewed below.

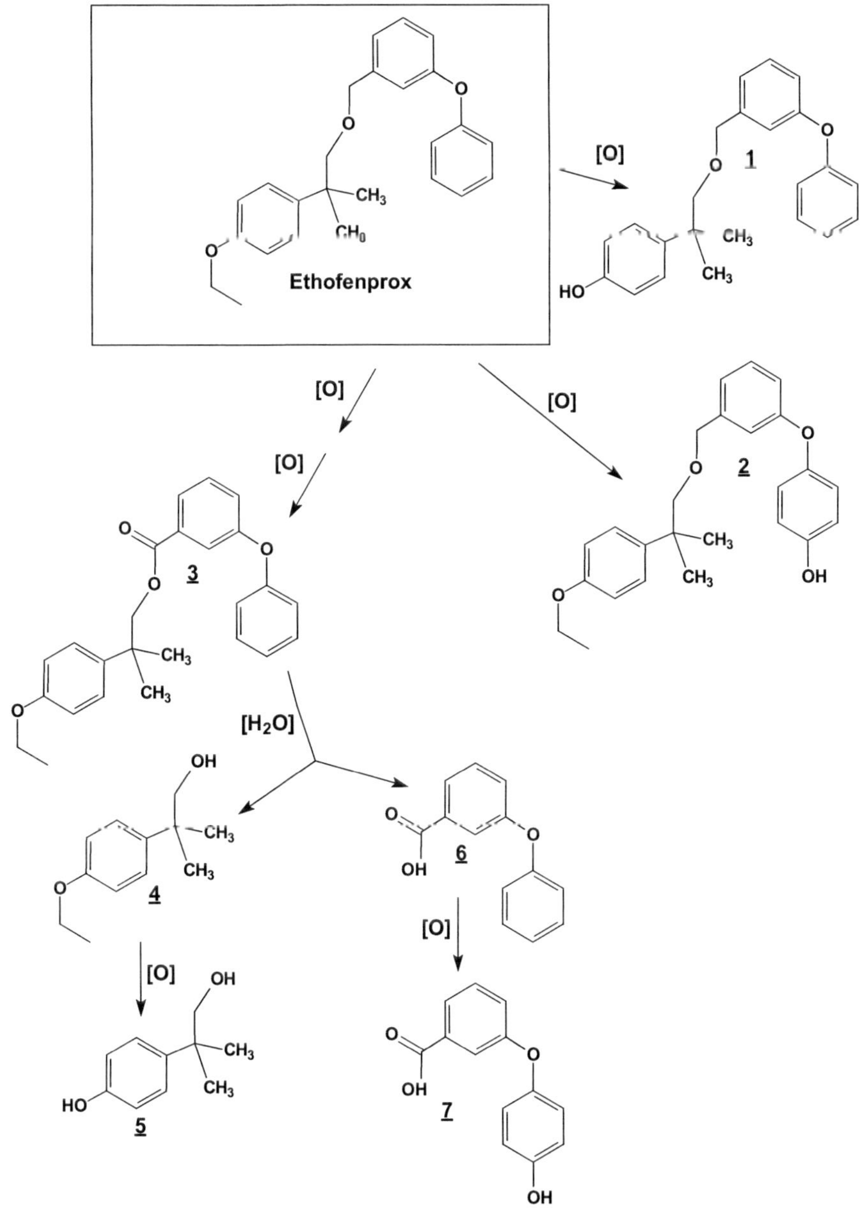

Figure 5.18 Proposed phase 1 metabolic pathways of ethofenprox in rats.

5.3.1 Voltage-Gated Sodium Channel (VGSC)

Unlike insects, which possess only a single gene coding for a VGSC protein, mammals have nine α subunit cDNA sequences (Nav1.1 to Nav1.9) and four β cDNA sequences.[62] The α channel proteins are capable of serving as ion channels alone or in the presence of one or more β channels. Whereas the α proteins appear to contain the ion pore and associated filters, the β proteins are associated with channel gating.

Breckenridge *et al.*[63] described the effects of 11 pyrethroids on rat Nav1.8 channels, expressed in *Xenopus* oocytes. This VGSC isoform is restricted to the peripheral nervous system and, unlike the insect VGSC,[64] is insensitive to blockage by TTX. However, unlike other VGSC isoforms tested in preliminary studies, Nav1.8 appeared to be sensitive to a broad range of pyrethroid structures. Under voltage-clamp conditions, pyrethroids caused a slowing of inactivation during depolarization (effect on peak current) as well as a slowing of deactivation following repolarization (prolongation and enlargement of Na^+ tail current at end of a depolarizing pulse). Type II pyrethroids also resulted in prolonged depolarization resulting in nerve conduction blockage. Factor and multivariate dissimilarity analysis were used to evaluate 56 functional observational battery (FOB) parameters[65] against 8 electrophysiological parameters associated with modification of Nav1.8. The neurotoxic responses of Type I (non-cyano) and Type II (cyano) pyrethroids separated into two groups. Two compounds, fenpropathrin and bifenthrin, fell between the two groups. This is consistent with the former having mixed Type I/II effects in mammals[10,11] and insects.[12] Bifenthrin failed to cause Type I discharges in the cockroach cercal sensory nerve assay, *in vivo*.[13,66] It is possible that the methyl group in the ortho position of the biphenyl alcohol moiety of bifenthrin is able to mimic the effect of the cyano group in the α position of pyrethroids, causing Type II effects with respect to VGSC effect(s). This possibility is supported by the finding that a pyrethroid with an α-ethynyl group in place of the α-cyano group of fenvalerate (Type II) exhibited Type I and II effects in the rat (tremors with salivation).[10]

The effects of selected pyrethroids have also been studied on other rat VGSC forms after expression in *Xenopus* oocytes.[18] For example Nav1.2, Nav1.3 and Nav1.6 have been found to have varying degrees of sensitivity to tefluthrin.[18] Co-expression of some αVGSCs with β1 and β2 VGSCs has been found to increase the pyrethroid potency, an indication that the gating mechanism may be especially susceptible to pyrethroid modification. Depending on the pyrethroid and the channel isoform, the pyrethroid effect may be increased with repeated depolarizations, indicating a pyrethroid preference for the open channel state.[67–68] In the latter study, cypermethrin (Type II) prolongation of the tail current was use-dependent whereas the (much reduced) effect of cismethrin (Type I) was not use-dependent. This study was conducted using the housefly Vssc1 VGSC expressed in *Xenopus* oocytes. In studies on Nav1.2 from rat brain expressed in *Xenopus* oocytes, the effects of cypermethrin were increased by around 20-fold by co-expression of rat brain β-1 subunit.[69] Another heterologous expression system for VGSCs, the CHO (Chinese hamster ovary) cell, has also been used to study pyrethroid effects. Esfenvalerate was found to cause a large increase in the (Na^+) tail current, typical of Type II pyrethroids, using human VGSC subtypes (ChanTest, Cleveland, OH, in preparation). In this study, 50 μM esfenvalerate was shown to maintain channels in an open state by removing fast inactivation during depolarization and slow channel closing (producing a long-lasting tail current) following repolarization in the following human VGSC α isoforms: Nav1.2, 1.6 and 1.8 (coupled to β3). These are found in the (rat) central nervous system (CNS) and peripheral nervous system (PNS). Esfenvalerate also caused similar effects on Nav1.4 (skeletal muscle) and Nav1.5 (cardiac muscle). However, it was ineffective or much less effective on Nav1.1, 1.3 and 1.7. These isoforms are found in the CNS/PNS and it is noteworthy that Nav1.3 is only found in the CNS of fetal or very young rats; it is not expressed in adults. It is therefore possible that infants may be less sensitive than adults to esfenvalerate and other Type II pyrethroids. On the other hand, the Type I pyrethroid tefluthrin has been found to affect the Nav1.3 channel after expression in *Xenopus* oocytes, with the rat channel more sensitive than the human one.[70] Other pyrethroids are to be tested shortly using a range of VGSCs incorporated into CHO cells. A review of the function and distribution of VGSC isoforms is provided in Ref. 16.

VGSCs are also present in cardiac muscle in mammals. The effects of pyrethroids have also been described on Na^+ currents, action potentials and contractions in isolated rat and guinea-pig ventricular myocytes or rat perfused hearts.[71] Tefluthrin (Type I), α-cypermethrin (Type II) and fenpropathrin (Type I/II), but not tetramethrin (Type I), prolonged ventricular action potentials and evoked after-depolarizations. They also modified the time course of the Na^+ current, causing a large slowing down of inactivation, as well as altering its voltage dependence. On the whole-heart assay, these pyrethroids caused cardiac arrhythmogenic changes due to variability of intervals between heartbeats and in contraction amplitude.

5.3.2 Voltage-Gated Calcium Channel (VGCC)

The effects of pyrethroids on the N-type (Cav2.2) VGCC in rat brain synaptosomes have been studied extensively.[72,73] Three fluorescence assays enabled calcium influx, membrane depolarization and neurotransmitter (glutamate) release to be measured and the effects of pyrethroids to be recorded. The effects on these three parameters[63] were compared with the FOB effects reported using factor analysis,[65] similar to the process used for VGSC (Nav1.8) analysis. The Type I pyrethroids tefluthrin and bioallethrin formed a tight group or cluster with fenpropathrin (mixed Type I/II) whereas bifenthrin and cismethrin each occupied a separate space. The Type II pyrethroids formed a cluster, but including permethrin, a Type I structure. The effects of pyrethroids on VGCC isoforms from rat brain, expressed in *Xenopus* oocytes, have also been described. Type II pyrethroids stimulated calcium influx that was TTX-insensitive.[72,73] Another group has studied the effects of pyrethroids on L-type VGCCs in neocortical neurons from day 16 mouse embryos in primary culture.[74] Nine of eleven pyrethroids stimulated calcium influx, the exceptions being permethrin and resmethrin. *S*-bioallethrin, bifenthrin and cypermethrin had very little effect but deltamethrin, tefluthrin, λ-cyhalthrin, β-cyfluthrin, esfenvalerate and fenpropathrin all had relatively large effects. In this study the calcium influx was TTX-sensitive, in contrast to the findings for the N-type VGCC.

Using patch-clamped rat hippocampal neurons in culture, at 10 μM, permethrin (Type I) but not deltamethrin (Type II) increased spontaneous excitatory transmitter (glutamate) release.[75,76] The mEPSC frequency was increased without an effect on mEPSC amplitude, thus indicating a presynaptic effect. Because the experiments were conducted in the presence of bicuculline (to block $GABA_A$ receptors) and TTX (to block VGSCs) it was concluded that presynaptic VGCCs were involved in the glutamate release stimulated by permethrin.The permethrin effect was insensitive to ω-conotoxin, indicating that it was not an effect on N- or P/Q-type VGCCs.

In mouse GC-2spd (ts) cells, fenvalerate induced Ca^{2+} transients *via* intracellular and extracellular pathways.[77] The former were through inositol triphosphate and ryanodine receptors and the latter *via* a store-operated channel.

5.3.3 Voltage-Gated Chloride Channel (VGClC)

The patch-clamp technique was used to study pyrethroid effects on VGClCs in mouse neuroblastoma cells (NIE-115). These cells express a maxi-chloride channel with high conductance.[14,63,78] Open channel probability was measured for the 11 pyrethroids at 10^{-5} M. The five Type I pyrethroids did not affect open channel probability, with the exception of a slight effect for bioallethrin, whereas the six Type II pyrethroids reduced this parameter, with the exception of λ-cyhalothrin. In the latter case, however, the more toxicologically active γ-cyhalothrin was active in reducing open channel probability. The VGClC data were plotted against the FOB data, for the T and CS scores,[63] but less clear-cut clustering was obtained than for the Nav1.8 or Cav2.2 channels. However, bifenthrin was separate from the Type I or Type II pyrethroid clusters/groups.

5.3.4 Voltage-Gated Potassium Channel (VGKC)

The effects of the α- and θ-cypermethrin on delayed rectifier K^+ currents in rat hippocampal neurons, *in vitro*, were evaluated by Tian *et al.* using patch-clamp analysis.[79] This Type II pyrethroid reduced the steady-state (outward) current in these neurons, at 10^{-9} to 10^{-7} M, in a concentration-dependent manner. However, the same group[80] found that the transient (outward) K^+ current amplitude was increased by α-cypermethrin but reduced by θ-cypermethrin. Because both isomer mixtures are toxic, further work is needed to clarify pyrethroid effects on VGKC channel isoforms.

5.3.5 $GABA_A$-Gated Chloride Channel

The Type II pyrethroid deltamethrin was found to be a potent inhibitor of the binding of a tritiated picrotoxinin analogue to rat brain membranes.[81] This convulsant blocks the GABA-activated Cl^- channel. The benzodiazepine diazepam, which amplifies the $GABA_A$ response through acting at a different GABA receptor, was then found to delay the onset of clinical signs and nerve effects of Type II pyrethroids in the mouse and cockroach, while having no effects on Type I pyrethroids.[82] The crayfish claw opener muscle was then used to show that Type II pyrethroids, but not Type I pyrethroids or inactive isomers, could increase the input resistance of fibres exposed to GABA.[83] Further, benzodiazepines were able to reduce the pyrethroid effect on these GABA-activated Cl channels. Type II pyrethroids but not Type I pyrethroids or inactive isomers were also found to inhibit the specific binding of [^{35}S]-TBPS (t-butyl bicyclophosphorothionate) to mouse brain receptors.[84] This has a similar site of action to picrotoxinin. Similar receptor binding studies using rat brain membranes confirmed this effect of Type II pyrethroids.[85,86]

5.3.6 Peripheral Benzodiazepine (BZ) Receptor

Ro5-4864 or 4-chlorophenyl diazepam is a BZ with very low affinity for the $GABA_A$-linked CNS receptor in mammals. It binds specifically to sites outside the CNS (peripheral BZ receptor) and does not bind to the central BZ receptor. It is however, the most potent inhibitor of [^{3}H]-flunitrazepam, a ligand for the CNS BZ receptor in mammals, to the insect BZ receptor in housefly thorax.[87,88] It was found that Type II but not Type I pyrethroids inhibited the binding of [^{3}H]-Ro5–4864 to rat brain membranes.[66,89] Although the binding inhibition was stereospecific, it had relatively low potency. However, it remains possible that some of the clinical signs of Type II pyrethroids, in both insects and mammals, represent an interaction at this receptor.[13]

5.3.7 Nicotinic ACh-Related Receptors

Pyrethroids, with the exception of deltamethrin, did not affect the binding of [^{3}H]-acetylcholine (ACh) to the nicotinic ACh receptor in the electric organ of the electric ray, *Torpedo ocelleta*.[90] The receptors in this organ are similar to those at the mammalian skeletal neuromuscular junction. The binding of [^{3}H]-isodihydrohistrionicotoxin, a frog skin toxin that blocks the ion channel associated with the nicotinic ACh receptor, was increased by pyrethroids when tested in the unstimulated state, *i.e.* in the absence of the agonist carbamylcholine. In the presence of this agonist, 9 of 10 pyrethroids caused significant inhibition of [^{3}H]-HTX binding. The six Type II pyrethroids were generally less potent than the (four) Type I pyrethroids. Receptor-activated transport of $^{45}Ca^{2+}$ was inhibited by a subset of pyrethroids, with IC_{50} values of about 0.1 μM. An interesting finding concerned the effects of temperature: pyrethroids were more potent as inhibitors of [^{3}H]-HTX binding as temperature was reduced, in line with insecticidal activity. It thus appears that pyrethroids may have toxicologically significant effects at receptors associated with the nicotinic ACh receptor.

5.4 Neurotoxicology

5.4.1 Clinical Signs—Type I/II

The original description of multiple modes of action of pyrethroids in the rat found that, in general, the absence or presence of a cyano group in the α-position relative to the phenoxybenzyl alcohol moiety resulted in differences in clinical signs.[10] The T syndrome (tremors) was equivalent to the Type I syndrome of the non-cyano group and the CS syndrome (choreoathetosis salivation) was equivalent to the Type II syndrome of the α-CN group. This classification of two subgroups, Types I and II, has been found to hold good, with a handful of exceptions, in a variety of assays since 1980. The mouse[11] and the American cockroach,[12] for example, also showed two types of action, the latter adding differences in nerve effects. Fenpropathrin, although possessing an α-CN phenoxybenzyl group, showed a mixture of Type I and II effects in the rat, mouse and cockroach. Bifenthrin, which was not considered in these studies, has been described as Type I, based on lack of an α-CN group and clinical signs of tremors.[63,65,91] However, it did not cause typical Type I nerve discharges in the cockroach[13,66] and appears not to fit readily into either Type I or Type II in several factor and multivariate dissimilarity analyses.[63]

In reviewing the neurobehavioral toxicology of pyrethroids,[17] several endpoints were assessed in terms of their ability to quantify acute pyrethroid neurotoxicity. Effects on motor activity and the acoustic startle response were the only two that had been used consistently. Other FOB effects that have sometimes been used included coordination, neuromuscular response (grip strength, *etc.*), tremors, learning and memory, somatosensory response, social interactions and other descriptions. However, in general the available data available for endpoints associated with these FOB effects have not been sufficiently robust to provide consistently useful information, *e.g.* monotonic dose–response relationships. In the category of physiological toxicity, body temperature in the rat has also been considered. There is some evidence that this is increased by Type I and reduced by Type II pyrethroids. However, there is also a suggestion that the temperature response may be bimodal with opposite effects at low and high doses for the same pyrethroid.[17]

5.4.2 Motor Activity

Motor activity in the rat has typically been measured using photobeams, in a maze. It assesses ambulatory (locomotor) activity, rather than non-locomotor activity such as tremors. In the majority of studies with pyrethroids, the reduction in motor activity, reported at sublethal doses, is therefore a reduction of locomotor activity. This may result because dosed rats (with tremors) are sufficiently uncoordinated that they show less ambulatory activity. As such, motor activity reduction is a very apical endpoint which lacks specificity and does not provide information on the MOA of the pyrethroid. Nevertheless, it has been used extensively to quantify pyrethroid toxicity (*e.g.* Refs 91–94). The route of administration had a profound effect on pyrethroid toxicity[94] as did the choice of solvent and the relative dilution of the pyrethroid administered *via* the PO route.[92] This has been found for both deltamethrin[94] and bifenthrin,[92] with corn oil being consistently a 'more toxic' solvent than less lipophilic ones, such as methylcellulose, by several orders. It is likely that greater oral absorption takes place with solvents such as corn oil.

The clinical signs (Type I/II) appear to be largely independent of the route of administration. The route mainly affects the dose required and the latency period for the syndromes to develop. The early studies used parenteral methods, such as intravenous dosing in the rat[10] and intracerebral dosing in the mouse.[11] However, more recent studies have used enteral dosing methods, such as oral gavage in the rat.[91–96] After such dosing, it took 2–8 h for peak clinical signs to develop.[63,65] Dietary administration, as used in most subchronic and chronic studies, also results in different clinical signs, similar to those occurring after oral gavage dosing (Type I/II). However, larger doses and longer lag times (sometimes several days) are required to observe clinical signs in dietary studies. Some differences in clinical signs do appear to be route-specific, *e.g.* choreoathetosis in the mouse was not observed after intracerebral dosing, perhaps reflecting a spinal origin for this sign.[11]

A dose-additivity study examined the effects of mixtures of 11 pyrethroids on motor activity in the rat.[93] Mixtures were based on fractions of the ED_{30} for reduced motor activity and the results were consistent, mathematically, with a dose-additivity model rather than an effect-additivity model. In other words, all pyrethroids were considered to be interchangeable and to have the same MOA, using this system. However, because all pyrethroids (Types I and II) cause a reduction in motor activity in this system, this finding does not seem surprising, but it does not enable conclusions to be made about MOA. In addition, the large number of pyrethroids studied (simultaneously) complicates the evaluation of data from such mixture studies. A more relevant assay system to study additivity or interactions between pyrethroids may be the auditory startle response (ASR), which appears to distinguish between (some) Type I and II pyrethroids.

5.4.3 Auditory Startle Response (ASR)

Type I pyrethroids such as permethrin and RU 11679, along with DDT, increased the amplitude of the ASR in the rat, without an effect on latency of onset.[95,96] In contrast, in the same system, the Type II pyrethroids cypermethrin and cyfluthrin, along with flucythrinate (a probable Type II pyrethroid), decreased the ASR amplitude and increased the latency of onset. However, the Type II pyrethroid fenvalerate had mixed effects on the ASR: similar effects to Type I pyrethroids on amplitude and latency but increased ASR sensitization to background noise, unlike Type I or II pyrethroids. It was also found that cismethrin (Type I) and deltamethrin (Type II) had effects on the ASR that were consistent with those of permethrin and cypermethrin, respectively.[95] Deltamethrin also reduced sensitization to background noise. The ASR results for permethrin and deltamethrin were largely confirmed by Hijzen *et al.*[97,98] and Sheets *et al.*[99] Similarly, permethrin and cypermethrin were compared in several measures of sensory evoked response in the rat: touch- and approach-response scores were both increased with permethrin and reduced by cypermethrin, but click-evoked responses were increased by both pyrethroids.[100] In another study, the interactions of cismethrin and deltamethrin with pharmaceutical agents acting at the $GABA_A$ receptor complex were investigated using the ASR and motor activity.[86] The agents were picrotoxin (PTX, antagonist at the Cl channel site), muscimol (agonist at GABA binding site) and chlordiazepoxide (central BZ receptor agonist). The only clear interaction was dose-addition for deltamethrin with PTX for combined effects on ASR and motor activity. This indicates that deltamethrin and PTX act at the same receptor and cause similar effects, *i.e.* both are antagonists at this site. However, an unanticipated finding in these studies was that PTX and muscimol caused similar instead of opposite effects on each of the (four) parameters measured: ASR amplitude, latency, sensitization and locomotor activity. The toxicological significance of Type II pyrethroids on the $GABA_A$-activated Cl^- channel complex is therefore in need of further clarification.

5.5 Conclusions

Pyrethroids are a class of insecticides considered to have limited mammalian toxicity. They are highly lipophilic and poorly absorbed through the skin. Oral absorption is generally in the range of 40–60% using corn oil as solvent and is followed by metabolism by oxidases and esterases. Structures with an acid moiety based on chrysanthemic acid, as found in pyrethrin I, are poorly metabolized by esterases whenever the *cis* isomer is present, whereas *trans* isomers are readily metabolized by esterases. Similarly, pyrethroids containing a secondary alcohol moiety are also resistant to ester cleavage. Structures that are slowly cleaved by esterases are metabolized almost exclusively by oxidation. Target sites in the neuromuscular system include voltage-gated ion channels, such as Na^+, which has been studied *in vitro*. Several isoforms have been found to be susceptible to modification by pyrethroids, including the Nav1.8 α isoform. Channels tend to inactivate slowly and exhibit prolonged tail currents at the end of a depolarizing pulse applied to a voltage-clamped Na^+ channel. Other ion channel types, including voltage-gated Ca^{2+} and Cl^- channels, as well as GABA-gated Cl^- channels, are also affected by pyrethroids.

In vivo, pyrethroid clinical signs in rodents have been divided into two syndromes, Types I and II, also known as T and CS. A limited number of pyrethroids are classed as hybrids or unknowns. Reduced locomotor activity in the rat after oral gavage dosing has been used to quantify pyrethroid neurotoxicity. Unfortunately, the effect is non-specific and does not readily allow conclusions to be made about MOA. Other clinical signs, such as effects on the rat ASR, show greater specificity. It is probable that different clinical signs associated with Type I and II syndromes result from effects on different channel isoforms or combinations of them. Pharmacokinetic studies have found that effects on clinical signs, in the first few hours following dosing of the rat, generally correlate with peak blood plasma concentration. The AUC is more relevant for estimating internal doses associated with chronic effects.

Overall, Type I pyrethroids with a *trans* substituted acid moiety have much lower mammalian toxicity than the corresponding *cis* isomers. This is partly a result of more rapid ester hydrolysis for *trans* than for *cis* isomers. For Type II pyrethroids, there is much less difference in toxicity between *trans* and *cis* isomers. This may be an indication of similar target site potency of geometric isomers of Type II pyrethroids.

References

1. T. Narahashi, Effect of the insecticide allethrin on membrane potentials of cockroach giant axons, *J. Cell. Comp. Physiol.*, 1962, **59**, 61–65.
2. T. Narahashi, Nature of the negative after-potential increased by the insecticide allethrin in cockroach giant axons, *J. Cell. Comp. Physiol.*, 1962, **59**, 67–76.
3. K. Murayama, N. J. Abbott, T. Narahashi and B. I. Shapiro, Effects of allethrin and Condylactis toxin on the kinetics of sodium conductance of crayfish axon membrane, *Comp. Gen. Pharmacol.*, 1972, **3**, 391–400.
4. T. Narahashi and N. C. Anderson, Mechanism of excitation block by the insecticide allethrin applied externally and internally to squid giant axons, *Toxicol. Appl. Pharmacol.*, 1967, **10**, 529–547.
5. C. M. Wang, T. Narahashi and M. Scuka, Mechanism of negative temperature coefficient of nerve blocking action of allethrin, *J. Pharmacol. Exp. Ther.*, 1972, **182**, 442–453.
6. D. W. Gammon, Nervous effects of toxins on an intact insect: a method, *Pestic. Biochem. Physiol.*, 1977, **7**, 1–7.
7. D. W. Gammon, Neural effects of allethrin on the free-walking cockroach *Periplaneta americana*: an investigation using defined doses at 15 and 32 °C, *Pestic. Sci.*, 1978, **9**, 79–91.
8. D. W. Gammon, An analysis of the temperature-dependence of the toxicity of allethrin to the cockroach, in *Neurotoxicology of Insecticides and Pheromones*, ed. T. Narahashi, Plenum Press, New York, 1979, pp. 97–117.
9. M. Elliott, N. F. Janes, *et al.* Insecticidal activity of the pyrethrins and related compounds. Part XI. Relative potencies of isomeric cyano-substituted 3-phenoxybenzyl esters, *Pestic. Sci.*, 1978, **9**, 112–116.

10. R. D. Verschoyle and W. N. Aldridge, Structure-activity relationships of some pyrethroids in rats, *Arch. Toxicol.*, 1980, **45**, 325–329.
11. L. J. Lawrence and J. E. Casida, Pyrethroid toxicology: mouse intracerebral structure-activity relationships, *Pestic. Biochem. Physiol.*, 1982, **18**, 9–14.
12. D. W. Gammon, M. A. Brown and J. E. Casida, Two classes of pyrethroid action in the cockroach, *Pestic. Biochem. Physiol.*, 1981, **15**, 181–191.
13. D. W. Gammon, K. F. Leggett and J. M. Clark, Pyrethroid mode(s) of action in the context of Food Quality Protection Act (FQPA) regulation, *J. Agric. Food Chem.*, 2011, **59**, 2773–2785.
14. D. E. Ray, Pyrethroid insecticides: mechanisms of toxicity, systemic poisoning syndromes, paresthesia and therapy, in *Hayes' Handbook of Pesticide Toxicology*, Elsevier, London, 2nd edition, 2001, pp. 1289–1303.
15. D. M. Soderland, J. M. Clark, L. P. Sheets, L. S. Mullin, V. J. Picirillo, D. Sargent, J. T. Stevens and M. L. Weiner, Mechanisms of pyrethroid neurotoxicity: implications for cumulative risk assessment, *Toxicology*, 2002, **171**, 3–59.
16. T. J. Shafer, D. A. Meyer and K. M. Crofton, Developmental neurotoxicity of pyrethroid insecticides: critical review and future research needs, *Environ. Health Perspect.*, 2005, **113**, 123–136.
17. M. J. Wolansky and J. A. Harrill, Neurobehavioral toxicology of pyrethroid insecticides in adult animals: a critical review, *Neurotoxicol. Teratol.*, 2008, **30**, 55–78.
18. D. M. Soderland, Toxicology and mode of action of pyrethroid insecticides, in *Hayes' Handbook of Pesticide Toxicology*, Elsevier, London, 3rd edition, 2010, pp. 1665–1686.
19. D. E. Ray and J. R. Fry, A reassessment of the toxicity of pyrethroid insecticides, *Pharmacol. Ther.*, 2006, **111**, 174–193.
20. J. E. Casida and L. O. Ruzo, Metabolic chemistry of pyrethroid insecticides, *Pestic. Sci.*, 1980, **11**, 257–269.
21. H. Kaneko, Pyrethroid chemistry and metabolism, in *Hayes' Handbook of Pesticide Toxicology*, Elsevier, London, 3rd edition, 2010, pp. 1635–1663.
22. T. Roberts and D. Hutson, *Metabolic Pathways of Agrochemicals: Part 2, Insecticides and Fungicides*, Royal Society of Chemistry, Cambridge, 1999.
23. J. P. Demoute, A brief review of the environmental fate and metabolism of pyrethroids, *Pestic. Sci.*, 1989, **27**, 375–385.
24. H. Kaneko, Pyerthroids: mammalian metabolism and toxicity, *J. Agric. Food Chem.*, 2011, **59**, 2786–2791.
25. M. Elliott, N. F. Janes, E. C. Kimmel and J. E. Casida, Metabolic fate of pyrethrin I, pyrethrin II, and allethrin administered orally to rats, *J. Agric. Food Chem.*, 1972, **20**, 301.
26. I. Yamamoto, M. Elliott and J. E. Casida, The metabolic fate of pyrethrin I, pyrethrin II, and allethrin, *Bull. World Health Org.*, 1971, **44**, 347–8.
27. K. Ueda, L. C. Gaughan and J. E. Casida, Metabolism of (+)-trans- and (+)-cis-resmethrin in rats, *J. Agr. Food Chem.*, 1975, **23**, 106–115.

28. J. Miyamoto, T. Nishada and K. Ueda, Metabolic fate of resmethrin, 5-benzyl-3-furylmethyl DL-transchrysanthemate in the rat, *Pestic. Biochem. Physiol.*, 1971, **1**, 293–306.
29. J. E. Casida, K. Ueda, L. Gaughan, L. T. Jao and D. M. Soderlund, Structure-biodegradability relationships in pyrethroid insecticides, *Arch. Environ. Contam. Toxicol.*, 1976, **3**, 491–500.
30. R. Edwards, P. Millburn and D. H. Hutson, Comparative metabolism and disposition of [^{14}C-benzyl] cypermethrin in quail, rat and mouse, *Pestic. Sci.*, 1990, **30**, 159–181.
31. L. C. Gaughan, T. Unai and J. E. Casida, Permethrin metabolism in rats, *J. Agric. Food Chem.*, 1977, **25**, 9–17.
32. A. Anadon, M. R. Martinez-Larranage, M. J. Diaz and P. Bringas, Toxicokinetics of permethrin in the rat, *Toxicol. Appl. Pharmacol.*, 1990, **110**, 1–8.
33. T. Takaku, K. Mikata, M. Matsui, K. Nishioka, N. Isobe and H. Kaneko, *In vitro* metabolism of trans-permethrin and its major metabolites, PBacc and PBacid, in humans, *J. Agric. Food Chem.*, 2011, **59**, 5001–5005.
34. M. J. Crawford, A. Croucher and D. H. Hutson, Metabolism of cis- and trans-cypermethrin in rats. Balance and tissue retention study, *J. Agric. Food Chem.*, 1981, **29**, 130.
35. M. J. Crawford, A. Croucher and D. H. Hutson, The metabolism of the pyrethroid insecticide cypermethrin in rats; excreted metabolites, *Pestic. Sci.*, 1981, **12**, 399.
36. D. H. Hutson and C. J. Logan, The metabolic fate in rat the pyrethroid insecticide WL85871 a mixture of two isomers of cypermethrin, *Pestic. Sci.*, 1986, **17**, 548.
37. B. Wielgomas and J. Krechniak, Toxicokinetic interaction of α-cypermethrin and chlorpyrifos in rats, *Polish J. Environ. Stud.*, 2007, **16**, 267–274.
38. S. Selim, Absorption, distribution, and excretion of FMC 54800 in rats. FMC Report PC-0047, 1987; unpublished report, FMC Corporation, Ewing, NJ 08628, USA.
39. S. ElNaggar, *et al.* Metabolism of FMC 54800 in rats. Identification of products in excreta, FMC Report P-1439, 1986; unpublished report, FMC Corporation, Ewing, NJ 08628, USA.
40. S. Selim, Kinetics of FMC 54800 in the blood of rats following a single oral dose. FMC Report PC-0048, 1986; unpublished report, FMC Corporation, Ewing, NJ 08628, USA.
41. R. H. Tullman, Analysis of FMC 54800 in plasma from rats dosed orally with ^{14}C-FMC 54800. FMC Report P-1448, 1986; unpublished report, FMC Corporation, Ewing, NJ 08628, USA.
42. S. ElNaggar, *et al.* Metabolism study: quantitative estimates of urinary, fecal and biliary excretion of alcohol (Phenyl)-^{14}C-bifenthrin in the laboratory rat. FMC Report P-2570, 1991; unpublished Report. FMC Corporation, Ewing, NJ 08628, USA.

43. Tefluthrin, in: *Metabolic Pathways of Agrochemicals Part 2: Insecticides and Fungicides*, ed. T. R. Roberts and D. H. Hutson, Royal Society of Chemistry, Cambridge, 1999.
44. R. Amos and J. P. Leahey, in *British Crop Protection Conference: Proceedings of the Conference Held at Brighton Metropole, England*, BCPC Publications, 1986, pp. 812–828.
45. Tefluthrin; pesticide tolerance, *Federal Register*, 1997, **62**, 62954–62961.
46. D. W. Bewick, I. R. Hill, J. Pluckrose, J. E. B. Stevens and M. S. Weissler, in *British Crop Protection Conference: Proceedings of the Conference Held at Brighton Metropole, England*, BCPC Publications, 1986, pp. 459–468.
47. J. Heath and J. P. Leahey, The metabolism of tefluthrin in the goat, *Pestic. Sci.*, 1989, **25**, 375–389.
48. Pesticide Safety Directorate, Evaluation No. 42, MAFF, London, 1991.
49. H. Ohkawa, H. Kaneko, H. Tsuji and J. Miyamoto, Metabolism of fenvalerate (Sumicidin) in rats, *J. Pesticide Sci.*, 1979, **4**, 143–155.
50. N. Isobe, H. Kaneko, K. Shiba, K. Saito, S. Ito, N. Kakuta, A. Saito, A. Yoshitake and J. Miyamoto, Metabolism of esfenvalerate in rats and mice and effects of its isomers on metabolic fates of esfenvalerate, *J Pestic. Sci.*, 1990, **15**, 159–168.
51. P. W. Lee, S. M. Stearns and W. R. Powell, Rat metabolism of fenvalerate (Pydrin insecticide), *J. Agric. Food. Chem.*, 1995, **33**, 988–993.
52. S. Misra and M. S. Sharma, Metabolism and accumulation of fenvalerate and its metabolites in rat organs, *Biomed. Chromatogr.*, 1997, **11**, 50–53.
53. G. Leng, A. Leng, K. H. Kuhn, J. Lewalter and J. Pauluhn, Human dose-excretion studies with the pyrethroid insecticide cyfluthrin: Urinary metabolite profiles following inhalation, *Xenobiotica*, 1997, **27**, 1273–1283.
54. A. Anadon, M. R. Martinez-Laranaga, M. L. Fernandez-Cruz, M. J. Diaz, M. C. Fernandez and M. A. Martinez, Toxicokinetics of deltamethrin and its 4′-OH-metabolite in the rat, *Toxicol. Appl. Pharmacol.*, 1996, **141**, 8–16.
55. A. Mirfazaelian, K. B. Kim, S. S. Anand, H. J. Kim, R. Tornero-Velez, J. V. Bruckner and J. W. Fisher, Development of a physiologically based pharmacokinetic model for deltamethrin in the adult male Sprague-Dawley rat, *Toxicol. Sci.*, 2006, **93**, 432–442.
56. S. J. Godin, M. J. Devito, M. F. Hughes, D. G. Ross, E. J. Scollon, J. M. Starr, R. W. Setzer, R. B. Conolly and R. Tornero-Velez, Physiologically based pharmacokinetic modeling of deltamethrin: development of a rat and human diffusion-limited model, *Toxicol. Sci.*, 2010, **115**, 330–343.
57. K. B. Kim, S. S. Anand, H. J. Kim, C. A. White and J. V. Bruckner, Toxicokinetics and tissue-distribution of deltamethrin in adult Sprague-Dawley rats, *Toxicol. Sci.*, 2008, **101**, 197–205.
58. M. J. Crawford and D. H. Hutson, The metabolism of the pyrethroid insecticide (±)-α-cyano-3-phenoxybenzyl 2,2,3,3-tetramethyl-cyclopropanecarboxylate, WL 41706, in the rat, *Pestic. Sci.*, 1977, **8**, 579–599.

59. H. Kaneko, K. Shiba, A. Yoshitake and J. Miyamoto, Metabolism of fenpropathrin (S-3206) in rats, *J. Pestic. Sci.*, 1987, **12**, 385–395.
60. A. Anadon, M. Martinez, M. A. Martinez, M. J. Diaz and M. R. Martinez-Larrañaga, Lambda-cyhalothrin toxicokinetics in rats, *Toxicol. Lett.*, 2006, **165**, 47–56.
61. D. R. Hawkins, D. Kilpatrick, B. Eween, I. Midgley, S. R. Biggs and B. R. Whithy, The biokinetics and metabolism of ^{14}C-ethofenprox in the rat. Report no. HRC/MTC 68/8610, Huntington Research Centre, England, 1985. Submitted to WHO by Mitsui Toatsu Chemicals, Inc, Tokyo, Japan.
62. A. L.Goldin, R. L. Barchi, J. H. Caldwell, F. Hoffman, J. R. Howe and J. C. Hunter, *et al.*, Nomenclature of voltage-gated sodium channels, *Neuron*, 2000, **28**, 365–368.
63. C. B. Breckenridge, L. Holden, N. Sturgess, M. Weiner, L. Sheets, D. Sargent, D. M. Soderlund, J.-S. Choi, S. Symington, J. M. Clark, S. Burr and D. Ray, Evidence for a separate mechanism of toxicity for the Type I and the Type II pyrethroid insecticides, *Neurotoxicology*, 2009, **30**, S17–31.
64. D. W. Gammon, The action of tetrodotoxin on the cockroach *Periplaneta americana*: a toxicological and neurophysiological study, *Physiol. Entomol.*, 1978, **3**, 37–42.
65. M. L. Weiner, M. Nemec, L. Sheets, D. Sargent and C. B. Breckenridge, Comparative functional observational battery study of twelve commercial pyrethroid insecticides in male rats following acute oral exposure, *Neurotoxicology*, 2009, **30**, S1–16.
66. D. W. Gammon and G. Sander, Two mechanisms of pyrethroid action: electrophysiological and pharmacological evidence, *Neurotoxicology*, 1985, **6**, 63–86.
67. V. L. Salgado and T. Narahashi, Immobilization of sodium channel gating charge in crayfish giant axons by the insecticide fenvalerate, *Mol. Pharmacol.*, 1993, **43**, 626–634.
68. T. J. Smith, P. J. Ingles and D. M. Soderland, Actions of pyrethroid insecticides cismethrin and cypermethrin on house fly Vssc1 sodium channels expressed in *Xenopus* oocytes, *Arch. Insect Biochem. Physiol.*, 1998, **38**, 126–136.
69. T. J. Smith and D. M. Soderlund, Action of the pyrethroid insecticide cypermethrin on rat brain IIa sodium channels expressed in *Xenopus* oocytes, *Neurotoxicology*, 1998, **19**, 823–832.
70. J. Tan and D. M. Soderland, Human and rat Nav1.3 voltage-gated sodium channels differ in inactivation properties and sensitivity to the pyrethroid insecticide tefluthrin, *Neurotoxicology*, 2009, **30**, 81–89.
71. C. I. Spicer, K. H. Yuill, J. J. Borg, J. C. Hancox and R. Z. Kozlowski, Actions of pyrethroid insecticides on sodium currents, action potentials and contractile rhythm in isolated mammalian ventricular myocytes and perfused hearts, *J. Pharmacol. Exp. Ther.*, 2001, **298**, 1067–1082.
72. J. M. Clark and S. B. Symington, Neurotoxic implications of the agonistic action of CS-syndrome pyrethroids on the N-type $Ca_v2.2$ calcium channel, *Pest Manag. Sci.*, 2008, **64**, 628–638.

73. S. B. Symington, R. K. Frisbie and J. M. Clark, Characterization of 11 commercial pyrethroids on the functional attributes of rat brain synaptosomes, *Pestic. Biochem. Physiol.*, 2008, **92**, 61–69.
74. Z. Cao, T. J. Shafer and T. F. Murray, Mechanisms of pyrethroid insecticide-induced stimulation of calcium influx in neocortical neurons, *J. Pharmacol. Exp. Ther.*, 2011, **336**, 197–205.
75. D. A. Meyer and T. J. Shafer, Permethrin, but not deltamethrin, increases spontaneous glutamate release from hippocampal neurons in culture, *Neurotoxicology*, 2006, **27**, 594–603.
76. D. A. Meyer, J. M. Carter, A. F. M. Johnstone and T. J. Shafer, Pyrethroid modulation of spontaneous neuronal excitability and neurotransmission in hippocampal neurons in culture, *Neurotoxicology*, 2008, **29**, 213–225.
77. W. Jun, J. Lei, G. Xiaohua, D. Haixia, W. Qiang, C. Jie, G. Rong and X. Hang, Fenvalerate-induced Ca^{2+} transients via both intracellular and extracellular way in mouse GC-2spd (ts) cells, *Toxicology*, 2009, **259**, 122–132.
78. S. E. Burr and D. E. Ray, Structure-activity and interaction effects of 14 different pyrethroids on voltage-gated chloride ion channels, *Toxicol. Sci.*, 2004, **77**, 341–346.
79. Y.-T. Tian, Z.-W. Liu, Y. Yao, Z. Yang and T. Zhang, Effect of alpha-cypermethrin and theta-cypermethrin on delayed rectifier potassium currents in rat hippocampal neurons, *Neurotoxicology*, 2009, **30**, 269–273.
80. Y.-T. Tian, Z.-W. Liu, Y. Yao, T. Zhang and Y. Zhuo, Effects of alpha- and theta-cypermethrin insecticide on transient outward potassium current in rat hippocampal CA3 neurons, *Pestic. Biochem. Physiol.*, 2008, **90**, 1–7.
81. F. Leeb-Lundberg and R. W. Olsen, Picrotoxinin binding as a probe of the GABA postsynaptic membrane receptor-ionophore complex, in *Psychopharmacology and Biochemistry of Neurotransmitter Receptors*, ed. H. I. Yamamura, R. W.Olsen and E. Usdin, Elsevier, New York, 1980, pp. 593–606.
82. D. W. Gammon, L. J. Lawrence and J. E. Casida, Pyrethroid toxicology: protective effects of diazepam and phenobarbital in the mouse and the cockroach, *Toxicol. Appl. Pharmacol.*, 1982, **66**, 290–296.
83. D. W. Gammon and J. E. Casida, Pyrethroids of the most potent class antagonize GABA action at the crayfish neuromuscular junction, *Neurosci. Lett.*, 1983, **40**, 163–168.
84. L. J. Lawrence and J. E. Casida, Sterospecific action of pyrethroid insecticides on the γ-aminobutyric acid receptor-ionophore complex, *Science*, 1983, **221**, 1399–1401.
85. K. M. Crofton, R. B. Mailman and L. W. Reiter, Pyrethroid insecticides and radioligand binding at the GABA receptor-ionophore complex, *Toxicol. Lett.*, 1987, **35**, 183–190.
86. K. Crofton and L. Reiter, Pyrethroid insecticides and the γ-aminobutyric acid receptor complex: motor activity and the acoustic startle response in the rat, *J. Pharmacol. Exp. Ther.*, 1987, **243**, 946–954.

87. I. M. Abalis, M. E. Eldefrawi and A. T. Eldefrawi, Biochemical identification of putative GABA/benzodiazepine receptors in house fly thorax muscles, *Pestic. Biochem. Physiol.*, 1983, **20**, 39–48.
88. M. E. Eldefrawi, D. B. Gant, I. M. Abalis and A. T. Eldefrawi, Interactions of insecticides with GABA-operated and voltage-dependent chloride channels, in *Sites of Action of Neurotoxic Pesticides*, ed. R. M. Hollingworth and M. B. Green, ACS Symposium Series 356, 1986, pp. 107–121.
89. L. J. Lawrence, K. Gee and H. Yamamura, Interactions of pyrethroid insecticides with chloride ionophore-associated binding sites, *Neurotoxicology*, 1985, **6**, 87–98.
90. M. A. Abbassy, M. E. Eldefrawi and A. T. Eldefrawi, Influence of the alcohol moiety of pyrethroids on their interactions with the nicotinic acetylcholine receptor, *J. Toxicol. Environ. Health A*, 1983, **12**, 575–590.
91. M. J. Wolansky, C. Gennings and K. M. Crofton, Relative potencies for acute effects of pyrethroids on motor function in rats, *Toxicol. Sci.*, 2006, **89**, 271–277.
92. M. J. Wolansky, K. L. McDaniel, V. C. Moser and K. M. Crofton, Influence of dosing volume on the neurotoxicity of bifenthrin, *Neurotoxicol. Teratol.*, 2007, **29**, 377–384.
93. M. J. Wolansky, C. Gennings, M. J. DeVito and K. M. Crofton, Evidence for dose-additive effects of pyrethroids on motor activity in rats, *Environ. Health Perspect.*, 2009, **117**, 1563–1570.
94. K. M. Crofton, L. S. Kehn and M. E. Gilbert, Vehicle and route dependent effects of a pyrethroid insecticide, deltamethrin, on motor function in the rat, *Neurotoxicol.Teratol.*, 1995, **17**, 489–495.
95. K. Crofton and L. Reiter, Effects of two pyrethroids on motor activity and the acoustic startle response in the rat, *Toxicol. Appl. Pharmacol.*, 1984, **75**, 318–328.
96. K. Crofton and L. Reiter, The effects of type I and II pyrethroids on motor activity and the acoustic startle response in the rat, *Fundam. Appl. Toxicol.*, 1988, **10**, 624–634.
97. T. H. Hijzen and J. L. Slangen, Effects of type I and type II pyrethroids on the startle response in rats, *Toxicol. Lett.*, 1988, **40**, 141–152.
98. T. H. Hijzen, R. De Beun and J. L. Slangen, Effects of pyrethroids on the acoustic startle reflex in the rat, *Toxicology*, 1988, **49**, 271–276.
99. L. P. Sheets, J. D. Doherty, M. W. Law, L. W. Reiter and K. M. Crofton, Age-dependent differences in the susceptibility of rats to deltamethrin, *Toxicol. Appl. Pharmacol.*, 1994, **126**, 186–190.
100. K. L. McDaniel and V. C. Moser, Utility of a neurobehavioral screening battery for differentiating the effects of two pyrethroids, permethrin and cypermethrin, *Neurotoxicology*, 1993, **15**, 71–83.

CHAPTER 6

Nicotine and the Neonicotinoids

PATRICK H ROSE

JSC International Limited, Simpson House, Windsor Court, Clarence Drive, Harrogate, North Yorkshire, UK. Email: Patrick.Rose@jsci.co.uk

6.1 Introduction

Nicotine is the main biologically active alkaloid found in tobacco extract and has been used for many years as an insecticide. Nicotine consumption for insecticidal use exceeded 2500 tons worldwide after the Second World War and then gradually declined to less than 200 tons by the early 1980s as a broad range of more selective, effective and cheaper insecticides became established.[1] Although nicotine was not very effective as an insecticide and was toxic to mammals, research was undertaken in the 1970s and 1980s to discover potentially better insecticides based on the nicotine structure. The optimization of a lead compound in the 1970s led to the synthesis of nithiazine, a potent nitromethylene with a thiazine ring; however, it was photolabile and thus not developed as an agricultural insecticide.[2] Further optimization resulted in introduction of a chloropyridinylmethyl substituent and replacement of the nitromethylene group with a nitroguanidine or cyanoamidine moiety to yield photostable molecules with similar potency.[2] This new class of systemic insecticides became known as neonicotinoids, based on their similarity to nicotine in terms of structure and action. The key advantages of neonicotinoids are their selective toxicity to insects with low risks for humans and other mammals, and their utility in counteracting the problems of insect resistance to other modes of action.[3]

Issues in Toxicology No. 12
Mammalian Toxicology of Insecticides
Edited by Timothy C. Marrs

Published by the Royal Society of Chemistry, www.rsc.org

The mammalian toxicity of neonicotinoids will be discussed in the contexts of both nicotine toxicity and their mode of action.

6.2 Neonicotinoid Insecticides

The development of chloropyridinyl compounds represented the first generation of systemic neonicotinoids,including imidacloprid, nitenpyram, acetamiprid and thiacloprid. This was followed by second-generation chlorothiazolyl compounds, thiamethoxam and clothianidin, and a teterahydrofuryl compound, dinotefuran. The chemical structures of neonicotinoid insecticides and nicotine are shown in Figure 6.1.

Imidacloprid was the first commercially successful neonicotinoid insecticide and since its launch in 1991 it has been widely registered for use in agriculture and for use in animal and public health products.[4] The neonicotinoid class accounts for about 20% of the current global insecticide market.[5]

The discussion of the mammalian toxicity of neonicotinoids has an emphasis on imidacloprid, being probably the most studied of the class.

6.3 Mode of Action of Nicotine and Neonicotinoid Insecticides

The mode of action of nicotine and neonicotinoids on insect and vertebrate nervous systems has been reviewed extensively in the open literature. The discussion which follows has been taken from a selection of these reviews.

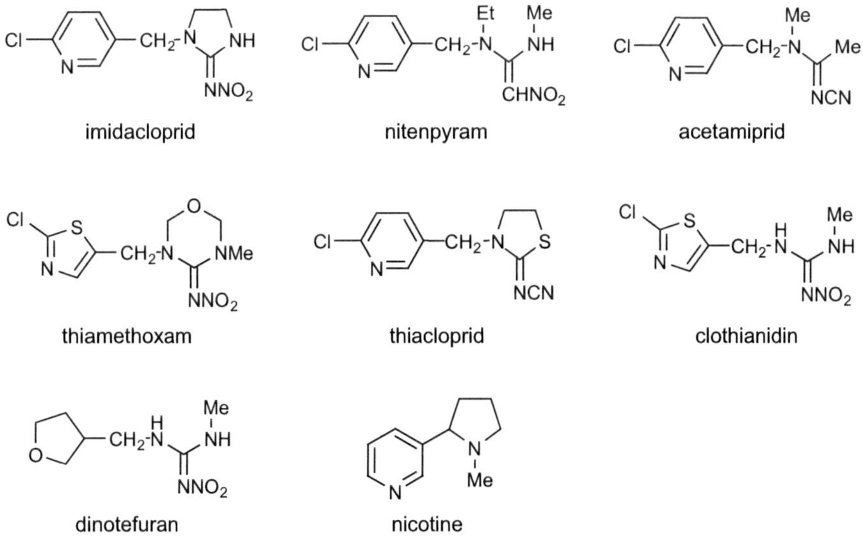

Figure 6.1 Chemical structures of neonicotinoids and nicotine.

Both nicotine and neonicotinoids exert their neurotoxic insecticidal action by binding to the nicotinic acetylcholine receptors (nAChRs) of the postsynaptic membrane of nerve cells in the central nervous system (CNS) and acting as nAChR agonists. They mimic the action of acetylcholine (ACh) by opening the ion channels which allows the entry of cations, including sodium and calcium, thus causing excitatory neurotransmission in the CNS.[2,3,6,7] Recent reviews have summarized the current state of knowledge.[2,8]

6.3.1 Insect and Mammalian Cholinergic Systems

There is a fundamental difference in the cholinergic system in insects compared with that in mammals which is a key factor in understanding the toxicity arising from exposure to nicotine or neonicotinoids. Insects possess a cholinergic system which is only located in the CNS, whereas mammals possess a cholinergic system located in both the CNS and peripheral nervous systems. Thus, in mammals toxicity can arise from effects of nicotine exposure on the central nAChRs in the CNS and from effects on the peripheral nAChRs in the ganglia and neuromuscular junctions.[2] There are also differences between insect and vertebrate nAChRs. The structure of the vertebrate nAChR is better understood than that of insects. The vertebrate nAChRs are large pentameric transmembrane allosteric proteins consisting of diverse subtypes assembled from five homologous subunits in combinations of ten α (α1–10), four β (β1–4), γ, δ, and ε subunits.[7,9] The expressed amounts of the various subtypes vary in different parts of the CNS and peripheral nervous systems. For example, the major population of vertebrate brain nAChRs appears to be of the α4β2 and α7 subtypes (α-bungaratoxin (BGT)-insensitive and α-BGT-sensitive, respectively).[10] The insect nAChRs have diverse subunits across many species, but the pentameric stoichiometries have not been fully resolved.[2,8] Nicotinoids and neonicotinoids vary in their affinity for different nAChR subtypes, which accounts in part for differences in toxicity between compounds and for differences in selectivity across insect species.

Non-neuronal cells such as those in the bronchial epithelium, endothelial cells and vascular smooth muscle also contain nAChRs, suggesting that these receptors have functions other than neurotransmission, as described later in section 6.4.

6.3.2 Selective Toxicity of Nicotine and Neonicotinoids

The selective toxicity of nicotine for vertebrates and of neonicotinoids for insects is attributed largely to the differential sensitivity of the insect and vertebrate nAChR subtypes.[2,6] Other factors, including bioavailability and metabolism of neonicotinoids, are also important.[3,6,8] In summary, neonicotinoids are significantly less toxic to vertebrates because of the combined effects of low affinity for vertebrate nAChRs and relatively poor penetration of the blood–brain barrier.

For nicotine to be effective in insects, free nicotine must penetrate the ion-impermeable barrier to the CNS whereupon it is ionized again in order to interact with the nAChR.[3] Although nicotine has a strong binding affinity for the insect nAChR, only a small fraction of nicotine is available in the non-ionized free form to penetrate the ion-impermeable barrier, thus limiting its insecticidal potency. The toxic moiety in nicotine and nicotinoids such as nornicotine and anabasine is 3-pyridylmethylamine, with the more toxic compounds having a highly basic nitrogen atom that is protonated at physiological pHs, as opposed to low toxicity compounds having a low basic nitrogen atom.[3]

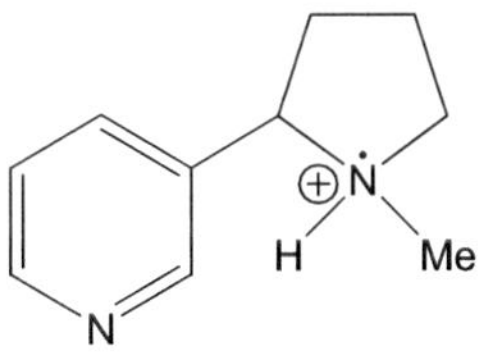

Ionized nicotine

The interactions of neonicotinoids with nAChRs have been studied extensively and much of the work involved studies with imidacloprid.[9] Both nicotine and imidacloprid interact with the α-BGT binding site of the nAChR.[3] It was found that neonicotinoids had a lower affinity for the mammalian nAChR, mainly the α-BGT-insensitive subtypes, compared with a strong affinity for the insect nAChR.[10] Although nicotinoids and some neonicotinoids contain the common 3-pyridylmethylamine moiety, the nitroguanidine or cyanoamidine groups are important for selectivity of the neonicotinoids.These groups confer a partial negative charge (δ^-) flow towards the tip of the molecule, *i.e.* an electronegative pharmacophore.[8] The negatively tipped nitro- or cyano-neonicotinoids interact with a proposed unique subsite consisting of cationic amino acid residues in the insect nAChR. This is in contrast to the strong affinity of ionized nicotine having cationic functionality for mammalian nAChRs; *i.e.* binding to anionic subsites.[2,3,8] A proposed binding model for ACh, nicotine and imidacloprid is shown in Figure 6.2.

It is thought that the high affinity for neonicotioids at the insect nAChR is related to a single dominant binding orientation, whereas relative insensitivity at vertebrate nAChRs is caused by multiple binding confirmations in the agonist-binding pocket.[8]

The liver cytochrome P450 enzymes and cytosolic aldehyde oxidase are involved in the activation and inactivation or detoxification of neonicotoids.[10,12] Studies of the metabolites of neonicotinoids have shown that they can be bioactive and act as nAChR agonists or cause secondary toxicity in mammals.[11] Nicotinoids can be formed as metabolites of neonicotinoids with greater selectivity for vertebrate nAChRs than for insect nAChRs.[11,13] For example, the desnitro metabolite of imidacloprid and the descyano metabolite of thiacloprid are protonated under physiological pH as with nicotinoids and the former is more potent than nicotine at selected vertebrate nAChRs.[2,8,13]

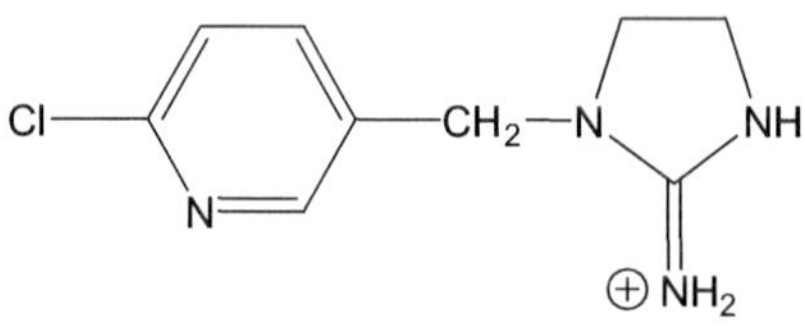

Desnitro-imidacloprid

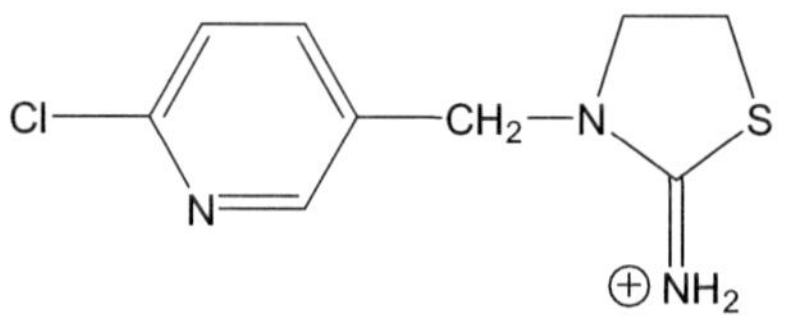

Descyano-thiacloprid

Other neonicotinoids may act as pro-insecticides such as derivatives of imidacloprid which are converted to imidacloprid. Although thiamethoxam yields clothianidin as a metabolite, both are reported to be insecticidally active rather than thiamethoxam being a pro-insecticide.[14]

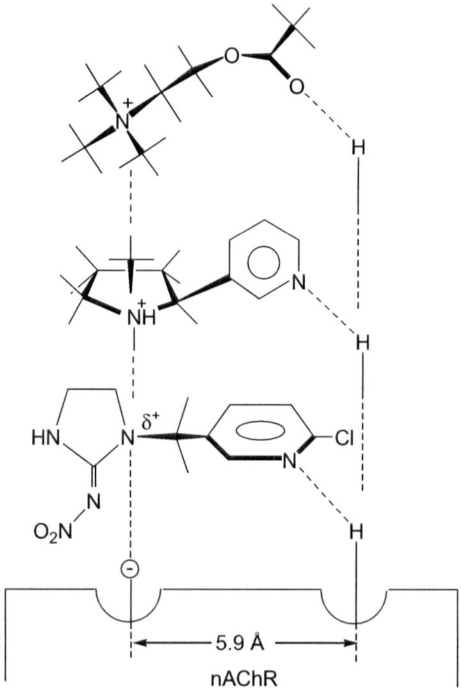

Figure 6.2 Binding model for acetylcholine, nicotine and imidacloprid to nAChR. (Reproduced with permission from ref. 5).

6.4 Toxicology of Nicotine

Nicotine is the common name for 3-[(2*S*)-1-methylpyrrolidin-2-yl]pyridine (IUPAC). The structural formula is:

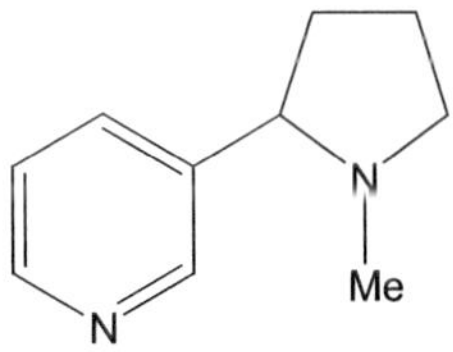

Nicotine is a pharmacologically active drug which has wide-ranging effects mainly involving the nervous, cardiovascular, respiratory and endocrine systems and the gastrointestinal tract. Nicotine has been extensively studied because of its role in diseases related to tobacco smoking. Although nicotine is usually associated with adverse effects on biological systems, it can provide beneficial effects at low doses. The pleasurable effects of nicotine are well known and lead to addiction. Complex pharmacodynamic effects involving initial behavioural stimulation followed by depression are related to the complexity of pharmacological actions and dose response, and rapid development of tolerance.[7,15] For example, smokers rapidly become tolerant to the low-dose effects of nicotine such as dizziness and nausea.

6.4.1 Metabolism and Toxicokinetics

The metabolism and toxicokinetics of nicotine is summarized from a number of reviews.[7,15–17] Nicotine can be absorbed by all routes of exposure. Alkaline pH increases the proportion of uncharged nicotine, which facilitates its absorption across mucous membranes and the skin. Once absorbed it is extensively distributed throughout the body and it rapidly reaches the CNS, easily penetrating the blood–brain barrier. It is about 69% ionized in the blood stream at pH 7.4. Most nicotine is metabolized in the liver, mainly by the microsomal enzyme CYP2A6, to the main metabolite cotinine which is itself extensively metabolized, the major metabolite being trans-3′-hydroxycotinine. N-oxidation of nicotine to 1′-N-oxide is another significant biotransformation pathway and concentrations in urine are about threefold lower than those of 3′-hydroxycotinine. Nicotine, cotinine and other phase 1 metabolites undergo phase 2 metabolism *via* glucuronidation to form a variety of glucuronide conjugates. Metabolites are excreted predominantly in urine, the rate of elimination being influenced by the pH of urine. The elimination half-life of nicotine is short, averaging 2 h.

6.4.2 Acute Toxicity

Nicotine is acutely toxic to humans and other mammals. The human median lethal dose (MLD) is estimated to be 0.5–1.0 mg kg^{-1} and the oral MLD or

LD_{50} values in rats, mice and dogs are 50, 3.3 and 9.2 mg kg^{-1} bw, respectively.[15] Signs of nicotine poisoning arise from effects on the gastrointestinal tract, respiratory and cardiovascular systems and from a variety of other centrally or peripherally mediated neurological responses.[7,15,17] Early signs of poisoning are characterized by symptoms including headache, dizziness, sweating, nausea and vomiting, increased blood pressure and heart rate. Cholinergic symptoms include salivation, lachrimation, increased bronchial secretions, miosis and later mydriasis. Later signs include diarrhoea, respiratory depression, bradycardia and hypotension. In cases of severe poisoning tremors and convulsions, prostration and death may occur from paralysis of respiratory muscles and/or central respiratory failure. Nicotine stimulation of central nAChRs results in the release of a variety of neurotransmitters, particularly dopamine, which play an important role in the behavioural effects of nicotine including addiction.[7,17]

6.4.3 Short-/Long-Term Toxicity/Carcinogenicity/Mutagenicity

The pharmacological actions of nicotine are believed to be associated with the promotion of cardiovascular disease and acute events such as constriction of coronary blood vessels and myocardial infarction. Chronic exposure to nicotine induces vascular endothelial disease which leads to a range of cardiovascular disorders including atherosclerosis, hypertension, heart failure, diabetes mellitus and stroke.[18] In addition, nicotine increases the release of catecholamines from the adrenal medulla and is thought to influence lipid metabolism by increasing the levels of free fatty acids, factors known to adversely affect the cardiovascular system.

Nicotine is probably not genotoxic on a weight of evidence evaluation, although it has been reported to cause genotoxicity in some *in vitro* assays.[15,17] It is not believed to be directly carcinogenic although there are concerns that nicotine may have tumour-promoting properties in certain circumstances.[15] Nicotine appears to inhibit apoptosis and increase cell proliferation in some cases and thus may be implicated in carcinogenesis.[17] It has been suggested that nicotine and some other agents in tobacco smoke may promote tumour growth *via* nAChRs.[19]

6.4.4 Reproductive Toxicity

Nicotine readily crosses the placenta and at high doses is toxic to the fetus.[15] Associations of smoking with impaired growth and development and possibly with malformations are well known, although the role of nicotine *per se* among the multitude of chemicals in tobacco smoke is not clear. Investigations in animal models have shown that nicotine is toxic to developing neurons during prenatal exposure through complex interactions involving nAChR binding, altered transmitter release and signalling, and adverse effects on neuronal growth.[20]

6.4.5 Neurotoxicity

High doses of nicotine are associated with neurotoxicity, and neuronal degeneration has been reported in rats.[20] On the other hand, nicotine is also reported to exert protective effects on developed neurons, mainly by interaction with nAChRs.[20] For example, nicotine appears to interact with specific CNS nAChRs (mainly α7 and /or α4/β2 subunits) which activate signalling pathways that act neuroprotectively in degenerative disorders such as Alzheimer's disease. Alternatively, nicotine can exert neuroprotective effects *via* other pathways such as an inhibition of apoptosis or an increase in the expression of neurotrophic factors, as demonstrated in some models of chemically induced Parkinsonism. The balance between neuroprotection and neurotoxicity of nicotine could be related to the dose-dependent levels of intracellular calcium arising from nicotine activated nAChRs, whereby high levels of calcium produce neurotoxicity. Nicotine also enhances permeability of the blood–brain barrier and this is thought to be related to interactions with specific nAChRs in brain endothelial cells.[19]

Longer-term exposure to nicotine leads to an increase in the density of binding sites for nAChRs in the brain.[21] Studies in mice have shown that sites labelled with L-^{3}H-nicotine increase in a time- and dose-related manner in some but not all brain regions. On withdrawal of treatment there is a corresponding decrease in sites which tends to mirror the changes in sensitivity to nicotine.[21] Thus nAChR up-regulation on chronic exposure may be related to the development of nicotine tolerance.

6.5 Toxicology of Imidacloprid

Imidacloprid is the ISO name for (*E*)-1-(6-chloro-3-pyridylmethyl)-*N*-nitroimidazolidin-2-ylideneamine (IUPAC). The structural formula is:

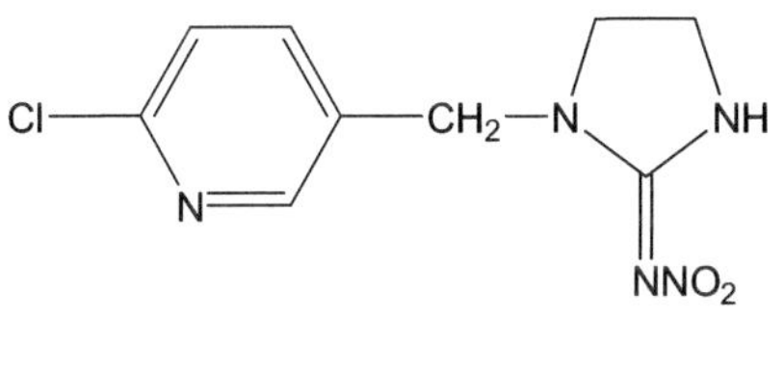

The following summary of the regulatory metabolism and toxicity studies on imidacloprid is taken from regulatory agency reviews.[22–26] Non-regulatory studies reported in the scientific literature are identified separately in the following discussion.

6.5.1 Metabolism and Toxicokinetics

In regulatory studies imidacloprid was rapidly and almost completely absorbed and widely distributed following oral administration in rats. Peak plasma levels (C_{max}) were achieved within 2 h and the initial and terminal half lives in plasma were about 3 h and 26–118 h, respectively. The highest tissue residues after 48 h were found in the liver, kidney, lung and skin. Low concentrations were found in the brain, indicating limited penetration of the blood–brain barrier in contrast to nicotine. Low concentrations were also found in fat, reflecting imidacloprid's relatively low lipophilicity. Imidacloprid was metabolized by two main pathways; oxidative cleavage of the methylene bridge to yield 6-chloronicotinic acid and nitroiminoimidazoline; and hydroxylation of the imidazolidine ring in the 4- or 5- position. Further biotransformation of these initial metabolites occurred in both pathways. The main metabolites in urine were 6-chloronicotinic acid and its glycine conjugate. Two hydroxylated metabolites of imidacloprid (5-hydroxyimidacloprid and 4-hydroxyimidacloprid) were also detected in urine. Imidacloprid and its metabolites were quickly excreted and more than 90% of the dose was eliminated within 24 h of administration. Approximately 75% of the dose was excreted in urine and most of the remaining dose found in faeces originated from biliary excretion. Less than 20% remained as unchanged parent in excreta. There was no evidence of bioaccumulation following oral dosing.

Metabolism studies in the mouse following single intraperitoneal injections of imidacloprid revealed a range of metabolites including those identified in the rat.[27] Imidacloprid, desnitroimidacloprid and nitrosoiminoimidacloprid were detected in the brain. It was noted earlier that desnitroimidacloprid is more potent than nicotine at vertebrate nAChR subtypes and thus may have a role in the toxicity expressed by imidacloprid.[8,13]

6.5.2 Acute Toxicity, Irritation and Sensitization

Imidacloprid exhibits moderate acute oral toxicity in rats with LD_{50} values in the range 379–648 mg kg^{-1} bw. Clinical signs of toxicity were observed typically within an hour of oral gavage dosing and signs at high doses include those indicative of neurological effects, such as staggering gait, decreased activity, salivation, trembling and spasms. The signs were transient and surviving animals usually recovered after the first few days of dosing. The neurological effects were attributed to imidacloprid activation of α-bungaratoxin-sensitive nAChRs.Imidacloprid was slightly more toxic in mice compared with rats with oral LD_{50} values of 130–170 mg kg^{-1} bw.

Imidacloprid was not acutely toxic by the dermal and inhalation routes. There were no deaths or signs of toxicity at a limit dose of 5000 mg kg^{-1} bw after a 24-h dermal exposure in rats. Minimal signs of toxicity were seen in a 4-h nose-only inhalation study in which groups of rats were exposed up to a maximum achievable dust concentration of 69 mg m^{-3} or to an aerosol (nebulized 2.5% solution) concentration of 5323 mg m^{-3}.

Imidacloprid is not irritating to eyes or skin and it is not a skin sensitizer.

6.5.3 Short-Term Toxicity

In regulatory 13-week studies in rats there were no clinical signs of toxicity at dietary concentrations producing average achieved imidacloprid doses of up to about 500 mg kg^{-1} bw day^{-1}. Dose levels of this magnitude are associated with neurological effects following an acute oral dose by gavage administration (see above). The absence of such effects in the dietary studies is considered to reflect toxicokinetic differences. Acute neurological effects are attributed to a C_{max} effect (*i.e.* transient peak circulating levels of imidacloprid) and they are less likely to occur during dietary administration owing to the gradual intake of imidacloprid over several hours. High doses of imidacloprid resulted in reduced bodyweight gain and the liver was identified as the principal target organ for toxicity. Hepatotoxicity in males was characterized by serum biochemical changes, lengthening of the blood clotting time and histopathological changes including cytoplasmic lesions and cellular necroses. Perturbation of liver function was also observed in females but there was no evidence of frank liver toxicity. The hepatic changes induced by imidacloprid were found to be reversible after a 4-week recovery period. Plasma, erythrocyte and brain cholinesterase activities were not affected.The overall no observed adverse effect level (NOAEL) was 14 mg kg^{-1} bw day^{-1} for males and 83 mg kg^{-1} bw day^{-1} for females.

In a 4-week inhalation study the highest atmospheric concentration of 191 mg m^{-3} produced similar toxicity to that seen in the oral studies. Induction of hepatic metabolizing enzymes was observed in response to imidacloprid exposure. Interestingly, plasma cholinesterase activities were decreased by about 27% in females at the higher concentrations. The reasons for this are not clear, since this would not be expected from the mode of action of imidacloprid and no effects on cholinesterases were observed in other regulatory studies. It was speculated that this may have been related to liver dysfunction.[20] The no observed adverse effect concentration (NOAEC) was 5.5 mg m^{-3}.

In a published non-regulatory 90-day study, groups of female rats were dosed orally by gavage with imidacloprid at dose levels of up to 20 mg kg^{-1} bw day^{-1}.[28] The highest dose level resulted in toxicity similar to that reported in the regulatory dietary subchronic studies. There were no clinical signs of toxicity or neurological effects although reduced locomotor activity was reported at 20 mg kg^{-1} bw day^{-1}. Serum and brain acetylcholinesterase (AChE) activities were inhibited in a dose-dependent manner by up to about 50% and 40% of control values, respectively. The authors provided no explanation for this finding and referred to a possible relationship to altered liver function as noted earlier in the summary of the 4-week inhalation study. Neuropathological findings in the brain of high-dose animals consisting of necrosis of Purkinje cells with loss of dendrites and granules in the granular layer of the cerebellum were attributed to imidacloprid exposure. The NOAEL of 10 mg kg^{-1} bw day^{-1} is lower than that reported for females in the regulatory 13-week dietary studies.

Regulatory studies in the dog revealed neurological effects such as tremors, trembling, ataxia and occasional vomiting at high doses of imidacloprid. There was evidence of induction of metabolic enzymes in the livers of high-dose dogs, manifest by slightly increased hepatic cytochrome P450 values and associated liver weight increases. There was no evidence of other target organ effects in studies of up to 1 year in the dog. An overall NOAEL for imidacloprid in the dog is established at 15 mg kg^{-1} bw day^{-1}.

6.5.4 Long-Term Toxicity/Carcinogenicity/Mutagenicity

A comprehensive range of regulatory *in vitro* and *in vivo* genotoxicity studies has been conducted on imidacloprid. The *in vitro* studies showed that imidacloprid is not mutagenic to bacterial and mammalian cells although clastogenic effects were observed at cytotoxic concentrations. No clastogenic effects were observed in the relevant *in vivo* studies and imidacloprid is considered to be non-genotoxic overall. This was confirmed in long-term bioassays for carcinogenicity conducted in the rat and mouse in which there was no evidence for a carcinogenic effect following chronic dietary exposure to imidacloprid for 2 years.

There were no remarkable chronic toxicity findings in the 2-year rat study. An exacerbation of normal age-related mineralization in the colloid of thyroid follicles was not associated with any apparent disturbance to thyroid function. There was no inhibition of plasma, erythrocyte or brain cholinesterases,which was consistent with regulatory subchronic dietary studies. The chronic toxicity NOAEL in the rat was 5.7 mg kg^{-1} bw day^{-1} for males and 25 mg kg^{-1} bw day^{-1} for females.

In the mouse study unusual vocalizations were noted at the highest dose level and increased mortality was observed after anaesthesia and blood withdrawal; this latter finding was also observed in the 15-week study. There was evidence for a slight adaptive effect in the liver and slight alteration to liver function but there was no evidence of chronic hepatotoxicity. An increased incidence of mineralization of the thalamic region of the brain at the highest dose is considered to be equivocal owing to difficulties in drawing a definitive conclusion from the study. The NOAELs of 208 and 274 mg kg^{-1} bw day^{-1} for males and females, respectively, were higher than those in the rat, although an alternative NOAEL of about 47 mg kg^{-1} bw day^{-1} has been suggested based on slightly reduced bodyweight in males at 208 mg kg^{-1} bw day^{-1} and an adjustment to the compound intake calculation.[25]

6.5.5 Reproductive Toxicity

In a regulatory reproduction study two generations of rats were exposed to imidacloprid up to achieved dose levels in the range 40–90 mg kg^{-1} bw day^{-1} except during lactation when they were increased to about 100 mg kg^{-1} bw day^{-1}. Fertility and other indices of reproductive performance of both generations were unaffected by imidacloprid. There were no effects on litter size although there was evidence of toxicity to offspring in the form of reduced bodyweight gain in high-dose pups of both generations during lactation. Slightly lower birth weights were also noted for high-dose second-generation litters of the second mating. The overall NOAEL for both parental and offspring toxicity was concluded to be 20 mg kg^{-1} bw day^{-1} during the pre-mating period and 40 mg kg^{-1} bw day^{-1} during lactation, respectively.

The potential effects of imidacloprid on the developing fetus were investigated in the rat and the rabbit. Imidacloprid was administered by oral gavage to groups of pregnant rats and rabbits from day 6 to day 15 or day 18 of gestation, respectively. The highest dose of 100 mg kg^{-1} bw day^{-1} in the rat and 72 mg kg^{-1} bw day^{-1} in the rabbit produced evidence of toxicity to the dams in the form of reduced bodyweight gain. The effects were more severe in the rabbit, with evidence of weight loss among does. One female rabbit aborted and total litter resorptions were observed in two others in the high-dose group, both findings potentially indicative of toxicity. Toxicity to the fetus was seen in the high-dose groups of both species, however there was no evidence of malformations associated with imidacloprid administration. Mild fetotoxicity in the rat was characterized by an increased incidence of wavy ribs, which is a reversible phenomenon indicative of delayed development. In the rabbit, slightly reduced fetal weights and an increased incidence of reduced ossification indicative of delayed development were observed. The fetoxicity seen in both species may be considered as secondary to maternal toxicity and thus overall imidacloprid is judged not to be a primary developmental toxicant. The NOAEL for both maternal toxicity and fetotoxicity in the rat was 30 mg kg^{-1} bw day^{-1}, whereas in the rabbit the NOAELs for maternal and fetotoxicity were 8 and 24 mg kg^{-1} bw day^{-1}, respectively.

6.5.6 Neurotoxicity

Detailed investigations of the neurotoxic effects of imidacloprid were undertaken in specific regulatory acute, 90-day and developmental neurotoxicity studies in the rat. These studies included comprehensive functional observations and behavioural tests, and detailed morphological examinations of the central and peripheral nervous systems.

In the acute neurotoxicity study, groups of male and female rats were administered single doses of imidacloprid by oral gavage up to the highest dose of 307 mg kg^{-1} which resulted in significant mortality. Clinical signs of toxicity at the highest dose were typical of those associated with severe generalized toxicity in addition to neurological effects such as tremors and uncoordinated gait. At the sublethal dose of 151 mg kg^{-1} there was little evidence of generalized toxicity and neurological effects were limited to tremors. These signs had resolved within 1–5 days of dosing. Decreases in grip strength, motor activity and locomotor activity at 151 mg kg^{-1} bw and above were fully reversible by the time of the next measurements after 7 days of dosing. There were no neuropathological findings related to imidacloprid administration. Therefore, neurotoxicity was confined to reversible neurological effects following a single acute dose. The NOAEL was 42 mg kg^{-1}, the lowest dose tested.

Dietary administration of imidacloprid in the 90-day neurotoxicity study up to an achieved average dose of approximately 200 mg kg^{-1} bw day^{-1} did not result in any mortality or clinical signs of toxicity, although bodyweight gain was reduced at high doses. The only significant change among the battery of functional observations was a decrease in forelimb grip strength; however, this correlated with the reduction in bodyweight and in the absence of any other significant neurological effects it was considered not to be indicative of neurotoxicity.[24] Locomotor activity was slightly reduced in high-dose females. The magnitude of the change is of debatable toxicological significance although it is consistent with the effects seen in the acute neurotoxicity study at a comparable dose level and in the published 90-day toxicity study summarized earlier.[28] Morphological examination of the central and peripheral nervous systems revealed no changes related to imidacloprid. The absence of any neuropathological changes in the brain is in contrast to the changes reported in the published 90-day toxicity study.[28] However, the protocol in the present study included a much more robust assessment of brain morphology. It was concluded that imidacloprid was not neurotoxic under the conditions of this study. The NOAEL for overall toxicity was 9.3 mg kg^{-1} bw day^{-1}.

In the developmental neurotoxicity study female rats received imidacloprid in the diet from the start of gestation through to 21 days *post partum* up to an achieved intake in the highest dose group of approximately 57 mg kg^{-1} bw day^{-1} during gestation and 80 to 155 mg kg^{-1} bw day^{-1} during lactation. Functional observations, learning and memory tasks and developmental landmarks (day of vaginal opening in females and balanopreputial separation in males) of pups were evaluated at intervals for up to 60 days *post partum*. Reduced food intake and slightly reduced bodyweight gain of dams in the high-dose group were indicative of slight maternal toxicity. Toxicity to the offspring observed at the highest dose consisted of reduced bodyweight gain by the end of the dosing period on postnatal day (PND) 21 although bodyweights had largely recovered by PND 60. There were no apparent behavioural changes in pups, with the possible exception of lower motor and locomotor activity in the high-dose group on PNDs 17 and 21 when pups were exposed to imidacloprid. This is consistent with findings in adult rats in the acute neurotoxicity study. There were no effects on PND 60 following cessation of dosing. Histological examination of the brain and selected neural tissues of pups on PND 11 revealed slight reductions in caudate putamen width and thickness of the corpus callosum of the brain in high-dose females. The former finding was also observed at termination of the study. These slight differences were not accompanied by neuropathological changes such as atrophy which is associated with decreased thickness of the corpus callosum.[29] The minor changes in brain morphometry were regarded as potentially significant by two regulatory agencies in the USA but not in Europe or in the authoritative review by the World Health Organization, Joint Meeting of Pesticides Residues (WHO JMPR).[22,23,25,26]

Overall, it may be considered that there was no significant developmental neurotoxicity attributed to imidacloprid exposure with the possible exception of the equivocal morphometric changes in the brain. The NOAEL for toxicity to dams and pups corresponded to maternal intakes of approximately 20–46 mg kg^{-1} bw day^{-1} during gestation and lactation.

The effects on the offspring of pregnant rats given a single high sublethal dose of imidacloprid (337 mg kg^{-1} bw) by intraperitoneal injection on gestation day 9 have been investigated in a published non-regulatory study.[30] Offspring were investigated on PND 30 using a range of behavioural, biochemical and neuropathological parameters. There were no clinical signs of toxicity in the dams or their offspring and no effects on litter parameters. Neurobehavioural tests revealed a decline in sensorimotor performance in the offspring. AChE activities in the midbrain, cortex and brainstem were increased by about 125–145% above control values. It was suggested by the authors that these changes may have contributed to the neurotoxic effects in offspring, but the increases are relatively small in terms of biological significance. Imidacloprid did not affect the ligand binding for $\alpha4\beta2$ subtype nAChRs (α-BGT-insensitive),which is consistent with other data.[8] However, it was found to increase the ligand binding density for the muscarinic acetylcholine receptor m2 mAChR in the cortex and midbrain of offspring. It was suggested that because normal functioning levels of m2 mAChR are involved in learning and memory that the increased ligand binding may lead to memory deficits in exposed individuals. There was an increased expression of glial fibrillary acidic protein (GFAP) in several regions of the brain indicative of glial activation which may be associated with neuronal degeneration. However, there was no evidence of neuronal degeneration in treated offspring. Although the results of this study are of academic interest the administration of a single high dose of imidacloprid by intraperitoneal injection is not representative of potential human exposure and the results need to be interpreted in this light.

6.5.7 Human Data

There are no published epidemiology studies on the health of workers or operators occupationally exposed to imidacloprid during its use as an insecticide; however, no adverse health effects have been reported.[22,23] Incidents of human poisoning largely *via* deliberate ingestion of formulated products have been summarized in a recent publication.[31] Despite the uncertainties of the doses ingested and complications arising from other factors such as the effects of co-exposure to other formulation components, it appears that high doses of imidacloprid can lead to symptoms partly resembling those seen with nicotine poisoning. In severe or fatal cases respiratory failure or aspiration pneumonia and CNS effects including agitation, confusion or coma were reported. In moderate cases of poisoning symptoms typically included nausea and vomiting, drowsiness, dizziness, lack of coordination and other occasional CNS effects reported in patients with severe poisoning. Cardiovascular effects including tachycardia, palpitation and ventricular fibrillation were also noted.

6.6 Toxicology of Thiacloprid

Thiacloprid is the ISO name for (*Z*)-3-(6-chloro-3-pyridylmethyl)-1,3-thiazolidin-2-ylidenecyanamide (IUPAC). The structural formula is:

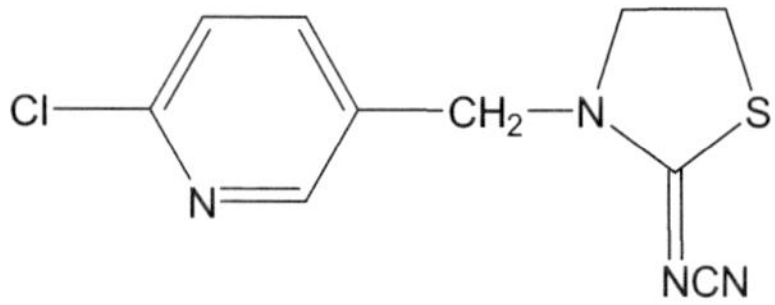

The following summary of the regulatory metabolism and toxicity studies on thiacloprid is taken from regulatory agency reviews.[32–34] A published study on metabolism in the mouse is reported separately in the following discussion.

6.6.1 Metabolism and Toxicokinetics

Thiacloprid was readily absorbed following single oral doses in rats and peak plasma concentrations were achieved within 1–3 h of a low dose and 3–4 h after a high dose. It was widely distributed in the body and residues remaining after 48 h were low. The highest residues were in the liver compared with minimal concentrations in tissues such as the brain and bone marrow. Thiacloprid was extensively metabolized, the main pathways being C- and N-hydroxylation, S-oxidation and methylation, oxidative ring cleavage and methylene bridge cleavage, followed by glucuronic acid, glycine and pentose sulphate conjugations.Thiacloprid was rapidly eliminated mainly in urine and to a lesser extent in the faeces, predominantly *via* biliary excretion. There was no evidence of bioaccumulation.

Chloronicotinic acid and seven of its derivatives were found to be common metabolites of thiacloprid and the other chloropyridinyl neonicotinoid insecticides in metabolism studies in the mouse.[27]

6.6.2 Acute Toxicity, Irritation and Sensitization

Thiacloprid was of moderate acute toxicity to rats by the oral route with LD_{50} values of 396–836 mg kg^{-1} bw and by the inhalation route with LC_{50} values of 1.2 mg L^{-1} (females) to greater than 2.5 mg L^{-1} (males). Clinical signs of toxicity at sublethal and lethal doses included neurological effects characterized typically by tremors, convulsions and uncoordinated gait. Thiacloprid was not acutely toxic by the dermal route in rabbits. It was a slight irritant to the skin and eye in rabbits, causing mild transient irritation. It is not a skin sensitizer in a guinea-pig test.

6.6.3 Short-Term Toxicity

Short-term studies in rodents showed that the liver was the main target organ for thiacloprid. There was evidence of adaptive liver growth characterized by hepatocellular hypertrophy and evidence of metabolic enzyme induction, but there was no significant liver toxicity. Adaptive changes were also observed in the thyroid gland of rats as a consequence of induction of hepatic thyroxine (T4)-uridine diphosphate-glucuronyltransferase (UDPGT), which enhances the metabolism and clearance of T4. The resulting perturbation of the hypothalamic–pituitary–thyroid axis causes the pituitary to release thyroid stimulating hormone (TSH) which leads to a compensatory increase in T4 production with associated enlargement of the thyroid gland. Increased thyroid weights and transiently increased circulating T4 and triiodothyronine (T3) levels were observed in 90-day studies. Increased TSH and evidence of increased thyroid follicular mitotic rate and follicular cell hypertrophy were observed in males in 14-day studies. There were no clinical signs of toxicity in the short-term studies up to dose levels causing general toxicity in the form of reduced bodyweight. Additional toxicity to the adrenal gland characterized by fatty vacuolation of the X-zone leading to hypertrophy was evident in mice after 90 days. The 90-day dietary NOAELs were 29 mg kg^{-1} bw day^{-1} in the rat and 20 mg kg^{-1} bw day^{-1} in the mouse.

The liver and the thyroid gland were also target organs in the subchronic and 1-year dog studies although the effects were less pronounced than in the rat. Slight changes in serum thyroid hormone levels were noted but there were no significant changes to thyroid morphology. Additional findings in the dog included increased prostate weights with moderate hypertrophy of the glandular epithelium. The overall NOAEL in the dog was approximately 9 mg kg^{-1} bw day^{-1}.

6.6.4 Long-Term Toxicity/Carcinogenicity/Mutagenicity

A comprehensive range of regulatory *in vitro* and *in vivo* genotoxicity studies demonstrated that thiacloprid is not genotoxic.

Carcinogenicity studies in the rat and mouse revealed increased frequencies of benign thyroid tumours and malignant uterine tumours in rats and of mainly benign ovarian tumours in mice at high doses.

The increased benign thyroid follicular adenomas were confined to males at the two higher dose levels and are likely to have been related to the effect of prolonged thyroid stimulation resulting from hepatic induction of UDPGT.[26] Male rats are particularly sensitive to the development of thyroid tumours by this mode of action.[35] Humans are far less susceptible to the induction of thyroid tumours and these findings in rats at high doses are usually considered not to be relevant for human exposure.[36] The modes of action of increased uterine tumours in rats and ovarian tumours in mice were not established, although mechanistic investigations suggested that they may have been related to induction of hepatic aromatase and consequent elevated plasma oestradiol concentrations. These increased tumour frequencies are unlikely to be attributed to a genotoxic mode of action given that thiacloprid is clearly non-genotoxic and that the increases only occurred at high doses.

The long-term studies reproduced the adaptive liver changes and associated secondary effects as seen in the shorter-term studies. There was some evidence of liver toxicity in addition to adaptive changes in males. The only other remarkable non-tumour-related pathology consisted of increased incidences of degenerative myelopathy in the nervous system with associated skeletal muscle atrophy, and increased retinal atrophy and lens degeneration in female rats at the mid and high dose levels. However, these differences in spontaneous age-related changes may have been attributed to differences in survival between control and treated rats rather than to a direct effect of thiacloprid. The NOAELs for chronic toxicity in rats and mice were 1.2 mg kg^{-1} bw day^{-1} and 5.7 mg kg^{-1} bw day^{-1}, respectively.

6.6.5 Reproductive Toxicity

Thiacloprid did not adversely affect fertility or mating performance in a two-generation study in rats; however, there was dystocia (difficulty in delivering at parturition) leading to death among some females of the mid- and high-dose groups in the first parental generation. Further studies confirmed this finding, and although a number of mechanistic studies were conducted, the mode of action for thiacloprid induced dystocia was not clearly established.Thiacloprid was shown to induce changes in metabolism of steroid hormones and a possible mechanism specific to rats involving perturbations of progesterone levels at the critical time of parturition was proposed. There were also increases in stillbirths but these were not always associated with dystocia. The overall NOAEL for reproductive effects and toxicity to pups was 3.5 mg kg^{-1} bw day^{-1}.

In studies of prenatal developmental toxicity in the rat and rabbit, increased embryo–fetal loss and delayed fetal development were observed at maternally toxic doses. Thiacloprid was not teratogenic and the embryo–fetal toxicity was considered to be secondary to maternal toxicity. The NOAELs for both maternal and developmental toxicity in the rat and rabbit were 10 mg kg^{-1} bw day^{-1} and 2 mg kg^{-1} bw day^{-1}, respectively.

6.6.6 Neurotoxicity

In studies of the acute neurotoxicity of thiacloprid there was no evidence of permanent neurotoxic effects following single sub lethal oral doses of up to 109 mg kg^{-1} bw. Transient neurological signs after dosing included tremors and ataxia. Neurobehavioural changes consisted of impaired motor and locomotor activity on the day of dosing. The NOAELs for acute neurotoxicity were 11 mg kg^{-1} bw in males and 3.1 mg kg^{-1} bw in females.

A dietary 90-day neurotoxicity study produced no clinical signs of neuro-toxicity or neuropathological changes in rats given thiacloprid up to dose levels causing slight general toxicity. The NOAEL for neurotoxicity was the highest dose level tested, equivalent to an achieved intake of 100–115 mg kg^{-1} bw day^{-1}. The absence of neurological effects at dose levels similar to that producing neurological signs in the acute study reflects differences in toxicokinetics between the two routes of administration.

There was no evidence of any specific developmental neurotoxicity in the offspring of rats given dietary concentrations of thiacloprid up to maternally toxic dose levels. Toxicity to pups manifest as reduced bodyweight and delayed sexual maturation in both sexes was attributed to the secondary consequences of maternal toxicity. The overall NOAEL for toxicity to both dams and offspring was equivalent to an achieved maternal intake of 4.4 mg kg^{-1} bw day^{-1}.

6.7 Toxicology of Acetamiprid

Acetamiprid is the ISO name for (*E*)-N^1-[(6-chloro-3-pyridyl)methyl]-N^2-cyano-N^1-methylacetamidine (IUPAC). The structural formula is:

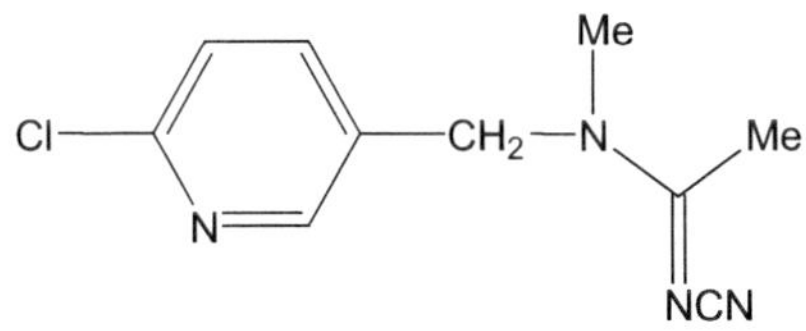

There are no published studies on the mammalian toxicity of acetamiprid and the following summary of the regulatory metabolism and toxicity studies on acetamiprid is taken from regulatory agency reviews.[37,38] A published study on metabolism in the mouse is reported separately in the following discussion.

6.7.1 Metabolism and Toxicokinetics

Acetamiprid is rapidly and almost completely absorbed after 24 h following single oral doses in rats. Peak plasma concentrations (C_{max}) were achieved within 0.5–7 h after single oral doses and intravenous administration. The highest residues were in the liver, adrenal, thyroid and kidney. Acetamiprid was extensively metabolized, mainly to chloronicotinic acid and demethylated compounds. Acetamiprid and its metabolites were rapidly eliminated mainly in urine after single and repeated oral administration and there was no potential for bioaccumulation.

In common with the other chloropyridinyl neonicotinoid insecticides, chloronicotinic acid and seven of its derivatives were found following administration of acetamiprid to the mouse.[27]

6.7.2 Acute Toxicity, Irritation and Sensitization

The acute oral toxicity of acetamiprid in the rat appears to have been assessed in two separate studies with LD_{50} values of 314–417 mg kg^{-1} bw and 146–217 mg kg^{-1} bw reported. Acetamiprid was moderately toxic by the inhalation route with an LC_{50} value of greater than 1.15 mg L^{-1}. Acetamiprid was not acutely toxic by the dermal route in rats. It is not irritating to the skin and eyes and it is not a skin sensitizer.

6.7.3 Short-Term Toxicity

Short-term regulatory studies in rodents and dogs produced generalized non-specific toxicity characterized by reduced bodyweight and food consumption at high dose; however, there was no evidence of significant target organ toxicity. Largely adaptive changes were observed in the liver of rodents manifest by hepatocellular hypertrophy in both rats and mice and hepatocellular vacuolation in the rat. Decreased glucose and cholesterol levels in mice may have been related to minor perturbations in liver function. Tremors were observed among high-dose female mice, but no other clinical signs of toxicity were reported. The NOAELs were approximately 12–15 mg kg^{-1} bw day^{-1} in rats, 106–129 mg kg^{-1} bw day^{-1} in mice and 20–21 mg kg^{-1} bw day^{-1} in dogs.

6.7.4 Long-Term Toxicity/Carcinogenicity/Mutagenicity

A comprehensive range of regulatory *in vitro* and *in vivo* genotoxicity studies demonstrated that acetamiprid is not genotoxic overall. Although a clastogenic effect was observed in an *in vitro* assay there was no evidence of clastogenicity in the corresponding *in vivo* study.

Acetamiprid was not oncogenic in lifetime bioassays in the rat and mouse up to dose levels producing generalized toxicity. Changes in the liver were observed in the rat, in common with the findings in the short-term studies. An increased incidence of microconcretions in the kidney papilla was noted in high-dose male rats. The incidence of mammary hyperplasia was increased in high-dose female rats but the toxicological significance was reported to be uncertain. Amyloidosis of multiple organs was increased in mice but no other evidence of toxicity was noted. The NOAELs for chronic toxicity were 7.1–8.8 mg kg^{-1} bw day^{-1} in rats and 20–76 mg kg^{-1} bw day^{-1} in mice.

6.7.5 Reproductive Toxicity

There was no adverse effect of acetamiprid on fertility or mating performance in a two-generation study in rats. Parental toxicity was characterized by reduced bodyweight and food consumption at high doses. Litter size and pup viability were reduced in the second-generation offspring and delayed sexual maturation was observed in the offspring of both sexes. Significant reductions in pup bodyweights were noted in both generations and it was reported that these effects were more severe than the parental effects. The NOAELs for reproductive effects and parental and offspring toxicity were 18–22 mg kg^{-1} bw day^{-1}.[37] A lower NOAEL of 6.5 mg kg^{-1} bw day^{-1} has also been reported.[38]

In developmental toxicity studies acetamiprid produced slight fetotoxicity in the rat at a maternally toxic dose, but there was no evidence of developmental toxicity in the rabbit in the presence of maternal toxicity. The NOAEL for both maternal and developmental toxicity in the rat was 16 mg kg^{-1} bw day^{-1}. The maternal NOAEL in the rabbit was 15 mg kg^{-1} bw day^{-1} and the developmental NOAEL was 30 mg kg^{-1} bw day^{-1}, the highest dose tested.

6.7.6 Neurotoxicity

In an acute neurotoxicity study, single oral sublethal doses of acetamiprid up to 100 mg kg^{-1} bw produced no evidence of neuropathology or other indications of permanent neurotoxicity. Neurological and behavioural effects were largely confined to the day of dosing. Neurological signs in high-dose animals included tremors, unsteady gait and dilated pupils. Decreased locomotor activity, forelimb grip strength and hindlimb foot splay were observed, mainly in high-dose animals. A slight decrease in the duration of movement persisted in mid and high dose males on days 7 and 14 after dosing. The acute neurotoxicity NOAEL was 10 mg kg^{-1} bw.

No clinical signs of neurotoxicity or neuropathological changes were observed in a 90-day dietary neurotoxicity study up to dose levels producing slight general toxicity. The NOAEL for neurotoxicity was at least 60 mg kg^{-1} bw day^{-1}.

In a developmental neurotoxicity study signs of toxicity consisting of reduced pup bodyweight and pup viability were observed at the highest dose tested (45 mg kg^{-1} bw day^{-1}). There was slight maternal toxicity at this dose level. The only apparent treatment related neurobehavioural change consisted of significantly decreased auditory startle response (ASR) amplitudes in high-dose males on PND 20 and PND 60. However, there were no neuropathological changes in the central or peripheral nervous systems and this isolated finding in one sex is probably of equivocal toxicological significance. The NOAEL for offspring toxicity was 10 mg kg^{-1} bw day^{-1} based on the maternal dose.

6.8 Toxicology of Nitenpyram

Nitenpyram is the ISO name for (*E*)-*N*-(6-chloro-3-pyridylmethyl)-*N*-ethyl-*N′*-methyl-2-nitrovinylidenediamine (IUPAC). The structural formula is:

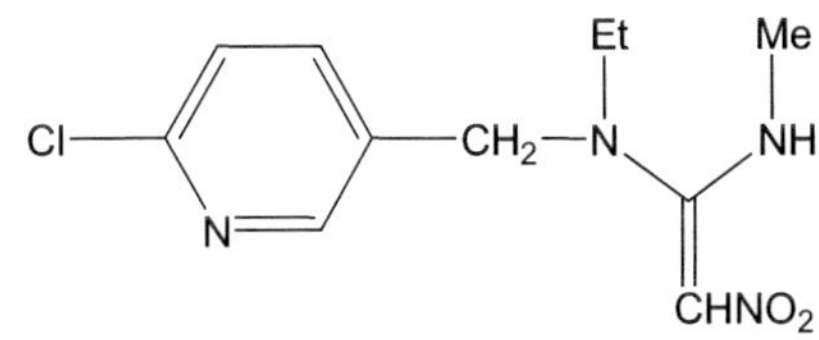

There are no published studies on the mammalian toxicity of nitenpyram and no regulatory agency reviews are publicly available. Limited toxicological information is available, including safety studies conducted on cats and dogs to support the registration of nitenpyram for the control of flea infestations.

6.8.1 Metabolism and Toxicokinetics

Nitenpyram was rapidly absorbed orally in cats and dogs with maximum blood levels attained within 0.6 and 1.2 h, respectively.[39] The half-life of nitenpyram in dogs (2.8 h) was shorter than that in cats (7.7 h).

Metabolism studies in mice given single intraperitoneal injections of nytenpyram showed that 7% of the dose excreted in urine was the N-demethylated derivative.[27] This metabolite was also detected in the brain, liver and plasma. About 40% of the dose excreted in urine was unchanged nitenpyram. Cleavage products included chloronicotinic acid and its seven derivatives.

6.8.2 Acute Toxicity, Irritation and Sensitization

Nitenpyram is of low acute toxicity. The acute oral LD_{50} of nitenpyram in rats is 1575–1680 mg kg^{-1} bw, the acute dermal LD_{50} is greater than 2000 mg kg^{-1} bw and the acute inhalation LC_{50} is greater than 5.8 mg L^{-1}. It is slightly irritating to the eye but it is not a skin irritant or a skin sensitizer.[40]

6.8.3 Short-Term Toxicity

Nitenpyram was administered orally in formulated tablets to cats and dogs at up to 10 times (10×) the minimum recommended dose of 1 mg kg^{-1} daily for 14–42 days. It was also administered as a single oral dose of 10× to dogs. No adverse effects were reported. Administration for 6 months at five times (5×) the recommended dose was also without adverse effects.[41]

6.8.4 Long-Term Toxicity/Carcinogenicity/Mutagenicity

Nitenpyram is not carcinogenic or mutagenic. The chronic toxicity NOEL (no observed effect level) in rats is 129 mg kg^{-1} bw day^{-1} for males and 54 mg kg^{-1} bw day^{-1} for females although the toxicity findings were not reported. The NOEL in a 1-year dog study was reported to be 60 mg kg^{-1} bw day^{-1}.[40]

6.8.5 Reproductive Toxicity

Nitenpyram does not adversely affect reproductive performance in rats and it is not teratogenic in rats and rabbits.[40]

The potential effects of nitenpyram on reproductive performance in cats and dogs were investigated in single-generation reproduction studies. Oral doses of either the recommended dose (1×) or three times (3×) the recommended dose were administered daily. There were no adverse effects on reproductive performance and there was no evidence of parental or neonatal toxicity.[41]

6.9 Toxicology of Thiamethoxam

Thiamethoxam is the ISO name for (*EZ*)-3-(2-chloro-1,3-thiazol-5-ylmethyl)-5-methyl-1,3,5-oxadiazinan-4-ylidene(nitro)amine (IUPAC). The structural formula is:

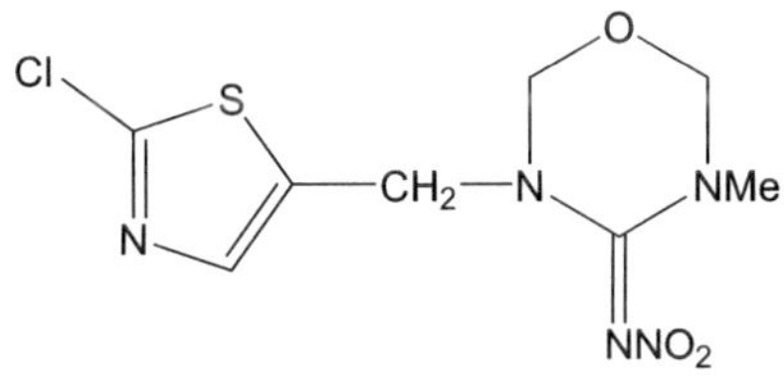

The following summary of the regulatory metabolism and toxicity studies on thiamethoxam is taken from regulatory agency reviews.[42–45] Studies reported in the scientific literature are identified separately in the following discussion.

6.9.1 Metabolism and Toxicokinetics

Thiamethoxam is rapidly and almost completely absorbed following single oral doses in the rat. It is widely distributed in the body and the highest tissue residues were found in the liver. Oral absorption in the mouse was reported to be lower than that in the rat, being about 75% of the administered dose. Thiamethoxam is poorly metabolized in the rat at the highest dose level (100 mg kg^{-1} bw) and 70–80% of the dose was eliminated as unchanged thiamethoxam. This is in contrast to complete metabolism at the lowest dose level (0.5 mg kg^{-1} bw). The predominant route of excretion was *via* urine for both unchanged thiamethoxam and its metabolites. About 95% of the dose was excreted in urine and 5% in faeces within 168 h. Tissue residues after 7 days were extremely low and there was no evidence of accumulation. In the mouse 30–60% of the administered dose was excreted as metabolites.

The metabolism of thiamethoxam follows the same major pathways in both the rat and the mouse as shown in Figure 6.3. However, there are significant species differences in quantitative metabolism which are discussed later in the context of species differences in hepatocarcinogenicity.[46,47]

The major biotransformation is cleavage of the oxadiazine ring to form the corresponding nitroguanidine CGA322704 (clothianidin). The *N*-demethylated nitroguanidine metabolite CGA265307 is formed either directly from clothianidin or *via* the intermediate *N*-demethylated thiamethoxam metabolite (CGA330050). Blood profiling of metabolites showed that CGA330050 and CGA265307 are present in 15-fold and 140-fold greater concentrations in the mouse compared with the rat. Comparative *in vitro* metabolism studies in rat, mouse and human liver microsomal fractions confirmed the species differences between rats and mice *in vivo* and indicated that the rates of metabolism by these pathways in humans would be equal to or lower than those in the rat.[47] Other studies in the mouse following single intraperitoneal injections of

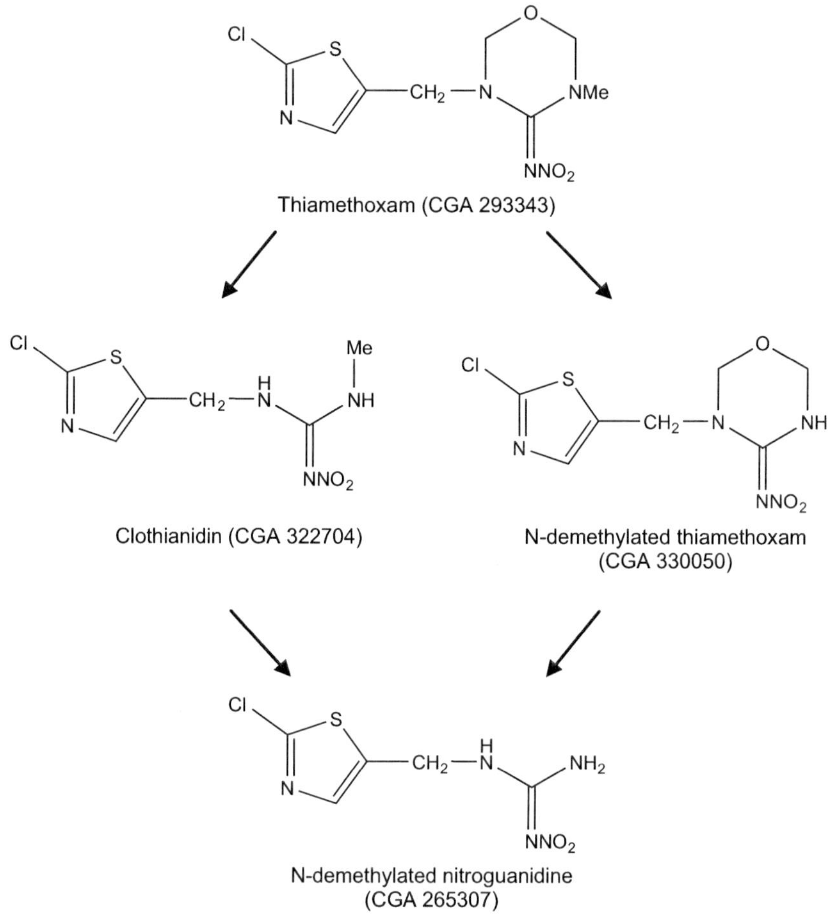

Figure 6.3 Thiamethoxam and its major metabolites in mice and rats. (Reproduced with permission from ref. 47).

thiamethoxam established that the major pathways of metabolism involved *N*-methyl and *O*-methylene hydroxylation and nitro reduction to form the metabolites identified above and further minor metabolites in urine.[48] Interestingly, the *N*-demethylated nitroguanidine metabolite of clothianidin (CGA265307) was a major metabolite in the brain and was remethylated in part to clothianidin. The metabolites formed from clothianidin were common to those found after dosing mice directly with clothianidin.

6.9.2 Acute Toxicity, Irritation and Sensitization

Thiamethoxam was of low acute acute oral toxicity in the rat (LD_{50} 1563 mg kg^{-1} bw) and the mouse (871 mg kg^{-1} bw). Neurological signs including tonic or clonic convulsions were observed. It was not acutely toxic by the dermal route in the rat ($LD_{50} > 2000\,mg\,kg^{-1}$ bw). There was no significant toxicity by inhalation up to the highest achievable concentration of $3.72\,mg\,m^{-3}$. Thiamethoxam is not a skin or eye irritant and it is not a skin sensitizer.

6.9.3 Short-Term Toxicity

The liver was identified as a target organ in rodents with evidence of both adaptive changes and toxicity. In the rat this was characterized principally by hepatocellular hypertrophy and inflammatory changes whilst in the mouse there was additional single-cell necrosis and apoptosis. The kidney was also a target organ in the rat although there was a sex difference in the nature of morphological changes and sensitivity. Males exhibited hyaline droplet accumulation in the renal tubular epithelium with associated acute and chronic tubular lesions which is characteristic of α-2-microglobulin nephropathy, a phenomenon specific to male rats that is not relevant to humans. In females there was evidence of nephrocalcinosis and chronic tubular lesions at higher doses. Other findings included fatty changes in the adrenal glands, changes in the spleen and follicular epithelial hypertrophy in the thyroid gland of rats. The latter finding may have been secondary to liver enzyme induction consistent with adaptive changes in the liver. The lowest relevant NOAELs in the rodent studies were $93\,mg\,kg^{-1}$ bw day^{-1} in rats (females) and $1.4\,mg\,kg^{-1}$ bw day^{-1} in mice (males).

Toxicity in the dog was characterized mainly by significantly lower weight gain, atrophy of the thymus and spleen and delayed maturation of the testes and ovaries. However, the changes in these organs may have been secondary to substantial growth retardation. Thc overall NOAEL in the dog was approximately $4\,mg\,kg^{-1}$ bw day^{-1}.

6.9.4 Long-Term Toxicity/Carcinogenicity/Mutagenicity

Thiamethoxam was not genotoxic in a comprehensive range of regulatory *in vitro* and *in vivo* genotoxicity studies.

Chronic studies revealed no evidence of an oncogenic response in rats, but there was an increased incidence of liver tumours in mice. Subsequent investigative studies established a non-genotoxic mode of action for hepatocarcinogenesis in the mouse which was shown not to be relevant for human exposure.[46,47,49] It was established that the metabolite *N*-demethylated thiamethoxam (CGA330050) was primarily responsible for the induction of hepatotoxicity and subsequent development of tumours (see Figure 6.3). Metabolite CGA265307 exacerbated the hepatoxicity induced by CGA330050 as a result of inhibition of inducible nitric oxide synthase, however, it was not directly hepatotoxic.[46] Comparative studies in the rat and mouse demonstrated that the absence of liver tumours in rats was related to substantially lower levels of CGA330050 in this species, which were insufficient to cause the hepatotoxicity characteristic of that seen in the mouse.[47] It was shown that potential human exposures to thiamethoxam would not generate sufficient levels of CGA330050 to cause hepatotoxicity and therefore it is highly unlikely that thiamethoxam would pose a carcinogenic hazard to humans.[47,49] Although clothianidin is a major metabolite of thiamethoxam it could not be converted into CGA330050, nor did it induce the characteristic hepatotoxicity observed with thiamethoxam in the mouse.[46] This correlated with the absence of liver tumours in chronic rodent bioassays on clothianidin (see section 6.10.4).

6.9.5 Reproductive Toxicity

There were no adverse effects on reproductive performance in two reproduction studies in the rat. Increased incidences of testicular tubules with germ cell loss/disorganization ± Sertoli cell vacuolation and sperm abnormalities were noted for high-dose males of the F1 generation although the effects were not consistent in both studies. These findings did not affect male reproductive function. Reduced pup weights were indicative of toxicity to offspring at high doses. The overall NOAEL for reproduction based on the findings in males was either 62 mg kg^{-1} bw day^{-1} or approximately 1 mg kg^{-1} bw day^{-1} depending on the interpretation of different agencies.[42–45] One agency in the USA considered that there were toxicologically significant effects at 3 mg kg^{-1} bw day^{-1} and the NOAEL was concluded to be 1.2 mg kg^{-1} bw day^{-1}.[44]

Developmental toxicity studies in the rat and rabbit revealed evidence of embryo-fetotoxicity at maternally toxic doses in both species characterized by reduced fetal weight and delayed ossification, and additionally in the rabbit by increased post-implantation loss. These findings were probably secondary to maternal toxicity. There was no evidence of teratogenicity associated with *in utero* exposure to thiamethoxam. The maternal and developmental NOAELs were 30 and 200 mg kg^{-1} bw day^{-1} in the rat and 50 mg kg^{-1} bw day^{-1} in both cases in the rabbit.

6.9.6 Neurotoxicity

Thiamethoxam administered to rats in an acute neurotoxicity study resulted in general clinical signs of severe toxicity at high lethal doses. Neurological effects at a sublethal dose of 500 mg kg^{-1} bw day^{-1} consisted of decreased motor activity and increased forelimb grip strength 2–3 h after dosing. There were no neuropathological changes and the acute neurotoxicity NOAEL was 100 mg kg^{-1} bw.

No clinical signs of neurotoxicity or neuropathological changes were observed in a 90-day neurotoxicity study. The NOAELs for neurotoxicity were 95 mg kg^{-1} bw day^{-1} in males and 216 mg kg^{-1} bw day^{-1} in females, the highest doses tested in each case.

In a developmental neurotoxicity study signs of toxicity consisting of reduced pup bodyweight and bodyweight gain were observed. Delayed sexual maturation in males may have been secondary to the reduced weight gain. A reduction in brain size and weight and significant differences in morphometric measurements in certain regions of the brain were reported in the offspring.

However, because no effect on brain weight relative to bodyweight or neuro-behavioural or neuropathological changes were observed the differences were considered secondary to reduced growth rather than a direct adverse effect of thiamethoxam. The overall NOAEL was 34.5 mg kg^{-1} bw day^{-1} based on toxicity to offspring.

In a non-regulatory study groups of adult rats were administered thiamethoxam by subcutaneous injection for 7 days at dose levels of up to 100 mg kg^{-1} bw day^{-1}.[50] There were no clinical signs of toxicity, although neuro behavioural investigations revealed a decrease in spontaneous motor activity and an apparent change in anxiety behaviour at the highest dose. It was suggested that a putative anxiogenic effect of thiamethoxam may have been mediated by nAChR activation and an associated increased release of serotonin in the dorsal hippocampus of the brain because this has been observed with high doses of nicotine. Measurements of AChE activity in the hippocampus, striatum and cortex showed a slight non-dose-related reduction in activity of 20–28% relative to control 2 h after the last dose at the mid and high doses. A slight reduction in AChE activity persisted at the highest dose 7 days after dosing. The authors noted that subchronic dosing of nicotine also reduced AChE activity in the rat brain. It is not known whether these effects of nicotine or thiamethoxam are the result of their interaction with nAChRs or a direct effect on AChE activity. Hippocampal high-affinity choline uptake (HACU) was also reduced 2 h after the last dose possibly in response to potentially altered intraterminal acetylcholine content.

6.10 Toxicology of Clothianidin

Clothianidin is the ISO name for (*E*)-1-(2-chloro-1,3-thiazol-5-ylmethyl)-3-methyl-2-nitroguanidine (IUPAC). The structural formula is:

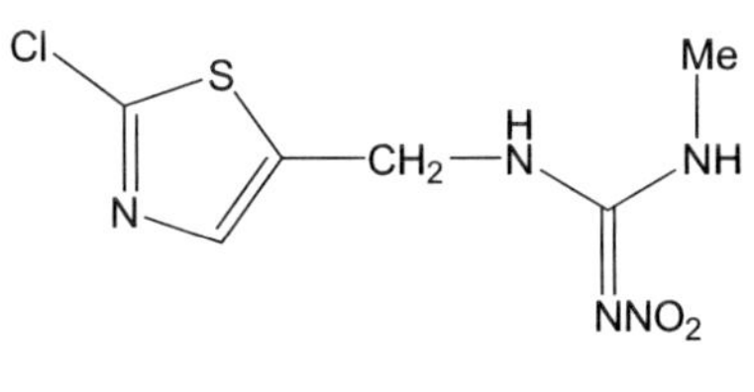

The following summary of the regulatory metabolism and toxicity studies on clothianidin is taken from regulatory agency reviews and manufacturer's information.[51–54] A published study on metabolism in the mouse is reported separately in the following discussion.

6.10.1 Metabolism and Toxicokinetics

Clothianidin was rapidly and almost completely absorbed (89–95%) after an oral dose in the rat. It was widely distributed up to 72 h and there was a slight preference for higher residues in the liver, kidney, adrenals and urinary bladder.

Clothianidin and its metabolites were rapidly excreted with 91% and 57% of the applied dose found in urine after 24 h in the low- and high-dose groups, respectively. Less than 7% was excreted in faeces and there was limited enterohepatic circulation. There was no potential for accumulation. Between 55% and 73% of administered clothianidin was excreted unchanged, demonstrating that clothianidin is poorly metabolized in the rat. The major biotransformation pathways are oxidative demethylation and oxidative cleavage of the methylene bridge.

Earlier it was shown that many of the metabolites of clothianidin are the same as those from thiamethoxam and that the metabolism of clothianidin follows the same major pathways of *N*-demethylation and nitro reduction.[48]

6.10.2 Acute Toxicity, Irritation and Sensitization

Clothianidin was not acutely toxic in the rat by oral, dermal and inhalation routes although it was of moderate acute oral toxicity in the mouse. The acute oral and dermal LD_{50} values in the rat were greater than 5000 mg kg^{-1} bw and greater than 2000 mg kg^{-1} bw, respectively, whereas the acute oral LD_{50} in the mouse was 425 mg kg^{-1} bw. The acute inhalation LC_{50} in the rat was greater than 6.1 mg L^{-1}. Clothianidin is not a skin and eye irritant and it is not a skin sensitizer.

6.10.3 Short-Term Toxicity

No specific target organ toxicity was reported for clothianidin in short-term oral toxicity studies in rodents and dogs apart from the haematopoietic system in dogs. Reduced bodyweight and weight gain indicative of non-specific general toxicity were observed in the rat and dog at high doses. There was clinical evidence of anaemia in some dogs at high doses and decreased circulating white blood cell counts, albumin and total protein levels were noted in females. The overall NOAELs were approximately 30 mg kg^{-1} bw day^{-1} in the rat and 40 mg kg^{-1} bw day^{-1} in the dog.

6.10.4 Long-Term Toxicity/Carcinogenicity/Mutagenicity

Clothianidin and its major metabolites were not genotoxic in a comprehensive range of regulatory *in vitro* and *in vivo* genotoxicity studies. Although there was a clastogenic response at cytotoxic concentrations of clothianidin in two *in vitro* assays these were not reproduced in the relevant *in vivo* study.

There was no evidence of oncogenicity associated with the long-term administration of clothianidin to rats and mice. Reduced bodyweight and weight gain were observed in the mouse study at high doses, but no specific target organ toxicity was reported other than a possible increased incidence of hepatocellular hypertrophy. In the 2-year rat study liver toxicity was characterized by altered hepatocellular eosinophilic foci in both sexes and mottled livers in males. Additional findings included interstitial gland hyperplasia and increased lymphohistiocytic infiltrates in the ovaries, and slightly increased incidences of pelvic mineralization and transitional hyperplasia in the kidneys. The overall NOAELs for chronic toxicity were 33 mg kg^{-1} bw day^{-1} (females) in rats and 65 mg kg^{-1} bw day^{-1} (females) in mice.

6.10.5 Reproductive Toxicity

There were no effects on fertility in rats administered clothianidin in a two-generation reproduction study, but there was an increase in stillbirths in both generations. Parental and offspring toxicity was manifest by reduced body-weight gain and reduced thymus weights. The latter finding may indicate an effect on the immune system; however, no evidence of immunotoxicity was reported in other studies. Delayed sexual maturation in first-generation offspring was secondary to reduced growth. Adverse effects on sperm characterized by decreased sperm motility and increased numbers of sperm with detached heads were not associated with any effects on male fertility. The overall NOAEL for reproductive and parental toxicity was 31 mg kg^{-1} bw day^{-1} and that for offspring toxicity was 10 mg kg^{-1} bw day^{-1}.

Clothianidin did not cause any adverse developmental effects following *in utero* exposure of rats up to maternally toxic doses. In a rabbit developmental toxicity study there was evidence of premature delivery and abortion at dose levels causing mortality and reduced weight gain. Reduced ossification of the sternal centra was indicative of fetotoxicity and there was an increased incidence of a missing lobe of the lung in fetuses. The adverse developmental effects in the rabbit were observed only in the presence of severe maternal toxicity which reduces the level of concern. The NOAELs for maternal and developmental toxicity were 10 mg kg^{-1} bw day^{-1} and 125 mg kg^{-1} bw day^{-1} (highest dose tested) in the rat and 25 mg kg^{-1} bw day^{-1} in both cases in the rabbit.

6.10.6 Neurotoxicity

Transient neurobehavioural effects consisting of decreased arousal and decreased motor activity were observed in an acute neurotoxicity study in rats. There were no indications of any specific or permanent neurotoxicity and the NOAEL was 60 mg kg^{-1} bw. There was no evidence of neurotoxicity in a subchronic study up to dose levels causing non-specific general toxicity.

Developmental neurobehavioural effects characterized by reduced motor activity and surface righting reflex in offspring were reported in a developmental neurotoxicity study in rats at doses causing general toxicity to offspring and dams. Brain morphometry findings were also noted but no details are available. The developmental NOAEL was 43 mg kg^{-1} bw day^{-1} based on maternal intake of clothianidin during gestation.

A neurotoxicity and pharmacology study in mice produced transient decreased motor activity, tremors and deep respirations at 50 mg kg^{-1} bw day^{-1}

and the NOAEL was 25 mg kg^{-1} bw day^{-1}. These findings suggest that the mouse is more sensitive than the rat, which is consistent with the differential sensitivity shown for acute oral toxicity.

6.11 Toxicology of Dinotefuran

Dinotefuran is the ISO name for (*EZ*)-(*RS*)-1-methyl-2-nitro-3-(tetrahydro-3-furylmethyl)guanidine (IUPAC). The structural formula is:

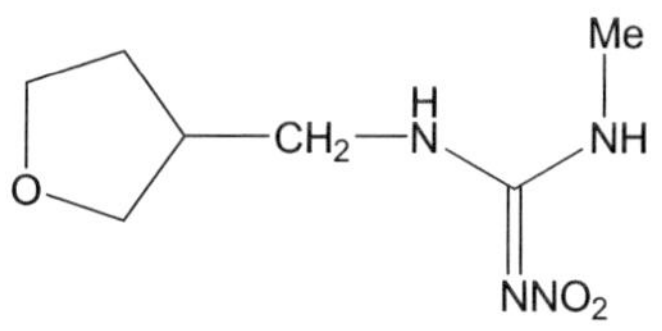

The following summary of the regulatory metabolism and toxicity studies on dinotefuran is taken from regulatory agency reviews and an independent review.[55–57] A published study on metabolism in the mouse is reported separately in the following discussion.

6.11.1 Metabolism and Toxicokinetics

Dinotefuran was rapidly and almost completely absorbed (>90%) after low and high oral doses in the rat. Maximum plasma concentrations were achieved between 0.5 and 2 h of dosing. Dinotefuran was widely distributed throughout the body and then completely eliminated within 168 h. The highest tissue residues were found in the stomach and kidneys. Almost all of the radioactivity was excreted *via* the urine and was up to 99% complete by 24 h after dosing. The vast majority of material was excreted unchanged following oral or intravenous dosing demonstrating that dinotefuran was poorly metabolized.

Studies in the mouse revealed that dinotefuran is metabolized by *N*-demethylation, nitro reduction, tetrahydrofuran hydroxylations and *N*-methylene hydroxylation and amine cleavage to give a complex metabolic pathway.[48] Some of the metabolites of dinotefuran were similar to those of clothianidin because of the structural similarity of both neonicotinoids whereby the easily metabolized tetrahydrofuranyl moiety replaces the chlorothiazolyl substituent in clothianidin (see Figure 6.1).

6.11.2 Acute Toxicity, Irritation and Sensitization

Dinotefuran exhibited very low acute oral toxicity to rats and mice (LD_{50} > 2000 mg kg^{-1} bw). It was not acutely toxic in the rat by the dermal (LD_{50} > 2000 mg kg^{-1} bw) or inhalation (LC_{50} > 4 mg L^{-1}) routes. Dinotefuran is a slight skin and eye irritant but it is not a skin sensitizer.

6.11.3 Short-Term Toxicity

Dinotefuran administration to rodents and dogs revealed mild toxicity in short-term studies. The mouse was particularly insensitive to dinotefuran. Reduced bodyweight and weight gain indicative of general non-specific toxicity were commonly observed at high doses. The only significant target organ effect in the rat was increased vacuolization of the zona glomerulosa and zona fasciculata of the adrenal cortex. No target organ effects were observed in the mouse. Thymus weights were reduced in the 1-year dog study but there were no associated histopathological changes. The NOAELs were 34 mg kg^{-1} bw day^{-1} in the rat, over 4000 mg kg^{-1} bw day^{-1} in the mouse and 22 mg kg^{-1} bw day^{-1} in the dog.

6.11.4 Long-Term Toxicity/Carcinogenicity/Mutagenicity

Dinotefuran was not genotoxic in a comprehensive range of regulatory *in vitro* and *in vivo* genotoxicity studies and it was not oncogenic in chronic rodent bioassays.

Chronic toxicity characterized by reduced weight gain and increased pelvic ulceration and mineralization of male kidneys was observed at the highest dose in rats. Although bodyweight gain was reduced in the mouse study there were no significant target organ effects. The NOAELs were 100 mg kg^{-1} bw day^{-1} in the rat and 345 mg kg^{-1} bw day^{-1} in the mouse.

6.11.5 Reproductive Toxicity

There were no adverse effects of dinotefuran on reproductive performance in a two-generation reproduction study in the rat. Toxicity to both parents and offspring was manifest by reduced bodyweight gain at high doses. Forelimb or hindlimb grip strength was reduced in the offspring of the first generation but no other neurobehavioural changes were noted. The most significant effect was an adverse change in sperm quality although the changes were inconsistent between generations and did not result in decreased male fertility. Reduced sperm motility, sperm count and altered sperm morphology were observed in first-generation parents whereas only sperm morphology was affected in second-generation parents. Minor changes were also reported in the female reproductive tract. The overall NOAEL was 241 mg kg^{-1} bw day^{-1}.

No evidence of developmental toxicity was apparent when dinotefuran was administered to pregnant rats and rabbits up to maternally toxic doses. The maternal NOAELs were 300 mg kg^{-1} bw day^{-1} in the rat and 52 mg kg^{-1} bw day^{-1} in the rabbit.

6.11.6 Neurotoxicity

There were no signs of significant neurotoxicity in both the acute and sub-chronic neurotoxicity studies. Transient decreases in motor activity were observed at the highest dose levels in both studies but there were no other neurobehavioural changes. The acute and subchronic NOAELs were 750 mg kg^{-1} bw and 327 mg kg^{-1} bw day^{-1}, respectively.

6.12 Discussion

Neonicotinoids are rapidly and readily absorbed following oral administration and they are widely distributed throughout the body. Absorption of nicotine is facilitated by alkaline pH and it is also widely distributed. However, in contrast to neonicotinoids it easily penetrates the blood–brain barrier to produce a wide range of centrally mediated neurological signs and symptoms. Nicotine is extensively metabolized and the same is true for the majority of neonicotinoids. Both nicotine and neonicotinoids are rapidly excreted and do not accumulate in the body. Nicotinoid metabolites derived from some neonicotinoids such as imidacloprid and thiacloprid are more potent vertebrate nAChR agonists than their parent molecules and are likely to be more toxic.

The toxicological characteristics of nicotine and neonicotinoids, while showing some similarities at acute neuropharmacologically active doses, generally display significant differences in overall toxicity. The toxicity profiles of nicotine and neonicotinoids are summarized in Table 6.1.

The acute reversible neurological effects observed for the majority of neonicotinoids tend to be associated with other signs of severe toxicity at high doses. This is consistent with the relative insensitivity of vertebrate nAChRs to neonicotinoids compared with nicotine. Acute and longer-term exposure to neonicotinoids is rarely associated with adverse morphological changes to the central or peripheral nervous systems, indicating that neurological effects are of pharmacological rather than pathological origin. The acute neurotoxic potency of neonicotinoids was variable, reflecting differences in their chemical structures, affinity for nAChRs and metabolism. Although cholinesterase inhibition has been occasionally reported for imidacloprid and thiamethoxam, the changes are relatively small and have not been reproduced in other studies. Inhibition of cholinesterase would not be expected from the known mode of action of neonicotinoids and the reasons for the reported effects were not established.

The liver is a common target organ among the neonicotinoids and hepatic microsomal enzyme induction accompanied by adaptive liver growth is seen with most neonicotinoids. Evidence of frank hepatotoxicity is observed with imidacloprid and thiamethoxam. The hepatotoxicity observed in the mouse with thiamethoxam led to the formation of liver tumours, but the mode of action was found to be specific for the mouse because of key species differences in metabolism. Although there was some evidence of liver toxicity in the chronic studies on clothianidin the nature of the histopathological changes was considerably different from those induced by a specific metabolite of thiamethoxam in the mouse. The neonicotinoids are not genotoxic and are generally not carcinogenic, the exceptions being thiamethoxam and thiacloprid for which non-genotoxic modes of action were established or presumed, respectively.

Table 6.1 Toxicity profiles of nicotine and neonicotinoids.

	Acute toxicity (rat)		*Short / long term toxicity*				
Compound	*Oral LD_{50}*[a]	*Neurotox. NOAEL*[a]	*Target organs*[b]	*NOAEL*[a]	*Carcinogenicity*	*Reproductive toxicity*	*Developmental toxicity*
Nicotine	50	No data	Cardiovascular, CNS, endocrine, g.i. tract	No data	Possible indirect effects	Negative	Fetotoxicity, CNS and possible malformations
Imidacloprid	379–648	42	Liver, CNS, thyroid	5.7	Negative	Negative	Fetotoxicity, CNS?
Thiacloprid	396–836	3.1	Liver, thyroid, adrenal	1.2	Thyroid and uterus (rat), ovaries (mouse)[c]	Dystocia, stillbirths	Fetotoxicity
Acetamiprid	146–417	10	Liver	7.1	Negative	Reduced litter size	Fetotoxicity
Nitenpyram	1575–1680	No data	No data	54	Negative	Negative	Negative
Thiamethoxam	1563	100	Liver, kidney, adrenal, thyroid	2.6	Liver (mouse)[d]	Testicular pathology, sperm changes[e]	Fetotoxicity, CNS?
Clothianidin	>5000	60	Liver, kidney, ovaries, heamatopoeitic system,	33	Negative	Stillbirths, sperm changes,[e] reduced thymus weight in offspring	Fetotoxicity, missing lung lobe (rabbit)
Dinotefuran	>2000	750	Adrenal, thymus	22	Negative	Sperm changes, minor female reproductive tract changes[e]	Negative

[a] mg kg^{-1} bw or mg kg^{-1} bw day^{-1}
[b] Other than pharmacological effects on the nervous system seen with both nicotine and neonicotinoids
[c] Non genotoxic mode of action presumed
[d] Non genotoxic mode of action established
[e] No associated adverse effects on fertility

The neonicotinoids display variable reproductive and developmental effects and there is no common theme among the class, apart from toxicity to the developing fetus at high doses which was usually secondary to maternal toxicity. Equivocal effects on development of the CNS were reported for imidacloprid and thiamethoxam and they were not associated with any persistent neurobehavioural changes or other evidence of neurotoxicity. In the case of thiamethoxam they were attributed to reduced growth rather than to a developmental adverse effect. Nicotine is also toxic to the fetus but in contrast to the majority of neonicotinoids it is associated with the potential to cause developmental malformations.

Overall, the neonicotinoids display characteristics of nicotine poisoning at high pharmacologically active doses consistent with agonist action and binding affinity at specific vertebrate nAChR subtypes. However, the much lower nAChR binding affinity and limited penetration of the blood-brain barrier compared with nicotine are the principal factors in their significantly lower potency. The toxicological profiles of neonicotinoids, while having some commonality, exhibit considerable variation across the range of toxicological endpoints studied. This is attributed to variations in chemical structures and associated toxicokinetic and toxicodynamic differences.

Acknowledgements

I am grateful to my colleagues at JSC International Limited for their support in preparing this chapter. I am particularly indebted to Peter Batten for drawing the chemical structures and for advice with the manuscript.

References

1. I. Ujváry, Nicotine and other insecticidal alkaloids, in *Nicotinoid Insecticides and the Nicotinic Acetylcholine Receptor*, ed. I. Yamamoto and J. E. Casida, Springer-Verlag, Tokyo, 1999, pp. 29–69.
2. M. Tomizawa and J. E. Casida, Molecular recognition of neonicotinoid insecticides: the determinants of life or death, *Acc. Chem. Res.*, 2009, **42**, 260–269.
3. I. Yamamoto, Nicotine to nicotinoids, in *Nicotinoid Insecticides and the Nicotinic Acetylcholine Receptor*, ed. I. Yamamoto and J. E. Casida, Springer-Verlag, Tokyo, 1999, pp. 3–27.
4. A. Elbert, M. Haas, B. Springer, W. Thielert and R. Nauen, Mini-review, Applied aspects of neonicotinoid uses in crop protection, *Pest Manag. Sci.*, 2008, **64**, 1099–1105.
5. S. Kagabu, Discovery of imidacloprid and further developments from strategic molecular designs, *J. Agric. Food Chem.*, 2011, **59**, 2887–2896.

6. P. Jeschke and R. Nauen, Nicotinic acetylcholine receptor agonists, target and selectivity aspects, in *Modern Crop Protection Chemicals*, ed. W. Krämer and U. Schirmer, Wiley-VCH Verlag, Weinheim, 2007, pp. 927–957.
7. N. L. Benowitz, Pharmacology of nicotine: addiction, smoking-induced disease, and therapeutics, *Annu. Rev. Pharmacol. Toxicol.*, 2009, **49**, 57–71.
8. M. Tomizawa and J. E. Casida, Unique neonicotinoid binding conformations conferring selective receptor interactions, *J. Agric. Food Chem.*, 2011, **59**, 2825–2828.
9. M. Tomizawa and J. E. Casida, Selective toxicity of neonicotinoids attributable to specificity of insect and mammalian nicotinic receptors, *Annu. Rev. Entomol.*, 2003, **48**, 339–364.
10. M. Tomizawa and J. E. Casida, Neonicotinoid insecticide toxicology: mechanisms of selective action, *Annu. Rev. Pharmacol. Toxicol.*, 2005, **45**, 247–268.
11. J. E. Casida, Neonicotinoid metabolism: compounds, substituents, pathways, enzymes, organisms, and relevance, *J. Agric. Food Chem.*, 2011, **59**, 2923–2931.
12. H. Honda, M. Tomizawa and J. E. Casida, Neonicotinoid metabolic activation and inactivation established with coupled nicotinic receptor-CYP3A4 and -aldehyde oxidase systems, *Toxicol. Lett.*, 2006, **161**, 108–114.
13. M. Tomizawa and J. E. Casida, Minor structural changes in nicotinoid insecticides confer differential subtype selectivity for mammalian nicotinic acetylcholine receptors, *Br. J. Pharmacol.*, 1999, **127**, 115–122.
14. P. Maienfisch, Chemical structural features of commercialized neonicotinoids—six-membered heterocycles—thiamethoxam, in *Modern Crop Protection Chemicals*, ed. W. Krämer and U. Schirmer, Wiley-VCH Verlag, Weinheim, 2007, pp. 994–1013.
15. J. Higa de Landoni, *Nicotine*, International Programme on Chemical Safety, World Health Organization, Geneva, 1991. http://www.inchem.org/documents/pims/chemical/nicotine.htm
16. J. Hukkanen, P. Jacob III and N. L. Benowitz, Metabolism and disposition kinetics of nicotine, *Pharmacol. Rev.*, 2005, **57**, 79–115.
17. D. Yildiz, Nicotine, its metabolism and an overview of its biological effects, *Toxicon*, 2004, **43**, 619–632.
18. P. Balakumar and J. Kaur, Is nicotine a key player or spectator in the induction and progression of cardiovascular disorders? *Pharmacol. Res.*, 2009, **60**, 361–368.
19. R. D. Egleton, K. C. Brown and P. Dasgupta, Angiogenic activity of nicotinic acetylcholine receptors: Implications in tobacco-related vascular diseases, *Pharmacol. Ther.*, 2009, **121**, 205–223.
20. S. Ferrea and G. Winterer, Neuroprotective and neurotoxic effects of nicotine, *Pharmacopsychiatry*, 2009, **42**, 255–265.
21. M. J. Marks, J. R. Pauly, S. D. Gross, E. S. Deneris, I. Hermans-Borgmeyer, S. F. Heinemann and A. C. Collins, Nicotine binding and nicotinic receptor subunit RNA after chronic nicotine treatment, *J. Neurosci.*, 1992, **12**, 2765–2784.

22. European Food Safety Authority, *Imidacloprid Draft Assessment Report*, 2008. http://www.efsa.europa.eu/en/praper/praperdareports.htm, http://dar.efsa.europa.eu/dar-web/provision.
23. R. Solecki, *Toxicological evaluations: imidacloprid.* International Programme on Chemical Safety/INCHEM/Joint Meeting on Pesticide Residues, 2001. http://www.inchem.org/documents/jmpr/jmpmono/2001pr07.htm
24. J. Doherty, *Memorandum: Imidacloprid. Review of the Series 81-8 Acute Neurotoxicity and 82-7 Subchronic Neurotoxicity Screen Studies*, U.S. Environmental Protection Agency, Washington, DC, 1995. http://www.epa.gov/pesticides/chemical/foia/cleared-reviews/reviews/129099/129099–076.pdf
25. California Environmental Protection Agency, *Imidacloprid: Risk Characterization Document Dietary and Drinking Water Exposure*, Department of Pesticide Regulation, Sacramento, CA, 2006. http://www.cdpr.ca.gov/docs/risk/rcd/imidacloprid.pdf
26. C. H. Bast, C. Forsyth and R. H. Ross, *Data Evaluation Record, Imidacloprid, HED Records Center Series 361 Science Reviews – File R055682*, U.S. Environmental Protection Agency, Washington DC, 2002, pp. 245–272. http://www.epa.gov/pesticides/chemical/foia/cleared-reviews/reviews/129099/129099-0000-00-00a.pdf
27. K. A. Ford and J. E. Casida, Chloropyridinyl neonicotinoid insecticides: diverse molecular substituents contribute to facile metabolism in mice, *Chem. Res. Toxicol.*, 2006, **19**, 944–951.
28. S. Bhardwaj, M. K. Srivastava, U. Kapoor and L. P. Srivastava, A 90 days oral toxicity of imidacloprid in female rats: Morphological, biochemical and histopathological evaluations, *Food Chem. Toxicol.*, 2010, **48**, 1185–1190.
29. H. Tomimoto, J.-X. Lin, A. Matsuo, M. Ihara, R. Ohtani, M. Shibata, Y. Miki and H. Shibasaki, Different mechanisms of corpus callosum atrophy in Alzheimer's disease and vascular dementia, *J. Neurol.*, 2004, **251**, 398–406.
30. M. B. Abou-Donia, L. B. Goldstein, S. Bullman, T. Tu, W. A. Khan, A. M. Dechkovskaia and A. A. Abdel-Rahman, Imidacloprid induces neuro behavioral deficits and increases expression of glial fibrillary acidic protein in the motor cortex and hippocampus in offspring rats following in utero exposure, *J. Toxicol. Environ. Health A*, 2008, **71**, 119–130.
31. D. H. Phua, C. C. Lin, M.-L. Wu, J.-F. Deng and C.-C. Yang, Neonicotinoid insecticides: an emerging cause of acute pesticide poisoning, *Clin. Toxicol.*, 2009, **47**, 336–341.
32. FAO, *Thiacloprid, Report of the Joint Meeting of the FAO Panel of Experts on Pesticide Residues in Food and the Environment and the WHO Core Assessment Group on Pesticide Residues, Rome, Italy, 3–12 October 2006*, World Health Organization, Geneva; Food and Agriculture Organization of the United Nations, Rome. http://www.fao.org/ag/AGP/AGPP/Pesticid/JMPR/DOWNLOAD/2006_rep/report2006jmpr.pdf
33. Chemicals Regulation Directorate. *Thiacloprid: Proposal for Harmonised Classification and Labelling*. Health and Safety Executive, Bootle,

Merseyside, UK, 2009. http://echa.europa.eu/doc/consultations/cl/clh_uk_axrep_thiacloprid.pdf

34. National Registration Authority for Agricultural and Veterinary Chemicals. *Evaluation of the New Active Thiacloprid in the New Product Calypso 480 SC Insecticide*, Australian Pesticides and Veterinary Medicines Authority, Canberra, Australia, 2001. http://www.apvma.gov.au/registration/assessment/docs/prs_thiacloprid.pdf
35. P. M. Hurley, R. N. Hill and R. J. Whiting, Mode of carcinogenic action of pesticides inducing thyroid follicular cell tumors in rodents, *Environ. Health Perspect.*, 1998, **106**, 437–445.
36. R. N. Hill, T. M. Crisp, P. M. Hurley, S. L. Rosenthal and D. V. Singh, Risk assessment of thyroid follicular cell tumors, *Environ. Health Perspect.*, 1998, **106**, 447–457.
37. Environmental Protection Agency, *Petition for Acetamiprid*, Office of Pesticide Programs, Washington, DC, 2009. http://nepis.epa.gov
38. European Commission, *Review Report for the Active Substance Acetamiprid*, Health and Consumer Protection Directorate General, Brussels, 2004. http://ec.europa.eu/food/plant/protection/evaluation/newactive/acetamiprid.pdf
39. R. Schenker, O. Tinembart, S. H. Barnett and S. T. Witte, A brief introduction to nitenpyram: a new systemic flea adulticide for cats and dogs, clinical advances, *Compend. Contin. Educ. Pract. Vet.*, 2001, **23** (Suppl. 3A), 4–6.
40. C. D. S. Tomlin (ed.), Nitenpyram, in *The Pesticide Manual*, British Crop Production Council, Alton, Hampshire, 15th edn, 2009, pp. 817–818.
41. Food and Drug Administration, *Freedom of Information Summary, CAPSTAR TM (Nitenpyram) Tablets, New Animal Drug Application (NADA)141-175*, Rockville, MD, 2000, pp. 1–28. http://www.fda.gov/downloads/AnimalVeterinary/Products/ApprovedAnimalDrugProducts/FOIADrugSummaries/ucm117258.pdf
42. European Commission, *Review Report For The Active Substance Thiamethoxam*, Health and Consumer Protection Directorate General, Brussels, 2006. http://ec.europa.eu/food/plant/protection/evaluation/newactive/thiamethoxam_en.pdf
43. Standing Committee on Biocidal Products, *Assessment Report: Thiamethoxam: Product-Type 8 (Wood Preservative)*, European Commission, Brussels, 2008. http://ecb.jrc.ec.europa.eu/documents/Biocides/ANNEX_I/ASSESSMENT_REPORTS/AnnexI_AR_153719-23-4_PT08_en.pdf
44. M. Doherty, M. Collantes and D. Smegal, *Memorandum: Thiamethoxam Human Health Risk Assessment for Proposed New Uses or Revised Uses on Artichoke, Barley, Brassica Vegetables, Bushberry, Caneberry, Cotton, Cranberry, Cucurbit Vegetables, Fruiting Vegetables, Hops, Juneberry, Leafy Vegetables, Legume Vegetables, Lingonberry, Mint, Oilseed Crops, Pecan, Pome Fruit, Potato Seed Pieces, Root Vegetables (Except Sugarbeet), Salal, Stone Fruit, Strawberry, Tobacco, Tuberous and Corm Vegetables, and Turf.* US Environmental Protection Agency, Washington, DC, 2007. http://www.regulations.gov/search/Regs/home.html#home

45. California Environmental Protection Agency, *Summary of Toxicology Data: Thiamethoxam*, Department of Pesticide Regulation, Sacramento, CA, 2008. http://www.cdpr.ca.gov/docs/risk/toxsums/pdfs/5598.pdf
46. T. Green, A. Toghill, R. Lee, F. Waechter, E. Weber and J. Noakes, Thiamethoxam induced mouse liver tumors and their relevance to humans. Part 1: mode of action studies in the mouse, *Toxicol. Sci.*, 2005, **86**, 36–47.
47. T. Green, A. Toghill, R. Lee, F. Waechter, E. Weber, R. Peffer, J. Noakes and M. Robinson, Thiamethoxam induced mouse liver tumors and their relevance to humans. Part 2: species differences in response, *Toxicol. Sci.*, 2005, **86**, 48–55.
48. K. A. Ford and J. E. Casida, Unique and common metabolites of thiamethoxam, clothianidin, and dinotefuran in mice, *Chem. Res. Toxicol.*, 2006, **19**, 1549–1556.
49. T. Pastoor, P. Rose, S. Lloyd, R. Peffer and T. Green, Case study: weight of evidence evaluation of the human health relevance of thiamethoxam-related mouse liver tumors, *Toxicol. Sci.*, 2005, **86**, 56–60.
50. K. J. A. Rodrigues, M. B. Santana, J. L. M. Do Nascimento, D. L. W. Picanço-Dintiz, L. A. L. Maués, S. N. Santos, V. M. M. Ferreira, M. Alfonso, R. Durán and L. R. F. Faro, Behavioral and biochemical effects of neonicotinoid thiamethoxam on the cholinergic system in rats, *Ecotoxicol. Environ. Saf.*, 2010, **73**, 101–107.
51. European Commission, *Review report for the Active Substance Clothianidin*, Health and Consumer Protection Directorate General, Brussels, 2005. http://ec.europa.eu/food/plant/protection/evaluation/newactive/list_clothianidin.pdf
52. Environmental Protection Agency, *Pesticide Fact Sheet*, Clothianidin, Office of Prevention, Pesticides and Toxic Substances, Washington, DC, 2003. http://www.epa.gov/opprd001/factsheets/clothianidin.pdf
53. California Environmental Protection Agency, *Summary of Toxicology Data: Clothianidin*, Department of Pesticide Regulation, Sacramento, CA, 2003. http://www.cdpr.ca.gov/docs/risk/toxsums/pdfs/5792.pdf
54. Food and Agriculture Organization of the United Nations, *Specifications and Evaluations for Agricultural pesticide: Clothianidin*, FAO, Rome, 2010. http://www.fao.org/fileadmin/templates/agphome/documents/Pests_Pesticides/Specs/Clothianidin2010.pdf
55. Environmental Protection Agency, *Pesticide Fact Sheet, Dinotefuran*, Office of Prevention, Pesticides and Toxic Substances, Washington DC, 2004. http://www.epa.gov/opprd001/factsheets/dunotefuran.pdf
56. California Environmental Protection Agency, *Summary of Toxicology Data: Dinotefuran*, Department of Pesticide Regulation, Sacramento, CA, 2003. http://www.cdpr.ca.gov/docs/risk/toxsums/pdfs/5822.pdf
57. P. R. Durkin, *Dinotefuran: Human Health and Ecological Risk Assessment,* Syracuse Environmental Research Associates Inc., Fayetteville, NY, 2009. http://www.fs.fed.us/foresthealth/pesticide/pdfs/0521803b_Dinotefuran.pdf

CHAPTER 7

Insecticides that Interfere with Insect Growth and Development

TIMOTHY C MARRS

Edentox Associates, Pinehurst, Four Elms Road, Edenbridge, Kent, UK.
Email: timothymarrs05@aol.com

7.1 Introduction

The attraction of insecticides that interfere with insect growth and development is that they have no direct target organ or system in mammals analogous to those that they target in insects, so that their toxicity tends to be unrelated to their insecticidal action, conferring considerable specificity towards killing insects and not mammals. This has an important consequence, in that the mammalian toxicity is generally rather non-specific and relatively high exposures are required to produce adverse effects. However, the compounds are not generally effective against adult insects. Some authorities classify all this group of compounds as insect growth regulators (IGRs), whereas others use the term IGR as a synonym for juvenile hormone (JH) analogues.

To some extent the ecdysone agonists (see below) are exceptions to the rule that interference with mammalian metabolism is unlikely, in that ecdysone is a steroid and structurally related to mammalian steroid hormones. Despite the structural dissimilarity of the three main groups of insecticides discussed in this chapter, curiously, many have effects on the mammalian haematological system. A widely accepted scheme for mechanism grouping has been proposed by the Insecticide Resistance Action Committee (IRAC);[1] this is used in this chapter with some modifications (see Table 7.1).

Issues in Toxicology No. 12
Mammalian Toxicology of Insecticides
Edited by Timothy C. Marrs

Published by the Royal Society of Chemistry, www.rsc.org

Table 7.1 Insecticides that interfere with insect growth and development.

Group	*Examples*
Insect growth regulators (juvenile hormone analogues)	Methoprene
	Hydroprene
	Cyromazine
	Dicyclanil
	Fenoxycarb
Chitin synthesis inhibitors	Diflubenzuron
	Flufenoxuron
	Lufenuron
	Hexaflumuron
Ecdysone agonists	Tebufenozide
	Methoxyfenozide

7.2 Insect Growth Regulators

JHs are a group of substances that regulate many aspects of insect physiology. JHs are acyclic sesquiterpenoids that control and regulate development and reproduction.[2] They are secreted by the corpora allata, which are a pair of endocrine glands behind the insect brain.

7.2.1 Actions of Insect Growth Regulators

The effect of IGRs (JH analogues) is to prevent reproduction by preventing metamorphosis of insect larvae into viable adults, when such insecticides are applied to the larvae.[3] The analogues used as insecticides are heterogeneous in structure: some are esters of long-chain fatty acids (methoprene and hydroprene) and these clearly act as JH analogues.[1] Cyromazine, which IRAC[1] separately classifies as a moulting disruptor, whose target protein is unknown or uncharacterized, is a triazine. Dicyclanil is a cyclopropylaminopyrimidine. Fenoxycarb is a carbamate whose target protein is unclear.[1] Although compounds similar to insect JHs exist in crustaceans, the juvenile hormone system is not present in mammals, and IGRs are generally of low acute mammalian toxicity (oral LD_{50} *c.*5 g kg^{-1}).

7.2.2 Methoprene

Methoprene is the ISO name for 1-methylethyl (E,E)-11-methoxy-3,7,11-trimethyl-2,4-dodecadienoate (IUPAC). The structural formula is:

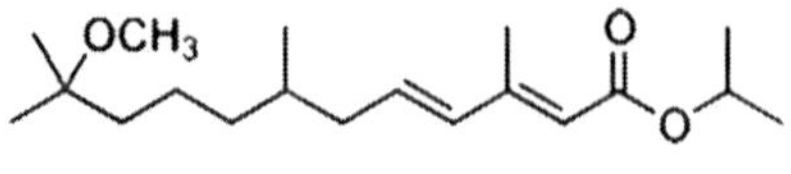

Methoprene is a racemic mixture of the *R* and *S* enantiomers in a 1:1 ratio, activity as a JH analogue being restricted to the *S* enantiomer.[4] Some studies have been carried out using *S*-methoprene while others have used the racemic mixture.

7.2.2.1 Absorption, Distribution, Metabolism and Excretion

Studies of the absorption, distribution, metabolism and excretion of methoprene after single doses have been undertaken in mice, rats, guinea-pigs, cows and chickens. Generally methoprene was rapidly absorbed and the radiolabel is excreted in the urine, faeces and expired air, mostly within 5 days. Methoprene was extensively metabolized by mammals and in guinea-pigs much of the label in urine was present as glucuronic acid conjugates.[5] In rats, concentrations of methoprene in fat were observed to decline very slowly.

7.2.2.2 Acute Toxicity, Irritancy and Sensitization

Methoprene and *S*-methoprene are of low mammalian toxicity;[5–7] the oral rat LD_{50} both of the racemate and *S*-methoprene is greater than 5 g kg^{-1} bw.[5] Methoprene was not irritating to the eye in two studies of the racemate in rabbits, while slight irritancy was observed in a third study using *S*-methoprene. Significant irritancy was not observed to the skin of rabbits, with racemic methoprene or *S*-methoprene. In a study in guinea-pigs, not to modern standards, the racemate appeared to have no sensitizing properties but in a study with a formulation containing 20% *S*-methoprene, skin sensitization was seen.

7.2.2.3 Repeated Dose Studies

In repeated dose studies, increased relative liver weight were frequently seen, not always accompanied by histopathological changes, but methoprene caused changes such as bile duct proliferation in the livers of rats during a 2-year study at doses of about 220 mg kg^{-1} bw day^{-1}.[5,7] In a 90-day study in dogs, using dietary concentrations giving intakes of up to 120 mg kg^{-1} bw day^{-1} of racemate, no deaths occurred, nor were there adverse effects on behaviour, bodyweight, food consumption or haematological endpoints. Serum alkaline phosphatase activity was increased in both sexes at the highest dose. At that concentration, at termination, the relative liver weight was increased in both sexes, but this finding was not accompanied by any histopathological abnormality.

7.2.2.4 Carcinogenicity and Mutagenicity

Methoprene was carcinogenic in neither rats nor mice, but bile duct proliferation in the livers of rats was seen during a 2-year study (see above). Methoprene did not appear to be mutagenic.[5] The range of tests was limited: Ames tests were carried out both on racemate and *S*-methoprene, while a test for chromosomal aberrations in Chinese hamster ovary (CHO) cells was performed on the racemate and tests for mitotic recombination, gene conversion and reverse mutation in *Saccharomyces cerevisiae* D7 were undertaken using *S*-methoprene.

7.2.2.5 *Reproductive and Developmental Toxicity*

There was no evidence of reproductive or developmental toxicity at doses below those causing maternal toxicity, except in a mouse study of developmental toxicity, where the dams were allowed to litter and suckle their pups: effects on organ weights were observed in pups at the top dose. A study on endocrine activity in mice was reviewed by the JMPR,[5] in which immature female mice (19–21 days of age) received racemic methoprene subcutaneously at doses of 0.015 or 0.15 mg kg^{-1} bw day^{-1} for 3 days. No increase in the uterine weight relative to bodyweight ratio was seen. When methoprene was given subcutaneously to castrated male rats at doses of 0.37 or 3.7 mg kg^{-1} bw day^{-1} for 7 days, no increase was seen in the seminal vesicle, ventral prostate or levator ani weights relative to bodyweight. In adrenalectomized male rats, 21–23 days old, subcutaneous injection of methoprene at 0.9 or 9 mg kg^{-1} bw day^{-1} for 6 days did not affect the thymus weight relative to bodyweight ratio. It was concluded that these studies suggested that methoprene had no oestrogenic, androgenic, anabolic or glucocorticoid activity.

7.2.2.6 *Reference Dose*

The JMPR[5] established an acceptable daily intake (ADI) of 0.09 mg kg^{-1} bw: this allowed for the purity (*c*.70%) of the racemate tested in the relevant study, which was the 90-day study in dogs. A safety factor of 100 was allocated. The effects observed at the lowest observed effect level (LOAEL) were increased liver weight and an increase in alkaline phosphatase activity. An acute reference dose (ARfD) was considered unnecessary.

7.2.3 Hydroprene

Hydroprene is the ISO name for ethyl (*E,E*)-(*RS*)-3,7,11-trimethyldodeca-2,4-dienoate (IUPAC). The structural formula is:

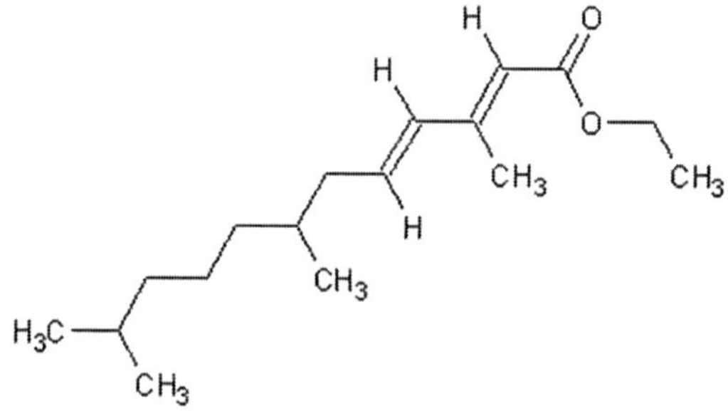

Hydroprene is similar in structure to methoprene. Like methoprene, it is a racemic mixture of the *R* and *S* enantiomers and activity as a JH analogue is limited to the *S* enantiomer. Both the racemate and *S*-hydroprene have a LD_{50}

greater than 5 g kg^{-1} bw in the rat. The toxicology, which was reviewed by the UK Advisory Committee on Pesticides (ACP),[8,9] is broadly similar to that of methoprene and will therefore be discussed in less detail. In a 4-week study in the rat, using the racemate, the main finding was increased liver weights (absolute and relative). In a 13-week rat study using *S*-hydroprene, there were disturbances in haematological and clinical chemistry parameters, including prothrombin time (dose-related decrease in males) (such changes were not seen with methoprene—see section 7.2.2.3). Increased liver, kidney and ovarian weights were observed at necropsy. Histopathogically, there was increased homogeneity of liver cytoplasm at high doses and in females, vacuolated interstitial cells in the ovaries. In a long-term/carcinogenicity study in rats using *S*-hydroprene, findings were similar and *S*-hydroprene was not carcinogenic. Neither hydroprene nor *S*-hydroprene appear to be mutagenic. The racemate was not fetotoxic or teratogenic in the rabbit nor was *S*-hydroprene in the rat. In a multigeneration study in rats, using *S*-hydroprene, there was some evidence of an effect on seminiferous tubules in males, associated with an effect on fertility; it was unclear whether these effects were test-material related. Hydroprene (racemate) had no oestrogenic activity in mice and no androgenic, anabolic or glucocorticoid activity in rats.[8,9] The ACP did not identify an ADI, as the evaluation was not for the use of hydroprene as a plant protection product, but the lowest NOAEL was 4.62 mg kg^{-1} bw day^{-1} in the rat long-term study on *S*-hydroprene based on ovarian lipid accumulation and regression of corpora lutea. However, the Australian ADI was 0.5 mg kg^{-1} bw.[10]

7.2.4 Cyromazine

Cyromazine is the ISO name and international non-proprietary name (INN) for *N*-cyclopropyl-1,3,5-triazine-2,4,6-triamine. The structural formula is:

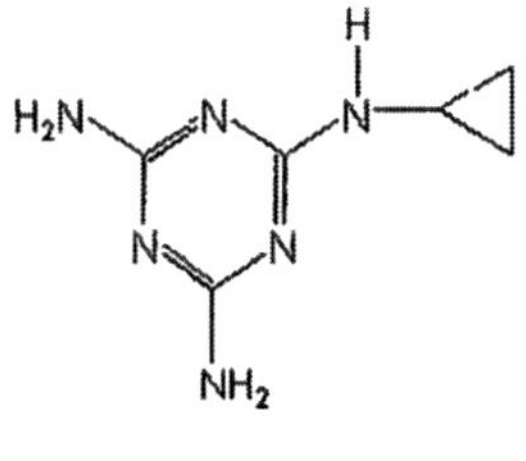

Cyromazine is used *inter alia* as a sheep ectoparasiticide. The absorption, distribution, metabolism, excretion and toxicology has been reviewed twice by the JMPR.[11,12]

7.2.4.1 *Absorption, Distribution, Metabolism and Excretion*

In most species (rat, sheep, goat, monkey), the majority is excreted within 24 h, as unchanged cyromazine. A higher proportion seems to be metabolized in the goat. Melamine (a bladder carcinogen) was a metabolite of cyromazine in mammals.[11,12]

7.2.4.2 *Acute Toxicity, Irritancy and Sensitization*

Cyromazine is of low acute toxicity (rat oral LD_{50} *c.*3–4g kg^{-1} bw). In rabbits, cyromazine was not an irritant to the skin or eyes, nor does this insecticide have sensitizing potential in the guinea-pig.[11,12]

7.2.4.3 *Repeated Dose Studies*

In short-term feeding studies in rats and dogs and in long-term studies in mice and rats, effects on bodyweight were seen. Red blood cell counts and haemoglobin levels were reduced in dogs at high dietary concentrations.[11] A 1-year dog study using diets containing cyromazine at concentrations giving intakes of up to 110 mg kg^{-1} bw day^{-1} was also undertaken.[12] Similar haematological effects (slight hypochromic and microcytic anaemia), were seen as in the earlier dog studies. Increased heart and liver weights (both absolute and relative) were seen at high doses. The effects on the blood were accompanied by hypercellularity of the bone marrow. In addition there were histopathological effects on the myocardium (chronic myocarditis) and kidney (focal chronic tubular epithelial regeneration) at high doses. The NOAEL of 5.7 mg kg^{-1} bw day^{-1} proved to be the critical NOAEL for establishing the JMPR ADI. This NOAEL was based upon haematological effects in males at the next highest dietary concentration.

7.2.4.4 *Carcinogenicity and Mutagenicity*

In long-term studies in mice and rats, dietary concentrations were used that gave intakes of up to 476 mg kg^{-1} bw day^{-1} in the former and 210 mg kg^{-1} bw day^{-1} in the latter. There was a slight increase in the number of females with mammary gland adenocarcinomas in the highest dose group in both species, but there was no clear dose relationship. The JMPR[11,12] considered that cyromazine was neither carcinogenic nor genotoxic. The NOAEL for the long-term rat study, which formed the basis of the 1990 JMPR ADI[11] (now superseded) was a dietary concentration giving an intake equal to 1.8 mg kg^{-1} bw day^{-1}, based on the bodyweight changes noted in females seen in the mid dose (the 2006 JMPR[12] considered the NOAEL for the rat study to be 15 mg kg^{-1} bw day^{-1}).

7.2.4.5 *Reproductive and Developmental Toxicity*

In a rat multigeneration study, cyromazine did not affect fertility, but there was increased perinatal pup mortality and reduced pup weight, at maternally toxic doses. A study of developmental in rats showed fetal toxicity at the highest dose

(600 mg kg^{-1} bw day^{-1}) and maternal toxicity at the next lower dose: teratogenic effects were absent. Several studies in rabbits were evaluated by the JMPR: abnormalities, including cyclopia with multiple head anomalies in two fetuses, were observed in one study, but there was no dose response and similar findings were not present in other rabbit studies. The JMPR did not consider cyromazine to be teratogenic.[11,12]

7.2.4.6 *Reference Doses*

The JMPR established an ADI of 0.06 mg kg^{-1} bw based on a NOAEL of 5.7 mg kg^{-1} bw day^{-1} for haematological effects in the 1-year study of toxicity in dogs, seen at 23 mg kg^{-1} bw day^{-1} in males. A safety factor of 100 was used. An ARfD of 0.1 mg kg^{-1} bw was based on the NOAEL of 10 mg kg^{-1} bw day^{-1}. This was based on bodyweight loss and decreased food consumption observed soon after the start of dosing at 25 mg kg^{-1} bw day^{-1} in dams in studies of developmental toxicity in rabbits treated by gavage, a safety factor of 100 being used.[12]

7.2.5 Dicyclanil

Dicyclanil is the ISO name and the INN for 4,6-diamino-2-cyclopropylaminopyrimidine-5-carbonitrile (IUPAC). The structural formula is:

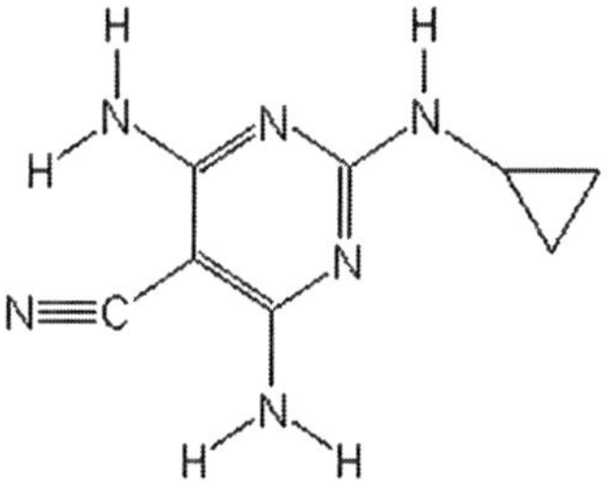

Dicyclanil is an insecticide also used on sheep as an ectoparasiticide.

7.2.5.1 *Absorption, Distribution, Metabolism and Excretion*

The absorption, distribution, metabolism, excretion and toxicology of dicyclanil has been reviewed by the Joint Expert Committee on Food Additives (JECFA).[13] When orally administered to rats for 7 days, label from radiolabelled dicyclanil was rapidly excreted (>90% within 24 h of the last dose) predominantly in the urine. Most of the label was present as metabolites. Biotransformation of dicyclanil in rats and sheep was limited to the cyclopropyl ring.

7.2.5.2 Acute Toxicity, Irritancy and Sensitization

The toxicology of dicyclanil has been reviewed, also by JECFA.[13] Dicyclanil is moderately hazardous by mouth (rat oral LD_{50} *c.*500 mg kg^{-1} bw), *i.e.* it is somewhat more acutely toxic than many of the other insecticides discussed in this chapter. In a skin irritancy test in rabbits, very slight erythema was observed, while in an eye irritancy test, also in rabbits, slight chemosis of the conjunctiva was observed in two out of three animals 1 h after instillation, but complete recovery was seen within 24 h. Dicyclanil was not a sensitizer in guinea-pigs.

7.2.5.3 Repeated Dose Studies

In a 28-day study in the rat, dietary concentrations were used which achieved intakes of up to 160 mg kg^{-1} bw. Piloerection was seen in both sexes at the top dose in the latter part of the study. Dose-dependent reductions in food consumption, bodyweight gain, and terminal bodyweight were noted in all treated groups. Increased incidences of erythrocytic anisocytosis and polychromasia were observed in males at the highest dose. Clinical chemistry disturbances at the highest dietary concentration were also seen in both sexes, notably raised blood urea and elevations in aminotransferase activities. At necropsy, reductions in the absolute and relative weights of the prostate in males at the mid and high doses were observed. Histopathologically, reduced spermatogenesis and accumulation of cellular debris in the epididymal duct in males at the high dose were noted. In a 3-month study in rats, including a 4-week recovery period, intakes of up to 34 mg kg^{-1} bw day^{-1} were achieved. Adverse clinical signs were not seen. Small reductions in bodyweight gain and food consumption were observed in animals of both sexes at high doses: bodyweights of animals at the high dose were comparable to those of the controls at the end of the recovery period. Higher organ:bodyweight ratios were observed for a number of organs (kidneys, brain, and testis in males at the high dose and for liver and brain in females at the highest dose) but these were reversible during the 4-week recovery period. No treatment-related adverse ophthalmological, haematological or histopathological findings were seen. In a 3-month study in dogs, intakes of up to 42 mg kg^{-1} bw day^{-1} were achieved. Clinical signs of neurotoxity (ataxia, raised tails and tremor) were noted at weeks 9–11 at the high dose. Vomiting was also seen as was, in the males, decreased food consumption and bodyweight gain. Changes in clinical chemistry (increased plasma cholesterol and phospholipid concentrations at higher doses) and haematology (reduced haemoglobin and haematocrit with microcytosis and erythrocytic hypochromasia at the top dose) were observed. Liver weights (absolute and relative) were increased at high doses in both sexes. The absolute and relative weights of the testes were decreased at the top dose. The main histopathological findings were minimal to moderate focal or multifocal subcapsular inflammation with fibrosis in the livers of some animals of both sexes at the highest dose. Enlarged oedematous hepatocytes were observed in some females at in the test groups. The testes in 3/4 males showed minimal tubular atrophy at the high dose and there was marked reduction in spermatogenesis in all males of that group. A 1-year dietary study of dicyclanil was undertaken in dogs with a 4 week recovery period, with 4 test groups and daily intakes of up to 23 mg kg^{-1} bw day^{-1}. One top-dose male was killed *in extremis* (vomiting and apathy) and one top-dose female died. Vomiting was observed in females at that dose. Adverse haematological effects were not seen. Throughout the treatment period, plasma cholesterol concentrations were increased in animals at the highest dose rate (in females not statistically significantly). This finding provided the basis for the NOAEL of the study, which was 0.71 mg kg^{-1} bw day^{-1}.

7.2.5.4 Carcinogenicity and Mutagenicity

Long-term studies were carried out in mice and rats, the former with intakes of up to 210 mg kg^{-1} bw day^{-1} and the latter up to 26 mg kg^{-1} bw day^{-1}. In the mice, there were clinical signs of toxicity and deaths in the high-dose groups, for which reason survivors were killed at weeks 58–59. There were findings at autopsy in a number of organs, the most important being the liver. Macroscopically, there were enlarged livers with masses and/or nodules of the liver at the higher doses. Microscopically, these findings were associated with Kupffer cell pigmentation, hepatocellular necrosis, increased in the number of hepatocellular mitotic figures and/or multinucleated hepatocytes. The incidence of hepatocellular adenomas was higher in females at the two highest doses than in controls. Additionally, the incidence of hepatocellular carcinomas was increased in females at thc highest dose. Pigmentation of the olfactory epithelium was observed at higher doses. Moto *et al.*[14,15] have studied the aetiology of the liver tumours seen in mice exposed to dicyclanil; they concluded that inhibition of apoptosis and DNA damage due to oxidative stress may be involved in the mechanism of hepatocarcinogenesis (see also Umemura *et al.*[16]). Moreover, the effect appears to be thresholded.[17] In the rat long-term study, there were effects on bodyweight, haematological and clinical chemistry endpoints; as in mice, pigmentation of the olfactory epithelium on both sexes was observed. Dicyclanil was not tumorigenic in the rat. Dicyclanil was not mutagenic in a variety of tests (including one test *in vivo*) and JECFA[13] considered that dicyclanil was neither genotoxic nor a carcinogenic risk to humans (see also Moto *et al.*).[18,19]

7.2.5.5 Reproductive and Developmental Toxicity

In a multigeneration study in rats, dicyclanil was not a reproductive toxin at the highest dietary concentration tested, while the NOAEL for pup toxicity (reduced bodyweight gain) was above the NOAEL for parental toxicity. Developmental toxicity studies were carried out in rats and rabbits, in the former using gavage doses of up to 75 mg kg^{-1} bw day^{-1}, and in the latter doses of up to 30 mg kg^{-1} bw day^{-1}. In rats, fetotoxicity was seen at doses above those that were maternally toxic (reduced fetal weight, increased renal pelvic dilatation, and increased skeletal anomalies and variations consistent with a slight delay in skeletal maturation). In the rabbit, dicyclanil was not teratogenic and was only fetotoxic at doses above those that were maternally toxic. Findings in the fetuses were consistent with delayed ossification (reduced fetal weight and increased skeletal variations).

7.2.5.6 *Other Studies*

Additional studies have been carried out on the pharmacological properties of dicyclanil both *in vitro* and *in vivo*. Dicyclanil at high concentrations had antagonistic effects on smooth muscle contractions induced by agonists. A single oral dose of 100 mg kg^{-1} bw (the highest dose tested) slightly inhibited both exploratory activity and the startle response in mice. Moreover, there were reductions in static activity and motor coordination. The same dose in rats affected the heart rate, and tidal and minute lung volume. Locomotor activity was also affected.

7.2.5.7 *Reference Dose*

JECFA[13] considered an ADI of 0.007 mg kg^{-1} bw appropriate. This was based on the NOAEL of 0.71 mg kg^{-1} bw day^{-1} in a 1-year study of toxicity in dogs concentrations and a safety factor of 100.

7.2.6 Fenoxycarb

Fenoxycarb if the ISO name for ethyl 2-(4-phenoxyphenoxy)ethylcarbamate (IUPAC). The structural formula is:

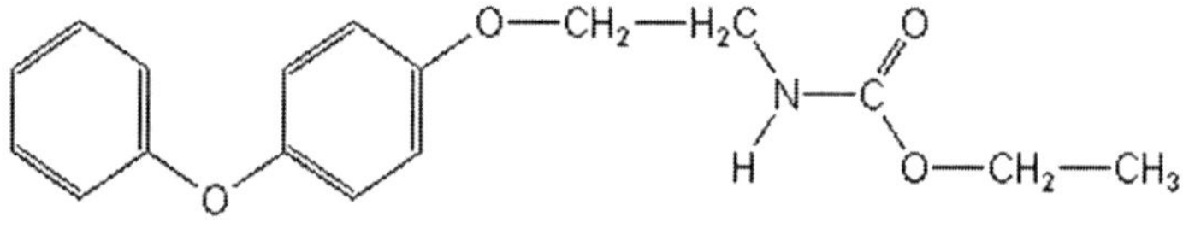

7.2.6.1 *Action*

In structure, fenoxycarb is a carbamate, without acetylcholinesterase-inhibiting properties and it exerts most of its insecticidal activity as a JH analogue. The absorption, distribution, metabolism and excretion and toxicology of fenoxycarb has been reviewed by a number of regulatory authorities, for example the USEPA and the UK PSD (now the Chemical Regulations Directorate, CRD).[20]

7.2.6.2 Absorption, Distribution, Metabolism and Excretion

In a study in rats using radiolabelled fenoxycarb, the majority of the label was excreted in 24 h mostly in the faeces. A study in bile duct cannulated rats showed that there was considerable excretion in the bile. In both single and repeated dose studies, concentration of the label in the liver was notable. Metabolism in the rat was complex involving loss of the ethylcarbamate side chain and hydroxylation. Sulfate conjugates were prominent in the urine.[20]

7.2.6.3 Acute Toxicity, Irritancy and Sensitization

Fenoxycarb was of low acute mammalian toxicity when administered to rats by mouth, *via* the skin or by inhalation.[20] The rat oral LD_{50} was greater than 10 g kg^{-1} bw. Fenoxycarb was not irritant to rabbit skin but is a mild eye irritant. Fenoxycarb was nor a sensitizer in the guinea-pig.

7.2.6.4 Repeated Dose Studies

In a 28-day rat study using gavage, using doses of up to 1000 mg kg^{-1} bw day^{-1}, there were no deaths or treatment-related clinical signs, nor were disturbances in bodyweight or food consumption seen. Slight but significant effects were seen at the highest dose on haematological parameters (decreased erythrocyte count, haemoglobin and haematocrit); in the same group increases in the alkaline phosphatase activity and calcium and albumin concentrations were seen. The most important finding was an increase in relative and absolute liver weights in all treated groups with a dose relationship in both sexes. Microscopically, there was hypertrophy and eosinophilia of hepatocytes, confined to the highest-dose group. Another microscopic finding was follicular hyperplasia of the thyroid, also confined to the highest-dose group. On electron microscopic examination, there was peroxisome proliferation in the livers of the high dose group, more marked in males. In a 13-week dietary study, with a 4-week reversibility group and using doses of up to 800 mg kg^{-1} bw day^{-1}, there were clinical signs (alopecia with some evidence of dose response, and diuresis and shaggy fur) at the top dose. Bodyweight and weight gain was reduced in both sexes at the top dose. Haematological disturbances were seen at the top dose and in some cases in the lower-dose treatment groups in females, the changes comprising decreased red blood cell counts, haemoglobin levels and packed cell volume. Anisoscytosis and poikilocytosis were also seen. Clinical chemistry changes at the high- and/or mid-dose group were seen: these included increased alanine aminotransferase, glutamate dehydrogenase, and alkaline phosphatase activities. Bilirubin levels were increased at the top two doses in both sexes. Total protein was increased in the top-dose males and the top two doses of females and the albumin:globulin ratio was increased in these groups. There was a dose-related increase in relative and absolute liver weights (all dose groups, both sexes). Increased thyroid weight was also observed at high doses. Histopathological changes observed included hepatocyte hypertrophy (mainly centrilobular), changes in the thyroids of females (small active follicles) and renal tubular changes. Most of these findings appeared to be reversible, but there was not a clear NOAEL for the study. Another rat study was conducted at lower doses (highest dose was 203 mg kg^{-1} bw day^{-1}), but otherwise using a similar design with a reversibility group. No clinical signs were seen at any dose. Bodyweight gain was decreased at the highest dose in both sexes (greater in females) and this was reversible only in the males. Top-dose females had reduced food consumption. No effects were seen on ophthalmoscopy or on haematological parameters. Changes in clinical chemistry parameters, generally confined to the top dose group were seen in aminotransferases (males) and alkaline phosphatase (both sexes). The top-dose males and females had increased albumin:globulin ratios. Reduced bilirubin concentrations were seen in the top-dose females. Absolute and relative liver weights were increased in both sexes at the top two doses. Absolute and relative thyroid weights were increased in females. Hepatocellular hypertrophy, largely centrilobular, was seen at the higher doses in females and the top dose in males. Hypertrophy of the thyroid follicular epithelium was seen in females at the top two doses and also in males at the top dose. A 13-week mouse study was undertaken, using dietary concentrations giving doses of up to 900 mg kg^{-1} bw day^{-1}. There were no adverse clinical signs or effects on food consumption, bodyweight, weight gain or food consumption. In males there were decreases in white blood cell counts and in females of thrombocytes. Dose-dependent changes in plasma proteins were seen in females. Significant increases in liver weights were seen in the top dose in both sexes. Periportal fatty change was seen in the liver, most notably in males. A capsule study in dogs was undertaken using doses of up to 260 mg kg^{-1} bw day^{-1}. Treatment-related clinical signs were not seen, but bodyweight gain was reduced in the high-dose group. No haematological effects were observed but there was an increase in relative liver weight at the high dose. No treatment-related histopathological effects were seen.

7.2.6.5 *Carcinogenicity and Mutagenicity*

Long-term/carcinogenicity studies in the mouse and rat were reviewed by ACP.[20] In the rat study, dietary concentrations which gave intakes up to about 104 mg kg^{-1} bw day^{-1} were used. There was a slight reduction in bodyweight at the top dose in males and in bodyweight at termination in females. Effects observed were broadly consistent with short-term rat studies, but haematological effects were minor. In males at the top dose and at the interim sacrifice, there was slight centrilobular hypertrophy of the liver and fibrosis was also observed. At termination, there were no test-material related effects on tumour incidence, but at the top dose effects on the liver were seen, including centrilobular hypertrophy and sometimes focal necrosis. Two mouse studies were available for review: the first was carried at dietary concentrations giving intakes of up to 56 mg kg^{-1} bw day^{-1}. The effects observed were mainly on the liver but fenoxycarb also increased the incidence of alveolar/bronchiolar and Harderian gland tumours. These findings were reviewed and while the presence of Harderian gland tumours was considered to be incidental, the lung tumours were considered to be significant. In the second study, which was carried out at higher dietary concentrations, an increase in lung tumours was also observed. Increased liver weights were seen as was enlargement and extramedullary haematopoiesis in the spleen. Fenoxycarb was non-mutagenic both *in vivo* and *in vitro*.

7.2.6.6 *Reproductive and Developmental Toxicity*

A rat multigeneration study was carried at dietary concentrations producing intakes of up to approximately 135 mg kg^{-1} bw day^{-1}. There were effects on liver weight in F_1 parental females and F_1 pups. Effects were seen on duration of gestation and pup weight. A NOAEL was not achieved for this study, but a derived no effect level (NOEL) was used to calculate the ADI. In a study of developmental toxicity in the rat, developmental toxicity, including teratogenicity was not seen even at the highest dose (500 mg kg^{-1} bw day^{-1}). In a study in rabbits of developmental toxicity, there was some evidence of teratogenicity at the highest dose (300 mg kg^{-1} bw day^{-1}). In a follow-up study using doses up to 200 mg kg^{-1} bw day^{-1} such findings were not seen, but it was concluded that it would be prudent to regard 200 mg kg^{-1} bw day^{-1} as a NOAEL for these effects. In a study of reproductive toxicity in sheep, fenoxycarb was administered at doses of 0, 0.69 or 1.38 mg kg^{-1} per day in gelatine capsules; nothing adverse was observed in the dosed ewes or rams or in their offspring up to 28 days.[21]

7.2.6.7 *Neurotoxicity*

Fenoxycarb is not thought to be an anticholinesterase, but Smulders and colleagues[22] observed that fenoxycarb inhibited rat neuronal $\alpha 4\beta 4$ nicotinic receptors, expressed in *Xenopus laevis* oocytes.

7.2.6.8 *Reference Dose*

The ACP proposed an ADI based on a derived NOAEL (considered to be 3 mg kg^{-1} bw day^{-1}) from the multigeneration study. The ADI was 0.03 mg kg^{-1} bw day^{-1}.[20]

7.3 Chitin Synthesis Inhibitors

Chitin is a modified polysaccharide which contains nitrogen and is found in many organisms. It is the main component of the exoskeletons of arthropods, including insects and crustaceans. Insects grow by repeatedly moulting and creating a new exoskeleton, and this requires the synthesis of chitin. Chitin synthesis inhibitors prevent this process. The action of chitin synthesis inhibitors such as diflubenzuron is very specific, in that the compounds do not appear to inhibit hexosamine transferases, which are responsible for connective tissue glycosaminoglycan formation in mammals.[23] The chitin synthesis inhibitors, which are benzoylphenylureas, are generally of low mammalian toxicity. Many of these compounds have haematological effects in mammals, probably because of metabolism to substituted anilines.

Many of the chitin synthesis inhibitors have uses in veterinary medicine as well as in plant protection.

7.3.1 Diflubenzuron

Diflubenzuron is the ISO name for 1-(4-chlorophenyl)-3-(2,6-difluorobenzoyl) urea (IUPAC). The structural formula is:

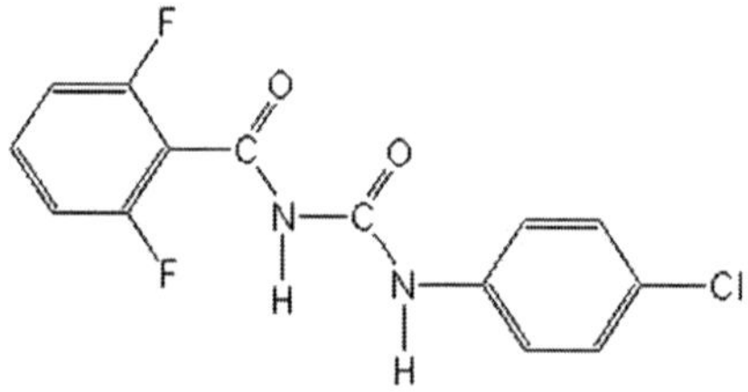

7.3.1.1 *Absorption, Distribution, Metabolism and Excretion*

The absorption, distribution, metabolism and excretion were reviewed by the JMPR.[5,23] Diflubenzuron administered orally was absorbed to a limited extent in mice, rats and cats, the rate of absorption being rapid. The extent of absorption decreased with increasing dose. The highest tissue concentrations of label in the rat were found initially in liver and fat. Diflubenzuron was rapidly excreted, most of the dose being excreted within 24 h. Absorbed diflubenzuron was excreted primarily in urine, with involvement of biliary excretion and enterohepatic circulation. Diflubenzuron (unchanged) was the only component excreted to a significant extent in the faeces. A number of metabolites were found in the urine. The primary metabolic steps were hydroxylation of the anilino ring, ureido bridge cleavage and sulfate conjugation.

7.3.1.2 *Acute Toxicity, Irritancy and Sensitization*

The acute toxicity of diflubenzuron is very low, the LD_{50} in mice and rats by gavage was greater than 4600 mg kg^{-1} bw.[5] Diflubenzuron is not a skin irritant or a skin sensitizer. It is marginally irritant to the eyes of rabbits.[24]

7.3.1.3 Repeated Dose Studies

As with many other insecticides discussed in this chapter, a notable toxicological effect of diflubenzuron in experimental animals is to cause changes in haematological parameters, such as methaemoglobinaemia, sulfhaemoglobinaemia and reticulocytosis, together with a fall in the red blood cell count and increased Heinz body formation. This may result from the metabolism of the compound to *inter alia* 4-chloroaniline.[23] Such changes were seen in rats, in a 7/13-week dietary study, but the JMPR[5] considered the findings not to be biologically significant, as they were sporadic. Increases in spleen and liver weight were seen. Histopathologically, there was, haemosiderosis and congestion of the spleen, erythroid hyperplasia of the bone marrow in all treated groups, and haemosiderosis in the liver. In a 13-week dietary study in the dog, achieved intakes were up to 6.4 mg kg^{-1} bw day^{-1}. Changes similar to those seen in rats were observed (reduction in haemoglobin concentration and erythrocyte count and an increase in methaemoglobin and free haemoglobin concentrations). In a study using dosing by capsule for a year, with doses of up to 250 mg kg^{-1} bw day^{-1}, there were signs of haemolytic anaemia, destruction of erythrocytes and compensatory regeneration of erythrocytes. Increases in methaemoglobin and sulfhaemoglobin concentrations were evident at some doses and the haematological effects appeared dose related. At autopsy, increased pigmentation of Kupffer cells and macrophages in the liver, was observed.

7.3.1.4 Carcinogenicity and Mutagenicity

In a long-term study in mice, diflubenzuron was given in the diet at concentrations producing intakes of up to 960 mg kg^{-1} bw day^{-1}. Increases in sulfhaemoglobin concentrations were noted and Heinz bodies were seen. Increased serum alkaline phosphatase and aspartate aminotransferase activities were also observed. There were elevations in liver and spleen weights, and histopathological changes were see in both organs. In the liver, these changes included hepatocyte enlargement and vacuolation and the presence of pigmented Kupffer cells. In the spleen evidence of extramedullary haematopoiesis was seen. No increase in tumour incidence was associated with administration of diflubenzuron. Two long-term studies in the rat were seen by the JMPR,[5] only one of which demonstrated an overall NOAEL. In the first study, dietary concentrations used were calculated to produce daily doses of up to 8 mg kg^{-1} bw. Increased methaemoglobin concentrations were seen. The study was characterized by poor survival and the range of tissues examined was limited. In the second rat study, much higher dietary concentrations were used and intakes up to 612 mg kg^{-1} bw day^{-1} were achieved. There were alterations in erythrocyte parameters (reductions in red blood cell count and haemoglobin, and increased methaemoglobin and reticulocyte count). Bone-marrow hyperplasia indicated a response to haemolysis. Splenomegaly was seen accompanied histopathologically by pigmented macrophages in the spleen and liver and erythroid hyperplasia of the bone marrow. No NOAEL was identified as increased concentrations of methaemoglobin and sulfhaemoglobin were seen at the lowest dietary concentration (equal to an intake of 7.1 mg kg^{-1} bw day^{-1}). In neither rat study was there evidence of carcinogenicity. Interestingly, the compound is reported to have antitumour activity in mice,[25] but it stimulates haematopoiesis in mice.[26] Diflubenzuron was not genotoxic *in vitro*, in a variety of systems and in the presence or absence of metabolic activation. The pesticide was not genotoxic *in vivo*.

7.3.1.5 Reproductive and Developmental Toxicity

In a two-generation study in rats seen by the JMPR,[5] the most notable effect was a reduction in pup bodyweight gain during lactation (F_1 pups only). Specific reproductive toxicology was not observed, even at the highest dietary concentration, which produced intakes of diflubenzuron of up to 4300 mg kg^{-1} bw day^{-1}. Haematological effects were observed in the parents at all doses. Developmental toxicity studies in rats and rabbits were available to the JMPR.[5] In both species the NOAELs for maternal toxicity, embryotoxicity, fetal toxicity and teratogenicity were considered to be the highest dose tested (1000 mg kg^{-1} bw day^{-1} for both species).

7.3.1.6 Reference Dose

The JMPR established an ADI of 0.02 mg kg^{-1} bw.[27] This was based on the NOAEL for haematological effects in the 2-year rat studies and the 1-year study in dogs and was later reaffirmed. An ARfD was deemed unwarrented.[5]

7.3.2 Flufenoxuron

Flufenoxuron is the ISO name for 1-[4-(2-chloro-α,α,α-trifluoro-*p*-tolyloxy)-2-fluorophenyl]-3-(2,6-difluorobenzoyl)urea (IUPAC). The structural formula is:

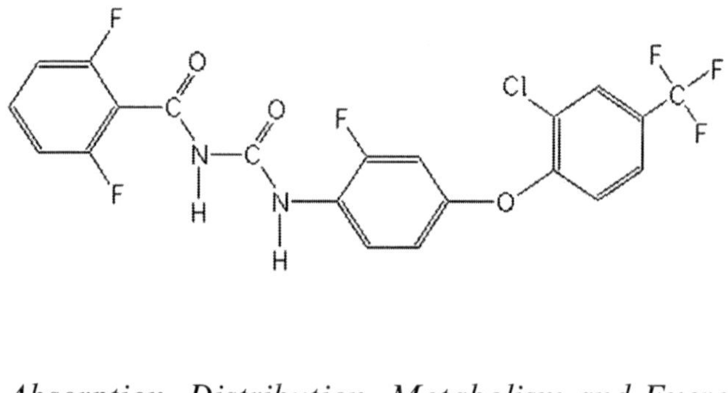

7.3.2.1 Absorption, Distribution, Metabolism and Excretion

The metabolism and toxicology of the compound has been reviewed.[28–30] The United States Environmental Protection Agency (EPA)[29] considered the major metabolic route in rats to be hydrolysis to benzoic acid, a substituted aniline derivative and a phenylurea metabolite. The major metabolite in the urine was benzoic acid. In rats at low doses the material was well absorbed but not at high doses, *i.e.* saturation occurred.

7.3.2.2 Acute Toxicity, Irritancy, and Sensitization

The rat oral LD^{50} is greater than 3 g kg^{-1} bw. The material is not a skin irritant in the rabbit and is only transiently and minimally irritant to the rabbit eye. EFSA considered the active substance to be non-sensitizing to guinea-pig skin.[30]

7.3.2.3 Repeated Dose Studies

The toxicology of flufenoxuron is similar to that of diflubenzuron and consequently it will be discussed in less detail. As with other pesticides of the group, the major effects in repeated dose studies are on the haematological system. These include decreases in red blood cell count, haemoglobin and haematocrit and the presence of methaemoglobin and sulfhaemoglobin, dogs being more susceptible than the rat.

7.3.2.4 Carcinogenicity and Mutagenicity

In one (out of two) mouse oncogenicity studies vascular tumours (haemangiosarcomas in the liver and spleen) were observed. This was considered to be a high dose effect and the EPA[29] considered flufenoxuron not likely to be carcinogenic to humans. Flufenoxuron was negative in a variety of genoxicity studies.

7.3.2.5 Reproductive and Developmental Toxicity

In a study of reproductive toxicity in the rat, parental toxicity was not seen and offspring toxicity consisted of reduced pup weight during lactation (the NOAEL for this effect was critical to the evaluation). Parental mating performance and fertility was unaffected. In a study of developmental toxicity in the rat, adverse effects were seen neither in the dams nor in the offspring. In rabbits, maternal toxicity was not seen, but there were effects on fetal growth at the highest dose.[29]

7.3.2.6 Neurotoxicity

There was no evidence of neurotoxicity in a 28-day rat neurotoxicity study.[29]

7.3.2.7 Human Data

There have been few case reports of human poisonings, although lactic acidosis was reported in one individual, who appeared to have ingested about 100 mL of a preparation containing flufenoxuron and other components including surfactants and solvents.[31]

7.3.2.8 Reference Dose

The EPA established a chronic reference dose (ADI) of 0.0375 mg kg^{-1} bw day^{-1}. This was based upon the NOAEL of 3.75 mg kg^{-1} bw day^{-1} in the two-generation study of reproductive toxicity in the rat and a uncertainty factor of 100.[29]

7.3.3 Lufenuron

Lufenuron is the ISO name for (*RS*)-1-[2,5-dichloro-4-(1,1,2,3,3,3-hexafluoropropoxy)phenyl]-3-(2,6-difluorobenzoyl)urea. The structural formula is:

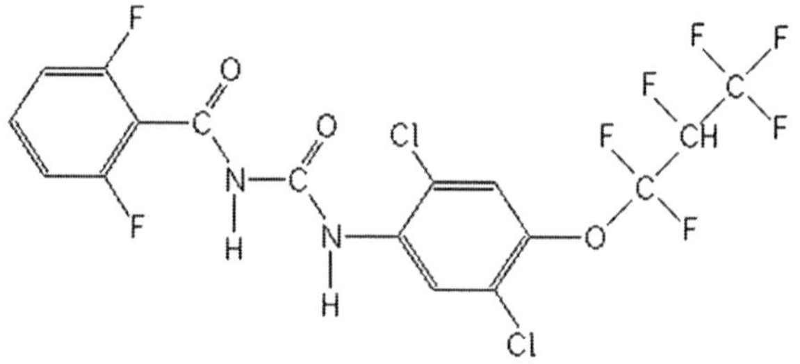

7.3.3.1 Absorption, Distribution, Metabolism and Excretion

The absorption, distribution and metabolism and toxicity of lufenuron has been reviewed.[32] Several studies of absorption, distribution and metabolism were available. The major route of excretion after gavage in rats with radiolabelled lufenuron was the faeces, particularly at high doses, and unchanged lufenuron was the only major component in the faeces. Elimination in the urine was very low at all doses. Following intravenous administration, the major route of elimination was again the faeces, but excretion in the first 24 h was much lower than after oral administration suggesting that, after oral administration, much of the lufenuron was simply not absorbed. The rate of elimination of the material which was absorbed or administered intravenously was slow. The highest tissue residue levels were in fat.

7.3.3.2 Acute Toxicity, Irritancy and Sensitization

The acute oral toxicity in mice and rats was greater than 2000 mg kg^{-1} bw. The material was non-irritant to the skin and eyes. However, lufenuron was a skin sensitizer in guinea-pigs.[32]

7.3.3.3 Repeated Dose Studies

Compared to other insecticides of its group, studies on lufenuron show a paucity of haematological signs but evidence of neurotoxicity (albeit at high doses). In 90-day study in rats, seizures were seen at very high doses (intakes of approximately 1 g kg^{-1} bw day^{-1}). Slightly increased haematocrits and prothrombin times were seen in females. Bodyweights were lower in males at high doses compared to controls. Seizures were also observed in range finding studies in mice and this effect appeared to be dose-related. In a 90-day study of the toxicity of lufenuron admixed into the diet of beagles, maximum intakes achieved were approximately 2 g kg^{-1} bw day^{-1}. Adverse clinical effects were not seen. A small decrease was seen in the highest-group males in red blood cell count and haematocrit: some changes in clinical biochemistry (decrease in potassium and phosphorus and increases in alkaline phosphatase and cholesterol) also occurred. Increased liver weight was observed at higher doses, and this was unaccompanied by histopathological changes. In a 1-year toxicity study in dogs, similar dietary concentrations were used. Neuromuscular signs and convulsions were observed at the higher doses starting at least 20 weeks after inception of the study. Increased adrenal weights were observed at higher doses. There was also a slight increase in thyroid weight (at higher doses), with follicular dilatation (at all doses) and an increase in liver weight (at higher doses) accompanied by histopathological alterations (hepatocyte hypertrophy and increased intracellular pigmentation). Changes were also seen in the adrenal cortex and lung at the higher doses. The rapporteur member state considered that no NOAEL could be derived for the study. A further dog study over 12 months was carried out at lower dietary concentrations (highest intakes were 31.8 mg kg^{-1} bw day^{-1}). One early death was seen at the highest dose and three in that dose group were killed after convulsions. Clinical signs seen at the highest dietary concentrations included convulsions, which were preceded by reduced activity, tremor and ataxia. Lower bodyweight was seen at the highest dietary concentration. There was an increase in liver weights at the higher doses and in adrenal weights in the high-dose males. Hypertrophy of hepatocytes with pigmentation of hepatocytes was seen at the higher doses. At the highest dose there were changes in the lymphoreticular tissue of the gastrointestinal tract (depletion of lymphocytes in Peyer's patches) and lungs (foamy histiocytes).

7.3.3.4 Carcinogenicity and Mutagenicity

In a long-term study in mice, the highest intakes achieved were approximately 63 mg kg^{-1} bw day^{-1}. Convulsions were seen at the higher doses, and at the highest dose the study was terminated early. Moreover, at the next to highest dose, there was excess mortality. At that dose, fatty liver was observed. In the females necrotic changes were seen in livers at the highest dose. Excess lung adenomas were seen. The relevance of these tumours was considered equivocal. Histopathological evidence of chronic inflammation of the prostate was seen at the highest dose. In a long-term oral toxicity study in the rat, dietary concentrations were used that produced daily intakes of up to 114 mg kg^{-1} bw. There were convulsions at high doses. Because of the clear evidence of toxicity, the animals of both sexes at the highest dose were killed at week 14. Some minor haematological changes were observed in these animals, but not the others in the study. There was no evidence of tumorigenesis. At higher doses there were pulmonary alveolar foam cells, ulcerative and/or inflammatory changes in the forestomach and focal inflammatory changes in the caecum and colon. Fatty change in the livers and urinary tract inflammation were seen in females. In a number of tests *in vitro* and *in vitro* there was no evidence of genotoxicity.

7.3.3.5 *Reproductive and Developmental Toxicity*

In a multigeneration study in the rat, the highest dietary concentration was 250 ppm, where intakes were approximately 20 mg kg^{-1} bw day^{-1}, except during lactation where intakes were higher. There was little evidence of reproductive toxicity: at the high dose, appearance of the righting reflex was delayed in both F_1 and F_2 pups. Bodyweights of the F_1 adults (both sexes) were increased by comparison with the controls. There was no histopathological evidence of effects related to treatment. In a rat study of reproductive toxicity, the highest dose being 1000 mg kg^{-1} bw day^{-1}, there was slight maternal toxicity at that dose (reductions in bodyweight and food consumption). No embryofetal toxicity was seen and there was no evidence of teratogenicity. A reproductive toxicity study was carried out in rabbits, using the same doses as in the rats. Maternal toxicity was not observed, There were no effects of treatment on fertility. Embryofetal toxicity and teratogencity was not observed at any dose.

7.3.3.6 *Neurotoxicity*

The neurotoxicity of lufenuron was investigated in male rats at four doses, with controls, in a 4-month feeding study (the section in the draft assessment report is incorrectly headed 'Delayed neurotoxicity', unlikely to be an issue with lufenuron). Daily test-material intakes achieved ranged up to 27 mg kg^{-1} bw. At the higher doses there was a 2-month recovery period. At the highest dose there was an episode of fasciculation in one rat and convulsions in another. Specific neurological examinations (motor activity, startle habituation, maze learning) were carried out, and the results did not differ between doses. Lufenuron, at the highest dose facilitated convulsions induced by pentalene tetrazole. No effects were seen of lufenuron in the central or peripheral nervous system on microscopy.

7.3.3.7 *Endocrine Effects*

In a study of the endocrine system of the rat, an increase in adrenocorticotropic hormone (ACTH) levels was seen at the highest dose: this was considered to be a stress response. Elevated prolactin and FSH levels were not considered to be biologically significant. No effects were observed on the female genital system or oestrous cycle length.

7.3.3.8 Reference Dose

EFSA[32] proposed an ADI of 0.015 mg kg^{-1} bw day^{-1} based on the 1-year dog study NOAEL of 1.5 mg kg^{-1} bw day^{-1} with a safety factor of 100.

7.3.4 Hexaflumuron

Hexaflumuron is the ISO name for 1-[3,5-dichloro-4-(1,1,2,2-tetrafluoroethoxy)phenyl]-3-(2,6-difluorobenzoyl)urea (IUPAC). The structural formula is:

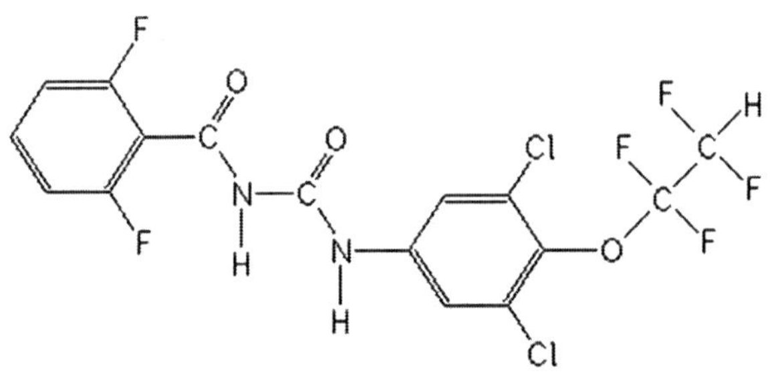

Hexaflumuron is a chitin synthesis inhibitor used predominantly as a termiticide, as are many other insecticides of this group.[33] It is of very low acute mammalian toxicity. As it is usually used in bait stations, exposure of the public is unlikely. The mammalian organ-specific toxicity is not dissimilar to that of diflubenzuron, with effects on the liver and haematological system.[34]

7.4 Ecdysone Agonists

Ecdysone is the precursor of 20-hydroxyecdysone, the major insect moulting hormone, which acts at the ecdysone receptor, a ligand-activated transcription factor. The ecdysone moulting system is a general one in arthropods, hence activity of the ecdysone agonists may be expected in other arthropod species such as crabs. Unlike the other insect hormonal systems discussed above, where the hormones are structurally unrelated to known mammalian hormones, ecdysone resembles hormones important in mammalian physiology. Both ecdysone and 20-hydroxy ecdysone are steroids and, unsurprisingly, are pharmacologically active in mammals (see review by Lafont and Dinan).[35]

The main insecticides in this group are tebufenozide and methoxyfenozide. They mimic the action of ecdysone and bind to the ecdysone receptor, which brings about an incomplete and premature unsuccessful moult, leading eventually to death.

7.4.1 Tebufenozide

Tebufenozide is the ISO name for *N-tert*-butyl-*N*′-(4-ethylbenzoyl)-3,5-dimethylbenzohydrazide (IUPAC). The structural formula is:

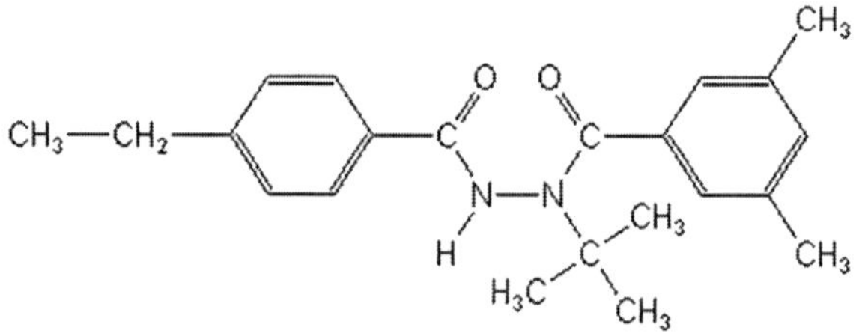

Tebufenozide is predominantly used to control lepidopteran pests in crops, including fruit and vegetables. Tebufenozide is fairly specific for lepidopteran insects.[36,37]

7.4.1.1 Absorption, Distribution, Metabolism and Excretion

The absorption, distribution and metabolism of tebufenozide has been reviewed.[38,39] In rats given a single dose by gavage of labelled tebufenozide, the profiles of excretion of radiolabel by males and females were similar. Absorption and excretion of label were rapid, with more than 70% of the administered dose eliminated within 48 h. The faeces was the main route of excretion, only minor amounts being excreted in the urine and trace amounts in expired air. Little radiolabel was retained in organs or tissues by 7 days after dosing. In a similar study in bile duct cannulated rats, most of the radiolabel (67–70%) was unabsorbed and was eliminated in the faeces. The major component in the faeces was unchanged tebufenozide, together with numerous metabolites. Parent tebufenozide was not found in the urine, but there were metabolites, as seen in the faeces. The extent of metabolism of tebufenozide was highly dose-dependent, being lower at higher doses. The major route of metabolism of tebufenozide appeared to be oxidation of benzylic carbons. In bile duct cannulated rats, parent tebufenozide was not seen in the bile.

7.4.1.2 Acute Toxicity, Irritancy and Sensitization

The toxicology of tebufenozide has been reviewed by the JMPR.[38–40] Tebufenozide is of low oral acute toxicity (lethal dose >5.0 g kg^{-1} bw in the rat). Dermal administration resulted only in transient and mild effects (erythema) at the site of application to rabbits. Tebufenozide was minimally irritating to the eyes of male rabbits and was not a skin sensitizer in guinea-pigs. A single-dose oral study was carried out in dogs to evaluate the NOAEL for haematotoxicity; this was used to set an ARfD (see below).

7.4.1.3 Repeated Dose Studies

In repeated dose studies, the primary target of tebufenozide is the red blood cell, resulting in increased erythrocytic turnover and a compensatory haematopoiesis. In short-term (2-week) toxicity studies in mice, using diets containing tebufenozide, giving intakes of up to 1100 mg kg^{-1} bw day^{-1}, increased liver weights were observed. In a 13-week study in mice using diets containing tebufenozide giving intakes of up to 4200 mg kg^{-1} bw day^{-1} reduced mean bodyweight gain was seen in males. At higher dietary concentrations haemolytic changes and a reduced mean myeloid:erythroid ratio in bone marrow were observed and there were dose-related increases in the weights of the spleen and liver. Histopathologically there was an increased frequency and/or severity of pigment deposition in the liver, spleen and kidney tubules, accompanied by increased extramedullary haematopoiesis in the spleen. In a 2-week study in rats, tebufenozide was admixed into the diet to provide intakes of up to 780 mg kg^{-1} bw day^{-1}. The only adverse effects seen were increased liver weights. In a 13-week rat study, using higher dietary concentration (maximum intake 1600 mg kg^{-1} bw day^{-1}), there were significant decreases in overall bodyweight gain and mean food consumption. As with mice, slight haemolytic anaemia, increased bone-marrow erythropoiesis and increased deposition of pigment in the spleen were observed. At the highest dose, additional effects included overt haemolytic changes, slightly and tubular 'nephrosis' of the kidney in four males. Several dietary studies in dogs were carried out. In a 2-week study maximum intakes achieved were 290 mg kg^{-1} bw day^{-1}, in the 6-week study 42 mg kg^{-1} bw day^{-1} and in the 90-day study, 200 mg kg^{-1} bw day^{-1}. In the 52-week study, the highest intake achieved was 56 mg kg^{-1} bw day^{-1}. Again the main effects seen were haematological: mild haemolytic anaemia (significantly reduced erythrocyte count, haemoglobin, and haematocrit) was observed. The 4-week study incorporated a recovery period after cessation of exposure and total recovery had occurred four weeks after cessation of treatment. In the 90-day study, Heinz bodies were observed together with elevated total bilirubin levels. Furthermore, there were increased absolute and relative spleen weights. Histopathologically, there was an increased incidence of haemosiderin deposition in the Kupffer cells of the liver and increased haematopoiesis and sinusoidal engorgement in the spleen.

At the highest dietary concentration, significant haemolytic changes and increased bone-marrow erythropoiesis (reduced mean myeloid:erythroid ratio) were seen at weeks 6 and 13 of treatment. In the 52-week study, slight but consistent haemolytic changes and a slightly elevated total plasma bilirubin level were seen. The mean absolute and relative spleen weights of females and the mean relative liver weight of males were increased, and there was an increased incidence of pigment deposition in the Kupffer cells of the liver.

Increased haematopoiesis in the spleen and splenic sinusoidal engorgement, and bone-marrow hyperplasia were also observed. The NOAEL for this study (1.8 mg kg^{-1} bw day^{-1}) provided one of the bases for the ADI.

7.4.1.4 *Carcinogenicity and Mutagenicity*

A study of long-term (18-month) toxicity was undertaken in mice, which included an interim sacrifice at 52 weeks. Maximum intakes achieved were 190 mg kg^{-1} bw day^{-1}. Survival rates were affected at the higher doses, but otherwise, the main effects seen were in the haematological system. There was mild haemolytic anaemia, accompanied by a small but significant increase in the blood level of methaemoglobinaemia and increased prevalence of polychromasia and echinocytosis in erythrocytes. Increases in splenic pigment deposition were observed. Extramedullary haematopoiesis in the spleen in females was noted. There was no evidence of carcinogenicity. Additionally, there was a rat 2-year study, with an interim sacrifice at about 1 year. Tebufenozide was admixed in the diet to produce intakes of test material up to 120 mg kg^{-1} bw day^{-1}. At high doses, there were effects on bodyweight and weight gain. Signs of mild haemolytic anaemia were seen during the first 52 weeks. Minor increases in the incidence and/or severity of pigment deposition (haemosiderin) were seen in the spleens of animals of both sexes at both the interim sacrifice and at termination. There was no evidence of carcinogenicity. In a battery of genotoxicity tests, mostly *in vitro* but including a test for chromosomal aberrations in rats *in vivo*, tebufenozide was non-genotoxic.

7.4.1.5 *Reproductive and Developmental Toxicity*

In two rat studies of reproductive toxicity, the main findings were haematological pigment deposition in the spleen and extramedullary haematopoiesis). In the earlier study, intakes achieved were up to 170 mg kg^{-1} bw day^{-1}. Effects were seen at high doses on parental weight gain. The number of implantation sites in F_1 females was decreased and gestational length in F_1 (but not F_0) dams was increased at the highest dose. The NOAEL was 0.8 mg kg^{-1} bw day^{-1}, providing one of the bases for the ADI. The findings were similar in the later study, but reproductive toxicity was not observed (this study involved dietary concentrations giving intakes up to 140 mg kg^{-1} bw day^{-1}). There was evidence of pup toxicity: significant reductions in mean bodyweight gain were seen in F_1 and F_2 pups at the highest dose between lactation days 14 and 21. In a rat study of developmental toxicity, initial slight, transient reductions in mean bodyweight gain and food consumption was observed in the dams. Embryotoxicity, fetotoxicity and teratogenicity were not observed. In a rabbit study of developmental toxicity, maternal toxicity, embryofetotoxicity and teratogenicity were not seen. In both rats and rabbits, the highest dose used was 1000 mg kg^{-1} bw day^{-1}.

7.4.1.6 Neurotoxicity

In a neurotoxicity test in rats, single doses of up to 2000 mg kg^{-1} bw were used. The study included a functional observational battery (FOB) testing and motor activity assessment before treatment and 1–3 h and 7 and 14 days after dosing. No effects referable to treatment were seen. At the end of the 14-day observation detailed neuropathological examination was carried out: treatment-related neuropathological changes were not found at any dose.

7.4.1.7 Reference Doses

The JMPR[38] set the ADI at 0.02 mg kg^{-1} bw based on the NOAEL from the 1-year dog study and one of the rat studies of reproduction, with a safety factor of 100. The JMPR[39] established an ARfD of 0.9 mg kg^{-1} bw based on the NOAEL of 89.4 mg kg^{-1} bw (the highest dose tested) in a single-dose oral study of haematotoxicity in dogs and a safety factor of 100.

7.4.2 Methoxyfenozide

Methoxyfenozide is the ISO name for *N*-*tert*-butyl-*N*′-(3-methoxy-*o*-toluoyl)-3,5-xylohydrazide (IUPAC). The structural formula is:

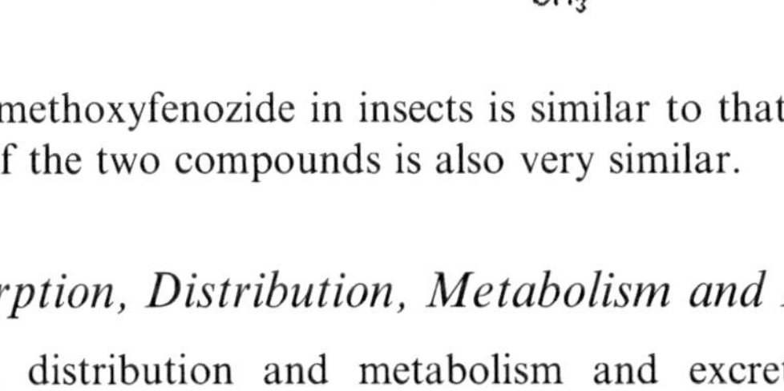

The action of methoxyfenozide in insects is similar to that of tebufenozide. The toxicology of the two compounds is also very similar.

7.4.2.1 Absorption, Distribution, Metabolism and Excretion

The absorption, distribution and metabolism and excretion of methoxyfenozide was reviewed by the JMPR.[39] When given orally to rats [^{14}C]-methoxyfenozide is absorbed rapidly, 58–77% of the administered dose being excreted within 24 h (mostly in the faeces). The peak plasma and blood concentrations of radioactivity (C_{max}) were attained approximately 15–30 min after dosing. Elimination of radiolabel from the plasma showed a biphasic pattern. Concentrations of radioactivity at C_{max} were highest in the liver, with concentrations in the adrenals and in the spleen also being higher than that in whole blood. Using methoxyfenozide radiolabelled at several sites, more than 30 metabolites of methoxyfenozide were identified in rat urine, faeces and bile, demethylation, glucuronidation and hydroxylation being the main metabolic processes. Methoxyfenozide was poorly absorbed (<4%) in the rat after dermal exposure.

7.4.2.2 Acute Toxicity, Irritancy and Sensitization

The toxicity of methoxyfenozide was reviewed by the JMPR.[39] Methoxyfenozide was of low acute toxicity when administered by the oral route to rats and mice (LD_{50} > 5 g kg^{-1} bw) and was of low toxicity in rats by the dermal and inhalation routes. Methoxyfenozide was not irritant to the skin of rabbits and produced only minimal, irritancy to the rabbit eye. Methoxyfenozide was not a sensitizer in guinea-pigs.

7.4.2.3 Repeated Dose Studies

In a 3-month study in mice, methoxyfenozide was given in diets at concentrations producing intakes up to 1742 mg kg^{-1} bw day^{-1}. Deaths were not seen nor were there treatment-related clinical signs of toxicity. Bodyweight gain after 4, 8 and 13 weeks was consistently reduced in both sexes at the highest dose. Substantial substance-related haematological effects were not seen, although a small increase in mean cell volume and mean cell haemoglobin concentration for females was seen at the highest dose. No important changes in any clinical chemistry parameter were seen. Clear effects were not noted on organ weights at any dose in either sex. A small increase in relative liver weight, probably secondary to reduced bodyweight was observed without histopathological correlates and there was no other evidence of hepatic toxicity in the study. No test-material related gross or microscopic pathological change findings was observed in any tissue. A 90-day dietary study of methoxyfenozide was undertaken in the rat, using 5 dietary concentrations and controls. Intakes achieved ranged up to 1531 mg kg^{-1} bw day^{-1}. No clinical signs of toxicity were noted and there was little effect on bodyweight, bodyweight gain, food consumption or ophthalmoscopic examination. Haematological findings indicative of mild anaemia were noted at the highest dose (decreased erythrocyte count and haemoglobin in females). Relative liver weights were increased in both sexes, and there was evidence of slight to moderate periportal hepatocyte hypertrophy in both sexes at the top two doses.

A total of five dog studies were reviewed by the JMPR.[39] They were two 2-week studies (highest intakes observed being 1225 and 1186 mg kg^{-1} bw day^{-1}), a 90-day study, a 1-year study and a study of reversibility of blood effects. In the 2-week studies, the main changes of interest were haematological (decreased erythrocyte counts and haemoglobin concentration, presence of Heinz bodies and an increase in methaemoglobin). Howell–Jolly bodies were also seen. There was an increase in plasma total bilirubin concentrations. Reticulocytosis was seen in one study, especially in females. Absolute and relative splenic weights were increased at high dietary concentrations. Minimal haemosiderin accumulation in Kupffer cells was observed in liver in one study (livers were not examined histopathologically in the other). In a 13-week dietary study in dogs, intakes achieved were up to 209 mg kg^{-1} bw day^{-1}. All dogs survived to termination and test-material-related clinical signs were not noted. Dogs from the lowest dose test group were dosed for a further 2 weeks at that dose and then for 6 weeks at a very high dietary concentration (intake about 460 mg kg^{-1} bw day^{-1}): there were no concurrent controls for that dose, which made interpretation difficult. Bodyweight and weight gain of females at the highest dose for which there were concurrent controls were decreased, but the significance of this was unclear. No effects were seen on food consumption by either sex during the study. Consistent effects on haematological parameters were not seen during the main 90-day study. There was no clear evidence of methaemoglobinaemia; moreover, there were no haematological changes in animals in the group in which the intake had been increased to approximately 600 mg kg^{-1} bw day^{-1}. There were no clear effects on clinical chemistry measurements. Ophthalmoscopic findings were similar in all groups. There were small rises in relative, but not absolute liver weights at some doses in both sexes compared with those of the concurrent controls, but as there was no clear dose–response relationship the significance of this is dubious. Absolute kidney weights were slightly decreased in males at all test doses, and this effect appeared to be dose-related. No test-material related morphological findings were recorded *post mortem* at the end of the main study or at the end of exposure to 460 mg kg^{-1} bw day^{-1}.

A 1-year dietary study in beagles was also undertaken, using dietary concentrations that gave intakes of up to 1199 mg kg^{-1} bw day^{-1}. Mortality was not observed, nor were there clinical signs of toxicity. Bodyweight gain was reduced in males at the top dose in the early part of the study, while in females, bodyweight gain was reduced at the top dose during the later part of the study. Food consumption was similar in all groups. There was some evidence of methaemoglobinaemia at the top dose in both sexes compared with controls and the top-dose group pretreatment. It should be noted that these measurements were on fasting blood samples and therefore may well have underestimated peak methaemoglobin values. Decreases were seen in erythrocyte counts at the top two dietary concentrations and nucleated erythrocytes were detected in both sexes at the top dose. Platelet numbers were increased in both sexes at the top two doses. There were no effects on leukocytes. Total bilirubin in both blood and urine was increased in both sexes at the top two doses. Ophthalmological examinations were all normal. The NOAEL was equal to 9.8 mg kg^{-1} bw day^{-1}, based on changes in the blood and liver.

A study to investigate the reversibility of haematological effects seen in the 1-year dog study was undertaken. There were two groups, controls and a test group similar to the high dose group in the 1-year study, the animals (all male) being dosed for 4 weeks and retained for 4 weeks untreated. Blood samples were collected from all dogs for haematology analysis before the start of the study, after 4 weeks of treatment and after 2 and 4 weeks of the period during which the dogs were not treated. There were no mortalities nor were adverse clinical signs observed. Methoxyfenozide produced a range of haematological effects after the 4-week period of dosing with methoxyfenozide, such as reduced haemoglobin and red blood cell count and elevated methaemoglobin. There was substantial evidence of recovery during the reversibility period.

7.4.2.4 Carcinogenicity and Mutagenicity

The JMPR[39] reviewed long-term studies in mice and rats. Dietary concentrations used in mice produced intakes of up to 1354 mg kg^{-1} bw day^{-1}. There were no effects on survival and treatment-related clinical signs of toxicity were not seen. Minor increases in the prevalence of extramedullary haematopoiesis in the spleen were observed at the highest dose in both sexes. The incidence of bronchiolo-alveolar adenoma or carcinoma was increased in females at the top two doses; however, the increase was small and was not considered to be treatment-related, as the prevalence was within the historical control range. The NOAEL for carcinogenicity and for other effects was the highest dietary concentration. In rats, the dietary concentrations used gave intakes of up to 1248 mg kg^{-1} day^{-1}. There was some evidence of reduced survival in the highest dietary concentration group, but survival was poor in all groups, which limited the power of the study. No test-material related clinical signs of toxicity were seen. Decreased bodyweight gain was seen at the highest-dose group in females towards the end of the study. Minor haematological changes were observed at the top two dietary concentrations (reduced red blood cell counts and haemoglobin and a small degree of methaemoglobinaemia). At the top two dietary concentrations, there was an increase in γ-glutamyltransferase activity. Increased liver weights (absolute and relative) were seen in both sexes at the top two dietary concentrations, and increased adrenal weights in females at the highest dietary concentration. Histopathological changes were seen in liver (periportal hepatocellular hypertrophy) at the top two dietary concentrations in both sexes and kidney (glomerular nephropathy) at the highest dietary concentration in females only. Follicular hyperplasia of the thyroid was observed at the top two dietary concentrations in males and the top dietary concentration in females. Changes were observed at the highest dose in females in the forestomach. There was a small increase in the incidence of hepatocellular adenoma in females at the highest dose, although the prevalence was within the historical control range. There were two instances of statistically significant increases in tumour incidence (C-cell adenoma of the thyroid in males and mammary gland tumours in females), but in neither case was there any dose relationship. The study authors considered there to be no clear oncogenic response. The overall NOAEL was equal to 10 mg kg^{-1} bw day^{-1}. This was on the basis of haematological changes and histopathological changes in the liver and thyroid at the next highest dietary concentration. Methoxyfenozide was non-genotoxic both *in vitro* and *in vivo*.

7.4.2.5 Reproductive and Developmental Toxicity

The JMPR[39] reviewed a two-generation study, which included sperm tests, oestrous cycling and developmental milestones. There were no deaths or clinical signs of toxicity in the parents or pups during the lactation period. Bodyweight gain was decreased at the highest dose in both generations of parental males but not females. Test-material-related effects on oestrous cycling, sperm parameters, mating performance, litter size and pup bodyweight and viability were not observed. There was an increased in the number of F_1 stillbirths but this was not seen in the F_2 offspring. A significant but small delay in vaginal opening was noted in F_1 female pups, but this was not significant in the F_2 generation. No effect was observed on reproductive success. There were no test-material-related effects on reproductive organs. In parental rats at the highest dose, there was an increase in absolute and relative liver weights, accompanied by hepatocellular hypertrophy, with pigmentation of Kupffer cells. Minimal effects on the liver at the mid dietary concentration were not considered to be an adverse effect. The NOAEL for toxicity was 143 mg kg^{-1} bw day^{-1}. This was on the basis of liver toxicity. The NOAEL for reproductive toxicity was 1474 mg kg^{-1} bw day^{-1}, the highest dose tested. The NOAEL for pup toxicity was 143 mg kg^{-1} bw day^{-1} because of minor delays in vaginal opening. In a study of developmental toxicity, mated female rats were given methoxyfenozide (purity 99.2%) at a doses of up to 1000 mg kg^{-1} bw day^{-1} on days 6–15 of gestation. Dosing was by gavage, dams being sacrificed at day 20 of gestation. No deaths occurred and clinical signs of toxicity were not seen. There were no treatment-related effects on the dams' bodyweights, weight gain or food consumption. There was no evidence of developmental toxicity or of treatment related-effects on the fetuses. There was no evidence of teratogenicity. The NOAELs for maternal toxicity and developmental toxicity were was 1000 mg kg^{-1} bw day^{-1}. Mated rabbits were given methoxyfenozide at doses of up to 1000 mg kg^{-1} bw day^{-1}. No test-material-related adverse clinical signs were observed or effects on reproductive or litter parameters were seen. The NOAELs for maternal toxicity and developmental toxicity were 1000 mg kg^{-1} bw day^{-1}. There was no evidence of teratogenicity.

7.4.2.6 Neurotoxicity

An acute neurotoxicity study was undertaken in the rat at single doses of up to 2000 mg kg^{-1} bw by gavage . The study included an FOB. No deaths were observed, nor were there test-material-related clinical signs of systemic toxicity or effects on bodyweight. Hindlimb grip strength for males at the highest dose was reduced, but it was considered that overall, there was no convincing evidence for a substance-related effect on grip strength or motor activity. Gross necropsy and histopathology did not reveal any treatment-related findings in the central or peripheral nervous systems. The NOAEL was 2000 mg kg^{-1} (the highest dose). A 13-week dietary study in rats at feed concentrations giving intakes of to up to 1577 mg kg^{-1} bw day^{-1} was undertaken. This study included an FOB. Deaths were not seen and there were no test-material-related adverse clinical signs. Moreover, effects on bodyweight were not seen. No test-material-related effects were seen either in the FOB or on motor activity. There was no evidence of neurotoxic or neuropathic effects in rats receiving diets at any test material concentration, therefore the NOAEL was the highest dietary concentration.

7.4.2.7 Reference Doses

The JMPR established an ADI of 0.1 mg kg^{-1} bw based on the NOAELs from the long-term study in rats, and the 1-year study in dogs, and a 100-fold safety factor; an ARfD of 0.9 mg kg^{-1} bw was established based on a 1-day dog study with tebufenozide. It was considered that this was a conservative approach because tebufenozide was more potent than methoxyfenozide in producing adverse effects on red blood cells.[39]

7.5 Conclusions

The insecticides discussed in this chapter are generally of low toxicity, but are not knock-down insecticides and do not usually treat established infestations of adult insects.

References

1. Insecticide Resistance Action Committee. *IRAC Mode of Action Classification*, September 2010. http://www.irac-online.org/wp-content/uploads/2009/09/MoA-classification_v7.0.4–5Oct10.pdf
2. L. M. Riddiford, Juvenile hormone action: a 2007 perspective, *J. Insect Physiol.*, 2008, **54**, 895–901.
3. T. G. Wilson, The molecular site of action of juvenile hormone and juvenile hormone insecticides during metamorphosis: how these compounds kill insects, *J. Insect Physiol.*, 2004, **50**, 111–121.
4. C. A. Henrick, Methoprene, *J. Am. Mosq. Control Assoc.*, 2007, **23** (2 Suppl.), 225–239.
5. FAO/WHO, *Pesticide Residues in Food—2001. Evaluations.* Joint Meeting of the FAO Panel of Experts on Pesticide Residues in Food and the Environment and the WHO Core Assessment Group, Geneva, 17–26 September 2001. World Health Organization, Geneva, 2002. http://www.inchem.org/documents/jmpr/jmpmono/2001pr09.htm#2.1.1
6. J. E. Wright, Environmental and toxicological aspects of insect growth regulators, *Environ. Health Perspect.*, 1976, **14**, 127–132.
7. FAO/WHO, *Pesticide Residues in Food—1984. Evaluations.* Joint Meeting Proceedings: Rome 24 September–3 October 1984. Food and Agricultural Organization of the United Nations, Rome; World Health Organization, Geneva, 1985. http://www.inchem.org/documents/jmpr/jmpmono/v84pr61.htm
8. ACP, *Evaluation on Hydroprene*, Advisory Committee on Pesticides, Chemicals Regulation Directorate, Health and Safety Executive, York, UK, July 1991.

9. *Evaluation on Hydroprene,* Advisory Committee on Pesticides, Chemicals Regulation Directorate, Health and Safety Executive, York, UK, January 1996.
10. ADHA, *Acceptable Daily Intakes for Agricultural and Veterinary Chemicals*. Australian Department of Health and Ageing, Office of Chemical Safety and Environmental Health, Canberra, 2007. http://www.health.gov.au/internet/main/publishing.nsf/content/ocs-adi-list.htm
11. FAO/WHO, *Pesticide Residues in Food—1990*. Joint Meeting of the FAO Panel of Experts on Pesticide Residues in Food and the Environment and the WHO Expert Group on Pesticide Residues, Rome, 17–26 September 1990, World Health Organization, Geneva, 1991. http://www.inchem.org/documents/jmpr/jmpmono/v90pr06.htm
12. FAO/WHO, *Pesticide residues in food—2006*. Joint Meeting of the FAO Panel of Experts on Pesticide Residues on Food and the Environment and the WHO Core Assessment Group, Rome, 3–12 October. World Health Organization, Geneva, Switzerland, 2007. http://www.inchem.org/documents/jmpr/jmpmono/v2006pr01.pdf
13. FAO/WHO, *Toxicological Evaluation of Certain Veterinary Drug Residues in Food.* WHO Food Additives Series 45. Prepared by the 54th Meeting of the Joint FAO/WHO Expert Committee on Food Additives (JECFA). World Health Organization Geneva, 2000. http://www.inchem.org/documents/jecfa/jecmono/v45je04.htm
14. M. Moto, M. Okamura, T. Muto, Y. Kashida, N. Machida and K. Mistumori, Molecular pathological analysis on the mechanism of liver carcinogenesis in dicyclanil-treated mice, *Toxicology*, 2005, **207**, 419–436.
15. M. Moto, T. Umemura, M. Okamura, M. Muguruma, T. Ito, M. Jin, Y. Kashida and K. Mitsumori, Possible involvement of oxidative stress in dicyclanil-induced hepatocarcinogenesis in mice, *Arch. Toxicol.*, 2006, **80**, 694–702.
16. T. Umemura, Y. Kuroiwa, M. Tasaki, T. Okamura, Y. Ishii, Y. Kodama, T. Nohmi, K. Mitsumori, A. Nishikawa and M. Hirose, Detection of oxidative DNA damage, cell proliferation and *in vivo* mutagenicity induced by dicyclanil, a non-genotoxic carcinogen, using *gpt* delta mice, *Mutat. Res.*, 2007, **633**, 46–54.
17. M. Jin, Y. Dewa, M. Kawai, J. Nishimura, Y. Saegusa, S. Kemmochi, T. Harada, M. Shibutani and K. Mitsumori, The threshold dose for liver tumor promoting effects of dicyclanil in ICR mice, *J. Toxicol. Sci.*, 2010, **35**, 69–78.
18. M. Moto, Y. F. Sasaki, M. Okamura, M. Fujita, Y. Kashida, N. Machida and K. Mitsumori, Absence of *in vivo* genotoxicity and liver initiation activity of dicyclanil, *J. Toxicol. Sci.*, 2003, **28**, 173–179.
19. M. Moto, M. Okamura, M. Muguruma, T. Ito, M. Jin, Y. Kashida and K. Mitsumori., Gene expression analysis on the dicyclanil-induced hepatocellular tumors in mice, *Toxicol Pathol.*, 2006, **34**, 744–51.
20. Advisory Committee on Pesticides, *Evaluation of Fenoxycarb*, Chemicals Regulation Directorate, Health and Safety Executive, York, UK, February 1997.

21. A. C. Barr, B. Abbitt, R. A. Fiske, J. T. Jaques, H. R. Maynard and J. C. Reagor, Reproductive effects of fenoxycarb on sheep, *J. Vet. Diagn. Invest.*, 1997, **9**, 401–406.
22. C. J. Smulders, T. J. Bueters, R. G. Van Kleef and H. P. Vijverberg, Selective effects of carbamate pesticides on rat neuronal nicotinic acetylcholine receptors and rat brain acetylcholinesterase, *Toxicol. Appl. Pharmacol.*, 2003, **193**, 139–146.
23. FAO/WHO, *Pesticide Residues in Food—1981. Evaluations 1981.* Joint Meeting of the FAO Panel of Experts on Pesticide Residues in Food and the Environment and the WHO Expert Group on Pesticide Residues, Geneva, 23 November–2 December 1981. World Health Organization, Geneva, 1982. http://www.inchem.org/documents/jmpr/jmpmono/v81pr13.htm
24. WHO/FAO, *Diflubenzuron*, Data Sheets on Pesticides. No. 77, World Health Organization, Geneva, July 1996. http://www.inchem.org/documents/pds/pds/pest77_e.htm
25. V. K. Jenkins, R. T. Mayer and R. R. Perry, Effects of diflubenzuron on growth of malignant melanoma and skin carcinoma tumors in mice, *Invest. New Drugs*, 1984, **2**, 19–27.
26. V. K. Jenkins, H. S. Juneja, K. Ives, S. Lee and R. R. Perry, Effects of diflubenzuron and clanfenur on mouse bone marrow cells, *Invest. New Drugs*, 1993, **11**, 279–289.
27. FAO/WHO, *Pesticide Residues in Food—1985. Evaluations 1985, Part II – Toxicology.* Joint meeting of the FAO Panel of Experts on Pesticide Residues in Food and the Environment and the WHO Expert Group on Pesticide Residues, Geneva, 23 September–2 October 1985. World Health Organization,Geneva, Switzerland, 1987. http://www.inchem.org/documents/jmpr/jmpmono/v85pr01.htm
28. ACP, *Review of Flufenoxuron.* Advisory Committee on Pesticides, Chemicals Regulation Directorate, Health and Safety Executive, York, UK, 1995.
29. USEPA, *Flufenoxuron: Human Health Risk Assessment for Proposed Tolerance on Apples, Pears, Grapes, Oranges and Livestock Commodities.* U.S. Environmetal Protection Agency, Washington, DC, 2006. http://www.epa.gov/opprd001/factsheets/flufenoxuranHHRA.pdf
30. EFSA, *Draft Assessment Report and Additional Report to the DAR. Risk Assessment Provided by the Rapporteur Member State France for the Existing Active Substance Flufenoxuron*, European Food Safety Agency, Parma, February 2010.
31. J. Jeong, S. Yeom, J. Ryu, S. K. Han, S. J. Cho and J. Kim, A case of human poisoning with a flufenoxuron-containing insecticide, *Clin. Toxicol. (Phila.)*, 2010, **48**, 87–89.
32. EFSA, *Draft Assessment Report. Risk Assessment Provided by the Rapporteur Member State Portugal for the Existing Active Substance Lufenuron*, European Food Safety Agency, Parma, July 2007.
33. J. L. Lewis and B. T. Forschler, Impact of five commercial baits containing chitin synthesis inhibitors on the protist community in *Reticulitermes flavipes* (Isoptera: Rhinotermitidae), *Environ. Entomol.*, 2010, **39**, 98–104.

34. ADHA, *Hexaflumuron*. Australian Department of Health and Ageing, Canberra, 2007. http://www.health.gov.au/internet/main/publishing.nsf/Content/termite-protection.htm ~ termite-protection-app-chemicals.htm ~ termite-protection-app-chemicals-hex.htm
35. R. Lafont and L. Dinan, Practical uses for ecdysteroids in mammals including humans: an update, *J. Insect Sci.*, 2003, **3**, 1–30.
36. M. Sundaram, S. R. Palli, P. J. Krell, S. S. Sohi, T. S. Dhadialla and A. Retnakaran, Basis for selective action of a synthetic molting hormone agonist, RH-5992 on lepidopteran insects, *Insect Biochem. Mol. Biol.*, 1998, **28**, 693–704.
37. T. S. Dhadialla, G. R. Carlson and D. P. Le., New insecticides with ecdysteroidal and juvenile hormone activity, *Annu. Rev. Entomol.*, 1998, **43**, 545–569.
38. FAO/WHO, *Pesticide Residues in Food—1996. Evaluations 1996. Toxicology*. Joint meeting of the FAO Panel of Experts on Pesticide Residues in Food and the Environment and the WHO Core Assessment Group, Rome, 15–25 September 1996. World Health Organization, Geneva, 1997. http://www.inchem.org/documents/jmpr/jmpmono/v96pr11.htm
39. FAO/WHO, *Pesticide Residues in Food—2003*. Joint Meeting of the FAO Panel of Experts on Pesticide Residues on Food and the Environment and the WHO Core Assessment Group, Geneva, Switzerland, 15–24 September 2003. World Health Organization, Geneva, 2004. http://www.inchem.org/documents/jmpr/jmpmono/v2003pr01.htm
40. FAO, *Pesticide Residues in Food—1996*. Joint Meeting of the FAO Panel of Experts on Pesticide Residues in Food and the Environment and the WHO Core Assessment Group, Rome, 15–25 September 1996. Food and Agricultural Organization, Rome, 1996. http://www.fao.org/docrep/w3727e/w3727e0s.htm

CHAPTER 8

Insecticides of Natural Origin, Other than Pyrethrum and Nicotine

ROLAND SOLECKI* AND LARS NIEMANN

Federal Institute for Risk Assessment, Berlin, Germany.
*Email: Roland.Solecki@bfr.bund.de

8.1 Introduction

Plants have been the most important sources of natural insecticides for centuries. Long before the advent of synthetic insecticides, materials of natural origin provided means for controlling insects affecting the human population both directly and indirectly. Insecticides of natural origin are obtained from animals, plants, bacteria or certain minerals.[1] There are several natural plant insecticides that have been widely used, although, compared with modern synthetic insecticides, their activities are relatively weak. More than 1500 species of plants have been reported to have insecticidal properties, and many more probably exist; however, only a few products are economically important. The body of scientific literature documenting effects of natural pesticides continues to expand, yet only a handful of botanicals are currently used in agriculture in the industrialized world. More recently, the immense potential of bacteria and other microorganisms for the production of biologically active insecticides was realized and many new pesticides commercialized since the middle of the 20th century are of microbial origin.[2] Clarification of the mode of action of these compounds at the receptor level has been made possible by advances in molecular biology during the past four decades. Spinosins, mectins

Issues in Toxicology No. 12
Mammalian Toxicology of Insecticides
Edited by Timothy C. Marrs

Published by the Royal Society of Chemistry, www.rsc.org

and neem extracts are well established commercially, while the use of others, *e.g.* rotenone, appears to be waning. A number of plant substances have been considered for use as insect antifeedants or repellents, but apart from some natural mosquito repellents, little commercial success has ensued for plant substances that modify arthropod behaviour.[3]

Insecticides of natural origin have long been touted as attractive alternatives to synthetic chemical insecticides for pest management because these substances reputedly pose little threat to the environment or to human health.[3] One benefit of plant insecticides is that many of them are readily biodegradable. However, insecticides of natural origin have certain disadvantages in that they are often mixtures of active and inactive components and the active ingredient content may be low, depending on origin, harvest, storage conditions and manufacturing process. Contamination of plant products with mycotoxins or other hazardous substances may occur. The biological variability may result in different toxic properties. This in turn makes the toxicological characterization of the active substances difficult and very often less reproducible than is the case with single synthetic active ingredients. Thus, an essential prerequisite for toxicological evaluation is that identity and quality are subject to permanent and strict control. To some extent, this problem has been addressed in recent decades by the production of naturally based substances in larger amounts with consistent technical specifications but, currently, the most successful insecticides based on naturally occurring substances are synthetic analogues of biological neurotoxins and growth regulators.[1]

This chapter focuses on insecticides of natural origin which have practical and historical importance. Table 8.1 summarizes the active substances of the insecticides discussed in this chapter.

8.2 Rotenone

Rotenone is the ISO name for (2*R*,6a*S*,12a*S*)-1,2,6,6a,12,12a-hexahydro-2-isopropenyl-8,9-dimethoxychromeno[3,4-b]furo(2,3-h)chromen-6-one (IUPAC). The structural formula is shown in Figure 8.1.

Rotenone is an example of a biologically active isoflavonoid. More than 60 plant species of the family *Leguminosae* are known to produce rotenone and

Table 8.1 Biological insecticides and their natural origin.

Active substance	*Natural origin*
Rotenone	Plant species of the *Leguminosae* family
Neem tree extracts	Neem tree preparations
Avermectins	Macrocyclic lactone antibiotics in the fermentation broth of soil microorganisms
Spinosyn	Natural fermentation product produced by an actinomycete bacterium
Quassia extract	Plant extracts derived from the wood of tropical trees
Anabasine	Pyridine alkaloid found in the wild tobacco tree plant

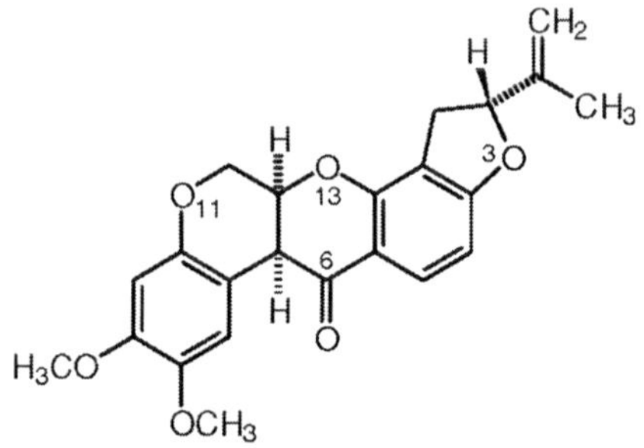

Molecular Formula:	$C_{23}H_{22}O_6$
IUPAC name:	2R,6aS,12aS)-1,2,6,6a,12,12a-hexahydro-2-isopropenyl-8,9-dimethoxychromeno[3,4-b]furo[2,3-h]chromen-6-one
CAS #	83-79-4

Figure 8.1 Chemical structure of rotenone.

other rotenoids, occurring chiefly in the roots. Rotenone is found in resin ducts, which occur in the phloem and xylem in ranges from 3% to 11%. Six rotenoid esters occur naturally and are isolated from the plant *Derris eliptica* found in South-east Asia, or from the plants *Lonchocarpus utilis* or *L. urucu* that are native to South America. The South American rotenone-bearing leguminous plant is no longer known in the wild, and now a days *Lonchocarpus* is hand-cultivated in tropical regions of Brazil and Peru. In South-east Asia, particularly in Java and Sumatra, the closely related legume, *Derris elliptica*, was used as an arrow poison. *Derris*, which has been grown commercially in Puerto Rico, gives lower yields of rotenone than does *Lonchocarpus*. Rotenone also occurs in the legume genus *Tephrosia* but this source is less important for extraction of rotenone for commercial purposes than the genera *Derris* and *Lonchocarpus*.[1–3]

Before the First World War, industrialized nations were ignorant of the plants that contain rotenone. Rotenone was a mysterious and unidentified fish poison of the deep forests of South America, where natives collected roots of a viney shrub, *Lonchocarpus* sp., and threw the crushed roots into small streams and pools. The chemical in the root stunned the fish and caused them to float to the surface where they could be easily collected. Humans were not poisoned by consuming rotenone. The use of rotenone as a fish poison became widespread in the 20th century. Although this use of rotenone-bearing plants as fish poisons was previously reported from the early 18th century, applications as insecticides are just over a century old, mainly because of the rapid photodegradation by ultraviolet light that has limited the commercial suitability of rotenone.[1] However, at the end of the 20th century, it was widely applied as an agricultural and household garden insecticide.[4] It is also applied as a piscicide directly to water to manage fish populations. Recently, the commercial use of rotenone appears to have declined.

The structure of rotenone (Figure 8.1) was established in 1932. Additional notable rotenoids are deguelin, ellipton, malaccol, sumatrol, tephrosin, and α-toxicarol as well as their recently identified oxahomologues.[2] Rotenone is relatively harmless to plants, highly toxic to fish and many insects, moderately toxic to mammals and leaves no harmful residue on vegetable crops. It can be applied as a spray on fruits and row crops, even several times before harvesting, because the chemical residues do not linger. It is a slow-acting contact and stomach poison in insects. In the presence of light and heat, its effectiveness and almost all toxicity can be lost after 2–3 days during the summer. Rotenone dusts and sprays have been used for years to control aphids, certain beetles and caterpillars on plants, as well as fleas and lice on animals. It is a potentially lethal toxin for aphids, cockroaches, houseflies, corn borers, Mexican bean beetles and mosquitoes.

Rotenone is a mitochondrial poison and its insecticidal activity is based on inhibition of mitochondrial oxidation. The critical effects in insects are related to blocking electron transport in mitochondria from complex I to ubiquinone, thus inhibiting oxidation linked to $NADH_2$, which results in nerve conduction blockade.[5] The insecticidal activity may also be related to interference with glutamate oxidation in nerves and muscles even at low concentrations.[6] The anaesthetic-like action on nerves appears to be related to this ability of rotenone.

8.2.1 Absorption, Distribution, Excretion and Metabolism

Rotenone is metabolized largely in the liver by hydroxylation at carbons 7 and 24. The acute toxicity of the metabolites hydroxyrotenone and rotelone I seems to be comparable to that of rotenone, whereas dihydroxyrotenone and rotelone II are significantly less toxic. One mechanisms of detoxification was found to be 3-*O*-demethylation. Within 24 h, approximately 20% of radioactivity was excreted in urine in rats and mice, respectively.[7]

8.2.2 Acute Toxicity, Irritancy and Sensitization

Acute poisoning in animals is characterized by an initial respiratory stimulation followed by respiratory depression, ataxia, convulsions, and death by respiratory arrest.[8] The acute toxicity appears to vary considerably between species, with oral LD_{50} values ranging from 25 to 3000 mg kg^{-1} bodyweight (bw). Factors other than species variation, which probably influences the oral absorption and consequently also the LD_{50} values, were the concentration and particle size of rotenone in the plant powders of various rotenone-bearing plants, and the diluents used for administration of the material. Several studies indicated that the finer the particle size the more toxic the preparation. The use of oil as a diluent also increases the toxicity. Thus, the LD_{50} in rats was estimated to be between 25 mg kg^{-1} bw and 200 mg kg^{-1} bw, when dissolved in olive oil and administered as suspension in vegetable gums, respectively.

When rotenone was injected into animals, tremors, vomiting, incoordination, convulsions, and respiratory arrest were observed.[9]

Derris powder was not irritant to animals and not sensitizing to human skin, but the dust produced intense eye irritation in rabbits.[10] When inhaled, derris caused severe pulmonary irritation in animals and was more toxic than pure rotenone, indicating higher toxicity of other plant constituents.[2]

8.2.3 Repeated Dose Toxicity

In short-term toxicity studies, clinical signs such as emesis and diarrhoea occurred and mean bodyweights were reduced. Changes observed in haematological and clinical chemistry parameters were decreased hematocrit and hemoglobin as well as lower cholesterol, total lipids and glucose levels. Furthermore, fatty changes in the liver and kidneys were observed.

In a long-term rat study, effects included a decrease in bodyweight, reduced food consumption, lower total protein and albumin levels in the blood, increased blood urea nitrogen levels and increased incidences of adrenal gland angiectasis and haemorrhage.[11]

8.2.4 Carcinogenicity and Mutagenicity

Whether rotenone is carcinogenic is still a matter of controversy. Some authors reported increased incidences of mammary tumours in rats, but these findings were not supported by other investigators.[12,13] No evidence of carcinogenic activity was observed in mice.[14] Recent reports have shown that rotenone inhibits spontaneously and chemically induced hepatic tumorigenesis in rodents through a decrease in hepatic focal proliferation and an increase in focal apoptosis.[15]

Rotenone is a potential spindle poison, which is comparable to colchicine. Microscopic investigations with cultured mammalian cells showed that rotenone delayed all phases of the cell cycle and reversibly inhibited the development of mitotic spindle microtubules.[1] In line with that, rotenone induced aneuploidy and polyploidy in Chinese hamster ovary (CHO) cells whereas induction of sister-chromatid exchanges and chromosome aberrations was not observed in this cell line. Also, rotenone induced forward mutations in the mouse lymphoma assay, increased the frequency of binucleated cells and caused a delay in cell division. In contrast, the substance proved negative in reverse mutation and unscheduled DNA synthesis assays. The mutagenic effects of rotenone were considered to depend on the inhibition of both microtubule assembly and cell respiration. The latter is assumed to occur by blocking electron transport.[16,17]

Chronic exposure to rotenone in rats has reproduced the anatomical, neurochemical, behavioural and neuropathological features of Parkinson's disease by systemic inhibition of mitochondrial complex I, which caused highly selective nigrostratial dopaminergic degeneration.[18,19] However, a new model based on daily inhalation exposure of neurotoxins in mice was used to assess the potential danger of toxins as risk factors for development of Parkinson's disease. In contrast to other complex I inhibitors, rotenone-treated mice or rats were asymptomatic.[11]

8.2.5 Effects on Reproduction and Development

In a two-generation study on rats, litter sizes and pup weights were reduced at doses toxic to the parents, as demonstrated by reduced bodyweight and bodyweight gain. Diets containing 0, 7.5, 37.5, or 75 ppm rotenone, equivalent to 0, 0.38, 1.88, or 3.8 mg kg^{-1} day^{-1}, were given to groups of 15 male and 25 male Charles River rats through two generations. The first parental generation animals were 6 weeks old at the beginning of the test, and they were given test diets for 105 days prior to mating. Parental rats were selected from pups 21 days after birth for the second generation mating and were given test diets for a period of 120 days before they were mated. Test diets were also administered during gestation and lactation for both generations. Litter sizes were reduced in the 75 ppm dose group in both generations indicating a reproductive effect at 75 ppm. Pup weights were reduced in both generations during lactation for the 37.5 and 75 ppm dose groups. Body weights and body weight gains in adult rats were reduced during the two generations also. Based on these results, the lowest effect level for reproductive toxicity was 37.5 ppm and the NOAEL was 7.5 ppm, equivalent to 0.38 mg kg^{-1} day^{-1}.[20]

Developmental studies in rats have shown increased resorptions and reductions in fetal weights. Delayed skeletal ossification, including increased incidences of extra ribs and unossified sternebrae, as well as renal pelvic cavitation and distended ureters occurred at dose levels in which decreased body weights were usually also present in the dams, suggesting maternal toxicity.[2]

8.2.6 Effects in Humans

Neither deaths nor systemic poisonings resulting from exposure to rotenone products have been reported in relation to ordinary use over many decades. Numbness of oral mucous membranes, dermatitis and respiratory tract irritation have been reported after inhalation of dust from powdered derris root in occupationally exposed humans. Dermatitis and respiratory tract irritation have also been reported in occupationally exposed persons.[9]

The United States Environmental Protection Agency (EPA) reviewed reported incidents from piscicide applications as well as agricultural and residential uses. The most common symptom reported was eye irritation, which was four times more prevalent than any other symptoms. Other symptoms reported included dermal irritation, throat irritation, nausea and cough/choke. Most incidents appeared to be caused by rotenone's irritant properties. Few neurological symptoms other than headache and dizziness were reported, though there were several reports of peripheral neuropathy, numbness, or tremor.[21]

8.2.7 Reference Doses

No acceptable daily intake (ADI) or acute reference dose (ARfD) published by the European regulatory bodies was obtainable. The toxicological endpoints and reference values published by the EPA[21] are shown in Table 8.2.

Table 8.2 Summary of toxicological endpoints for rotenone.[21]

Exposure scenario	*Dose used in risk assessment, uncertainty factor (UF)*	*Level of concern for risk assessment*	*Study and toxicological effects*
Acute dietary (females 13–49)	NOAEL = 15 mg kg^{-1} bw day^{-1} UF = 1000 ARfD = 0.015 mg kg^{-1} bw	Acute PAD: 0.015 mg kg^{-1} bw	Developmental toxicity study in mouse LOAEL = 24 mg kg^{-1} day^{-1} based on increased resorptions
Acute dietary (all populations)	An appropriate endpoint attributable to a single dose was not identified in the available studies, including the developmental toxicity studies		
Chronic dietary (all populations)	NOAEL = 0.375 mg kg^{-1} bw day^{-1} UF = 1000 cRfD = 0.0004 mg kg^{-1} bw day^{-1}	Chronic PAD: 0.0004 mg kg^{-1} bw day^{-1}	Chronic/oncogenicity study in rat LOAEL = 1.9 mg kg^{-1} bw day^{-1} based on decreased body weight and food consumption in both males and females
Incidental oral Short-term (1–30 days) Intermediate-term (1–6 months)	NOAEL = 0.5 mg kg^{-1} bw day^{-1}	Residential MOE = 1000	Reproductive toxicity study in rat LOAEL = 2.4/3.0 mg kg^{-1} bw day^{-1} based on decreased parental (male and female) body weight and body weight gain
Dermal short-, intermediate-, and long-term	NOAEL = 0.5 mg kg^{-1} bw day^{-1} 10% dermal absorption factor	Residential MOE = 1000 Worker MOE = 1000	Reproductive toxicity study in rat LOAEL = 2.4/3.0 mg kg^{-1} bw day^{-1}
Cancer	Classification: No evidence of carcinogenicity		

aPAD, acute population adjusted dose; cPAD, chronic population adjusted dose; LOAEL, lowest observed adverse effect level; MOE, margin of exposure; NA, not applicable; NOAEL, no observed adverse effect level; RfD, reference dose; UF, uncertainty factor.

8.3 Neem Tree Products

The neem tree (*Azadirachta indica*) was first described in botany by Laurent de Jussieu in 1830. It belongs to the mahogany family, *i.e.* the Meliaceae, and has its origin in South Asia, probably in India and Burma (Myanmar). It is now widespread in many tropical and subtropical regions all over the world, presumably because of its many useful properties.[22] Thus, from about 25 million neem trees in India alone, some hundred thousand tons of seeds are annually collected and about 100 000 tons of neem oil produced. Among the reasons for large-scale plantation of neem trees are their ecological advantages, in addition to the economic ones. Because of its high photosynthesis rate and oxygen liberation, the fast-growing and evergreen neem tree may help to clean the air. In Africa, the neem tree is an important plant in afforestation programmes that are conducted in the Sahel zone in particular to combat desertification. Moreover, neem cake is widely used as a fertilizer and as animal feed.[23]

The neem tree contains, in its roots, bark, leaves, fruit, flowers and seed kernels, a number of substances which possess, alone or in combination, remarkable insecticidal, 'antifeedant' or insect-repellent properties that may be successfully used in plant protection and for protection of stored food but also in veterinary medicine and public health. The major chemical constituents of interest as insecticides belong either to the limonoids or the triterpenoids.[23–27] The best known of them is, without doubt, the tetranortriterpenoid azadirachtin (or azadirachtin A, see Figure 8.2). This substance is often considered the 'active principle' of neem products. However this is not likely, taking into consideration the low concentration, the biological activity of other ingredients, the variety of effects and the presumably complex mode of action.[24,27,28] Azadirachtin A may be used for analytical purposes as a 'lead substance', *e.g.* to monitor residues after application of neem-based plant protection products

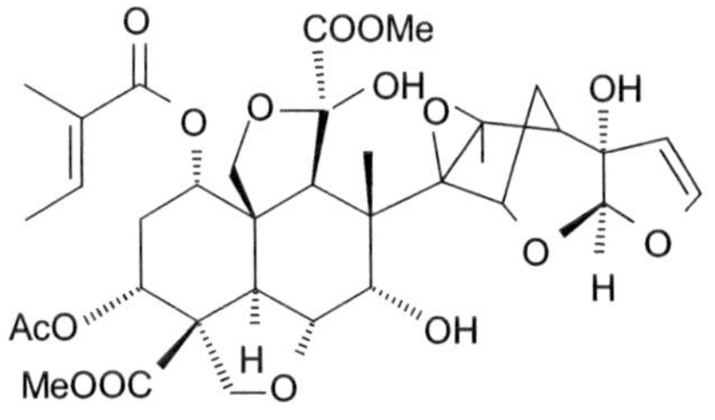

Molecular Formula:	$C_{35}H_{44}O_{16}$
IUPAC name:	dimethyl (2a*R*,3*S*,4*S*,*R*,*S*,7a*S*,8*S*,10*R*,10a*S*,10b*R*)- 10-(acetyloxy)- 3,5-dihydroxy- 4-[(1*S*,2*S*,6*S*,8*S*,9*R*,11*S*)- 2-hydroxy- 11-methyl- 5,7,10-trioxatetracyclo[6.3.1.0^{2,6}.0^{9,11}]dodec- 3-en- 9-yl]- 4-methyl- 8-{[(2*E*)- 2-methylbut- 2-enoyl]oxy}octahydro- 1*H*-furo[3',4':4,4a]naphtho[1,8-*bc*]furan- 5,10a(8*H*)-dicarboxylate
CAS #	11141-17-6

Figure 8.2 Chemical structure of azadirachtin A.

in field crops. Furthermore, it has been proposed to adjust reference values for certain neem kernel extracts to their azadirachtin A contents to allow better comparison of toxicities among different products.

Neem seed oil or extracts from kernels or leaves have been used for centuries in traditional Indian medicine for treatment of a wide range of diseases including fevers, hepatitis, respiratory diseases, gastrointestinal disturbances and helminthoses, or, more general, as 'health strengtheners'. Sometimes, it has been argued that, because of this 'familiarity', safety of neem-derived products may be taken as proven and further toxicological testing would not be needed, but this argument should not be accepted without proof. In fact, blood glucose lowering, anti-inflammatory, antiulcer, antiparasitic and hepatoprotective properties are ascribed to certain neem ingredients. However, no therapeutic use for these purposes has been established in contemporary Western medicine. Precise mechanisms or modes of action are unknown and the role of individual ingredients, either isolated or combined, still have to be elucidated. The presumed protective effect on the liver is assumed to be due to a stimulation of glutathione-*S*-transferase production potentially resulting in enhanced detoxification processes.[23,28,29]

Azadirachtin and the neem kernel extracts are very toxic to aquatic organisms and proper risk assessment will be needed before authorizations as plant protection products can be granted. Populations of sensitive (non-target) arthropods will be reduced, but are expected to recover with recolonization of treated areas within 1 year. The possibility of risk to mammalian wildlife or birds as well as to bees was considered to be low.[30]

Neem ingredients are also contained in various cosmetic preparations, and in particular neem-based toothpaste is widely used in India and Europe. Furthermore, mattresses are treated to reduce the population of house dust mites.

Neem-derived products such as seed oil and leaf extracts are reported to exhibit contraceptive effects due to spermicidal and anti-implantation activity when applied in the female genital tract but also due to inhibition of spermatogenesis after oral intake in males. On the other hand, these findings, together with the potential to cause abortion in primates, may raise concern of reproductive safety of insecticides containing neem ingredients and make comprehensive testing for reproduction and developmental toxicity necessary.[28,29,31]

A comprehensive toxicological database is available only for the two insecticides NeemAzal and Fortune AZA, *i.e.*, two extracts that are obtained from neem seed kernel by means of chemical solvents. In the EU, the intended representative use was control of Colorado beetles on potatoes. In the following subsection, the available studies with these extracts are described in greater detail.[28,30,32] Taking into account that different parts of the tree may be used, the findings and conclusions including reference doses for NeemAzal or Fortune AZA do not apply to other extracts from the leaves or neem oil. Neither are they applicable to seed kernel extracts that were obtained by other extraction methods. Accordingly, the studies with the two extracts cannot be used to predict toxic effects, demonstrate safety or derive reference values for other products collected from the neem tree.

Toxicological information on other neem-derived products or certain ingredients other than azadirachtin, such as the bitter substance nimbidin, at least so far publicly available, is scarce and more or less anecdotal and certainly does not allow scientifically sound evaluation.[33] A comprehensive review on what is known about the toxicity of diverse materials collected from the neem tree and some processed products was published in 2004.[29]

8.3.1 Toxicity of the Insecticidal Neem Seed Kernel Extracts NeemAzal and Fortune AZA

Azadirachtin A is considered to be the main active ingredient of kernel extracts. However, in the course of the EU evaluation process, it was acknowledged '... to consider the active substance as the sum of all biologically active identified compounds in the specification'.[30] The test substances in the toxicological studies were the extracts and reference values were directly set on the basis of appropriate NOAELs, *i.e.* without adjusting for azadirachtin A content. These extracts contain azadirachtin A at variable concentrations of 250–500 g kg^{-1} for NeemAzal or 111–180 mg kg^{-1} for Fortune AZA in the concentrates, depending on the batch of seed kernels provided as raw material and the manufacturing process. Despite these apparent differences, both extracts were considered toxicologically equivalent, mainly based on similar findings at effect doses and NOAELs of the same magnitude in feeding studies in rats that were performed under nearly identical conditions in the same laboratory following the same experimental design.

A formulation NeemAzal-F 5% was not developed further due to more serious adverse health effects, and not marketed, but a number of toxicological studies are still available that may be partly used for worst-case considerations.

Contamination with aflatoxins is quite common with neem products and must be strictly monitored. In line with current FAO specifications, the aflatoxin content (sum of aflatoxins B1, B2, G1 and G2) must not exceed 300 µg kg^{-1}azadirachtin, *i.e.* 0.00003% of the azadirachtin content.[30]

8.3.1.1 Absorption, Distribution, Excretion and Metabolism

As for most plant extracts, no data on absorption, distribution, excretion and metabolism are available. For technical reasons, it was not feasible to perform such studies with azadirachtin (A) because not enough technical material in sufficiently pure form could be synthesized and radiolabelled. Furthermore, it would represent only a fraction of the extract to be applied and information about absorption, distribution, metabolism and elimination of all the other ingredients was lacking. It is not known if the insecticidal activity or the observed toxic effects were due to the azadirachtin fraction or if and to which extent other components might have contributed. Thus, even if it had been technically feasible, the scientific value of such a study would be quite limited. A dermal absorption rate of 10% has been estimated.

8.3.1.2 Acute Toxicity

The acute oral and dermal toxicity of neem extracts in rats is very low, with LD_{50} values greater than 5000 mg kg^{-1} bw for oral application or greater than 2000 mg kg^{-1} bw for dermal application. Neither mortality nor gross pathological abnormalities at necropsy were evident at these limit doses. Clinical signs were confined to transient observations of apathy and piloerection. For NeemAzal, this low oral toxicity was confirmed in the mouse and low dermal toxicity was seen in the rabbit. In contrast, the acute oral toxicity of the formulation NeemAzal-F 5% was markedly higher, with LD_{50} values of 765 mg kg^{-1} bw in the rat and 1570 mg kg^{-1} bw in the mouse with a wide range of clinical signs, some deaths and multiple pathological findings in different organs clearly pointing to an impact of ingredients in the formulation other than azadirachtin A on these endpoints. It should be emphasized that no signs and no pathological findings in particular in liver and brain were noted in these studies that would resemble the observations in children following neem oil poisoning. In contrast, such effects were seen in rats and mice after experimental administration of the oil with the objective of reproducing a Reye-like syndrome. Thus, this specific toxicity seems related to neem oil but not to the extracts under investigation.

Inhalation toxicity of NeemAzal to rats was also low with no deaths and only weak and transient respiratory signs to be observed at the highest technically available concentrations of 0.72 mg L^{-1} on 4-h whole-body exposure. In a similar experiment with Fortune AZA, the LC_{50} was greater than the highest concentration of 2.45 mg L^{-1} but, this time, a single death was observed in this high-dose group whereas the surviving animals had recovered by day 2 after the 4-h exposure.

The extracts proved not irritating, either to the skin or the eyes, but there was evidence of skin sensitization at least by higher concentration products in various tests.

8.3.1.3 Repeated Dose Toxicity

Subchronic toxicity of the neem kernel extracts was investigated in oral feeding studies in rats over 90 days. These experiments were performed in the UK under Good Laboratory Practice conditions and in line with OECD requirements for studies of this type. Liver and thyroid proved the target organs. The effects comprised higher organ weights, histopathological findings such as degenerative liver changes and follicular epithelial hypertrophy in the thyroid and alterations in clinical chemistry parameters. At the higher dose levels, bodyweight and food consumption were decreased and some haematological parameters altered. In particular, Fortune AZA caused a number of further toxic effects such as hair loss, degeneration of the sciatic nerve or reduced number of corpora lutea and slightly reduced ovary weight. However, the NOAELs were virtually the same.

In contrast, a 2-year feeding study with NeemAzal was not fully acceptable due to many deficiencies. No toxic effects were noted up to the highest dose level of 6400 ppm, equivalent to 450 mg kg^{-1} bw for male rats and 630 mg kg^{-1} bw for females. This dietary concentration was clearly above the effect doses in the well-conducted 90-day studies but this might point to a lower sensitivity of the Wistar rat strain that was employed in the long-term study. The only available long-term study in mice is only supportive since the test material was the formulation NeemAzal-F with only about 5% azadirachtin A content. In this study, the NOAEL was 63 mg kg^{-1} bw day^{-1}, *i.e.* the highest dose tested.

8.3.1.4 Carcinogenicity and Mutagenicity

In the above mentioned long-term studies with NeemAzal or NeemAzal-F in rats and mice, gross and histopathological examination did not reveal a higher tumour rate. Thus, these studies provide some evidence that this extract was not carcinogenic. For Fortune AZA, no such data are available.

Both neem kernel extracts proved negative in a number of mutagenicity tests, but both elicited a positive response in a chromosome aberration assay in cultured human lymphocytes. However, negative results in appropriate *in vivo* studies such as the micronucleus assay in mouse bone marrow contravened these findings and suggested that NeemAzal and Fortune AZA had no clastogenic potential of relevance to humans.

8.3.1.5 Effects on Reproduction and Development

Dietary administration of NeemAzal had no impact on fertility and reproductive outcome in rats which were fed over two generations at dose levels of up to 750 ppm, equivalent to a mean daily dose of 50 mg kg^{-1} bw.

In developmental studies in rats, no evidence of teratogenicity was observed with both extracts up to the limit dose level of 1000 mg kg^{-1} bw day^{-1}. However, a lower bodyweight gain of the dams was observed at this dose and, with NeemAzal but not with Fortune AZA, a higher number of fetuses with supernumerary ribs was seen. This finding is considered a variation but not a malformation. The developmental NOAEL was the next lower dose of 225 mg kg^{-1} bw day^{-1}. A study in rabbits was performed with a third extract (ATI 720) that cannot be considered as equivalent to NeemAzal or Fortune AZA. This study revealed a lower number of viable fetuses per dam due to an increase in intrauterine deaths at a maternally toxic dose. Maternal toxicity was apparent, as shown by a lower body weight gain. The NOAEL was 20 mg kg^{-1} bw day^{-1} for maternal toxicity and 100 mg kg^{-1} bw day^{-1} for developmental toxicity.

8.3.2 Toxicity of Neem-Derived Products to Humans

Side effects of traditional remedies based on neem, mostly neem seed oil, have been reported mainly from India and Malaysia, reflecting their widespread use in these countries. In particular, severe intoxications and even fatalities have been observed in children. About 60 cases of presumed or confirmed cases of neem oil poisoning in children were treated in a single hospital in India during a 5-year period in the 1970s, but this number must be seen against the background of frequent use.[33–35] Typical symptoms of these poisoning incidents were vomiting and seizures occurring within minutes or hours following oral application of neem oil to children for the treatment of febrile illnesses. Toxicological assessment of these cases is difficult because the oil was of unknown purity and of generally ill-defined origin. Furthermore, the oil was administered in variable amounts with estimated volumes ranging from 5 to 30 mL. Subsequently, these acute signs may be followed by metabolic acidosis, coma, and sometimes, death. Autopsies revealed liver damage with fatty degeneration of hepatocytes and, on the subcellular level, alterations of mitochondrial integrity. Possibly in line with that, activity of microsomal liver enzymes may be decreased. In some cases, encephalopathy and brain oedema were noted in addition to hepatotoxicity. In survivors, neurological deficits have been noticed in some cases. It is not known to which components of neem oil the apparent toxicity was due. Thus, a so far unknown toxic principle must be assumed. It is striking that most confirmed cases of intoxications in humans occurred in children who were already ill, and perhaps this situation will be not properly reflected when the test substances are administered to laboratory animals. There is some similarity to Reye's syndrome.[28,29,34,35] This clinical entity occurs in rare cases in children who have fallen ill with an influenza-like illness or chickenpox and have been treated with aspirin (acetylsalicylic acid). Indeed, a 'Reye-like' syndrome could be experimentally produced by application of neem oil to rats and mice.[36]

Reports on poisoning incidents in adults are rare. However, quite recently, a case of attempted suicide by a 35-year old woman was described. She ingested 250 mL of the plant protection product NeemAzal-T/S obtained from an Indian manufacturer containing only 1% azadirachtin but 51% vegetable oil and 45% sodium lauryl sulphate as tenside. Intensive care with intubation and mechanical ventilation became necessary because of neurotoxic symptoms and coma but she recovered very soon without long-term complications. Poisoning was not verified by biochemical methods and, again, the contribution of azadirachtin is not clear.[37]

All these intoxications were apparently related to oral ingestion. In contrast, there is no evidence of adverse effects of topical application of neem-derived drugs or other products for control of ectoparasite infestations (*e.g.* fleas or lice) in humans or pets. These medical uses are the most common.

8.3.3 Reference Doses for NeemAzal and Fortune AZA

For both extracts, an ADI (Acceptable Daily Intake) of 0.1 mg kg^{-1} bw has been published by the European Commission in the EU pesticide database (http://ec.europa.eu/sanco_pesticides/public/index.cfm?event=activesubstance.detail) that is based on the NOAEL of 32 mg kg^{-1} bw day^{-1} in the oral 90-day study in rats. A higher safety factor of 300 was applied because a fully valid long-term study and a developmental toxicity study in a second species were not submitted.

The ARfD of 0.75 mg kg^{-1} bw was derived from the NOAEL of 225 mg kg^{-1} bw day^{-1} for maternal toxicity in the developmental study in rats. Again, a higher safety factor of 300 was needed because developmental toxicity and teratogenicity were not investigated in a second species.

The acceptable operator exposure level (AOEL), is numerically the same as the ADI and was established on the same experimental basis. The use of the safety factor of 300 was again justified by the absence of the developmental toxicity study in a second species but also by the absence of data on oral absorption and systemic bioavailability.[30]

8.4 Avermectins

Avermectins are complex disaccharides that were discovered in the 1970s in the Japanese Kitasato Institute as a new class of macrocyclic lactone antibiotics in the fermentation broth of the soil microorganism *Streptomyces avermitilis*. At least eight natural avermectin compounds, (A1a, A1b, A2a and b, B1a, B1b, B2a and b) have been identified.[38,39] Lacking strong antibacterial or antifungal activities, the avermectins have found their major field of application in the control of arthropods (insects and mites) and helminths (nematodes), due to their broad-spectrum antiparasitic acitvity and a prolonged post-treatment effect. In general, all compounds from this group share common pharmacological and toxicological mechanisms although they differ in potency and safety. The intended toxic effect on target organisms is mediated by activation of glutamate-gated chloride channels in nerve and muscle cells of invertebrates, resulting in the relatively slow but irreversible opening for chloride ions. The subsequent hyperpolarization or depolarization of the target cells makes them unable to respond to excitatory stimuli and eventually leads to paralysis that is considered the main cause for death of the target species.[38–42] In the case of the drug ivermectin, there is also evidence of a reprotoxic effect on the eggs and in particular the larvae of target species at very low dose levels. Furthermore, ivermectin may suppress egg production and inhibit larval maturation in ticks.[38] At concentrations above the therapeutic range, however, avermectins additionally bind to γ-aminobutyric acid (GABA) receptors in peripheral neurons of nematodes and neuromuscular synapses of arthropods but also in the brain of vertebrates. This mechanism may contribute to the antiparasitic efficacy but is certainly behind the toxicity to non-target organisms such as humans, farm animals and pets. It should be clearly understood that avermectin toxicity to mammals mainly depends on the ability of avermectins to permeate the blood–brain barrier and on the concentrations that are reached in the brain. Accordingly, individuals or subpopulations (such as neonates in some species or certain strains or breeds) with no intact or an insufficiently mature blood–brain barrier are at higher risk of experiencing adverse effects of avermectins. This is well in line with reports from poisoning incidents as well as with the outcome of toxicological studies. Mechanistic studies have shown that sensitivity to avermectins is related to underexpression of P-glycoprotein, which is crucial for development of the blood–brain barrier. This protein is part of the cell membrane and associated with multidrug resistance. While newborn children usually produce an adequate amount of P-glycoprotein, it is partly lacking in certain mouse strains (CF-1 mice), neonatal rats and quite frequently in certain breeds of dogs such as collies, Shetland sheepdogs, Australian shepherd dogs, Border collies, occasionally also in whippets or German shepherds, making the blood-brain barrier in these animals more readily permeable. The high vulnerability of collies as compared to other dogs is considered to result from a generally rare but in this breed widespread mutation in the *MDR1a* gene which is responsible for P-glycoprotein expression in the brain, whereas expression in the liver (regulated by the *MDR2* gene) was not altered. Differences in toxicity of avermectins to humans might be also explained by the rather new evidence that polymorphisms of the *MDR1* gene occur within the human population, resulting in either over- or underexpression of P-glycoprotein. Those with underexpression would be at greater risk but, for the time being, it is assumed that the 10-fold UF for interindividual variability as part of the overall 100-fold safety (uncertainty) factor that is applied to derive reference values will be sufficient to cover these differences.[39–43]

The most widely used and best-known substances from that group are ivermectin and abamectin (Figure 8.3). Ivermectin is a mixture of 22,23-dihydroavermectin B1a (80%) and 22,23-dihydroavermectin B1b (20%). Since its introduction in the early 1980s, it has very quickly become a popular drug in human and veterinary medicine.[38,44] Today, ivermectin is the most important remedy for onchocerciasis (river blindness) in humans, *i.e.* the infestation with the nematode *Onchocerca volvulus* that is endemic in parts of Africa and Latin America and is a major cause of vision loss there. It is also effective in treatment of some less common diseases caused by nematodes and of ectoparasitic infestations such as scabies or head lice (see Chapters 12 and 13). In animals, it is widely used to control both nematodes and ectoparasites in many species such as cattle, horses, camels, cats and dogs but also in rabbits, guinea-pigs, hedgehogs or rodent colonies and even in birds and reptiles. Application is mostly oral, but parenteral and topical routes are also common. In many cases, a single dose will be sufficient to get rid of the parasites. Development of resistance to ivermectin in the target organisms has seldom been reported and seems to be confined to very few nematode species in farm animals such as goat or sheep. Most likely, it is due to molecular alterations in glutamate-gated chloride channels.[38,39,44] The toxicity of ivermectin is further discussed in section 8.4.2.

In contrast, abamectin was primarily developed for agricultural use as an insecticide and acaricide, in greenhouses as well as in open fields. It acts as a contact and stomach poison in the target organisms and is used for control of the motile stages of mites, leaf miners, suckers or Colorado beetles in many crops such as cotton, citrus fruit, nuts and pome fruit, potatoes, vegetables such as lettuce and tomatoes, as well as ornamentals. Biocidal applications include

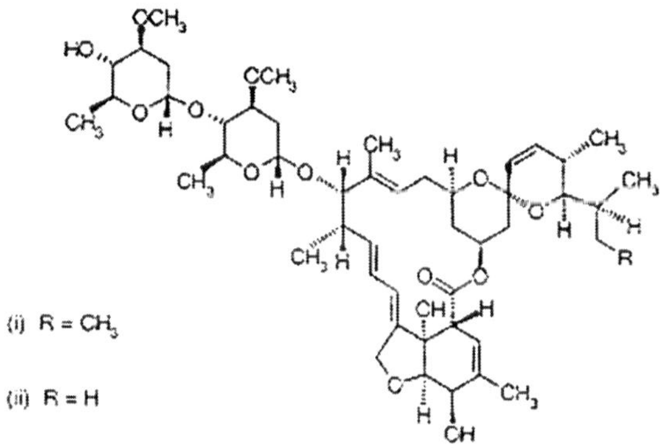

Molecular Formula:	Unspecified (in ChemIDplus), $C_{48}H_{72}O_{14}$ (B1a) und $C_{47}H_{70}O_{14}$ (B1b) (in Wikepedia)
IUPAC name:	*mixture of:* (10E,14E,16E)-(1R,4S,5′S,6S,6′R,8R,12S,13S,20R,21R,24S)-6′-[(S)-sec-butyl]-21,24-dihydroxy-5′,11,13,22-tetramethyl-2-oxo-(3,7,19-trioxatetracyclo[15.6.1.14,8.020,24]pentacosa-10,14,16,22-tetraene)-6-spiro-2′-(5′,6′-dihydro-2′H-pyran)-12-yl 2,6-dideoxy-4-O-(2,6-dideoxy-3-O-methyl-α-L-arabino-hexopyranosyl)-3-O-methyl-α-L-arabino-hexopyranoside *and* (10E,14E,16E)-(1R,4S,5′S,6S,6′R,8R,12S,13S,20R,21R,24S)-21,22-dihydroxy-6′-isopropyl-5′,11,13,22-tetramethyl-2-oxo-(3,7,19-trioxatetracyclo[15.6.1.14,8.020,24]pentacosa-10,14,16,22-tetraene)-6-spiro-2′-(5′,6′-dihydro-2′H-pyran)-12-yl 2,6-dideoxy-4-O-(2,6-dideoxy-3-O-methyl-α-L-arabino-hexopyranosyl)-3-O-methyl-α-L-arabino-hexopyranoside
CAS #	71751-41-2

Figure 8.3 Chemical structure of abamectin.

control of cockroaches, fire ants and pharaoh ants. In addition, there is a very limited use in veterinary medicine for nematode treatment in ruminants. No applications in human medicine are known to date. Abamectin is composed of 80% (minimum) avermectin B1a (CAS 65195-55-3), and 20% (maximum) avermectin B1b (CAS 65195-56-4). The structural difference between avermectins B1a and B1b is that B1a has a ethyl group at the 25-C position in one of the ring structures rather than a methyl group in the B1b form. Thus, the chemical structure is very closely related to ivermectin except that the hydrogen atoms bound to the 22-C and 23-C positions are missing.[40,43,45,46]

Whereas the risk of abamectin to birds and terrestrial mammals is considered as generally low, the substance is very toxic to aquatic organisms, bees and other non-target arthropods. Turtles are known to be particularly vulnerable to toxic side effects.[2,6]

Several toxicity studies suggest that abamectin and ivermectin have a comparable level of toxicity in mice, rats, rabbits and dogs.

Doramectin is an antiparasitic drug in veterinary medicine that is closely related to ivermectin. An additional cyclohexyl group results in a longer plasma half-life than that of ivermectin and, accordingly, a prolonged phase of therapeutic activity. Dogs with known or presumed *MDR1* gene deficiency must not be treated.[39]

The new agricultural insecticide emamectin (benzoate) is less frequently used than abamectin but apparently of similar toxicity.

8.4.1 Toxicology of Abamectin

In the following, toxicokinetics and toxicology of abamectin are described in greater detail as an representative example for the whole group. This compilation of data is based on evaluations by the Joint Expert Meeting of the FAO Panel of Experts on Pesticide Residues in Food and the Environment and the WHO Core Assessment Group on Pesticide Residues (JMPR),[40] the European Medicines Agency (EMEA),[41] the former UK Pesticide Safety Directorate (now part of Chemicals Regulation Directorate, CRD),[45] and by the EU.[43] Subsequently, effects in humans are summarized for both abamectin and ivermectin.

8.4.1.1 Absorption, Distribution, Excretion and Metabolism

Following oral intake of low doses of abamectin, 80–90% was rapidly absorbed from the gastrointestinal tract of rats and widely distributed throughout the body. Maximum blood levels were achieved within 4–8 h after administration. Highest residues were found in fat, kidney, liver, and muscle but there was no evidence of bioaccumulation. More than 95% of the dose had been eliminated within 7 days after dosing. Excretion was mainly via the faeces with significant enterohepatic circulation occurring. Only small amounts (about 1%) were detected in urine. Approximately one half of the administered dose was metabolized, mainly by demethylation and hydroxylation but also by cleavage of the oleandrosyl ring and oxidation, whereas 40–55% was excreted as unchanged compound. The main metabolites, 3′-desmethyl abamectin and 3′-hydroxymethyl abamectin, accounted for around 25% and 5–10% of the dose, respectively. In total, more than 10 metabolites have been isolated.

Dermal absorption can be expected to be low considering the high molecular mass (>859 Da) and the high octanol/water partition coefficient (log P_{OW} 4.4). This assumption was confirmed in an *in vivo* study on rhesus monkeys and in an *in vitro* experiment on human skin that revealed a dermal absorption rate of not more than 1% for local concentrations of 18 and 180 mg cm^{-2} of a commercial formulation.

When abamectin is applied to plants, the δ-8,9-isomer of avermectin B1a is formed as a major photometabolite that accounts for up to 20% of the applied amount but is not present in mammals. This plant metabolite exhibits toxicological activity, including neurotoxicity and teratogenicity, that is very similar to that of the parent and the same reference values may be used for risk assessment.

8.4.1.2 Acute Toxicity

The acute oral toxicity is high, with LD_{50} values around 10 mg kg^{-1} bw in the rat. Ataxia, whole body tremors and prostration are the predominant clinical signs. Deaths occurred in the absence of histological changes. It should be noted that sesame oil was the vehicle in that study. When water was used, toxicity was lower with LD_{50} values of 214 or 232 mg kg^{-1} bw in female and male rats, suggesting that toxicity may depend partly on the carrier substance, most probably because of a different oral absorption rate after hydrolysis.

Abamectin proved also very toxic by the inhalation route with an LC_{50} value that is higher than 0.051 but below 0.21 mg L^{-1} in male rats with a 4-h nose-only exposure. Females were even more vulnerable with an LC_{50} in the range from 0.034 to 0.051 mg L^{-1}.

As expected when the low dermal absorption is taken into account, dermal toxicity was rather low with LD_{50} values in excess of 330 mg kg^{-1} bw in a limit test in rats and even greater than 2000 mg kg^{-1} bw in rabbits. No deaths occurred, but progressive weight loss was observed throughout the postobservation period.

Abamectin was not irritant to the skin or the eyes, nor was it a skin sensitizer.

8.4.1.3 Repeated Dose Toxicity

Subchronic toxicity via the oral route was investigated in beagles giving an overall NOAEL of 0.25 mg kg^{-1} bw day^{-1} in studies over 18 weeks and 1 year with mortality first occurring already at the next higher dose level of 0.5 mg kg^{-1} bw day^{-1}. In a 90-day neurotoxicity study, rats proved less vulnerable with an NOAEL of 1.6 mg kg^{-1} bw day^{-1}. In both species, neurotoxic signs such a mydriasis, tremor, ataxia, clonic convulsions or salivation were observed, together with vomiting in dogs. Body weight gain was consistently compromised. Histopathological examination revealed some liver changes in dogs and stomach erosions in rats. Subacute inhalation exposure (30 days) of rats to abamectin resulted in clinical signs and reduced motor activity with a NOAEC of 0.577 μg L^{-1}. The calculated corresponding systemic value of 0.11mg kg^{-1} bw day^{-1} suggests higher toxicity via the inhalation route as compared to the oral.

Similar findings, but also an increase in mortality, were reported in long-term feeding studies in rats and mice with NOAELs of 1.5 and 4 mg kg^{-1} bw day^{-1}. In mice only, extramedullary haematopoiesis was noted in the spleen suggesting red blood cells as a potential additional target.

8.4.1.4 Carcinogenicity and Mutagenicity

There was no evidence of carcinogenicity of abamectin in long-term studies in rodents and no indications for genotoxicity were obtained in a standard test battery.

8.4.1.5 Effects on Reproduction and Development

Fertility and reproductive performance were not altered by dietary administration of abamectin to rats in a two-generation study. However, in the absence of parental toxicity, there were clear adverse effects on the pups at the dose level of 0.4 mg kg^{-1} bw day^{-1} since postnatal mortality was increased and body-weight gain delayed. These observations suggested a particular vulnerability of neonates that would be in line with lower P-glycoprotein in newborn rats as compared to adult animals. In humans, in contrast, development of the blood–brain barrier is mainly prenatal.

Abamectin proved teratogenic in rats, mice and rabbits, with CF-1 mice being particularly vulnerable. Teratological findings comprised cleft palates in all three species and, in addition, omphaloceles and clubbed forefeet only in rabbits and exencephaly in CF-1 mice. Furthermore, skeletal variations concerning the ribs were seen in rat fetuses and delayed ossification in rabbits. In rats, the NOAEL for developmental toxicity (0.8 mg kg^{-1} bw day^{-1}) was below that for maternal effects (1.6 mg kg^{-1} bw day^{-1}).

8.4.1.6 Neurotoxicity

Abamectin exhibits a specific neurotoxic potential by binding to GABA receptors in different tissues and opening of GABA-controlled chloride channels (see 8.4 above). Accordingly, neurotoxic signs were seen in all types of toxicological studies in laboratory animals by the oral and inhalation routes and were the main cause of mortality if deaths occurred. In an acute oral neurotoxicity study in rats, neurotoxic signs appeared from a dose level of 1.5 mg kg^{-1} bw onwards. The NOAEL was 0.5 mg kg^{-1} bw. As for developmental effects, a subpopulation of CF-1 mice proved extremely sensitive to acute toxicity and specific neurotoxicity not only of abamectin but of the avermectins in general. This high sensitivity is considered to be associated with a deficiency in the expression of P-glycoprotein in the small intestine and brain capillary epithelium in this strain which results in higher concentrations of the compounds in brain and plasma. In fact, about 17% of the CF-1 mice were highly sensitive and these animals had low P-glycoprotein levels in the cerebral cortex, cerebellum and jejunum. These findings were assumed to reflect the distribution of three different P-glycoprotein genotypes in this strain. Because of the species-specific poor P-glycoprotein expression and genetic heterogenicity of CF-1 mice, this strain is considered to be an inappropriate model for studying the toxicity of avermectins.

8.4.1.7 Reference Doses

The offspring NOAEL of 0.12 mg kg^{-1} bw day^{-1} from the multigeneration study in rats was used for setting the ADI by the JMPR in 1997. Because of the hypersusceptibility of neonatal rats, a reduced safety factor of only 50 was applied. The resulting value of 0.002 mg kg^{-1} bw was supported by the NOAEL of 0.24 mg kg^{-1} bw from the 1-year dog study.[40]

The Committee for Veterinary Medicinal Products of the EMEA in 2002 proposed an ADI 0.0025 mg kg^{-1} bw on the basis of a 1-year study in dogs: the committee considered the offspring effects in the two-generation rat study to be not relevant to humans.[41]

In 2005, the EPA established a chronic reference dose (cRfD) of 0.00012 mg kg^{-1} bw day^{-1} on the basis of the NOAEL for offspring effects in that two-generation study and an ARfD of 0.00025 mg kg^{-1} bw that was derived from the NOAEL in the 1-year dog study. In both cases, a UF of 1000 was applied because of severity of effects, steepness of the dose–response curve in several studies and lack of a developmental neurotoxicity study.[42]

As result of its joint evaluation process, in 2008 the EU set an ADI of 0.0025 mg kg^{-1} bw with the same figure as AOEL. This was based on the NOAEL of 0.25 mg kg^{-1} bw day^{-1} in the subchronic dog studies, employing the usual safety factor of 100. In addition, an ARfD of 0.005 mg kg^{-1} bw was established that was derived from the NOAEL for occurrence of neurotoxic signs in an acute neurotoxic study in rats.[43]

8.4.2 Toxicity of Ivermectin and Abamectin to Humans

In humans, more than 50 000 000 doses of ivermectin had been administered worldwide up to the end of the 1990s for treatment of various diseases caused by ecto- and endoparasites, with no report of toxicity directly attributable to the drug. The main adverse effects noted in patients treated with ivermectin have been those arising from the death of the parasites, the so-called Mazzotti reaction. This is characterized by arthralgia, pruritus, fever, hypertension, tachycardia, headache and ocular change. Another effect was a minor degree of hypersensitivity seen in some cases. The available data on reproductive toxicity in humans is very limited, but suggest that ivermectin does not increase the incidence of birth defects. The adverse effects experienced by the small number of persons who had been accidentally exposed to ivermectin by self-injection or oral ingestion included pain at the injection site, variable blood pressure, nausea, paresthesia, urticaria, mydriasis, vomiting, tachycardia, or somnolence.[38,44]

Severe intoxications with abamectin in humans are rare and nearly always due to suicide attempts. The symptoms differ from those in laboratory animals and consist of a comatose state occurring within 3 h after ingestion, shock and respiratory and/or multiple organ failure sometimes resulting in death, after ingestion of about 40 mg kg^{-1} bw or more.[39,45]

8.5 Spinosyn Products

Spinosad (Figure 8.4) is a natural fermentation product produced by the Gram-positive soil bacterium *Saccharopolyspora spinosa*, a species from the order Actinomycetales. It has insecticidal activity and its structure consists of a large complex hydrophobic ring, a basic amine group, and two sugar moieties. Spinosad is composed of numerous spinosyns, but nearly all of the insecticidal activity of spinosad is produced by the two closely related compounds spinosyns A and D, in a ratio of approximately 7:1. These two spinosyns differ from each other only in the substitution of hydrogen by a methyl group and

Spinosyn A

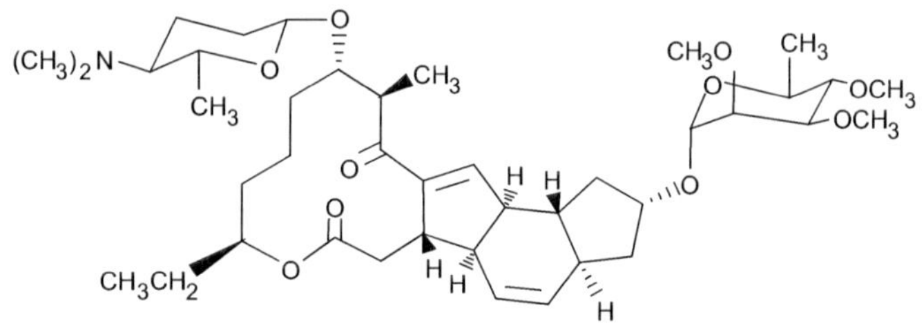

Empirical Formula: $C_{41}H_{65}NO_{10}$

Molecular Mass: 731.98

Spinosyn D

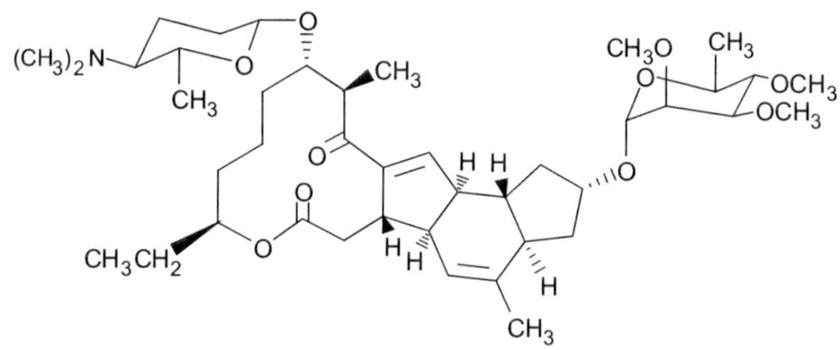

Empirical Formula: $C_{42}H_{67}NO_{10}$

Molecular Mass: 746.00

Molecular Formula:	Spinosyn A: $C_{41}H_{65}NO_{10}$, Spinosyn D : $C_{42}H_{67}NO_{10}$
IUPAC name:	Not found
CAS #	Spinosyn A: 131929-60-7, Spinosyn D: 131929-63-0

Figure 8.4 Chemical structure of spinosad.

represent about 88% of the composition of spinosad. The remaining components in spinosad consist of a number of additional spinosyns, which have other minor substitutions at various locations in the molecule, and impurities consisting of inorganic salts, carbohydrates, and proteinaceous materials that may be expected from the fermentation process.[47]

Insects exposed to spinosad exhibit classical symptoms of neurotoxicity, including lack of coordination, prostration, tremors, and other involuntary muscle contractions leading to paralysis and death. Spinosad kills susceptible species by causing rapid excitation of the insect nervous system. Spinosad has proven effective in controlling many chewing insect pests and possesses high selectivity since more than 70% of beneficial insects and predatory wasps are left unharmed. The insect control spectrum includes Lepidoptera, Diptera, Hymenoptera, Siphonaptera, Thysanoptera and certain Coleoptera, but it is relatively inactive on sucking insects, predatory insects and mites. Spinosad is unique amongst biologically derived insecticides in having an activity spectrum and level comparable to some modern synthetic insecticides.[48]

In the following, toxicokinetics and toxicology of spinosad are described as an representative example for the spinosyn products. This compilation of data is based on evaluations by the JMPR,[49] the EPA[50] and the EU.[51]

8.5.1 Absorption, Distribution, Excretion and Metabolism

Following oral gavage, 80% of spinosyn A and 66% of spinosyn D was rapidly absorbed. Peak blood concentrations of radiolabel were achieved 1–6 h after administration. The residues were initially widely distributed with highest residues in perirenal fat, liver, kidneys, and lymph nodes. In the thyroid gland, a slow rate of decline was observed. This resulted in higher concentrations in the thyroid than in other tissues. However, the absolute tissue levels were very low, 90% of the administered radiolabel being excreted within 48 h after single oral dose application. Spinosad is excreted primarily in the faeces. Most of the faecal radioactivity originates from biliary excretion. Urine and faecal excretion was almost completed at 48 h after dosing. The routes and rates of excretion were not greatly altered by repeated administration, as compared to single-dose administration.

Spinosad is extensively metabolized, primarily *via* *O*-demethylation and/or glutathione conjugation. There are no major differences in the bioavailability, routes or rates or excretion, or metabolism of spinosyn A or spinosyn D following oral administration in rats.

The absorption, distribution, metabolism, and excretion of radiolabelled spinosyn A showed no relevant differences based on sex, dose or repeated administration.

A rat *in vivo* study has shown a dermal absorption of approximately 1%. The comparison of the *in vitro* studies on rat and human skin has indicated a twofold higher penetration in rats for the concentrate but a tenfold higher penetration for the dilution.

8.5.2 Acute Toxicity

Spinosad is of low acute toxicity after oral or dermal administration and by inhalation. The oral LD_{50} is greater than 2000 mg kg^{-1} bw for rats and mice. The dermal LD_{50} is greater than 5000 mg kg^{-1} bw for rabbits and the inhalation LC_{50} is greater than 5.18 mg $L^{-1}$of air for male and female rats. Spinosad is mildly irritating to eyes but is not irritating to the skin of rabbits and non-sensitizing to the skin of guinea-pigs.

8.5.3 Repeated Dose Toxicity

The main effect associated with repeated exposure to spinosad in all test species was observed histologically as cellular vacuolation, inflammatory changes including necrosis, histiocytosis, and regenerative and degenerative changes in a wide range of tissues. Electron microscopy of selected tissues from rats and mice has shown that the cytoplasm of affected cells contained clear vacuoles that consisted of variable numbers of secondary lysosomes, which contained concentric cytoplasmic lamellar inclusion bodies, reflecting a lysosomal storage disorder. While such disorders may arise through a variety of mechanisms, which prevent degradation of cell constituents that are usually processed in the lysosomes, spinosad probably acts through a largely physicochemical mechanism associated with its cationic amphophilic structure.[49] The vacuolation is largely reversible on withdrawal of treatment. A number of other effects have been seen in subchronic studies on rats and dogs, including decreases in body weights and feed consumption, increased spleen, thyroid, and liver weights, altered haematology and clinical chemistry parameters, resulting in microcytic hypochromic anaemia and increased serum activity alanine aminotransferase, alkaline phosphatase, aspartate aminotransferase, and creatinine phosphokinase.

Spinosad did not cause specific neurotoxicity in rats in acute, subchronic or chronic toxicity studies.

A comparison of spinosad, spinosyn A and spinosyn D revealed notable differences in the toxicological profiles. Whereas the toxicological effects of spinosyn A were closely similar to those of spinosad, spinosyn D failed to produce most of the haematological and clinical chemical alterations seen with spinosad or spinosyn A.

8.5.4 Carcinogenicity and Mutagenicity

In carcinogenicity studies in mice and rats, histological effects consisted of vacuolation, particularly in the thyroid and kidneys, degeneration and inflammatory lesions. However, there was no treatment-related increase in the incidence of neoplasms in any tissue.

None of the genotoxicity studies showed mutagenic activity associated with spinosad.

8.5.5 Effects on Reproduction and Development

In a two-generation study, decreased bodyweights, vacuolation of the thyroid gland, and degenerative or inflammatory lesions in other tissues were observed in parental animals. Only at these parental toxic dose levels, litter size and pup bodyweights were decreased. In developmental studies, spinosad caused decreased bodyweights in pregnant rats and rabbits, but this maternal toxicity was not accompanied by embryo–fetal effects or teratogenicity.

8.5.6 Effects in Humans

No poisonings or adverse reactions in humans have been reported.

8.5.7 Reference Doses

The most sensitive overall toxicological endpoint was thyroid vacuolation in rats treated in the diet in a two-year study. Based on the NOAEL of 2.4 mg kg^{-1} bw day^{-1}, an ADI of 0.02 mg kg^{-1} was derived by the JMPR applying a 100-fold safety factor.[49] In the EU an ADI of 0.024 mg kg^{-1} was derived (*i.e.* the two organizations performed separate evaluations, which came to similar, but not identical, conclusions on the same toxicological basis).[51] Because of the low acute toxicity and the absence of toxicological alerts in repeated dose studies, there is no need for the establishment of an ARfD.

8.6 Quassin

Quassin is the generic term for plant extracts derived from the wood of tropical quassia trees, *e.g. Quassia amara* or *Picrasma excelsa*. It should be noted that extracts of quassia are often simply referred to as 'quassin' and may appear commercially under this name.[52] Quassia is the dried stem wood of *Quassia amara* L. or of *Picrasma excelsa* (Sw) Planch (family Simarubaceae). Commercial 'quassin' from *Quassia amara* is known to contain a mixture of bitter principles (quassinoids), such as quassin, neoquassin, and 18-hydroxyquassin,[53] while *P. excelsa* contains isoquassin, also known as picrasmin, instead of quassin as the major quassinoid.[54] These plant extracts are mixtures of several compounds of the chemical class of quassinoids and other naturally occurring plant ingredients (Figure 8.5). Depending on the origin of the wood, on environmental conditions during growth and on the extraction procedure, the contents of the single constituents may vary. Traditional uses include remedies for infestations of lice or worms, anorexia and dyspepsia. Quassin has also demonstrated antilarval activity and was effective for this purpose at concentrations of 6 ppm. One mechanism of this larvicidal activity may be due to inhibition of cuticle development.

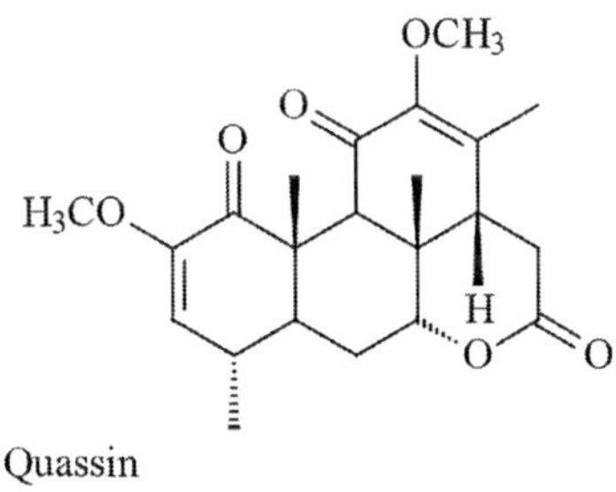

Quassin

Molecular Formula:	Unspecified
IUPAC name:	Not found
CAS #	68915-32-2

Figure 8.5 Chemical structure of quassin.

From the results of experimental studies on quassinoids, it has been suggested that several of these compounds might have an anticarcinogenic, antiprotozoal, antiviral or amoebicidal potential.[55–60] Several quassinoids, including quassin, have been shown to possess antifeedant and insecticidal properties.[52]

Because of its extremely bitter taste, with a bitter threshold of 0.08 ppm, quassia may act as a repellent against mammals and other species. Quassia is mostly used in orchards against sawflies and in ornamentals against aphids. In addition to the repellent effect, it shows oral toxicity, although a contact effect could not be excluded. Interestingly, quassia was found to act quite differently in comparison to most conventional insecticides, which may be important for users. Quassia is described in the literature as a slow-acting larvicide, but eggs are not affected. The newly hatched larvae of some insect species undergo a flaccid paralysis without prior convulsions.[61]

Quassin is also used in traditional Chines medicine and as an additive in soft drinks. In humans, both contact and stomach poisoning have been observed.

8.6.1 Absorption, Distribution, Excretion and Metabolism

No studies are available on absorption, distribution, metabolism or excretion.

8.6.2 Acute Toxicity

There are only few toxicological studies on quassin toxicity and most of the available studies are of poor quality. No sign of acute toxicity was observed at any doses given orally to albino rats and mice up to 1000 mg kg^{-1} bw of aqueous quassia extract. The quassin content was not given.[62] From the set of studies presented, it is concluded that quassia has only a low potency of toxicity. Dermal exposure did not reveal any acute toxicity. In acute inhalation studies at the highest technically attainable concentration, rats did not show adverse effects. Quassia extract also exhibits no irritant properties for the skin and was found to be a mild and transient irritant to the eyes of rabbits.

8.6.3 Repeated Dose Toxicity

There are only few repeated dose toxicity studies on quassin which are of very poor quality.

8.6.4 Carcinogenicity and Mutagenicity

No data are available and no indication for genotoxicity is recognized from other sources.

8.6.5 Effects on Reproduction and Development

Quassin was shown to inhibit steroidogenesis in rat Leydig cells *in vitro* and *ex vivo* in a concentration/dose-related manner.[52] However, even despite the fact that the data present were published in refereed international journals, the protocol schedules (*e.g.* treatment timing and route, type of analysis, lack of evaluation of toxic effects in other organs) do not permit an adequate evaluation of effects on male fertility. The effect of *Quassia amara* L. on steroidogenesis in rat Leydig cells was studied in an *in vitro* system.[63] Both the basal and the LH-stimulated testosterone production by the Leydig cells were inhibited in a dose-related manner with doses from 5 ng mL^{-1} onwards, up to 25 ng mL^{-1} of the isolated quassin. The inhibition of testosterone production was shown not to be caused by cytotoxic effects of the quassia extract or of the isolated quassin. An extensive *in vivo* study demonstrated that the crude methanol extract of the stem wood of *Quassia amara* L. significantly reduced the weight of the testis, epididymis and seminal vesicle and significantly increased that of the anterior pituitary gland.[64] Epididymal sperm counts and serum levels of testosterone, luteinizing hormone (LH) and follicle stimulating hormone (FSH) were significantly reduced when the rats were treated with the extract. All these changes proved reversible because they were restored completely 8 weeks after withdrawal from the 8 weeks of treatments. Furthermore, the basal and LH-stimulated testosterone secretion from Leydig cells isolated from rats pretreated with the extract was inhibited. But this study had a number of unusual features; for example, the viability of the Leydig cells was unchanged after the treatment, all the effects were shown at all three dose levels and no dose relationship was demonstrated. Moreover, data on fertility, obtained by mating treated males with females, were not produced. As far as the overall evaluation of the toxicity on reproduction is concerned, the complete lack of information on female fertility, as well as the lack of multigeneration and teratogenic tests, does not allow a comprehensive toxicological evaluation.

8.6.6 Effects in Humans

Little information can be found in the literature regarding poisoning effects of quassia in humans. Some dermatosic effects were observed when two farmers were working with quassia wood in one case and with an extract of quassia in another.[65] Parenteral administration of quassin is toxic, leading to cardiac irregularities, tremors, and paralysis. Large amounts given orally have been known to irritate the mucus membrane in the stomach and may lead to vomiting.[66] In the USA, quassia (plants of origin: *Quassia amara* or *Picrasma excelsa*) is listed as 'Generally Recognized As Safe' (GRAS) and can be used as a food additive. In humans, oral intake of quassia is useful in failure of appetite due to gastric debility, and in overdoses is capable of sufficiently irritating the stomach to produce vomiting.

8.6.7 Reference Doses

No chronic or acute reference doses published by any regulatory body were obtainable.

Quassin was classified as an active principle by the Committee of Experts on Flavouring Substances of the Council of Europe (CEFS) in 1981 with limits of 5 mg kg^{-1} in beverages and food, except alcoholic beverages where 50 mg kg^{-1} is allowed. In 1991, CEFS proposed to remove quassin from the list of active principles as 'there is little evidence of quassin toxicity although most of the studies available are of poor quality'. In the USA, quassia extract may be used in beverages (3.4 mg kg^{-1}), alcoholic beverages (3.4 mg kg^{-1}) and baked goods (50 mg kg^{-1}).[52]

8.7 Anabasine

Anabasine (Figure 8.6) also known as neonicotine or nicotimine, is an alkaloid with *R* and *S* enantiomers, which occurs naturally in the wild tobacco plant *Nicotinana glauca* and may also be detected in tobacco smoke.[67,68] It was first isolated in the 1930s by Russian scientists from the shrub plant *Anabasis aphylla* L., which is indigenous to Mongolia and of which it is the major alkaloid.[69,70] It is chemically related to nicotine and may produce similar neurotoxicity based on its activity as an acetylcholine agonist that binds to nicotinic receptors. As an insecticide it is, or was, mainly used in Russia and other countries of the former Soviet Union, but no details on the crops treated or application rates are available.[69]

No information about comprehensive toxicological testing or evaluation of anabasine is available. Published data is scarce and partly of poor quality, limited to certain toxicological endpoints. Very often, plant material from *N. glauca* was administered instead of anabasine itself. In these cases, the anabasine content is unknown and the applied doses cannot be quantified.

8.7.1 Absorption, Distribution, Excretion and Metabolism

Because of its structural similarity to nicotine, the toxicokinetics of anabasine are assumed to be similar.[71] Therefore, this alkaloid is also expected be become readily available following exposure by all routes with a high degree of

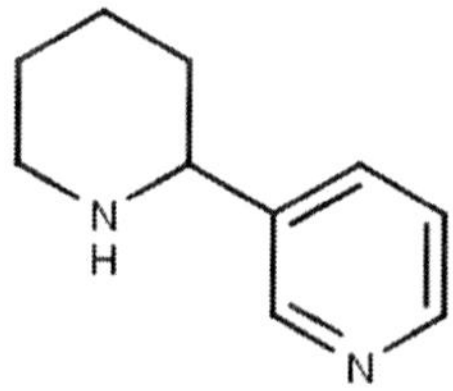

Molecular Formula:	$C_{10}H_{14}N_2$
IUPAC name:	3-[(2*S*)-2-piperidyl]pyridine
CAS #	494-52-0

Figure 8.6 Chemical structure of anabasine.

first-pass metabolism occurring. It must be anticipated that anabasine, as well as nicotine, will cross the blood–brain barrier as well as the placental barrier and will be excreted in breast milk. The main elimination route will be *via* urine.

8.7.2 Acute Toxicity

The acute toxicity of anabasine was investigated in mice but apparently only by a highly artificial route, namely intravenous injection, revealing an LD_{50} of 11 or 16 mg kg^{-1} bw, depending on the enantiomer used. The similar substance anabaseine, which is reported to be part of some animal venoms, proved much more potent with an intravenous LD_{50} as low as 0.58 mg kg^{-1} bw.[72] Fatal human poisoning was due to cardiac arrest.[68] Another fatal case was reported recently from Israel where an older woman died after accidental intake of cooked tobacco leaves which were mistaken for wild spinach. Acute signs comprised dizziness, nausea, and vomiting, followed by loss of consciousness, dilation of pupils and extreme bradycardia. In spite of intensive medical care, the victim died after 20 days because of multi-organ failure. Anabasine intoxication was verified by laboratory methods but no information about the ingested dose was available.[73]

8.7.3 Repeated Dose Toxicity

Based on the body of knowledge about toxicity mechanisms of nicotine, anabasine is also assumed to act by binding to nicotinic cholinergic receptors. These sodium-gated receptors are located mainly in the central and peripheral nervous system (including autonomic ganglia), in neuromuscular junctions, and in the adrenals. In the nervous system, they may be found at pre- and postsynaptic sites.[71] Stimulation of these receptors at the beginning of intoxication with abdominal pain, vomiting, tachypnoea, tachycardia and hypertension, ataxia or confusion as key symptoms may be followed by a prolonged 'depression' phase that is characterized by diarrhoea, respiratory depression or dyspnoea, bradycardia and hypotension, and eventually by shock, lethargy, paralysis and coma.[71] This anticipated course of a poisoning incident seems to be in line with what was described in case reports.[68,73]

Anabasine may moderately inhibit acetylcholinesterases.[74]

8.7.4 Carcinogenicity and Mutagenicity

Data on carcinogenicity are not available but the related nitrosated compound *N′*-nitrosoanabasine proved a weak carcinogen when given for 30 weeks to rats *via* drinking-water because an oesophageal tumour was observed in 1 out of 20 male Fischer rats after administration of a total dose of 630 mg over the study period.[75] In contrast to other tobacco ingredients, however, *N′*-nitrosoanabasine did not increase tumour incidence in Syrian golden hamsters in 25-week study with subcutaneous injection three times a week in which a total dose of 375 mg was administered.[76]

No mutagenicity data is available and no indication for genotoxicity was obtained from other sources.

8.7.5 Effects on Reproduction and Development

Teratogenicity was demonstrated in goats by occurrence of cleft palates and multiple flexion contractures in the joints after oral application of *N. glauca* plant material or anabasine-rich extracts. Developmental effects were less pronounced in sheep although maternal toxicity was similar in both ruminant species.[77] Similar findings were reported in piglets after feeding the sows *N. glauca* mainly during the first quarter of the gestation period and were attributed to the anabasine content.[78] Unfortunately, these data do not allow characterization of dose–response or to derive a threshold value for teratogenicity.

8.7.6 Effects in Humans

No poisonings or adverse reactions in humans have been reported.

8.7.7 Reference Doses

No chronic or acute reference doses published by any regulatory body were obtainable.

References

1. R. Solecki, Toxicology of miscellaneous insecticides, in *Pesticide Toxicology, Part 1: Insecticides*, ed. T. C. Marrs and B. Ballantyne, John Wiley & Sons, London, 2004, pp. 159–192.
2. I. Ujvary, Pest control agents from natural products, in *Handbook of Pesticide Toxicology*, ed. R. Krieger, Academic Press, Elsevier, Amsterdam, 3rd edn, 2010, pp. 119–230.
3. M. B. Isman, Botanical insecticides, deterrents, and repellents, in *Modern Agriculture and an Increasingly Regulated World, Annu. Rev. Entomol.*, 2006, **51**, 45–66.
4. C. D. S. Tomlin, *The Pesticide Manual*, British Crop Protection Council, Alton, Hants, 11th edn, 1999.
5. J. R. Corbett, *The Biochemical Mode of Action of Pesticides*, Academic Press, New York, 1974.
6. R. L. Rose, E. Hodgson and R. M. Roe, *Pesticides in Toxicology*, ed. H. Marquardt, S. G. Schäfer, R. O. McClellan and F. Welsch, Academic Press, San Diego, CA, 1999, pp. 663–696.

7. J. Fukami, T. Shishido, K. Fukunaga and J. E. Casida, Oxidative metabolism of rotenone in mammals, fish and insects and its relation to selective toxicology, *J. Agric. Food Chem.*, 1969, **17**, 1217–1226.
8. M. B. Shimkim and N. N. Anderson, Acute toxicities of rotenone and mixed pyrethrins in mammals, *Proc. Soc. Exp. Biol. Med.*, 1936, **34**, 135–138.
9. J. R. Reigart and J. R. Roberts, *Recognition and Management of Pesticide Poisonings*, OPP US-EPA, 5th edn, 1999. http://www.epa.gov/pesticides/safety/heathcare
10. T. J. Hayley, A review of the literature of rotenone 1,2,12,12a-tetrahydro-8,9-dimethoxy-(2-(1-methyl-ethenyl)-1-benozopyrano[3,5-B]fluro[2,3-H][1]-benzopyran-6(6h)-one, *J. Environ. Pathol. Toxicol.*, 1978, **1**, 315–337.
11. A. I. Rojo, C. Cavada, M. R. de Sagarra and A. Cuadrado, Chronic inhalation of rotenone or paraquat does not induce Parkinson's disease symptoms in mice or rats, *Exp. Neurol.*, 2007, **208**, 120–126.
12. H. Marquardt, S. G. Schäfer, R. O. McClellan and F. Welsch, *Toxicology*, Academic Press, San Diego, CA, 1999, pp. 3 and 1005.
13. M. Gosálvez, Carcinogenesis with the insecticide rotenone, *Life Sci.*, 1983, **32**, 809–816.
14. J. R. M. Innes, B. M. Ulland, M. G. Valerio, M. Petucelli, I. Fishbein, E. R. Hart, A. J. Pallota, R. R. Bates, H. L. Falk, J. J. Gart, M. Klein, I. Mitchell and J. Peters, Bioassay of pesticides and industrial chemicals for tumorigenicity in mice, a preliminary note, *J. Natl. Cancer Inst.*, 1969, **42**, 1101–1114.
15. J. S. Isenberg and J. E. Klaunig, Role of the mitochondrial membrane permeability transition (MPT) in rotenone-induced apoptosis in liver cells, *Toxicol. Sci.*, 2000, **53**, 340–351.
16. S. S. Barham and B. R. Brinkley, Action of rotenone and related respiratory inhibitors on mammalian cell division. 1. Cell kinetics and biochemical aspects, *Cytobios*, 1976, **15**, 85–96.
17. S. S. Barham and B. R. Brinkley, Action of rotenone and related respiratory inhibitors on mammalian cell division. 2. Ultrastructural studies, *Cytobios*, 1976, **15**, 97–109.
18. R. Betarbet, T. B. Sherer, G. McKenzie, M. Garcia-Osuna, A. V. Panov and J. T. Greenamyre, Chronic systemic pesticide exposure reproduces features of Parkinson's disease, *Nat. Neurosci.*, 2000, **3**, 1301–1306.
19. B. I. Giasson and V. M. Y. Lee, A new link between pesticides and Parkinson's disease, *Nat. Neurosci.*, 2000, **3**, 1227–1228.
20. U.S. EPA, Review of studies submitted to support registration of rotenone (Reg. No. 6704-Q). Tox Chem. No. 725. Tox Project Nos. 1582 and 7-0865. MRIDs 00156739, 401798-01 thru 06.Tox review 006537, in *Index of Cleared Science Reviews for Rotenone*, 1987. http://www.epa.gov/pesticides/chemical/foia/cleared-reviews/reviews/071003/071003.htm
21. U.S. EPA, *Reregistration Eligibility Decision for Rotenone*, List A Case No. 0255, 2007. www.epa.gov/oppsrrd1/REDs/rotenone_red.pdf

22. H. Schmutterer, The tree and its characteristics, in *The Neem Tree (Azadirachta indica A. Juss.) and other Meliaceous Plants*, ed. H. Schmutterer, Neem Foundation, Vithalnager, Mumbai (Bombay), 2nd edn, 2002, pp. 1–37.
23. I. P. Ogbuewu, V. U. Odoemenam, H. O. Obikaonu, M. N. Opara, O. O. Emenalom, M. C. Uchegbu, I. C. Okoli, B. O. Esonu and M. U. Iloeje, The growing importance of neem (*Azadirachta indica* A. Juss) in agriculture, industry, medicine and environment, a review, *Res. J. Med. Plant*, 2011, **5**, 230–245.
24. W. Kraus, Azadirachtin and other triterpenoids—Parts I and II, in *The Neem Tree (Azadirachta indica A. Juss.) and other Meliaceous Plants*, ed. H. Schmutterer, Neem Foundation, Vithalnager, Mumbai (Bombay), 2nd edn, 2002, pp. 39–110.
25. M. S. Mulla and T. Su, Activity and biological effects of neem products against arthropods of medial and veterinary importance, *J. Am. Mosq. Control Assoc.*, 1999, **15**, 133–152.
26. A. J. Mordue and A. Blackwell, Azadirachtin, an update, *J. Insect. Physiol.*, 1993, **39**, 903–924.
27. S. V. Ley, A. Denholm and A. Wood, The chemistry of azadirachtin, *Nat. Prod. Rep.*, 1993, **10**, 109–157.
28. L. Niemann and V. Hilbig, Die gesundheitliche Bewertung des Einsatzes von Naturstoffen im Pflanzenschutz am Beispiel von Neemkernextrakten, *Gesunde Pflanzen*, 2000, **52**, 135–141.
29. S. Boeke, M. G. Boersma, G. M. Alink, J. J. A. van Loon, A. van Huis, M. Dicke and I. M. C. M. Rietjens, Safety evaluation of neem (*Azadirachta indica*) derived pesticides, *J. Ethnopharmacol.*, 2004, **94**, 25–41.
30. European Food Safety Authority, Conclusion on the peer review of the pesticide risk assessment of the active substance azadirachtin, *EFSA J.*, 2010, **8**, 1858–1935.
31. G. P. Talwar, P. Raghuvanshi and M. Jacobson, Neem for control of fertility and sexually transmitted infections of the reproductive tract, in *The Neem Tree (Azadirachta indica A. Juss.) and other Meliaceous Plants* ed. H. Schmutterer, Neem Foundation, Vithalnager, Mumbai (Bombay), 2nd edn, 2002, pp. 666–677.
32. EFSA, *Draft Assessment Report (DAR) on the Active Substance Azadirachtin*, Prepared by the Rapporteur Member State Germany in the Framework of Directive 91/414/EEC, November 2007 (unpublished report but available on request in electronic format from EFSA, with the exception of some confidential parts dealing with details of the manufacturing process and impurity profile), European Food Safety Authority, 2007.
33. L. Niemann, S. Stinchcombe and V. Hilbig, Toxicity to mammals including humans, in *The Neem Tree (Azadirachta indica A. Juss.) and other Meliaceous Plants*, ed. H. Schmutterer, Neem Foundation, Vithalnager, Mumbai (Bombay), 2nd edn, 2002, pp. 607–622.

34. N. Sundaravalli, B. Bhaskar Raju and K. A. Krishnamoorthy, Neem oil poisoning, *Indian J. Pediatr.*, 1982, **49**, 357–359.
35. D. Sinniah, G. Baskaran, B. Viyavalakshmi and N. Sundaravalli, Margosa oil poisoning in India and Malaysia, *Trans. R. Soc. Trop. Med. Hyg.*, 1981, **75**, 903–904.
36. D. Sinniah, P. H. Schwartz, R. A. Mitchell and E. L. Arcinue, Investigation of an animal model of a Reye-like syndrome caused by margosa oil, *Pediatr. Res.*, 1985, **19**, 1346–1355.
37. R. Iyyadurai, V. Surekha, S. Sathyendra, B. P. Wilson and K. G. Gopinath, Azadirachtin poisoning, a case report, *Clin. Toxicol.*, 2010, **48**, 857–858.
38. S. Ömura and A. Crump, The life and times of ivermectin—a success Story, *Nat. Rev. Microbiol.*, 2004, **2**, 984–989.
39. IVPT, *CliniPharm—CliniTox*, Informationssystem des Institutes für Veterinärpharmakologie und -toxikologie, University of Zurich, Switzerland. www.vetpharm.uzh.ch/index.htm
40. JMPR, Abamectin (addendum), in *Pesticide Residues in Food—1997*, Joint Meeting of the FAO Panel of Experts on Pesticide Residues in Food and the Environment and the WHO Core Assessment Group, Lyon, 1997, World Health Organization, Geneva, 1998.
41. EMEA, *Abamectin—Revision of the Acceptable Daily Intake in Summary Report 4*, EMEA/MRL/838/02-Final, May 2002.
42. EPA, Avermectin B1 and its delta-8,9-isome—pesticide tolerance, *Environmental Protection Agency, Federal Register (Rules and Regulations)*, 2005, **70**, 7876–7886.
43. EFSA, Conclusion on the peer review of abamectin, *EFSA Scientific Report*, 2008, **147**, 1–106.
44. P. Del Giudice, O. Chosidow and E. Caumes, Ivermectin in Dermatology, *J. Drugs Dermatol.*, 2003, **2**, 13–21.
45. PSD, *Abamectin, Evaluation of Fully Approved or Provisionally Approved Products*, Pesticide Safety Directorate, Department for Environment, Food and Rural Affairs, London, 1992.
46. EU, *Review Report for the Active Substance Abamectin*, European Commission Health & Consumers Directorate-General, SANCO/138/08—final, 11 July 2008.
47. R. Solecki, Toxicology of miscellaneous insecticides, in *Pesticide Toxicology, Part 1: Insecticides*, ed. T. C. Marrs and B. Ballantyne, John Wiley & Sons, London, 2004, pp. 164–166.
48. C. V. DeAmicis, J. E. Dripps, C. J. Hatton and L. L. Karr, Physical and biological properties of the spinosyns, novel macrolide pest control agents from fermentation, in *Phytochemicals for Pest Control*, ed. P. A. Hedin, American Chemical Society, Washington, DC, 1997, pp. 144–154.
49. JMPR, Spinosad, in *Pesticide Residues in Food—2001*. Report of the joint meeting of the FAO Panel of Experts on Pesticide Residues in Food and the Environment and the WHO Core Assessment Group on Pesticide

Residues, Geneva, Switzerland, 17–26 September 2001, FAO plant production and protection paper **167**.

50. U.S. EPA, Spinosad, *Pesticide Fact Sheet*, U.S. Environmental Protection Agency, Washington, DC, 1997. http://www.epa.gov/opprd001/factsheets/spinosad.pdf
51. EU, *Review Report for the Active Substance Spinosad*, European Commission Health & Consumers Directorate-General, SANCO/1428/2001—rev. final, 14 July 2006. http://ec.europa.eu/food/plant/protection/evaluation/newactive/spinosad_in_en.pdf
52. EU, *Opinion of the Scientific Committee on Food on Quassin*, Health & Consumer Protection Directorate-General, Scientific Committee on Food, SCF/CS/FLAV/FLAVOUR/29 Final, 25 July 2002. http://ec.europa.eu/food/fs/sc/scf/out134_en.pdf
53. J. R. Robins, M. R. A. Morgan, M. J. C. Rhodes and J. M. Furze, An enzyme-linked immuno-sorbent assay for quassin and closely related metabolites, *Annal. Biochem.*, 1984, **136**, 145–156.
54. A. Y. Leung, S. Fosters, Quassia, in *Encyclopedia of Common Natural Ingredients Used in Food, Drugs and Cosmetics*, John Wiley & Sons, London, 1996, pp. 430–431.
55. M. Alvarez, K. Paull, A. Monks, C. Hose, J. S. Lee, J. Weinstein, M. Grever, S. Bates and T. Fojo, Generation of a drug resistance profile by quantitation of mdr-1/P-glycoprotein in the cell lines of the National Cancer Institute Anticancer Drug screen, *J. Clin. Invest.*, 1995, **95**, 2205–2214.
56. F. D. Gillin and D. S. Reiner, In vitro activity of certain quassinoid antitumor agents against, *Entamoeba histolytica, Arch. Invest. Med.*, 1982, **13**(Suppl. 3), 43–49.
57. K. H. Lee, Novel antitumor agents from higher plants, *Med. Res. Rev.*, 1999, **19**, 569–596.
58. A. Pierré, M. Robert-Géro, C. Tempete and J. Polonsky, Structural requirements of quassinoids for the inhibition of cell transformation, *Biochem. Biophys. Res. Commun.*, 1980, **93**, 675–686.
59. W. Trager and J. Polonsky, Antimalarial activity of quassinoids against chloroquineresistant *Plasmodium falciparum* in vitro, *Am. J. Trop. Med. Hyg.*, 1981, **30**, 531–537.
60. E. O. Ajaiyeoba, U. I. Abalugo, H. C. Krebs and A. M. J. Oduola, *In vivo* antimalarial activities of *Quassia amara* and *Quassia undulata* plant extracts in mice, *J. Ethnopharmacol.*, 1999, **67**, 321–325.
61. M. Jacobson and D. G. Crosby, *Naturally Occurring Insecticides*, Dekker, New York, 1971.
62. G. M. Garcia, S. M. C. Gonzalez and L. S. Pazos, Pharmacologic activity of the aqueous wood extract from Quassia amara (Simurabaceae) on albino rats and mice, *Rev. Biol. Trop.*, 1997, **44–45**, 47–50.
63. V. C. O. Njar, T. O. Alao, J. I. Okugun, Y. Raji, A. F. Bolarinwa and E. U. Nduka, Antifertility activity of *Quassia amara*: quassin inhibits the steroidogenesis in rat Leydig cells in vitro, *Planta Med.*, 1995, **61**, 180–182.

64. Y. Raji and A. F. Bolarinwa, Antifertility activity of *Quassia amara* in male rats—in vivo study, *Life Sci.*, 1997, **61**, 1067–1074.
65. M. Artom, Some occupational dermatoses of agricultural workers not applicable, *Minerva Derm.*, 1959, **34**, 77–79.
66. V. Schulz, R. Haensel and V. E. Tyler, *Rational Phytotherapy: A Physician's Guide to Herbal Medicine*, Springer-Verlag, Berlin, 2001, p. 207.
67. L. Jacob, L. Yu, A. T. Shulgin and N. L. Benowitz, Minor tobacco alkaloids as biomarkers for tobacco use: comparison of users of cigarettes, smokeless tobacco, cigars, and pipes, *Am. J. Public Health*, 1999, **89**, 731–736.
68. N. Mizrachi, S. Levy and Z. Goren, Fatal poisoning from *Nicotiana glauca* leaves: identification of anabasine by gas chromatography/mass spectrometry, *J. Forensic Sci.*, 2000, **45**, 736–741.
69. S. Dev and O. Koul, *Insecticides of Natural Origin, Section 1: General Introduction*, Harwood Academic Publishers, Amsterdam, 1997.
70. ACS, *Molecule of the week*, American Chemical Society website, 29/12/2008.
71. S. T. Lee, K. Wildeboer, K. E. Panter, W. Kem, D. R. Gardner, R. J. Molyneux, C. T. Chang, F. Soti and J. A. Pfister, Relative toxicities and neuromuscular nicotinic receptor agonistic potencies of anabasine enantiomers and anabaseine, *Neurotoxicol. Teratol.*, 2006, **28**, 220–228.
72. V. Furer, M. Hersch, N. Silvetzki, G. S. Breuer and S. Zevin, *Nicotiana glauca* (tree tobacco) intoxication—two cases in one family, *J. Med. Toxicol.*, 2011, **7**, 47–51.
73. L. J. Schep, R. J. Slaughter and D. M. G. Beasley, Nicotinic plant poisoning, *Clin. Toxicol.*, 2009, **47**, 771–781.
74. N. Karadsheh, P. Kussie and D. S. Linthicum, Inhibition of acetylcholinesterase by caffeine, anabasine, ethyl pyrrolidine and their derivatives, *Toxicol. Lett.*, 1991, **55**, 335–342.
75. D. Hoffmann, R. Rainieri, S. S. Hecht, R. Maronpot and E. L. Wynder, A study of tobacco carcinogenesis, XIV. Effects of *N'*-nitrosonornicotine and *N'*-nitrosoanabasine in rats, *J. Natl. Cancer Inst.*, 1975, **55**, 977–981.
76. J. Hilfrich, S. S. Hecht and D. Hoffman, A study of tobacco carcinogenesis. XV. Effects of *N'*-nitrosonornicotine and *N'*-nitrosoanabasine in Syrian golden hamsters, *Cancer Lett.*, 1977, **2**, 169–175.
77. K. E. Panter, J. Weinzweig, D. R. Gardner, B. L. Stegelmeier and L. F. James, Comparison of cleft palate induction by *Nicotiana glauca* in goats and sheep, *Teratology*, 2000, **61**, 203–210.
78. R. F. Keeler and M. W. Crowe, Congenital deformities in swine induced by wild tree tobacco, *Nicotiana glauca, J. Toxicol. Clin. Toxicol.*, 1983, **20**, 47–58.

CHAPTER 9

Toxicology of Some Insecticides Not Discussed Elsewhere

TIMOTHY C MARRS[a] AND IAN C DEWHURST[b]

[a] Edentox Associates, Pinehurst, Four Elms Road, Edenbridge, Kent, UK. Email: timothymarrs05@aol.com; [b] Chemicals Regulation Directorate, Health and Safety Executive, Mallard House, York, UK

9.1 Phenylpyrazole GABA Antagonists

The phenylpyrazoles are antagonists at the γ-aminobutyric acid A ($GABA_A$) receptor.[1] This receptor is a gated chloride channel. Because the GABA neurotransmission system in adult mammals (and insects) is inhibitory, the effects are excitatory. The main insecticides of this group are fipronil and ethiprole. In insects, fipronil also affects glutamate-gated chloride channels (see review by Narahashi and colleagues).[2]

9.1.1 Fipronil

Fipronil is the ISO name for (*RS*)-5-amino-1-[2,6-dichloro-4-(trifluoromethyl) phenyl]-4-(trifluoromethylsulfinyl)-^{1}H-pyrazole-3-carbonitrile (IUPAC). The structural formulae for the two fipronil enantiomers are:

Issues in Toxicology No. 12
Mammalian Toxicology of Insecticides
Edited by Timothy C. Marrs

Published by the Royal Society of Chemistry, www.rsc.org

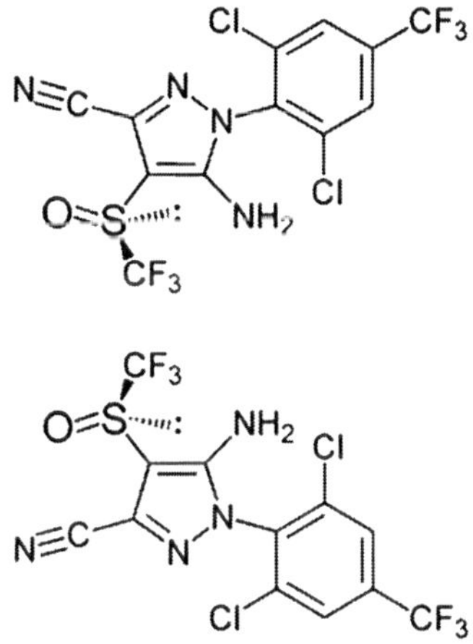

Fipronil is an antagonist at the $GABA_A$ receptor; specificity against insects and not against mammals may be achieved by some specificity for the insect $GABA_A$ receptor. Also glutamate-gated chloride channels, where fipronil also acts (see above) do not appear to exist in mammals. The absorption, distribution, excretion and metabolism of fipronil have been summarized by the Joint Expert Meeting on Pesticide Residues (JMPR).[3] Fipronil is discussed in detail in Chapter 12.

9.1.2 Ethiprole

Ethiprole is a compound with a similar mode of action to fipronil, and has a similar structure. Ethiprole is the ISO name for 5-amino-1-(2,6-dichloro-α,α,α-trifluoro-*p*-tolyl)-4-ethylsulfinylpyrazole-3-carbonitrile (IUPAC). The structural formula is:

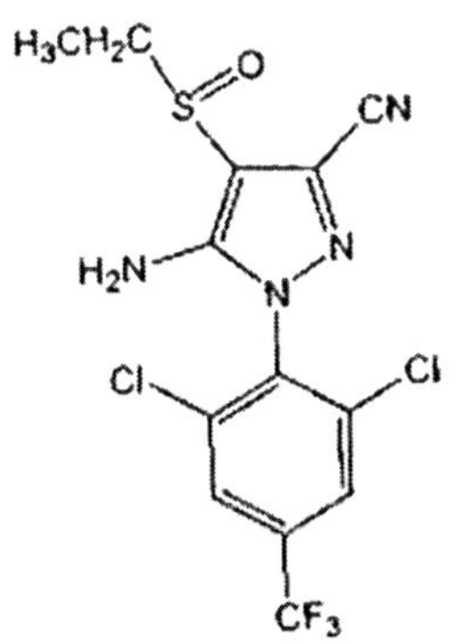

Differences between fipronil and ethiprole, principally from the point of view of photochemistry, insect and mammalian metabolism, action at the GABA receptor, and insecticidal potency were reviewed by Caboni and colleagues.[4]

9.1.2.1 Absorption, Distribution, Metabolism and Excretion

The absorption, distribution, metabolism, excretion and toxicology has been summarized in Japan.[5] Using radiolabelled ethiprole, after single doses in rats, the radioactivity was eliminated rapidly, largely in the faeces. Numerous metabolites were found in the urine and faeces.

9.1.2.2 Toxicology

The acute oral toxicity in rats was low (acute oral $LD_{50} > 7$ g kg^{-1} bw in both sexes). The insecticide was not an irritant either on rabbit skin or in the rabbit eye. Ethiprole was not a sensitizer in a guinea-pig maximization test.

In a 90-day study in rats, using dietary administration, lethalities were observed at the top two doses. The main changes observed were piloerection, effects on motor activity, decreased bodyweight and decreases in food consumption. Effects on haematological and clinical chemistry parameters were seen and these included effects on thyroid hormones. Necrosis of hepatocytes was observed in males and pigmentation in the kidneys in females. Hypertrophy/hyperplasia of the thyroid follicular epithelium was seen. The lethalities at the top dose were attributed to coagulation failure secondary to liver damage. The no observed adverse effect level (NOAEL) for the study was a dietary concentration of ethiprole giving intakes of 1.2 mg kg^{-1} bw day^{-1}. In a 90-day dietary study in dogs, the main changes were decreased bodyweight gain, decreased prostate and testis weight and an absence of epididymal sperm, centrilobular hypertrophy of hepatocytes and thymic atrophy. The NOAEL was 1.0 mg kg^{-1} bw day^{-1}. Effects on bodyweight were seen in a 1-year study on dogs and the NOAEL was a dietary concentration giving an intake of 0.70 mg kg^{-1} bw day^{-1}.

Long-term dietary studies were undertaken in rats and mice. In the rat, there were effects on the thyroid (follicular cell hyperplasia and adenoma) possibly *via* an effect on biliary T4 excretion: T4 levels were decreased, while levels of thyroid stimulating hormone (TSH) were increased in both sexes at higher doses. At the top dose, in males, there were effects on the liver (increased relative liver weight with basophilic altered foci) and progressive chronic nephropathy. In the females at the same dose, haematological changes were observed together with centrilobular hypertrophy of hepatocytes and bile duct fibrosis. Focal sinusoidal dilation in the liver and renal arteritis and periarteritis were also noted. The NOAEL was 0.85 mg kg^{-1} bw day^{-1}. In mice, there were increases in alanine transaminase (ALT) and the relative liver weight and altered foci and fatty change in the livers of males at the top dose, while an increase in incidence of liver adenomas was observed in females at the top dose. The NOAEL for the study was 12.5 mg kg^{-1} bw day^{-1}. Ethiprole was not genotoxic in a variety of tests, both *in vitro* and *in vivo*. In a two-generation study of reproductive toxicity in rats, the NOAELs for parents and offspring were the same dietary concentration. The lowest test material intake was seen in parental (F_0) males (4.77 mg kg^{-1} bw day^{-1}) and specific reproductive toxicity was not seen. In studies of developmental toxicity in rats and rabbits teratogenicity was not seen and the NOAEL for fetotoxicity (10 mg kg^{-1} bw day^{-1}) was higher than that for maternal toxicity (3 mg kg^{-1} bw day^{-1}) in rats: in rabbits the NOAELs for maternal and fetal toxicity were the same (0.5 mg kg^{-1} bw day^{-1}).

9.1.2.3 Reference Dose

The acceptable daily intake (ADI) established in Japan was 0.005 mg kg^{-1} bw day^{-1} based on maternal and fetal toxicity in a rabbit study developmental toxicity, using a safety factor of 100.[5]

9.2 Indoxacarb

Indoxacarb is the ISO name for methyl 7-chloro-2,5-dihydro-2-[[(methoxy-carbonyl)[4-(trifluoromethoxy)phenyl]amino]carbonyl]indeno[1,2-e][1,3,4]oxa-diazine-4a(3H)-carboxylate (IUPAC). Indoxacarb is an oxadiazine insecticide. Initially indoxacarb was a racemate containing the enantiomers (*S*:*R*), the insecticidal activity residing in the *S* enantiomer, to which the ISO name correctly applies. The structural formula is:

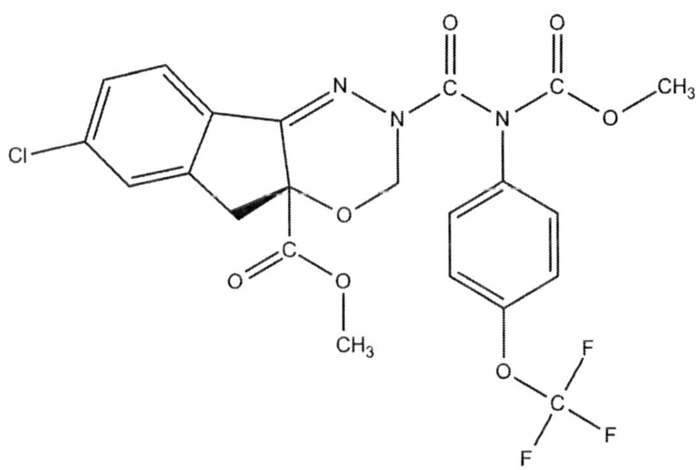

Indoxacarb exerts its insecticidal action at voltage-dependent sodium channels.

Some studies of absorption, distribution, metabolism and excretion toxicological studies used the racemate. Later studies used preparations containing about 75% of the insecticidally active *S* enantiomer.

9.2.1 Absorption, Distribution, Metabolism and Excretion

The absorption, distribution, metabolism and excretion of indoxacarb were reviewed by the 2005 JMPR.[6]

In rats the kinetics and metabolism of racemate and *S*-enantiomer-enriched preparation were similar. Indoxacarb administered by gavage at low doses (5 mg kg^{-1} bw) was slowly but extensively absorbed at low doses, but at higher doses there was saturation kinetics and the proportion absorbed was small. The time to C_{max} was shorter in males than females. Label was found predominantly in the fat or erythrocytes depending on the position of the label. Substantial amounts of label were eliminated both in the faeces and the urine, the precise proportions depending on the position of the label. Indoxacarb was extensively metabolized in rats to numerous compounds and in urine and erythrocytes, 4-trifluoromethoxyaniline, whose *N*-hydroxy derivative is thought to be responsible for haemolytic effects observed with indoxacarb, was found.

9.2.2 Toxicology

9.2.2.1 Acute Toxicity, Irritancy and Sensitization

In male rats, indoxacarb had low acute oral toxicity (LD_{50} = 1730 mg kg^{-1} bw), while in females the insecticide was of moderate oral toxicity (LD_{50} = 268 mg kg^{-1} bw). Indoxacarb was not a skin irritant in rabbits but was a moderate eye irritant. Indoxacarb was a skin sensitizing agent in the guinea-pig maximization test. These studies of acute toxicity, irritancy and sensitization were carried out using the *S*-enantiomer-enriched preparation.

9.2.2.2 Repeated Dose Studies

A 28-day dietary study in mice using dietary concentrations giving doses of up to 35.3 mg kg^{-1} bw day^{-1} and a 90 d study using dietary concentration giving doses of up to 30 mg kg^{-1} bw day^{-1}, were undertaken. Similar studies were carried out in rats using dietary concentration giving intakes of up to 23.5 mg kg^{-1} bw day^{-1} (28-day study) and 16 mg kg^{-1} bw (90-day study). In dogs 90-day and 1-year studies were carried out, using dietary concentrations giving intakes of up to 18 mg kg^{-1} bw day^{-1} in the former and up to 36.1 mg kg^{-1} bw day^{-1} in the latter. These studies in rodents and dogs all used the racemic mixture. The two main toxicological findings after repeated dosing with indoxacarb in all three species were reduced bodyweight gain and mild haemolysis. The reduction in bodyweight gain was usually associated with a concomitant decrease in food intake. Heinz bodies were seen in the erythrocytes of mice at high doses and reticulocytosis was seen in rats and mice. In rats, there was evidence of haemosiderin deposition in the livers and/or spleens. Neurotoxicity was seen in the mice at high doses. In the 90-day study in the dog, sporadic bodyweight loss and/or reduced bodyweight gain was observed. Haemolytic anaemia with a regenerative response and slight reticulocytosis were present. Heinz bodies were seen in red blood cells and serum bilirubin concentrations were increased at higher doses. Haemosiderin was present in the spleen and livers of both sexes at all test doses. Findings were similar in the 1-year study in dogs. NOAELs for the mouse studies were 10.8 mg kg^{-1} bw day^{-1} (28-day) and 5.5 mg kg^{-1} bw day^{-1} (90-day). Corresponding NOAELs for the rat studies were 2.61 mg kg^{-1} bw day^{-1} (28-day), 2.3 mg kg^{-1} bw day^{-1} (90-day). The NOAEL for the 90-day dog study was 3 mg kg^{-1} bw day^{-1}, while that for the 1-year study was 1.1 mg kg^{-1} bw day^{-1}.

A second rat 90-day study was carried out with the *S*-enantiomer-enriched preparation using doses of up to 15.0 mg kg^{-1} bw day^{-1}. In the females there were early deaths, otherwise the findings were similar to the 90-day rat study using the racemate. The NOAEL for this study was 2.1 mg kg^{-1} bw day^{-1}, based on reduced bodyweight gain and haematological changes.

9.2.2.3 *Carcinogenicity and Mutagenicity*

In a long-term study in mice, using dietary concentrations of racemate, giving daily intakes of up to 30.8 mg kg^{-1} bw, increased decedency was observed at the highest dietary concentration, so that was reduced twice in the course of the study. Clinical signs suggesting neurotoxicity, reduced bodyweight, weight gain and food consumption were observed at higher doses. Haematological effects were not seen. In some males, at the highest dietary concentration, necrotic, haemorrhagic and inflammatory heart lesions, accompanied by red fluid in the pleural cavity, were noted. No treatment-related neoplastic changes were seen in either sex in any group. The NOAEL for the study was 2.63 mg kg^{-1} bw day^{-1} based upon decreased bodyweight gain at higher dietary concentrations. In a long-term study in rats, dietary concentrations used gave intakes of up to 10 mg kg^{-1} bw day^{-1}, the racemate being used. Decedency was not seen, but at high doses there were decreased bodyweight gain and food consumption. Decreased red blood cell count was seen, most notably at 6 months. There was some evidence of a regenerative response, as mild reticulocytosis was seen. Increased pigment in splenic macrophages and/or liver Kupffer cells were seen. Treatment-related neoplastic changes were not noted. The NOAEL was 2.1 mg kg^{-1} bw day^{-1}, because of decreased bodyweight gain, and food consumption and effects on the haematological system. An adequate battery of tests *in vitro* and *in vivo* indicated that indoxacarb (enriched with the insecticidally active *S* enantiomer) was not genotoxic.

9.2.2.4 *Reproductive and Developmental Toxicity*

In a multigeneration study, rats were fed diets containing indoxacarb (racemate) at dietary concentrations giving intakes of up to 6.7 mg kg^{-1} bw day^{-1}. No notable clinical signs were seen but reductions in bodyweight gain and food consumption indicative of parental toxicity in the F_0 generation were seen at higher doses. No effects on gonadal function, oestrous cycling or mating behaviour were observed in either generation of parents. Test-material-related effects on pup survival were not observed. The weights of F_1 pups at higher dietary concentrations were reduced during lactation, but no such effect was noted in the F_2 litters. An increase in absolute and relative spleen weights of 56%, 13%, 37% and 63% was noted in F_0 males and females and F_1 females, and in those animals examined microscopically *post mortem* there was extramedullary haematopoiesis and haemosiderin pigmentation in the spleen. The NOAEL for parental toxicity was 1.3 mg kg^{-1} bw day^{-1}, this being based on lower bodyweights and food intake at higher doses. The NOAEL for the offspring development was 1.3 mg kg^{-1} bw day^{-1}, based on lower bodyweights of F_1 pups during lactation at higher doses. No effects were noted on the reproductive performance of parental animals at the top dose used in the study. After two pilot studies, a study of developmental toxicity in the rat using dosing by gavage was undertaken. Doses used were up to 4 mg kg^{-1} bw day^{-1} of the *S*-enantiomer-enriched preparation. There were no-test material-related maternal deaths, but maternal bodyweight gain and food consumption was considerably reduced at the top dose. Treatment was without effects on the number of resorptions, fetal viability or sex ratio. Fetal mortality was not affected by treatment, but there was a significant decrease in mean fetal weight compared with controls at the top dose. The prevalence of fetal malformations was not increased by the treatment. NOAELs for maternal and fetotoxicity were thus 2 mg kg^{-1} bw day^{-1}, based on effects on maternal and fetal bodyweight at the top dose. A study of developmental toxicity was undertaken in groups of mated rabbits were given indoxacarb (racemate) technical at doses of up to 1000 mg kg^{-1} day^{-1}. Test-material-related maternal deaths were not seen but maternal bodyweight gain was substantially reduced at the highest dose, as was food consumption. Moreover, green-coloured stools were observed at that dose. Treatment had no effect on the fetal viability or sex ratio, but mean fetal weights were reduced and there was an increase in retarded sternebral ossification, both at the highest dose. The NOAEL for maternal and developmental toxicity was 500 mg kg^{-1} bw day^{-1} on the basis of lower maternal and fetal bodyweight and retarded fetal ossification at the top dose (see also USEPA[7]).

9.2.2.5 Neurotoxicity

In a study of acute neurotoxicity in rats, using gavage with up to 200 mg kg^{-1} bw in males and 100 mg kg^{-1} bw in females, reduced bodyweight gain and food consumption occurred at high doses. In males, reduced forelimb grip strength and decreased foot splay was observed at 200 mg kg^{-1} bw. In females, evidence of neurotoxicity, including reduced motor activity, was observed at the highest dose used. The NOAEL in males was 100 mg kg^{-1} bw for both systemic toxicity and neurotoxicity. The NOAELs in females were 12.5 mg kg^{-1} bw for systemic toxicity and 50 mg kg^{-1} bw for neurotoxicity. Indoxacarb enriched with the insecticidally active *S* enantiomer was used in this study.

Zhao and colleagues studied the effects of indoxacarb on tetrodotoxin-sensitive and tetrodotoxin-resistant sodium channels, using neurons from the rat dorsal root ganglion. Indoxacarb slowly and irreversibly blocked both tetrodotoxin-sensitive and tetrodotoxin-resistant sodium channels.[8]

9.2.2.6 Human Data

Methaemoglobinaemia was reported in a 68-year-old healthy man: he presented with generalized 'cyanosis' after ingesting indoxacarb. He recovered after receiving treatment with methylene blue, without sequelae.[9]

9.2.2.7 Reference Doses

The JMPR[6] established an ADI of 0.01 mg kg^{-1} bw day^{-1} calculated from the NOAEL of 1.1 mg kg^{-1} bw day^{-1} in the 1-year dietary study in dogs, based on erythrocyte damage and the secondary increase in haematopoiesis in the spleen and liver, at higher doses. A 100-fold safety factor was used. This was supported by a NOAEL of 1.3 mg kg^{-1} bw day^{-1} in the rat two-generation reproductive study. The JMPR also established an acute reference dose (ARfD) of 0.1 mg kg^{-1} bw based on the NOAEL of 12.5 mg kg^{-1} bw after administration of a single dose of indoxacarb in a rat of neurotoxicity study, with a 100-fold safety factor. This NOAEL was based upon reductions in bodyweight gain and food intake.

9.3 Amitraz

Amitraz is the ISO name for *N,N'*-[(methylimino)dimethylidyne]di-2,4-xylidine (IUPAC). The structural formula is:

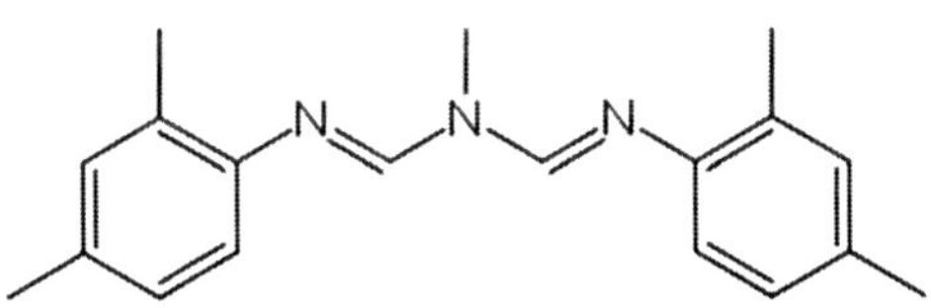

Amitraz is an amidine acaricide and insecticide used on crops and farm animals. It is thought to act *via* the octopamine receptors in insects, in which octopamine is a neurotransmitter.[1] Octopamine is present in mammals and is closely related to noradrenaline (norepinephrine) but its functions in humans are uncertain. The toxicity in mammals is probably due to α_2-adrenoreceptor stimulation[10,11] and octopamine has sympathomimetic effects.

The toxicity of amitraz has been reviewed in detail by JMPR[12] in 1998 and also by the United States Environmental Protection Agency (EPA).[13]

9.3.1 Absorption, Distribution, Metabolism and Excretion

In rats, mice, dogs, baboons and humans, orally administered amitraz is extensively absorbed, with a C_{max} in dogs at 8 h, and excreted quickly (*c.* 80%

in 24 h) predominantly in the urine. Dermal absorption has been studied in rats, dogs and pigs, with about 10% of the applied dose being absorbed under varying conditions. The initial step in metabolism is cleavage at the tertiary nitrogen followed by conjugation or sequential oxidation reactions to give toluic acid derivatives. No unmetabolized amitraz was detected in urine.

9.3.2 Toxicology

9.3.2.1 *Acute Toxicity, Irritancy and Sensitization*

Amitraz is of moderate acute toxicity by the oral route but low acute toxicity *via* dermal and inhalation routes. Oral LD_{50} values ranged from 100 mg kg^{-1} in dogs to 1600 mg kg^{-1} bw in mice; the inhalation LC_{50} was reported to be 65 mg L^{-1} in one study in rats but 2.4 mg L^{-1} in another (cited by Proudfoot[11] and the JMPR[12] respectively). Signs seen in oral studies varied across species. Mice, rats and guinea-pigs showing hyperexcitability, ataxia and tremors, while rabbits and dogs showed signs of central nervous system (CNS) depression, muscle weakness, decreased temperature and decreased pulse rate. In dogs, clear effects were seen after a single oral dose of 20 mg kg^{-1} bw with minimal effects on pulse and temperature at 4 mg kg^{-1} bw.

Two male human volunteers who received a single oral dose of 0.25 mg kg^{-1} bw reported symptoms within 4 h including dry mouth, drowsiness, light-headedness, reduced pulse rate and hypotension. Minimal, transient effects, considered to be non-adverse, were seen at 0.125 mg kg^{-1} bw in a separate study of six male volunteers.

In rabbits, amitraz was not irritating to the skin but was a mild eye irritant. In the guinea-pig, amitraz exhibited skin sensitizing potential in a maximization protocol but not in a Buehler protocol.

9.3.2.2 *Repeated Dose Studies*

In mice and rats the most sensitive endpoint following oral exposure for up to 90 days was reduced bodyweight gain at $\geq$12 mg kg^{-1} bw day^{-1}, with irritability/aggressive behaviour seen at around 50 mg kg^{-1} bw day^{-1}. Similar signs were reported in rats exposed to amitraz dusts ($\geq$0.1 mg L^{-1}) for 6 h. In dogs, the predominant effects following gavage or capsule dosing at $\geq$1 mg kg^{-1} bw day^{-1} related to transient CNS depression that reversed between doses. Dermal exposure of rabbits at $\geq$50 mg kg^{-1} bw day^{-1} resulted in transient sedative effects and reduced food consumption and bodyweight.

9.3.2.3 *Carcinogenicity and Mutagenicity*

In chronic studies in rats and mice, irritable/aggressive behaviour was seen at $\geq$10 mg kg^{-1} bw day^{-1}. There was no increase in tumours in male mice or rats of either sex. Amitraz produced an increase in hepatocellular tumours in female B6C3F1 mice receiving 400 ppm in the diet, a level above the maximum tolerated dose as survival was reduced. In CFLP mice lymphoreticular tumours were increased in females receiving 400 ppm. Amitraz was not genotoxic in a battery of *in vitro* and *in vivo* assays.

9.3.2.4 Reproductive and Developmental Toxicity

In a reproduction study in rats, reductions in litter size and pup survival were seen in all three generations at parental dietary exposures equivalent to about 5 mg kg^{-1} bw day^{-1}, the NOAEL for parental toxicity. Extended oestrous cycling was seen in rats receiving amitraz in the diet at 20 mg kg^{-1} bw day^{-1} for 18 weeks but not in mice receiving up to 60 mg kg^{-1} bw day^{-1} for 28 weeks. Amitraz exhibited no teratogenic effects in rats or rabbits at dose levels producing maternal toxicity. Findings of fetotoxicity were consistent with, and probably secondary to, the observed effects in the dams.

9.3.2.5 Human Data

Human experimental studies are discussed in section 9.3.2.1. There are data from a large number (*c.*50) of case reports or series of intentional or accidental ingestion[14,15] (see reviews by Proudfoot[11] and Yimlaz and Yildizdas).[16] It is not always clear which findings are associated with the toxicity of amitraz itself, of the hydrocarbon solvents in which it is commonly dissolved or of other products taken concurrently. Signs of amitraz poisoning are consistent with stimulation of α_2-adrenoceptors resulting in general CNS depression, impairment of consciousness, respiratory depression, convulsions, disorientation, miosis, vomiting, bradycardia, hypotension, hypothermia and abnormal glucose levels (increased or decreased). Initial signs are seen within 1 h hour of exposure. The prognosis is good, with even the most severely poisoned patients recovering following nothing more than intensive care, including mechanical ventilation. Antidotal studies in animals indicate that the α_2-adrenoceptor antagonists yohimbine and atipamezole can reverse amitraz-induced toxicity, but they have not been assessed in poisoned humans.[11]

Findings in the human volunteer studies at 0.25 mg kg^{-1} bw indicate that humans are more sensitive than dogs to amitraz, but exhibit a similar pattern of effects. In rodents the pattern of effects (*e.g.* aggressive behaviour) is different from those reported in human poisoning cases. Humans are significantly more sensitive than rodents, where the overall NOAEL is 3 mg kg^{-1} bw day^{-1} for general toxicity.

9.3.2.6 Reference Doses

The JMPR[12] based its ADI on the NOAEL of 1.3 mg kg^{-1} bw day^{-1} in the study of reproductive toxicity in rats, using a safety factor of 100. This gave an ADI of 0.01 mg kg^{-1} bw. An ARfD of 0.01 mg kg^{-1} bw was established on the basis of a study in humans, in which there was a NOAEL of 0.13 mg kg^{-1} bw day^{-1}, a safety factor of 10 being used.

9.4 Pyridalyl

Pyridalyl is the ISO name for 2,6-dichloro-4-(3,3-dichloroallyloxy)phenyl 3-[5-(trifluoromethyl)-2-pyridyloxy]propyl ether (IUPAC). The structural formula is:

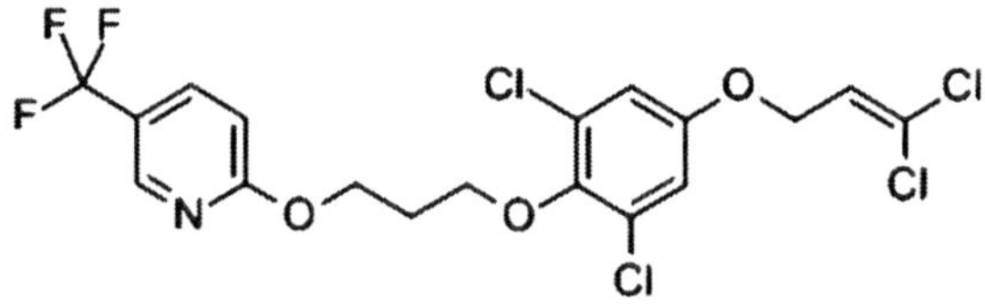

The Insecticide Resistance Action Committee (IRAC) deemed the mode of action of pyridalyl to be uncertain.[1] Powell and colleagues considered that cytochrome P450 action leads to an active pyridalyl metabolite in insects as action requires cytochrome P450 activity.[17] *In vitro*, pyridalyl exhibited high selectivity in cytotoxicity between an insect and mammalian cell line,[18] possibly by a selective inhibition of cellular protein synthesis.[19] The absorption, distribution, metabolism and excretion, and the toxicology of pyridalyl has been examined by the EPA.[20]

9.4.1 Absorption, Distribution, Metabolism and Excretion

In rats, more than 90% of the amount of the radiolabel from labelled pyridalyl was eliminated in the faeces within 72 h of treatment.[20] After a single oral administration of [^{14}C]-labelled pyridalyl, the blood ^{14}C concentration peaked between 2 and 10 h. Thereafter, the ^{14}C concentration in blood fell rapidly with a biological half-life of around 11–12 h. The major route of elimination of label was *via* the faeces, with smaller amounts being eliminated *via* the urine and expired air. One major urinary and faecal metabolite was noted, produced by *O*-dealkylation.[21] A metabolic pathway has been proposed.[22]

9.4.2 Toxicology

Pyridalyl was of low acute toxicity with a rat oral LD_{50} greater than 5000 mg kg^{-1}.[20] The insecticide was not an eye or dermal irritant in rabbits, but was a skin sensitizer in guinea-pigs (both Buehler and Magnusson and Kligman maximization test). In 90-day studies in rats and mice, decreased bodyweight gain was observed. In rats alveolar foamy cells were seen in the lungs in both sexes and vacuolation of interstitial gland cells in the ovary in females. In mice, pigmentation in the adrenal gland was observed in the males and ovarian atrophy in the females. In a 90-day oral study in the dog, vacuolation of cells in the adrenal cortex was present. In a 12-month oral study with dogs, pyridalyl produced alterations in blood biochemistry, while liver weights were increased. In long-term studies in rats and mice, decreased bodyweight gain was seen.

There was no evidence of carcinogenicity in either species and in an adequate range of tests *in vitro* and *in vivo*, there was no evidence of mutagenicity. In a two-generation study of reproductive toxicity in rats,increased ovarian weight and microscopic lesions in the ovaries of both generations of adults were seen. Decreased bodyweight and weight gain in both sexes and food consumption in males. Lesions in the thyroid, consisting of an increase in small-sized follicles, were also noted in females. Delayed vaginal opening in the offspring in F_1 and F_2 offspring was seen. Additionally, decreased thymus weights were observed in the offspring. In a study of developmental toxicity in the rat, maternal toxicity consisted of reduced bodyweight gain, while developmental toxicity (decreased incidence of thymus remnants) was only observed at higher doses than those causing maternal toxicity. In a study of developmental toxicity in rabbits, death, abortion/premature delivery, and decreased bodyweight gain and food consumption was observed in the does, and decreased fetal bodyweight. In a steroid hormone study in rats no treatment-related effects were found on sex hormones, uterine weight, or the oestrus cycle.

9.4.3 Reference Dose

The EPA identified a chronic reference dose (≡ADI) of 0.034 mg kg^{-1} day^{-1} and it was considered that an ARfD was not needed.[20] This chronic reference dose was based upon a NOAEL of 3.4 mg kg^{-1} day^{-1}, decreased bodyweights, weight gain and food efficiency being seen at the next higher dose.

References

1. Insecticide Resistance Action Committee. *IRAC Mode of Action Classification*, September 2010. http://www.irac-online.org/wp-content/uploads/2009/09/MoA-classification_v7.0.4–5Oct10.pdf
2. T. Narahashi, X. Zhao, T. Ikeda, K. Nagata and J. Z. Yeh, Differential actions of insecticides on target sites: basis for selective toxicity, *Hum. Exp. Toxicol.*, 2007, **26**, 361–366.
3. FAO/WHO, *Pesticide Residues in Food—1997. Evaluations 1997 Toxicology*. Joint Meeting of the FAO Panel of Experts on Pesticide Residues in Food and the Environment and the WHO Core Assessment Group: Lyon, 22 September–1 October 1997. World Health Organization, Geneva, 1998. http://www.inchem.org/documents/jmpr/jmpmono/v097pr09.htm
4. P. Caboni, R. E. Sammelson and J. E. Casida, Phenylpyrazole insecticide photochemistry, metabolism, and GABAergic action: ethiprole compared with fipronil, *J. Agric. Food Chem.*, 2003, **24**, 7055–7061.
5. Pesticides Expert Committee, *Evaluation Report, Ethiprole, July 21 2004*, Food Safety Commission, Tokyo, 2004.
6. FAO/WHO, *Pesticide Residues in Food—2005. Toxicological Evaluations.* Joint Meeting of the FAO Panel of Experts on Pesticide Residues in Food

and the Environment and the WHO Core Assessment Group: Geneva, Switzerland, 20–29 September 2005. World Health Organization, Geneva, 2006. http://www.inchem.org/documents/jmpr/jmpmono/v2005pr13.pdf
7. U.S. EPA, *Indoxacarb: Conditional Registration October 30, 2000*. Pesticide Fact Sheet, Office of Prevention, Pesticides, and Toxic Substances, U.S. Environmental Protection Agency, Washington, DC, 2000. http://www.epa.gov/opprd001/factsheets/indoxacarb.pdf
8. X. Zhao, T. Ikeda, J. Z. Yeh and T. Narahashi, Voltage-dependent block of sodium channels in mammalian neurons by the oxadiazine insecticide indoxacarb and its metabolite DCJW, *Neurotoxicology*, 2003, **24**, 83–96.
9. Y. J. Wu, Y. L. Lin, H. Y. Huang and B. G. Hsu, Methemoglobinemia induced by indoxacarb intoxication, *Clin. Toxicol. (Phila.)*, 2010, **48**, 766–767.
10. S. G. Lalchandani, L. Longping, W. Zheng, M. M. Suni, R. M. Moore, S. B. Liggett, D. D. Miller and D. R. Feller, Yohimbine dimers exhibiting selectivity for the human alpha2c-adrenoreceptor subtype, *J. Pharmacol. Exp. Ther.*, 2002, **303**, 979–984.
11. A. T. Proudfoot, Poisoning with amitraz, *Toxicol. Rev.*, 2003, **22**, 71–74.
12. FAO/WHO. *Pesticide residues in food—1998. Toxicological evaluations.* Joint Meeting of the FAO Panel of Experts on Pesticide Residues in Food and the Environment and the WHO Core Assessment Group, Rome, 21–30 September 1998. World Health Organization, Geneva, 1999. http://www.inchem.org/documents/jmpr/jmpmono/v098pr01.htm
13. U.S. EPA. *Reregistration Eligibility Decision (RED). Amitraz. List A Case 0234*. Office of Pesticide Programs, Special Review and Reregistration Division, U.S. Environmental Protection Agency, Washington DC, 1995. http://www.epa.gov/oppsrrd1/REDs/0234red.pdf
14. L. Avsarogullari, I. Ikizceli, M. Sungur, E. Sözüer, O. Akdur and M. Yücei, Acute amitraz poisoning in adults: clinical features, laboratory findings, and management, *Clin. Toxicol.*, 2006, **44**, 19–23.
15. Y. Demirel, A. Yilmaz, Y. Gursoy, K. Kaygusuz and C. Mimaroglu, Acute amitraz intoxication: retrospective analysis of 45 cases, *Hum. Exp. Toxicol.*, 2006, **25**, 613–617.
16. H. L. Yilmaz and D. R. Yildizdas, Amitraz poisoning, an emerging problem: epidemiology, clinical features, management, and preventive strategies, *Arch. Dis. Child.*, 2003, **88**, 130–134.
17. G. F. Powell, D. A. Ward, M. C. Prescott, D. G. Spiller, M. R. White, P. C. Turner, F. G. Earley, J. Phillips and H. H. Rees, The molecular action of the novel insecticide, Pyridalyl. *Insect Biochem. Mol. Biol.*, 2011, **41**, 459–469.
18. S. Isayama, S. Saito, K. Kuroda, K. Umeda and K. Kasamatsu, Pyridalyl, a novel insecticide: potency and insecticidal selectivity, *Arch. Insect. Biochem. Physiol.*, 2005, **58**, 226–233.
19. K. Moriya, S. Hirakura, J. Kobayashi, Y. Ozoe, S. Saito and T. Utsumi, Pyridalyl inhibits cellular protein synthesis in insect, but not mammalian, cell lines, *Arch. Insect Biochem. Physiol.*, 2008, **69**, 22–31.

20. U.S. EPA. *Pesticide Fact Sheet Pyridalyl, Date Issued: April 24, 2008*. Office of Prevention, Pesticides and Toxic Substances, United States Environmental Protection Agency, Washington, DC, 2008. http://www.epa.gov/opprd001/factsheets/pyridalyl.pdf
21. H. Nagahori, K. Saito, Y. Tomigahara, N. Isobe and H. Kaneko, Metabolism of pyridalyl in rats, *Drug Metab. Dispos.*, 2009, **37**, 2284–2289.
22. H. Nagahori, Y. Tomigahara, N. Isobe and H. J. Kaneko, Metabolism of pyridalyl in rats: excretion, distribution, and biotransformation of dichloropropenyl-labeled pyridalyl, *Agric. Food Chem.*, 2009, **57**, 10845–10851.

CHAPTER 10
Toxicology of Biological Insecticides

IAN C DEWHURST

Chemicals Regulation Directorate, Health and Safety Executive, Mallard House, York, UK. Email: ian.dewhurst@hse.gsi.gov.uk

10.1 Introduction

Biological insecticides are a heterogeneous group of compounds and entities/organisms that can be used to control insect pests. There is a small but increasing market for biological insecticides. They can consist of extracts of plant material, purified chemicals obtained from fermentation, or viable organisms. Examples of plant extracts include azadirachtin from the neem tree (see Chapter 8), nicotine from *Nicotiana* species (see Chapter 6) and pyrethrins from chrysanthemums (see Chapter 5). Purified chemicals from fermentations include avermectins from *Streptomyces avermitilis* (see Chapter 8) and spinosyns from *Saccharopolyspora spinosa* (see Chapter 8).

The majority of viable organisms used as insecticides are highly specific for the target pests and are considered to have minimal toxicity to non-target species. Because of the specificity and relatively low level of use, the data available are often very limited. Basic information on a wide range of biological pesticides can be obtained from the US Environmental Protection Agency (EPA) biopesticide fact sheets.[1]

This chapter focuses on preparations based on viable organisms, known as microbial pest control agents (MPCAs), and in particular the most widely used

Issues in Toxicology No. 12
Mammalian Toxicology of Insecticides
Edited by Timothy C. Marrs

Published by the Royal Society of Chemistry, www.rsc.org

biological insecticides, those based on *Bacillus thuringiensis* and its insecticidally active components.

MPCAs have the ability to multiply and/or secrete secondary metabolites after application; some of the secreted metabolites have subsequently been developed into purified chemical insecticides, *e.g.* spinosyns (Chapter 8).

10.2 *Bacillus thuringiensis*

Bacillus thuringiensis (Bt) is a spore-forming, aerobic, Gram-positive bacterium that is closely related to *B. cereus*. Bt can be differentiated from *B. cereus* in that Bt produces crystalline protein inclusion bodies in its spores. These inclusion bodies have toxicity to insects, nematodes, protozoa and mites. In other respects Bt and *B. cereus* are so closely related that some workers have considered them to be the same species, and work with RNA probes has identified cases of misclassification. Using 16S rRNA probes, Giffel[2] showed that 6 of 20 isolates previously identified as *B. cereus* by classical biochemical techniques were more correctly Bt. Jackson and colleagues[3] also reported that cases of gastroenteritis attributed to *B. cereus* might have been due to Bt. In neither of these studies were the serotypes identified, so it is not clear if the incidents were linked with insecticidal use of Bt. The properties of the different strains and variants of Bt differ, and it is important to relate data to specific strain or variant of Bt wherever possible.

The crystal proteins (Cry) produced by Bt are responsible for the insecticidal activity and have been identified by the term δ-endotoxin. The δ-endotoxin is coded for by plasmids that can be exchanged with other bacilli. Different strains of Bt have activity against different classes of insects and this activity can be related to differences in crystal shape and composition. The different crystals are identified by the code 'Cry #'. The δ-endotoxin in Bt spores is in the form of a protoxin that becomes active only after it is ingested, solubilized by the alkaline conditions and activated by enzymes in the insect gut. In mammals the gut is acidic to weakly alkaline and the activation of the endotoxin does not occur. The C-terminal of the protein, which binds to receptors on the gut epithelial cells, is variable and provides target species specificity. The N-terminal is conserved between strains of Bt and produces a pore in the gut epithelial cells, leading to cell lysis and death. The protein crystals are reported to be readily degraded by UV light, though a wide range of half-lives have been estimated, from a few hours to 10 days.[4] The development of genetically modified crop plants (maize, corn, cotton) that can synthesize Bt δ-endotoxins (Cry proteins) confers insecticidal activity without the potential risks associated with the use, generally by broadcast spraying, of live microorganisms (the toxicity of Cry proteins is addressed below). Genetic modification of other bacteria to produce large amounts of the Cry proteins has been achieved, *e.g.* Cry 1A and Cry 1C have been produced in *Pseudomonas fluorescens*.

Commercial products containing Bt have a range of uses, differing with the strain or variant used.[5] Bt *aizawa* has uses against lepidopteran larvae, Bt *israelensis* is active against mosquito and gnat larvae on plants and in aquatic situations, Bt *tenebrionis* is used to control beetle larvae and Bt *kurstaki* is active against cabbage moths.

Some strains of Bt also produce other toxins. The 'β-exotoxin' is a heat-stable nucleotide that inhibits RNA polymerases and is toxic to most animal phyla including insects and mammals. An enterotoxin similar to that produced by *B. cereus* is thought to be responsible for producing diarrhoea and vomiting in mammals. Several other poorly characterized toxins have also been reported.[4] The Bt strains supplied for insecticidal use are reported to be β-exotoxin negative and are tested for this on a batch basis.

Data from animal studies on toxicity, pathogenicity and infectivity, plus a range of additional studies,are available on Bt strains used as MPCAs. Much of the data on Bt as the organism were generated in the 1980s and 1990s, with more recent work focusing on the toxicity of the Cry proteins (see below). These indicate that use of Bt as a pesticide is without significant risk to human health (the data have been reviewed by *inter alia* the International Programme of Chemical Safety (IPCS),[4] McClintock *et al.*[6] and Siegel[7]).

A review of data on Bt by the IPCS[4] cites data on commercial preparations of Bt showing that administration (by oral, inhalational or parenteral routes) of more than 10^7 colony forming units (CFU)/animal did not produce deaths or evidence of pathogenicity or infectivity. Mice which had been immunosuppressed by corticosteroids or thymectomy were not killed by an intraperitoneal administration of around 10^7 spores of Bt *israelensis* (H14) and showed evidence of a slow but sustained clearance of the organism. Artificially activated crystal protein from Bt *israelensis* was lethal when injected into mice and was cytolytic against human erythrocytes, mouse fibroblasts and pig lymphocytes.[4] A review of data available to the EPA was published by McClintock *et al.*[6] This confirmed the generally safe nature of Bt strains developed for pesticidal uses, with no infectivity or mortality seen at exposures in the region of 10^7 spores by various routes. When several isolates of Bt *kurstaki* and Bt *israelensis* (including those registered as pesticides) were administered intraperitoneally to mice at 10^8 spores/animal, mortality varied from 0 to 100% depending on the isolate. A report of mortality following intraperitoneal injection of 3.4×10^7 CFU of Bt *israelensis* into athymic mice may be preparation specific, as a different commercial formulation produced no deaths when administered intraperitoneally to athymic or euthymic mice at 2.5×10^7 CFU. A range of studies on different Bt strains and life stages showed that toxicity and mortality could be produced by intraperitoneal or intranasal administration and that effects appeared to be related to the vegetative growth phase rather than the crystal protein, flagella or spores. Bt preparations had been found to be mild eye irritants but there were no confirmed cases of hypersensitivity.[6]

Following two seasons of aerial spraying of an area in Oregon to control gypsy moth, an epidemiology study covering approximately 80 000 potentially exposed subjects, was conducted to investigate any link between Bt and infections.[8] The study looked at cases of infection and any associated Bt isolates. Out of a total of 95 bacillus-positive cultures, 55 contained Bt. Of these, 52 were considered to have been associated with adventitious contamination of skin, tissue, surfaces or culture plates. In the 3 remaining cases it was not possible to dismiss the presence of Bt as being involved, but there were also confounding factors which prevented firm conclusions being drawn and in all three cases the immune system might have been compromised.

An epidemiological investigation was performed to monitor any health effects relating to the widespread use of Bt *kurstaki* (H3a3b) to eradicate the Asian gypsy moth in British Columbia.[9] Both aerial and ground spraying techniques were used in four operations between mid-April and July. The study is slightly compromised as few comparative data were obtained prior to spraying and 'cases' were classed by post code of residence rather than location at time of spraying. There was an increase in symptoms in exposed ground spray workers consistent with exposure to an irritant (consistent with the known properties of the formulation): eye, nose and throat irritation, dry skin and chapped lips. There was no increase in gastrointestinal problems nor in days off work. Nasal swabs were positive for 104 of the 112 workers sampled; though clearance was generally rapid, some of the highly exposed workers still had Bt-positive swabs after 4 weeks. Analysis of phone line reports, GP cases and casualty admissions produced nothing indicating a clear effect of exposure to Bt *kurstaki*, with a similar pattern of complaints seen in the 'control' groups not resident in the main spray zone. Swabs taken by GPs showed 10–15% of samples to be positive for Bt *kurstaki*, with a higher frequency in the spray zone. In none of the cases did the presence of Bt *kurstaki* meet predetermined criteria indicating it was associated with infection or disease, although there was an indication of a weak link to genitourinary infections and eye contamination. Food and environment samples indicated Bt *kurstaki* was present in the environment independent of the spray programme, as positive and negative samples were found in produce from the same stores independent of site of production and time of sampling (up to 6 weeks after the spraying operation). The authors concluded that though there was clear evidence of exposure to Bt *kurstaki* in both workers and the general population, there was no case indicating pathogenicity or infection. The report authors noted the potential for immunosuppressed people and those with HIV to be at extra risk but did not comment specifically on whether such people were present in the study population or as subjects with Bt-positive swabs.

An investigation of the effects of spraying Foray 48B (containing Bt *kurstaki* H3a3b) from aircraft over an area of New Zealand to control tussock moth in 1996/97 was performed. The authors concluded that there was no evidence that Bt *kurstaki* would produce: (1) infection of, or compromise, the respiratory tract; (2) gastrointestinal infection; (3) primary or secondary wound infection; (4) corneal ulcers or (5) infection or illness in conjunction with other bacteria leading to respiratory infection or miscarriage.[10]

10.2.1 Adverse Findings

A number of adverse findings relating to Bt have been reported, but it is not always clear how these reports relate to the actual use of *Bt* strains as MPCAs.

A French soldier injured by a land mine was found to have Bt sero-type H34 in necrotic soft tissue from the site of injury. Initial studies with this strain in immunosuppressed mice showed it to induce myonecrosis following cutaneous administration.[11] The same group of workers then performed studies with immunocompetent mice (5/group, BALB/c) on cultures of Bt H34, Bt H12 (obtained from a clinical specimen but considered clinically irrelevant), and two serotypes used in pesticides (H14 and H3a3b).[12] H34 was found to have the most potent haemolytic activity and to be the most toxic; all mice administered 10^8 spores intranasally died within 8 h, the lungs showing oedema, alveolar damage, nucleophilic infiltrate and ulceration. There were no deaths in animals receiving H12, just an inflammatory response. Administration of 10^8 spores of H14 produced deaths in 40% of the mice, while 80% of animals receiving 10^8 spores of H3a3b died; there were no deaths at lower inoculation rates. Lesions were similar to those produced by H34. H14 and H3a3b also had significant haemolytic activity associated with a heat-labile fraction of more than 30 kDa molecular weight identical to a haemolysin derived from *B. cereus* HG-6A associated with a case of food poisoning. The authors note that the exposures used in this study are very high compared to likely exposures from pesticidal use but show the two pesticidal strains to produce a haemolytic toxin with the potential to produce lung lesions. No information has been found relating to the potential for Bt strains to produce haemolytic toxins under typical conditions associated with pesticidal use.

Four strains of Bt (Bt *israelensis* ONR60A & O2-72; Bt *kurstaki* HD1 and HD73) closely related to those used in pesticide formulations were found to give positive results in tests for *B. cereus* enterotoxin. All four strains expressed no or minimal haemolytic activity.[13]

Investigations into the levels of *Bacillus* diarrhoeal enterotoxin in one strain of *B. cereus* (positive control) and 10 strains of Bt (one pure strain HD-1; an isolate from a case of bovine mastitis and 8 commercial preparations) were reported by Damgaard.[14] The enterotoxin titres in the vegetative forms of the Bt strains varied between 15 and 242, with the *B. cereus* strain having a titre of 1629. The author expressed concerns over the validity of the testing performed on Bt products, which were performed on spores, rather than the vegetative forms which can produce toxins.[14] This is a valid point and applies to all MCPAs that are formulated as spores. Subsequently the same group isolated a number of Bt strains from food items, but it was not possible to confirm whether contamination was due to natural occurrence or pesticidal treatment of raw materials.[15] They reported the presence of four strains of Bt associated with burns patients; none of the strains had insecticidal activity.

10.2.2 *Bacillus thuringiensis* Cry Proteins

The potential use of Bt *kurstaki* δ-endotoxin (Cry1Ac) as an adjuvant in vaccination was investigated by Vázquez-Padrón *et al.*[16] These workers showed that intraperitoneal or gavage administration of 100 μg δ-endotoxin per mouse in solution or as crystals could increase anti-Cry1Ac immunoglobulin titres in serum, faeces or the gut. Exposures to lower doses (<10 μg per mouse) did not produce any consistent pattern or statistically significant increase in antibodies. However, due to inadequate control investigations (*e.g.* mice administered bovine serum albumin (BSA) were investigated for anti-Cry1Ac immunoglobulins and not anti-BSA) it is not clear whether the findings were due to anything other than a general response to a foreign protein. The authors concluded that Cry1Ac is a potent systemic and mucosal immunogen and its ingestion may potentially produce a response typical of mucosal inflammatory diseases such as Crohn's disease. Though the basis for this conclusion is uncertain as dosing included an acid neutralizing vehicle, the no-effect dose of 10 μg per mouse (bodyweight *c.* 40 g) equates to 2.5 mg for a 10-kg infant and other work with Cry proteins has shown no evidence of allergenic potential.[17]

A number of specific evaluations of Cry proteins (1Ab, 1Ac, 1F, 3A, 3Bb1, 34Ab1 and 35Ab1), have been performed by the EPA as part of the evaluations of genetically modified crops (corn, cotton, potato) containing Cry proteins.[18] The assessment process included a consideration of the DNA inserted into the plant cells, its inheritance and stability, together with data on Cry protein expression in different parts of the plant. The vast majority of proteins are of low toxicity with the few exceptions showing toxicity at particularly low doses.[19] Another concern is allergenicity, with most allergenic proteins being glycosylated and showing resistance to degradation in gastric juice. The potential hazards of the Cry proteins were addressed by a basic data set based on an acute oral toxicity test in mice, degradation rate in gastric juice and searching for sequence homology with known allergenic proteins. In order to obtain sufficient Cry protein for the acute toxicity studies, Cry protein expressed in bacteria (*E. coli* or *P. fluorescens*) was used following confirmation of similarity with proteins expressed in plants. In the acute toxicity studies in mice, no deaths were seen at dose levels ranging from 576 mg kg^{-1} bw (Cry1F) to 5200 mg kg^{-1} bw (Cry3A), the only consistent clinical sign was transient reduction in bodyweight gain. None of the Cry proteins are glycosylated or showed any homologous sequences with known allergenic proteins and degradation in synthetic gastric fluid, *in vitro*, occurred in less than 7 min with the exception of Cry 34Ab1 which took approximately 20 min.[20]

10.2.3 Conclusion

The majority of data available indicate pesticidal strains of Bt to be non-pathogenic, non-infective and of low toxicity, though sporadic effects have been associated with individual preparations. The latter highlights the need for reliable quality control procedures to be operative during the production process. Epidemiology studies following widespread aerial and ground spraying of Bt in the USA, Canada and New Zealand did not find any clear evidence of adverse effects. A number of reports indicate that some strains of Bt, including some used as pesticides, have the potential to produce toxins in addition to the δ-endotoxin associated with insecticidal activity. These other toxins can produce symptoms typical of *B. cereus* poisoning (diarrhoea and vomiting). There are clear differences between different strains or variants of Bt, reinforcing the need for those producing commercial, insecticidal strains of Bt to adopt adequate quality control procedures. Investigations of specific types of the δ-endotoxin (Cry proteins) indicate they present a low hazard to mammals.

10.3 Granulovirus and Nuclear Polyhedrosis Virus Preparations

Granuloviruses and nuclear polyhedrosis viruses are members of the family of baculoviruses, which are naturally occurring pathogens of arthropods. These are complex, rod-shaped viruses with a double-stranded DNA genome, protected by a protein overcoat forming an occlusion body containing numerous virions. A number of these, including *Adoxephes orana* granulosis virus, *Cydia pomella* granulosis virus, *Mamestra brassica* nuclear polyhedrosis virus, *Neodiprion sertifer* nuclear polyhedrosis virus and *Spodoptera* spp. nuclear polyhedrosis virus have commercial insecticidal uses, primarily against moths.[5,21] Susceptible insect larvae ingest the occlusion bodies. In the alkaline ($pH > 9$) larval gut, the protein overcoat quickly disintegrates and releases the viral particles, which proceed to infect cells and interfere with the function of several larval organs, including food absorption in the gut. Larvae start to show signs of infection after about 4 days and are dead within 10–15 days.

Baculoviruses occur naturally and are considered to be arthropod specific and to present no known risks to humans.[21] When tested in mammalian cell lines, although there was cellular uptake, there was no evidence of viral replication or cytopathological effects. Exposure to mammals by the oral route produced no specific immunological response indicative of viral replication. Tests in a range of mammals with nuclear polyhedrosis viruses at up to 10^{12} occlusion bodies per animal found no evidence of adverse effects or immunological responses following oral or intraperitoneal administration.[22] No evidence of teratogenicity or carcinogenicity was reported in tests with *Helicovera* nuclear polyhedrosis virus.[22]

10.4 *Beauveria bassiana*

Beauveria bassiana is a fungus that grows naturally in soils and has activity against a range of insect pests. Four strains are registered with the EPA (HF23, GHA, 447 and ATCC 74040). A common mode of action involves growth of the fungus on the insect, secretion of enzymes to weaken the insect cuticle, allowing infection and subsequent growth inside the insect. In some strains the production of a depsipeptide secondary metabolite, beauvericin, is possibly related to the insecticidal activity. Pesticidal strains of *Beauveria bassiana* have shown no significant pathogenicity or toxicity *via* oral, dermal, intraperitoneal or intratracheal routes. The time to clear the organism following intratracheal or intraperitoneal doses is less than 15 days for ATCC 74040 and less than 3 days for HF23. An inflammatory response was noted in the lungs of animals dosed intratracheally with ATCC 74040 but this is a normal reaction to the presence of large titres of foreign material.[23,24]

References

1. U.S. EPA, *Index Page for Biopesticide Fact Sheets*, U.S. Environmental Protection Agency, Washington DC, 2011. http://www.epa.gov/pesticides/biopesticides/ingredients/index.htm#1A
2. M. Giffel, Discrimination between *Bacillus cereus* and *Bacillus thuringiensis* using specific DNA probes based on the variable regions of 16S rRNA, in *Isolation and Identification and Characterisation of Bacillus cereus from the Dairy Environment*. Dissertation, Wageningen University, The Netherlands, 1997.
3. S. G. Jackson, R. B. Goodbrand, R. Ahmed and S. Kasatiya, *Bacillus cereus* and *Bacillus thuringiensis* isolated in a gastroenteritis outbreak investigation, *Lett. Appl. Microbiol.*, 1995, **21**, 103–105.
4. IPCS, *Environmental Health Criteria No. 217. Bacillus thuringiensis*. International Programme on Chemical Safety, World Health Organization, Geneva, 1999.
5. European Union, *Pesticides database,* Main active substance search page, 2011. http://ec.europa.eu/sanco_pesticides/public/index.cfm?event=activesubstance.selection
6. J. T. McClintock, C. R. Schaffer and R. D. Sjoblad, A comparative review of the mammalian toxicity of *Bacillus thuringiensis* based pesticides, *Pesticide Sci.*, 1995, **45**, 95–105.
7. J. P. Siegel, The mammalian safety of *Bacillus thuringiensis*-based insecticides, *J. Invertebrate Pathol.*, 2001, **77**, 13–21.
8. M. Green, M. Heumann, R. Sokolow, L. R. Foster, R. Bryant and M. Skeels, Public health implications of the microbial pesticide *Bacillus thuringiensis*: an epidemiological study in Oregon 1985–86, *Am. J. Public Health*, 1990, **80**, 848–852.
9. M. A. Noble, P. D. Riben and G. J. Cook, *Microbiological and epidemiological Surveillance Programme to Monitor the Health Effects of Foray 48B Btk Spray*, Dept of Pathology (Medical Microbiology) and Health Care and Epidemiology, Vancouver, BC, Canada. Ministry of Forests, Province of British Columbia, Victoria, BC, Canada, 30 September 1992.
10. Operation Ever Green, *Health Risk Assessment of the Possible Future Control Programme for the White Spotted Tussock Moth (Orgyia Thyellina) in the Eastern Suburbs of Auckland, and the Environmental Impact Assessment of Aerial Spraying Btk in New Zealand*, Auckland, 1997.
11. E. Hernandez, F. Ramisse, J. P. Ducoureau, T. Cruel and J. D. Cavallo, *Bacillus thuringiensis* serovar H34–*konkukian* superinfection: report of one case and experimental evidence of pathogenicity in immunosuppressed mice, *J. Clin. Microbiol.*, 1998, **36**, 2138–2139.

12. E. Hernandez, F. Ramisse, T. Cruel, R. le Vagueresse and J. D. Cavallo, *Bacillus thuringiensis* serotype H34 isolated from human, and insecticidal strains serotypes 3a3b and H14 can lead to death of immunocompetent mice after pulmonary infection, *FEMS Immunol. Med. Microbiol.*, 1999, **24**, 43–47.
13. C. R. Carlson, D. A. Caugant and A.-B. Kolsto, Genotypic diversity among *Bacillus cereus* and *Bacillus thuringiensis* strains, *Appl. Environ. Microbiol.*, 1994, **60**, 1719–1725.
14. P. H. Damgaard, Diarrhoeal enterotoxin production by strains of *Bacillus thuringiensis* isolated from commercial *Bacillus thuringiensis*-based insecticides, *FEMS Immunol. Med. Microbiol.*, 1995, **12**, 245–250.
15. P. H. Damgaard, H. D. Larsen, B. M. Hansen, J. Bresciani and K. Jørgensen, Enterotoxin-producing strains of *Bacillus thuringiensis* isolated from food, *Lett. Appl. Microbiol.*, 1996, **23**, 146–150.
16. R. I. Vázquez-Padrón, L. Moreno-Fierros, L. Neri-Bazán, G. A. de la Riva and R. López-Revilla, Intragastric and intraperitoneal administration of Cry1Ac prooxin from *Bacillus thuringiensis* induces systemic and mucosal antibody responses in mice, *Life Sci.*, 1999, **64**, 1897–1912.
17. U.S. EPA, *Biopesticides Registration Action Document (BRAD)—Bacillus thuringiensis Plant-Incorporated Protectants.* U.S. Environmental Protection Agency, Washington, DC, 2001. http://www.epa.gov/pesticides/biopesticides/pips/bt_brad.htm
18. U.S. EPA, *Index of Bt Cry protein information—Plant-Incorporated Protectant Fact Sheets*, U.S. Environmental Protection Agency, Washington DC, 2011. http://www.epa.gov/pesticides/biopesticides/ingredients/factsheets/factsheet_plant_pesticides.htm#006430
19. R. D. Sjoblad, J. T. McClintock and R. Engler, Toxicological considerations for protein components of biological pesticide products, *Reg. Toxicol. Pharm.*, 1992, **15**, 3–9.
20. U.S. EPA, *Biopesticides Registration Action Document (BRAD)—Bacillus thuringiensis Cry34Ab1 and Cry35Ab1 Proteins and the Genetic Material Necessary for Their Production (PHP17662 T-DNA) in Event DAS-59122-7 Corn*, U.S. Environmental Protection Agency, Washington DC, 2010. http://www.epa.gov/pesticides/biopesticides/pips/cry3435ab1-brad.pdf
21. U.S. EPA, *Occlusion Bodies of Nucleopolyhedrovirus (NPV) of Various Moths (PC Codes 107300; 107302; 107303; 127885; 129078). Occlusion Bodies of Granulovirus (GV) of Codling Moth and Indian Meal Moth (PC Codes 129090 and 108896, Respectively)*, Biopesticides fact sheet, U.S. Environmental Protection Agency, Washington, DC, 22 October 2002. http://www.epa.gov/oppbppd1/biopesticides/ingredients/factsheets/factsheet_107300.htm
22. OECD, *Consensus Document on Information used in the Assessment of Environmental Applications Involving Baculoviruses. ENV/JM/MONO(2002)1*, Series on Harmonization of Regulatory Oversight in Biotechnology, No. 20.

Organisation for Economic Co-operation and Development, Paris, 8 January, 2002. http://www.oecd.org/dataoecd/16/59/46815698.pdf

23. U.S. EPA, *Biopesticide Fact Sheet for Beauveria bassiana ATCC 74040*, U.S. Environmental Protection Agency, Washington, DC, 2000. http://www.epa.gov/pesticides/biopesticides/ingredients/factsheets/factsheet_128818.htm
24. U.S. EPA, *Biopesticide Registration Action Document for Beauveria bassiana HF23*, U.S. Environmental Protection Agency, Washington, DC, 2006. http://www.epa.gov/pesticides/biopesticides/ingredients/tech_docs/brad_090305.pdf

CHAPTER 11

Clinical Toxicology of Insecticides

ALLISTER VALE,[1,2,3,*] SALLY BRADBERRY[1,2,3] AND ALEX PROUDFOOT[1]

[1] National Poisons Information Service (Birmingham Unit), City Hospital, Birmingham, UK. *Email: allistervale@npis.org; [2] West Midlands Poisons Unit, City Hospital, Birmingham, UK; [3] School of Biosciences and College of Medical and Dental Sciences, University of Birmingham, Birmingham, UK

11.1 Introduction

This chapter reviews the clinical features and management of exposure to organophosphorus and carbamate insecticides, neonicotinoids, phosphides and pyrethroids, the major insecticides responsible for pesticide-induced morbidity and mortality internationally.

11.2 Organophosphorus (OP) Insecticides

The toxicokinetic and toxicodynamic aspects of OP insecticides which are pertinent to the development of toxicity have been reviewed by us previously[1] and we have also reviewed the features and management of poisoning.[2]

11.2.1 Clinical Features

Eddleston *et al.*[3] have stressed that OP poisoning is not a single entity, as the clinical course, response to oximes and outcome vary substantially from one

Issues in Toxicology No. 12
Mammalian Toxicology of Insecticides
Edited by Timothy C. Marrs

Published by the Royal Society of Chemistry, www.rsc.org

OP to another. The clinical presentation and severity of OP poisoning is dependent not only on the pesticide and the magnitude of exposure but on factors including the route of exposure, the age of the patient, whether exposure was a suicidal attempt (when a substantial ingestion is likely) and the solvent in the formulation. Not only may skin absorption of the OP be enhanced by the solvent but ingestion of solvent may induce vomiting with risk of aspiration, pneumonia and depressed consciousness. There is increasing evidence that the solvents in formulations are responsible for the high morbidity and mortality.[4]

The onset and severity of toxicity depends on the speed and degree of depression of acetylcholinesterase (AChE) activity. In addition, some OP compounds, for example the phosphorothioates, require biotransformation to become biologically active and, as a result, signs of intoxication may be delayed. Furthermore, extensive occupational misuse of OP compounds may cause progressive depletion of AChE activity until toxic effects occur.

The typical features of OP poisoning are those of acetylcholine (ACh) excess. Symptoms can present within 5 min of massive ingestion and almost always occur within 12 h. Muscarinic features generally appear first and characterize mild to moderate poisoning but are not always present. In one study, no single feature was noted in more than 60% of cases.[5] For example, miosis, although the most prevalent specific sign, was found in only 44% of cases.[5] In other studies, miosis was observed in 82% of 61 cases[6] and in 83% of 23 cases.[7] Muscarinic symptoms rarely present more than 24 h after ingestion, although it must be remembered that they may reappear if therapy with atropine and an oxime is discontinued too early.

The first symptom of poisoning is often a feeling of exhaustion and weakness, particularly in those occupationally exposed. Vomiting, cramp-like abdominal pain, sweating and hypersalivation may follow. Miosis and tightness in the chest during inspiration may occur at an early stage but are not reliable indicators of the severity of systemic poisoning because they may be caused by local anticholinesterase effects of spray mist on the eye or bronchi.

In more severely poisoned patients, nicotinic features predominate at an early stage. Muscle twitching affects the eyelids, tongue, face and calf muscles; respiratory muscles then become involved and generalized weakness ensues. There is a decrease in dynamic lung compliance, an increase in total pulmonary resistance and an increased alveolar-arterial oxygen gradient. Convulsions may also occur.[7] Sinus tachycardia is likely to be present (although some patients may be bradycardic) and mydriasis may be observed,[8] particularly if atropine is given. Later effects may include diarrhoea, tenesmus, incontinence, ataxia and confusion. Cardiac effects include A-V block, S-T changes, peaked T waves and QT prolongation. Ventricular arrhythmias are a common cause of death[9] and tachyarrhythmias of the *torsades des pointes* type[10,11] may progress to ventricular fibrillation and/or asystole. Bronchial hypersecretion (bronchorrhoea) with bronchoconstriction, cyanosis, respiratory depression and coma supervene in severe cases and death may then follow from respiratory failure. A triad of muscarinic, nicotinic and central nervous system (CNS) symptoms will be apparent in severe cases and was found in 17% of patients in one series,[5] although in a study of severely poisoned patients requiring ventilatory support, it occurred in 60%.[12]

11.2.1.1 Intermediate Syndrome

Relapse after apparent resolution of cholinergic symptoms has been reported, particularly in patients who have ingested highly lipophilic OP insecticides, and is termed the intermediate syndrome.[13,14] Paralysis of limb muscles, neck flexors and cranial nerves develops some 24–96 h after exposure and probably represents the nicotinic effects of ACh.

11.2.1.2 Delayed Neuropathy

OP-induced delayed neuropathy is a rare complication of acute exposure to some OP insecticides.[15] It is initiated by phosphorylation and subsequent aging of at least 70% of neuropathy target esterase (NTE) in peripheral nerves, a two-step process that occurs within hours of poisoning.[16] While there is compelling evidence of the involvement of NTE in the initiation of OP-induced delayed neuropathy, its role in axonal degradation remains obscure.

Lotti and Moretto[15] have reviewed the features resulting from distal degeneration of some axons in the peripheral nervous system and CNS occurring 1–4 weeks after single or short-term exposures. Cramping muscle pain in the lower limbs, distal numbness and paraesthesiae are followed by progressive weakness, depression of deep tendon reflexes in the lower limbs and, in severe cases, in the upper limbs. Signs include a high-stepping gait from bilateral foot drop and, in severe cases, quadriplegia with foot and wrist drop as well as pyramidal signs. In time, there might be significant recovery of peripheral nerve function but, depending on the degree of pyramidal involvement, spastic ataxia may be permanent.[15]

11.2.1.3 Other Delayed Sequelae

Since acute poisoning with OP insecticides can cause convulsions, respiratory failure and cardiac arrhythmias, all of which can result in cerebral anoxia, it would be surprising if severe poisoning was not associated with long-term neurological sequelae and there is now evidence that such complications occur.[17–25] Despite some shortcomings in methodology, these studies show that severe acute OP poisoning can damage the central and peripheral nervous systems subclinically.

11.2.1.4 Neurological Effects After Low-Level Exposures

There is controversy as to whether low-level exposures to OP insecticides (those without features of acute poisoning) can result in peripheral nerve dysfunction. Lotti[26] is of the view that 'A single pattern of subclinical disturbances that relates low-level OP exposure to human peripheral neuropathy has not been identified' and gives cogent reasons for his conclusion.

11.2.2 Confirmation of the Diagnosis

The activity of two enzymes may be measured to confirm a diagnosis of OP poisoning. These are red cell AChE and plasma butyrylcholinesterase (plasma cholinesterase; plasma pseudocholinesterase). Both are surrogates for AChE activity in the CNS and peripheral nervous system. A number of factors need to be borne in mind when interpreting the results. AChE and butyryl cholinesterase are different gene products, although there are similarities in their structures.[27] The kinetics of the inhibition of the two enzymes by OPs, and the kinetics of aging and reactivation of the inhibited enzymes differ.[28] For these reasons plasma butyrylcholinesterase is a poorer surrogate than red cell AChE for the activity of AChE in nervous tissue.

11.2.3 Management

All cases of OP poisoning should be dealt with as an emergency and all patients with more than minor symptoms should be admitted to a critical care unit as quickly as possible. Ventilation must be maintained. Bronchorrhoea requires prompt relief with intravenous (IV) atropine (see below) and supplemental oxygen should be given to maintain P_aO_2 higher than 10 kPa. If these measures fail, the patient should be intubated and mechanical ventilation (with positive end-expiratory pressure) should be instituted. Careful attention must be given to fluid and electrolyte balance and adjustments to infusion fluids made as necessary. Heart rate, blood pressure, electrocardiograph (ECG) and arterial blood gases should be monitored routinely.

Once the patient is stabilized, and if dermal exposure was the major route of exposure, contaminated clothes should be removed and stored safely, without rescuers/carers contaminating themselves. Contaminated skin should be washed thoroughly with soap and cold water.

Gastric lavage may be considered in all potentially serious cases if ingestion has occurred less than 1 h previously, although its value is unproven.[29] It should be performed with care and with an endotracheal tube *in situ* if consciousness is depressed, as hydrocarbons are present in many OP insecticide formulations. Induced emesis should be avoided as it is dangerous in a patient whose level of consciousness might deteriorate; aspiration pneumonia is a well-recognized hazard and can also occur in conscious patients given emetics when hydrocarbons are present in the pesticide formulation. The capacity of activated charcoal to adsorb most OP compounds has not yet been demonstrated. In a study in Sri Lanka involving OP and carbamate insecticide poisoning, no benefit from single- or multiple-dose charcoal was demonstrated.[30]

11.2.3.1 Appropriate use of Atropine, Oxime and Diazepam

Atropine competes with ACh and other muscarinic agonists for a common binding site on the muscarinic receptor, thus effectively antagonizing their actions, including increased tracheobronchial and salivary secretions, bronchoconstriction and bradycardia. Clinically, the main benefit of oximes is to reverse cholinergic effects at peripheral nicotinic sites so that, for example, muscle strength may improve within 10–30 min. Oximes are much less effective than atropine at peripheral muscarinic sites and their effects on CNS-mediated symptoms and signs may not be clinically important. While the therapeutic combination of oxime and atropine is well established in the treatment of OP pesticide poisoning, diazepam may also be of benefit by reducing anxiety, restlessness and muscle fasciculation, suppressing seizures, and reducing morbidity and mortality when used in conjunction with pralidoxime and atropine.

The evidence base for these interventions has been reviewed recently.[31]

11.2.3.1.1 Atropine. The peripheral antimuscarinic effects of atropine may not be the only antidotal property of the drug; it may also be of value in treating the acute dystonic reactions occasionally observed.[32–35]

Atropine sulfate 2 mg IV (0.02–0.1 mg kg^{-1} in a child) should be given as soon as possible in moderately and severely poisoned adults and may be life-saving in those with rhinorrhoea and bronchorrhoea. If the initial dose produces only a partial response, it should be doubled and doubled again if there is only a limited clinical response.[36] The dose should be titrated to control rhinorrhoea and bronchorrhoea, to raise the pulse above 80 bpm, and restore systolic blood pressure to more than 80 mmHg. In severe cases, several hundred milligrams may be required, though in a study in Sri Lanka the mean dose was 23.4 mg (standard deviation 22.0, range 1–75 mg).[36]

11.2.3.1.2 Rationale for Oximes. The fundamental action of the pyridinium oximes is to reactivate AChE inhibited by OPs thus allowing ACh to be hydrolysed in the normal way and resumption of normal cholinergic neurotransmission. It is usually considered that the beneficial effects of oximes in OP poisoning are confined to peripheral nicotinic sites and that CNS effects are clinically insignificant,[37] although there is some evidence that pralidoxime can cross the blood–brain barrier.[38] This means that the beneficial effects will mainly be upon neuromuscular transmission and that there will be little action on parasympathetic effects, such as bronchorrhoea, bronchoconstriction, rhinorrhoea, or on central nervous effects.

The reactivation of inhibited AChE and clinical improvement depends on:

- The chemical form of inhibited AChE. While spontaneous reactivation of dimethyl phosphorylated AChE proceeds rapidly, in the case of diethyl

phosphorylated AChE it is very slow. Reactivation of diethyl phosphorylated AChE can be accelerated by oximes.
- The plasma concentration of the OP insecticide.[39,40]
- Aging, which results from monodealkylation of dialkoxyphosphorylated AChE;[41] aged enzyme reactivates neither spontaneously nor under the influence of oximes so that recovery depends on synthesis *de novo* of more enzyme.
- The plasma oxime concentration and duration of oxime therapy.

With the typical dimethoxy and diethoxy structure of pesticides, aging rates are such that the phenomenon only becomes a problem when treatment is delayed and/or exposure prolonged. It is commonly, but erroneously, believed that one day after intoxication with an OP insecticide virtually all the phosphorylated enzyme will be aged so that oxime therapy will be ineffective by this time. However, this interpretation derives from *in vitro* studies in which AChE is rapidly inhibited and is maintained fully inhibited by an excess of inhibitor and in the absence of oxime throughout. These experiments do not represent the case *in vivo* and should not be used as a reason to abandon the use of oxime therapy after 24 h.

De Silva *et al.*[42] performed a study in which 21 patients received atropine alone, and 24 atropine plus pralidoxime chloride. The mortality in both groups was 29%. The authors concluded that nothing is to be gained by the addition of an oxime to the standard regimen of atropine plus mechanical ventilation for severe OP insecticide poisoning.

The value of oximes has also been challenged on the basis of a systematic review of clinical trials.[43]

Eddleston *et al.*[44] performed a double-blind, randomized, placebo controlled trial of pralidoxime chloride (2 g loading dose, followed by a constant infusion of 0.5 g/h for up to 7 days) versus saline in patients with OP insecticide self-poisoning. Despite clear reactivation of red cell AChE activity in diethyl OP pesticide poisoned patients, there was no evidence that this regimen improved survival or reduced the need for intubation.[44] The reason for this failure was not apparent, though there is increasing evidence that the solvents in formulations are primarily responsible for the high morbidity and mortality, which could explain the apparent lack of benefit of oximes.[4]

Pralidoxime. Willems *et al.*[39] reported that parathion and parathion-methyl poisoning could be treated effectively with pralidoxime methylsulfate (plasma concentrations approximately 4 mg L^{-1}) and atropine provided that the pesticide concentrations were less than 30 μg L^{-1}. In more severe cases, plasma pralidoxime concentrations of 14.6 mg L^{-1} did not produce improvement. In dimethoate poisoning, omethoate-inhibited enzyme could not be reactivated at concentrations of up to 20 mg L^{-1}. Data from a case report suggests that pralidoxime concentrations greater than 40 mg L^{-1} may be required to produce reactivation of inhibited AChE.[45,46] In a further case, a continuous pralidoxime infusion (with atropine sulfate) successfully controlled the nicotinic signs and symptoms of chlorpyrifos poisoning.[47]

These studies support the recommendation that plasma oxime concentrations of far more than 4 mg L^{-1} are required to produce AChE reactivation and substantial clinical improvement in OP poisoning. Thus, the modest doses of pralidoxime that have often been recommended in the past (to achieve plasma oxime concentrations of approximately 4 mg L^{-1}) will be insufficient to produce not only reactivation of phosphorylated enzyme but also lasting clinical improvement, unless the patient is only mildly poisoned. Moreover, clinically effective pralidoxime concentrations need to be maintained as long as inhibitory oxons are circulating.

Based on clinical studies,[39,45,48] it has been recommended that pralidoxime chloride 30 mg kg^{-1} IV should be administered as soon as possible in any severe case. Repeat doses at intervals of 4–6 h or preferably an IV infusion of 8–10 mg kg^{-1} h^{-1} in an adult may then be necessary. Pralidoxime should be continued for as long as atropine is required, that is, until clear, irreversible clinical improvement is achieved, a process which may take many days while residual insecticide is cleared from body stores.

Obidoxime. Thiermann *et al.*[49] reported that in parathion poisoning, obidoxime 250 mg IV as a bolus followed by 750 mg day^{-1} by infusion was effective, but that in severe poisoning reactivation did not occur until the concentration of inhibitor in the plasma had declined. The same regimen was ineffective with oxydemeton-methyl when oxime therapy was delayed more than 1 day after poisoning. In another series, Thiermann *et al.*[50] found that in parathion poisoning, reactivation was possible 7 days after poisoning, whereas with oxydemeton-methyl, response was only seen when obidoxime therapy was instituted soon after poisoning. Similarly, Zilker *et al.*[51] reported that obidoxime 750 mg day^{-1} by infusion drastically reduced the need for atropine in parathion poisoning but that demeton-S-methyl poisoning only responded to obidoxime if therapy was instituted shortly after intoxication.

HI-6. Human experience of oximes other than pralidoxime salts and obidoxime in the management of OP pesticide poisoning is limited. An exception is a case series reported by Kusić *et al.*[52] The oxime HI-6, administered four times daily as a single intramuscular injection of 500 mg, accompanied by atropine and diazepam therapy, was studied. Oxime treatment was started on admission and continued for a minimum of 48 h and a maximum of 7 days. HI-6 rapidly reactivated human red blood cell AChE inhibited by diethoxy OPs (phorate, pyridaphenthion and quinalphos), as well as that inhibited by dichlorvos. With the exception of dichlorvos, enzyme inhibited by dimethoxy OPs (dimethoate, phosphamidon), was generally resistant to treatment by HI-6. The general clinical improvement with HI-6 which was sometimes more rapid than the rise in AChE activity, suggested that direct pharmacological effects were occurring in addition to the presumed mechanism of oxime reactivation of inhibited AChE. No adverse side effects were noted when plasma concentrations of HI-6 were maintained above therapeutic concentrations for up to 7 days.

11.2.3.1.3 Diazepam. Data from experimental nerve agent poisoning[53,54] suggest that benzodiazepines ameliorate or prevent the development of pathological changes in the CNS. From a variety of pharmacodynamic studies in experimental animals, it can be concluded that diazepam adds to the benefits of atropine and combinations of atropine and oximes.[55–61] In experimental animal models diazepam can prevent, stop and/or ameliorate convulsions due to OP insecticide poisoning and render less severe or even prevent structural changes in the brain. However, the effects on lethality are less clear.[62]

There are many case reports and series of apparently successful treatment of OP insecticide poisoning with diazepam as adjunctive therapy.[41,52,63–71] In most, the indication for the use of diazepam was convulsions, but control of muscle fasciculation[64] and agitation have also been reasons. In the case of convulsions, the adult dose is 10–20 mg IV, while that in children is 0.2–0.3 mg kg^{-1} IV. If large doses of diazepam are required to suppress seizure activity, phenytoin should be considered as an alternative.[72]

In the absence of convulsions, diazepam administration at doses of 5–10 mg IV has also been recommended for poisoning accompanied by anxiety and restlessness.[72]

11.3 Carbamate Insecticides

11.3.1 Clinical Features

Although carbamate insecticide poisoning is generally less severe than OP insecticide poisoning, carbamates should not be considered toxicologically benign as acute poisoning with them can be severe and even fatal.[73–77]

Systemic toxicity may occur following ingestion, inhalation or skin contamination; transplacental transfer has been reported with propoxur.[78] Most cases of severe carbamate poisoning occur following ingestion, which may be accidental[79,80] or deliberate.[76,79,81,82]

11.3.1.1 Ingestion

Cholinergic symptoms usually develop within a few minutes[82,83] and in the most severe cases muscle twitching, profound weakness, profuse sweating, incontinence, mental confusion and progressive cardiac and respiratory failure ensue.[80,82,84] In less severe cases, cholinergic symptoms are usually evident within 2 h and typically resolve within 24 h.[85] There is some evidence that CNS depression in moderate carbamate poisoning is more common in young children than in adults, possibly due to a greater blood–brain barrier permeability in infants.[85] Coma may persist for 18–24 h.[82]

Seizures[83] are relatively uncommon as a primary complication in severe carbamate poisoning since carbamate penetration into the CNS is limited, but they may occur secondary to hypoxia. Yang *et al.*[86] have reported a patient who developed a sensorimotor neuropathy following ingestion of carbofuran. Recovery commenced after 1 week and continued for 4 months. Pontine myelinolysis, confirmed by MRI, has been described in a child poisoned with carbaryl; it improved after 4 months and disappeared after 2 years.[87]

There has been only one report of the intermediate syndrome following carbamate (carbofuran) poisoning in a patient who abused alcohol chronically.[88] The diagnosis was based on the development of dyspnoea (respiratory rate of 40/minute; fall in oxygen saturation) and weakness of neck flexors in the absence of signs of pulmonary infection. As there was no analytical confirmation of the diagnosis it cannot be certain that the patient was poisoned with a carbamate.

The effect on heart rate is unpredictable and tachycardia occurs as commonly as the bradycardia that one would expect from the mechanism of action of carbamates. Arrhythmias, including *torsades des pointes*, are recognized.[9] Complete heart block has been reported following carbaryl.[89]

Death has occurred within a few hours in untreated cases[90] and is usually due to respiratory failure;[80,91] pulmonary oedema and evidence of hypoxic brain damage are the main findings at autopsy.[73,74,80,82,92] Organs have been transplanted from patients who died from carbamate insecticide poisoning.[93]

11.3.1.2 Dermal Exposure and Inhalation

Exposures in pesticide applicators are typically inhalational and/or dermal. Life-threatening poisoning in these circumstances is rare, though cholinergic symptoms have occurred following equipment malfunction or failure to observe adequate protective measures.[94–96] In these cases the rapid (often within 30 min) onset of features including nausea, headache, sweating, lacrimation, salivation, chest tightness, coughing, bronchorrhoea and blurred vision caused by constriction of one or both pupils, serves to alert affected individuals to remove themselves immediately from exposure.[94,95,97] Resolution of symptoms occurs typically within hours,[97] often without the need for atropinization.[95] If exposure continues, vomiting, diarrhoea, abdominal pain, muscle fasciculation and weakness develop.[95] Hypotension, sinus tachycardia or bradycardia and dyspnoea may also occur.[95] T-wave changes including inversion are recognized.[98,99] Stevens–Johnson syndrome has been reported.[99]

Severe sequelae following occupational carbamate exposure are rare, have often involved methomyl, and include several cases resulting from pilots of crop-spraying aircraft being exposed to spray that gained entry into the cockpit.[100–102] One such case[101] proved fatal as a result of the pilot crashing the plane.

11.3.1.3 Delayed Neurological Sequelae

Carbamate poisoning is generally held not to be associated with delayed polyneuropathy such as is seen with some OP insecticides, but three cases have been reported involving ingestion of carbaryl,[103] metolcarb (*m*-tolyl methylcarbamate)[104] and carbofuran.[86] The first developed a sensorimotor neuropathy;[103] the second had loss of large myelinated fibres and axonal degeneration on sural nerve biopsy.[104] The third patient[86] developed arm and foot sensory symptoms and muscle weakness on the sixth day after hospitalization; the sensory component resolved rapidly but motor features and associated abnormal electrophysiological studies persisted for several months. Lotti and Moretto[105] have concluded that high doses of carbamate may cause a sensorimotor polyneuropathy in humans similar to OP-induced delayed polyneuropathy. Moreover, carbamates may exacerbate polyneuropathy initiated by prior dosing with an OP insecticide.

Peripheral neuropathy was also seen in three out of 36 patients aged 14–70 years poisoned with carbamates, 80% of which involved carbofuran.[106] Though it is not clear whether this finding was carbamate-induced or predated the poisoning, other common causes of neuropathy, especially ethanol were carefully excluded leaving carbamates as the probable explanation.[106]

11.3.2 Analytical Confirmation of the Diagnosis

Most laboratories are not acquainted with the special procedure required to assay red blood cell cholinesterase activity in the presence of carbamates. Samples must be kept on ice or frozen at 20 °C during transportation to the laboratory prior to analysis. Spontaneous reversal of the inhibition is rapid and is accelerated by the time interval between sampling and analysis, the dilution of the sample, the addition of substrate (usually acetylcholine at high concentration which competes successfully for the enzymatic active site) and the duration of the assay. The latter should take less than three minutes and employ minimal dilution and minimal amounts of substrate.[107] Although the red cell cholinesterase activity correlates well with the severity of intoxication, its measurement adds little to management as the majority of patients recover rapidly.

Quantitative analysis of carbamate concentrations in body fluids is available only in specialist centres. Where blood carbamate concentrations have been measured, they correlate closely to the clinical picture.[84] Measurement of carbamate concentration may also prove useful in forensic practice where the cause of death is not known.[76]

11.3.3 Management

All cases of carbamate poisoning should be dealt with as an emergency and the patient admitted to hospital as quickly as possible. Severe cholinergic features necessitate supportive care, often in the setting of a critical care unit,[75,108] and the administration of atropine if significant morbidity and mortality is to be prevented.

Gastric lavage may be considered in all potentially serious cases if ingestion has occurred less than 1 h previously, although its value is unproven. Lavage should be performed with a cuffed endotracheal tube *in situ* if consciousness is depressed, as hydrocarbon solvents are often present in the formulations. Alternatively, the administration of oral activated charcoal, 50–100 g for an adult, may be considered in potentially serious cases if ingestion has occurred less than 1 h previously, though this again is of unproven value.

Bronchorrhoea requires removal of secretions by suction and prompt relief with intravenous atropine (see below). Supplemental oxygen should be given to maintain P_aO_2 greater than 10 kPa (75 mmHg). If these measures fail the patient should be intubated and mechanical ventilation instituted. Heart rate, blood pressure, ECG and arterial oxygen saturation or blood gas tensions should be monitored.

11.3.3.1 Atropine

Atropine antagonizes the effects of accumulated acetylcholine at muscarinic receptors. A dose of 2 mg IV (0.02 mg kg^{-1} bodyweight for a child) should be given as soon as possible in moderately or severely poisoned patients, particularly in those who develop rhinorrhoea and bronchorrhoea. Repeated injections will be required over the first few hours in those severely poisoned. If the initial dose produces only a partial response, it should be doubled and doubled again if there is only a limited clinical response. The dose should be titrated to control rhinorrhoea and bronchorrhoea, to raise the pulse above 80 bpm, and to restore systolic blood pressure to more than 80 mmHg.

11.3.3.2 Pralidoxime

Pralidoxime has been demonstrated in some animal studies[109–113] to increase the toxicity of the carbamate insecticide carbaryl. In contrast, Mecurio-Zappala *et al.*[114] showed that it protected against carbaryl poisoning in mice. Additionally, some human case reports have demonstrated the benefit of pralidoxime in carbamate poisoning,[74,115–118] though there has been at least one death despite the use of pralidoxime.[119] At present there is insufficient evidence to either recommend or prohibit the use of pralidoxime in severe poisoning with carbamate insecticides, though it should seldom need to be considered since carbamates have a shorter duration of action than OP insecticides. Pralidoxime chloride 30 mg kg^{-1} bodyweight by IV injection over 5–10 min, repeated 4–6 hourly, should be considered in cases of life-threatening intoxication unresponsive to atropine alone and supportive measures. Cases of mixed OP and carbamate poisoning should be treated as for OP poisoning (see above).

11.4 Neonicotinoid Insecticides

11.4.1 Clinical Features

Poisoning with neonicotinoids is characterized by the rapid onset of symptoms including nausea and vomiting[120–125] (occasionally haematemesis[120]), fever,[126–128] sweating,[120,123,129] increased salivation,[121,122,130] bronchorrhoea,[121,122,129] agitation,[124,129,131] lack of coordination,[121,122] disorientation,[126,129] muscle weakness,[125] seizures[125] and coma.[120–124,126–128,130] Breathlessness,[120–123,125,129] depressed respiration, cyanosis[123,127,129] and respiratory arrest[120,123,124,130,131] have been reported. Bradycardia[121,122,132] is sometimes present, but tachycardia[120,123,125,127,128,130,131] is observed more frequently; ventricular tachycardia/fibrillation has been reported occasionally,[123,127] as has cardiac arrest.[120,127] Severe hypotension and shock have occurred in those severely poisoned.[123–125,127,132] Miosis[121,122,124] was present in some cases but not the majority. Metabolic acidosis[132] and renal failure have been observed.[127,132]

Mohamed *et al.*[133] have analysed the patients reported by Phua *et al.*[120] and themselves[124] and some of the case reports published up to 2009. The median volume of imidacloprid ingested by 33 of 56 patients in their own study[124] was 15 mL (interquartile range (IQR) = 10–15 mL). In the severe group reported by Phua *et al.*[120] it was 75 mL (range 30–200 mL) and in the case reports, 150 mL (IQR = 65–250 mL). Thus, in interpreting the clinical reports published to date, the amount of the formulation ingested must be taken into account. In those ingesting substantial amounts, respiratory arrest and coma were observed frequently, though the solvent (*N*-methylpyrrolidone) present in many formulations may have contributed to the deaths of eight of these patients.

Mohamed *et al.*[124] collected data prospectively in 68 patients poisoned with imidacloprid and admitted to 3 hospitals in Sri Lanka; 61 cases followed ingestion and 7 were due to occupational exposure. Five reported co-ingestion with another pesticide and were excluded, thus leaving 56 patients for analysis. Imidacloprid ingestion was confirmed in 38 by HPLC/MSMS. All 7 patients with occupational exposure did not require specific treatment and were discharged within 24 h. In 23 of the 56 non-occupational cases, the amount ingested was unknown; in the remaining 35 the median amount was 15 mL (IQR 10–50) and the median time from ingestion to presentation at hospital was 4 h (IQR 2.3–6.0), at which time the median Glasgow Coma Score (GCS) was 15 (IQR 10–15). Fifty-four of the 56 patients developed only one of the following symptoms: nausea or vomiting, abdominal pain, diarrhoea, headache or dizziness.

Of the 70 patients reported by Phua *et al.*,[120] two were children who unintentionally ingested small amounts of imidacloprid. Forty-six ingested a neonicotinoid alone, 8 drank alcohol as well, 14 were co-exposed to other pesticides and 2 to other drugs and chemicals. Thirteen of the 70 patients were exposed by inhalation, dermally or a combination of both routes. Most patients ($n = 62$) had only mild–moderate symptoms while 8 developed major effects (*e.g.* respiratory failure and coma) and two died. All of the 10 patients with severe/fatal poisoning had ingested imidacloprid alone and were older than those less severely poisoned. Six of them developed respiratory failure and five aspiration pneumonia.

11.4.2 Management

The poisonings most likely to be associated with cholinergic features are those due to OP and carbamate insecticides (see above). It is therefore advisable to urgently measure erythrocyte AChE activity to exclude these diagnoses.

Gastric lavage may be considered if the patient presents within 1 h of ingestion of a substantial quantity of neonicotinoid, but there is no evidence that it alters the course of poisoning. Oral activated charcoal, 50–100 g for an adult, may be considered in those presenting within the same period as activated charcoal is known to bind imidacloprid *in vitro*.[134]

In patients who are unconscious a clear airway should be established and, if ventilation is impaired, assisted ventilation should be commenced. Hypotension and cardiac dysrrhythmias should be managed conventionally and acid–base and electrolyte balance corrected.

Since the clinically important features of neonicotinoid poisoning, notably bronchorrhoea, are caused by cholinergic overactivity, atropine is the logical antidote. Atropine sulfate 2 mg IV (in an adult) should be given and the dose repeated until the signs of atropinization are present (dry skin and sinus tachycardia).

11.5 Phosphides and Phosphine

Aluminium and zinc (and less commonly magnesium) phosphides are highly effective insecticides that are used widely to protect bulk commodities held in stores, the holds of ships, international shipping containers and railway wagons. They are formulated as tablets and pellets that usually also contain ammonium carbonate or carbamate to reduce the risk of spontaneous ignition. Phosphides interact with moisture in air (or with water or acid) to liberate phosphine, which diffuses readily and unaided and is the active pesticide. Inhalation of phosphine, however, is a much less frequent cause of human poisoning than ingestion of a phosphide, though the toxicity by the oral route is also due to phosphine liberated by contact of the salt with gut fluids. It is then absorbed through the alimentary mucosa and distributed to tissues where it depresses mitochondrial respiration by inhibiting cytochrome *c* oxidase,[135,136] succinate dehydrogenase and NADH dehydrogenase.

The toxicology of aluminium and zinc phosphides has been reviewed by us recently.[137]

11.5.1 Clinical Features of Phosphide Ingestion

The early features of phosphide ingestion include nausea, vomiting, retrosternal and epigastric pain and dyspnoea. Diarrhoea is less common and patients may be anxious and agitated while mentally clear. There is often a smell of garlic on the breath due to an impurity in many formulations. Aluminium phosphide tablets are some 18–20 mm in diameter and, if swallowed with an insufficient volume of fluid, may stick in the oesophagus causing release of phosphine and an exothermic reaction that results in mucosal damage.[138] Corrosive lesions of the stomach[139–141] have also been reported and haematemesis has occurred.[142–144]

Shock and peripheral circulatory failure develop rapidly in severe poisoning and appear to be due to cardiac congestion as a result of separation and fragmentation of myocardial fibres by oedema with focal necrosis and inflammatory cell infiltration. Creatine kinase MB activity may be increased. Echocardiography,[145] radionuclide ventriculography[146] and invasive haemodynamic investigations[147] frequently reveal hypokinesia causing severe hypotension and reduced cardiac output. Hypotension may be exacerbated by both tachycardia and bradycardia; arrhythmias such as supraventricular and ventricular tachycardias[148] and, rarely, ECG changes resembling those of myocardial ischaemia[149] or infarction have been reported.[150–153] Less serious ECG abnormalities including ST and T-wave changes are common and were present in 38% of 418 aluminium phosphide ingestions[144] and in 66% of 471 cases in another study.[154]

The lungs are also commonly damaged and pulmonary oedema results; a raised respiratory rate, breathlessness, crackles and wheeze are present on examination. The oedema fluid may be protein-rich and haemorrhagic.[155] It is not always clear whether the oedema is cardiogenic or non-cardiogenic in aetiology. On occasions adult respiratory distress syndrome has been diagnosed.[146,156–159] Whatever its true nature, pulmonary oedema tends to develop 4–48 h after ingestion.

Other complications of phosphide ingestion include atrial infarction,[160] subendocardial infarction,[161] acute pericarditis,[162,163] pleural effusion,[159,164–166] ascites,[159,165] skeletal muscle damage,[167] rhabdomyolysis,[168,169] malignant hyperthermia,[169] a bleeding diathesis,[170] adrenocortical congestion, haemorrhage and necrosis,[141] and acute pancreatitis.[171,172]

Intravascular haemolysis after ingestion of aluminium phosphide has been described[144] and a mean methaemoglobin concentration of 26.5 ± 19.9% on the second day of poisoning in non-survivors has been reported.[173] The mechanism by which these occur is uncertain, though phosphine is known to cause Heinz body formation and oxidize haemoglobin *in vitro*.[174,175] Oxygen free radical generation may be involved.[176] These, together with much more common factors such as refractory hypotension and hypoxaemia lead to acute renal failure[143,144,155,159,165] but the mortality from phosphides in the first 24 h after ingestion is so high that few patients survive to require renal replacement therapy.

A wide variety of laboratory abnormalities also complicates phosphide ingestion. Elevations of alanine and aspartate aminotransferase activities are not infrequent;[145,159,165,177–179] jaundice is much less common but was present in 12 out of 92 cases in one series.[157] Metabolic acidosis, hypokalaemia and hypocalcaemia[171] have been observed.[171] Hyperglycaemia is not uncommon[180] but hypoglycaemia[181] is very unusual. The most controversial issue is whether or not magnesium homeostasis is disturbed.[137]

The tragically small percentage of patients who survive phosphide ingestion usually recover completely. Oesophageal injury is the only cause of medium- and long-term complications and a considerable number of patients have developed oesophageal strictures,[138,142,182–195] the prevalence being 2% (7 out of 342) in one study[192] and 33% in another.[193]

11.5.2 Management of Phosphide Ingestion

Many patients poisoned with phosphide are likely to die despite optimum care. A low GCS, systolic hypotension, hyperglycaemia and acidosis have all been proposed as adverse prognostic factors.[196,197]

Gut decontamination is probably best avoided after ingestion as it might only increase the rate of disintegration of the pesticide and increase toxicity. The complete correction of metabolic acidosis appeared to improve prognosis in one series but the number of patients studied was small.[198] Shock frequently proves refractory to treatment with high-dose inotropes.[166,199] Intra-aortic balloon counterpulsation was ineffectual in one case[199] but successful in another.[166] The blood glucose concentration should be measured in every case and hypoglycaemia corrected if found. Similarly, hypokalaemia should be sought and supplements given if indicated. Currently, the evidence that magnesium administration is beneficial is not encouraging.[137]

11.5.3 Inhalation of Phosphine

Pure phosphine is a colourless and odourless gas up to toxic concentrations (200 ppm).[200–203] The literature on its inhalation has been reviewed recently.[204] Although it has been speculated that children are particularly vulnerable to inhaled phosphine,[205] adults are undoubtedly also at risk.

The initial features of phosphine intoxication are alimentary rather than respiratory with nausea, vomiting and diarrhoea sometimes so striking that acute gastroenteritis has been diagnosed. However, chest tightness, breathlessness and chest pain or soreness were reported by 52%, 34% and 29% of 67 men exposed during fumigation of bulk wheat with aluminium phosphide.[206] Consciousness is usually only mildly depressed.

Phosphine is cardiotoxic causing focal myocarditis and impairing myocardial contraction.[207] Acute heart failure and pulmonary oedema were considered to be the causes of the deaths of five girls aged 6–16 years with no history of ill-health but accidentally exposed to phosphine from aluminium phosphide in their homes or in the vicinity of grain stores. They all died suddenly in relationship to walking, running or bathing.[205] Similarly, sisters aged 6 and 9 years died from shock, respiratory distress and ventricular tachyarrhythmias allegedly due to phosphine arising from misuse of aluminium phosphide in the agricultural warehouse adjacent to their home. Myocardial congestion with interstitial oedema and pulmonary oedema were found at autopsy.[208] Myocardial depression, sometimes severe, has also featured in adults who have inhaled phosphine[209,210] and the resulting hypotension may result in acidaemia and hyperlactataemia.[210] Ventricular premature beats (>600/h) and bigeminy were present in one young man[211] and atrial fibrillation in another[212] and other complications include pericardial effusion,[210] hypomagnesaemia[211] and hyperglycaemia.[208]

11.5.4 Management of Phosphine Inhalation

Removal of victims of phosphine inhalation from the contaminated atmosphere is clearly the first priority. Oxygen may be given if necessary but further measures for airway control are unlikely to be required. Supportive measures should be implemented as required by clinical developments.

11.6 Pyrethroids

The clinical features and management have been reviewed by us previously.[213] The skin is the main route of occupational pyrethroid absorption, particularly when liquid formulations are used.[214] Inhalation is much less important[215–217] but becomes increasingly so when the pesticides are used in confined spaces and when employing products presented as dusts, powders, aerosols, sprays and total release foggers.[214] Ingestion of pyrethroids is occasionally accidental but much more commonly deliberate.

11.6.1 Clinical Features

11.6.1.1 Dermal Exposure

The main adverse effect of dermal exposure to pyrethroids is paraesthesiae presumed due to hyperactivity of cutaneous sensory nerve fibres. These occur at doses far lower than those required to cause systemic symptoms,[218] suggesting that concentrations of the pesticide at skin nerve terminals are higher with dermal than oral exposure. They generally start 0.5–2 h after exposure, peak at about 6 h and, resolve completely within 24 h.[219–223] The face is affected most commonly[224] and the paraesthesiae are exacerbated by sensory stimuli such as heat, sunlight and scratching,[220] sweating or application of water.[221] In two studies, they were more severe after deltamethrin and flucythrinate exposure, less after cypermethrin and fenvalerate, and least after permethrin.[220,225]

Pyrethroids also cause local skin irritation in up to 10% of workers.[226] Pruritus, tingling or pricking are the most common complaints though erythema, burning and blisters have been reported occasionally.[227–231] When used at recommended doses in the treatment of scabies and lice, however, pyrethroids appear to only rarely produce irritation.[227,232] It is possible, however, that the low frequency is due to the adverse effects being indistinguishable from the symptoms of the infestation.

Allergic reactions to pyrethroids are equally uncommon though there are isolated reports of contact sensitivity[233] and eczema.[234] In one case, possible exposure to spilt cypermethrin resulted in generalized urticaria the following day.[235]

Nineteen individuals who spent a week planting conifer seedlings that had been treated with cypermethrin did not develop adverse health effects despite urine concentrations of 3-phenoxybenzoic acid increasing.[236]

Since some permethrin formulations for medical use contain petroleum distillates there is a risk of thermal injury. A 10-year-old girl who applied phenothrin-containing lotion to her head to treat lice sustained partial thickness burns of her face and neck when her hair caught fire as she warmed herself.[237]

Systemic pyrethroid toxicity has been reported rarely after dermal exposure and generally occurs after inappropriate occupational handling, particularly of too concentrated solutions. A 23-year-old woman who had 'heavy' dermal exposure to deltamethrin for 3 days had headache, nausea, dizziness, anorexia, fatigue and muscle fasciculation. High amplitude, transient, slow and sharp waves were seen on her electroencephalogram (EEG) though she did not have seizures.[219] Acute polyarthralgia has also been observed after skin exposure to flumethrin[238] and a young man who applied some 1.5 g permethrin cream from neck to toe and did not wash for about 10 h developed severe torticollis which persisted unchanged for 24 h. No cause other than the pesticide was identified.[239]

11.6.1.2 Inhalational Exposure

Pyrethroids are not volatile but are inhaled as dust or aerosol droplets rather than vapour.[240] Nasal and respiratory tract irritation result and systemic effects may follow substantial exposures,[219] generally after inappropriate occupational handling such as using too concentrated solutions, prolonged exposure and spraying against the wind.[219] Fenvalerate, deltamethrin and cypermethrin are most commonly involved with systemic features starting 4–48 h after spraying.[219]

Of 12 workers who sprayed lambda-cyhalothrin indoors, 11 complained of nasal irritation and 6 of throat irritation,[241] while plant nursery workers treating conifer seedlings with pyrethroids experienced coughing, dyspnoea, increased nasal secretions and sneezing.[226] Cypermethrin, inadvertently introduced into the air-conditioning ducts of an office building, induced wheezing and shortness of breath which persisted for 7 months in three smokers and one non-smoker, none of whom had a history of such symptoms.[242] Similarly, persistent upper respiratory irritation compelled the parents of four children to evacuate the family home after it was oversprayed from an aircraft with granules of permethrin in kaolin. The symptoms recurred each time they attempted to re-occupy their home and 5 days after exposure the three oldest children developed urticaria.[243] The temporal relationship between illness and exposure strongly supports a causal link but it cannot be considered proven.

11.6.1.3 Ingestion

A strong chemical odour may emanate from those who have ingested pyrethroids[244] and the features that follow strongly indicate that the pesticides cause considerable direct irritation of the gastrointestinal tract. Often the throat becomes sore within minutes,[219,245] though in one case the onset of a buccal burning sensation was delayed for several hours after permethrin/pyrethrins were accidentally sprayed directly into the mouth.[246] Increased salivation, mouth ulceration and/or dysphagia may follow.[245] Erythema and swelling of the lips and buccal mucosa have also been features.[247] Nausea and vomiting occur in up to 57% of cases.[219,245,248]

One man who consumed an agricultural formulation of deltamethrin was found at autopsy to have a dark green substance coating his upper gastrointestinal mucosa; its removal revealed severe congestion and extravasation of blood, particularly in the stomach.[244] Similarly, oesophagitis, oesophageal ulceration, gastritis, gastric ulceration and, in one case, duodenal ulceration were present at endoscopy in five patients with permethrin, xylene and surfactant ingestion.[245] In addition, the Chinese literature includes a case of erosive gastritis with haematemesis following ingestion of 900 mL of a deltamethrin solution of unknown concentration.[249,250] Diarrhoea is a less frequent symptom but one man defecated loose, dark green material after consuming an agricultural formulation of deltamethrin[244] and another had diarrhoea that smelt of the pesticide.[251]

Dizziness, headache and fatigue are not uncommon, but palpitations, chest tightness and blurred vision are even less frequent.[219]

Coma[219,252–254] and convulsions[219,252–254] are the principal life-threatening features and may develop as soon as 20 min after substantial ingestions[219,254] and persist for days or weeks.[219,255] Consciousness was impaired in 16 of 48 patients who ingested a permethrin/xylene/surfactant mixture[245] and two brothers became unconscious and suffered convulsions following ingestion of one allethrin mosquito coil each; consciousness was regained after 36 h.[252] Continuous tonic-clonic seizures were also features of deltamethrin ingestion by a 16-year-old girl[256] and a young woman who consumed it in an anti-lice medication.[257] A combination of lambda-cyhalothrin and parathion-methyl induced convulsions in eight adults aged 38–62 years 4–12 h after ingestion.[258] Occasionally, the EEG may be abnormal.[219] Cerebral oedema was detected by CT scan in an adolescent who developed status epilepticus after ingesting deltamethrin[256] and was present at autopsy in an adult who had no seizures.[244]

Six of the eight patients in the series of combined lambda-cyhalothrin/parathion-methyl ingestions manifested choreoathetosis preceding coma and seizures and the pyrethroid was considered responsible rather than the OP insecticide, which we consider is improbable as OPs are known to cause choreoalthetosis.[32,258–261] Muscle fasciculation is also a feature of pyrethroid toxicity[262] and in such cases electromyography (EMG) may be abnormal.[219] Repetitive muscle discharges without denervation potentials were found in one case.[219] Less serious intoxications may feature muscle cramps, paraesthesiae, loss of sensation in the limbs, breathlessness and hypersalivation.[262,263]

The ECG changes reported in acute pyrethroid poisoning include ST and T-wave changes, sinus tachycardia, ventricular ectopics and (rarely) sinus bradycardia[219,245] that resolved over 2–14 days. Tefluthrin (a type I pyrethroid) prolonged mean action potential duration at 90% repolarization apparently due to persistent sodium influx and activated early after-depolarizations in isolated rat ventricular myocytes.[264]

A 45-year-old man died 3 h after eating beans and cheese prepared using cypermethrin 10% instead of oil. He developed symptoms within minutes and died from respiratory arrest after prolonged vomiting, abdominal colic, tenesmus, diarrhoea, convulsions and coma. Cypermethrin 0.7 g was found in his stomach at autopsy. Severe symptoms were also experienced by his family.[254]

Abnormal laboratory findings in 48 patients who ingested a formulation containing permethrin, xylene and a surfactant (5–650 mL) included leucocytosis in 16, metabolic acidosis in 5, increased creatinine concentrations in 3 and increased aspartate transaminase (AST) activity in three.[245] Raised transaminase activities were also found in 20 patients acutely poisoned with deltamethrin.[265]

11.6.2 Management

Following dermal exposure, pyrethroid-induced paraesthesiae usually resolve in 12–24 h, they do not generally require treatment. However, topical application of *dl*-α-tocopherol acetate (vitamin E) reduced the severity of skin reactions to fenvalerate,[221,266] flucythrinate, permethrin and cypermethrin.[225] In a controlled volunteer study, a commercial vitamin E oil preparation inhibited the cutaneous symptoms from fenvalerate by 98% when applied immediately after exposure;[267] at 4 h inhibition was only 50%.[268] The mechanism of this effect has not been clarified but may involve blockade of the sodium tail current.[269]

Systemic toxicity following pyrethroid exposure is very uncommon and most patients do not need treatment or require only symptomatic and supportive measures.

References

1. J. A. Vale, Toxicokinetic and toxicodynamic aspects of organophosphorus (OP) insecticide poisoning, *Toxicol. Lett.*, 1998, **102–103**, 649–52.
2. T. C. Marrs and J. A. Vale, Management of organophosphorus pesticide poisoning, in *Toxicology of Organophosphate and Carbamate Compounds*, ed. R. C. Gupta, Academic Press, Amsterdam, 2006, pp. 715–33.
3. M. Eddleston, P. Eyer, F. Worek, F. Mohamed, L. Senarathna, L. Von Meyer, E. Juszczak, A. Hittarage, S. Azhar, W. Dissanayake, M. H. R. Sheriff, L. Szinicz, A. H. Dawson and N. A. Buckley, Differences between organophosphorus insecticides in human self-poisoning: a prospective cohort study, *Lancet*, 2005, **366**, 1452–59.
4. M. Eddleston, J. B. Harris, I. Self, F. Worek, H. Thierman, A. J. Simpson and R. E. Clutton, The major solvent in agricultural dimethoate preparations is essential for toxicity in minipigs, *Clin. Toxicol.*, 2010, **48**, 297.

5. A. Hirschberg and Y. Lerman, Clinical problems in organophosphate insecticide poisoning the use of a computerised information system, *Fundam. Appl. Toxicol.*, 1984, **4**, S209–214.
6. P. G. Bardin, S. F. Van Eeden and J. R. Joubert, Intensive care management of acute organophosphate poisoning. A 7-year experience in the Western Cape, *S. Afr. Med. J.*, 1987, **72**, 593–97.
7. P. Lee and D. Y. H. Tai, Clinical features of patients with acute organophosphate poisoning requiring intensive care, *Intensive Care Med.*, 2001, **27**, 694–99.
8. D. W. Wyckoff, J. E. Davies, A. Barquet and J. H. Davies, Diagnostic and therapeutic problems of parathion poisonings, *Ann. Intern. Med.*, 1968, **68**, 875–82.
9. A. M. Saadeh, N. A. Farsakh and M. K. Al-Ali, Cardiac manifestations of acute carbamate and organophosphate poisoning, *Heart*, 1997, **77**, 461–64.
10. Z. Kiss and T. Fazekas, Organophosphates and *torsade de pointes* ventricular tachycardia, *J. R. Soc. Med.*, 1983, **76**, 984–85.
11. A. Ludomirsky, H. O. Klein, P. Sarelli, B. Becker, S. Hoffman, U. Taitelman, J. Barzilai, R. Lang, D. David, E. Disengni and E. Kaplinsky, Q-T Prolongation and polymorphous ('torsade de pointes') ventricular arrhythmias associated with organophosphorus insecticide poisoning, *Am. J. Cardiol.*, 1982, **49**, 1654–58.
12. Y. Finkelstein, A. Kushnir, B. Raikhlin-Eisenkraft and U. Taitelman, Antidotal therapy of severe acute organophosphate poisoning: a multihospital study, *Neurotoxicol. Teratol.*, 1989, **11**, 593–96.
13. N. Senanayake and L. Karalliedde, Neurotoxic effects of organophosphorus insecticides: an intermediate syndrome, *N. Engl. J. Med.*, 1987, **316**, 761–63.
14. P. Jayawardane, A. Dawson, N. Senanayake and V. Weerasinghe, Serial neurophysiological studies in 70 patients with organophosphate poisoning: early prediction of intermediate syndrome, *Clin. Toxicol.*, 2006, **44**, 729.
15. M. Lotti and A. Moretto, Organophosphate-induced delayed polyneuropathy, *Toxicol. Rev.*, 2005, **24**, 37–49.
16. M. K. Johnson, Organophosphates and delayed neuropathy—is NTE alive and well?, *Toxicol. Appl. Pharmacol.*, 1990, **102**, 385–99.
17. F. Sánchez-Santed, F. Cañadas, P. Flores, M. López-Grancha and D. Cardona, Long-term functional neurotoxicity of paraoxon and chlorpyrifos: behavioural and pharmacological evidence, *Neurotoxicol. Teratol.*, 2004, **26**, 305–17.
18. L. Rosenstock, M. Keifer, W. E. Daniell, R. McConnell and K. Claypoole, Pesticide Health Effects Study Group, Chronic central nervous system effects of acute organophosphate pesticide intoxication, *Lancet*, 1991, **338**, 223–27.
19. R. McConnell, M. Keifer and L. Rosenstock, Elevated quantitative vibrotactile threshold among workers previously poisoned with methamidophos and other organophosphate pesticides, *Am. J. Ind. Med.*, 1994, **25**, 325–34.

20. K. Steenland, B. Jenkins, R. G. Ames, M. O'Malley, D. Chrislip and J. Russo, Chronic neurological sequelae to organophosphate pesticide poisoning, *Am. J. Public Health*, 1994, **84**, 731–36.
21. E. P. Savage, T. J. Keefe, L. M. Mounce, R. K. Heaton, J. A. Lewis and P. J. Burcar, Chronic neurological sequelae of acute organophosphate pesticide poisoning, *Arch. Environ. Health*, 1988, **43**, 38–45.
22. J. Miranda, R. McConnell, E. Delgado, R. Cuadra, M. Keifer, C. Wesseling, E. Torres and I. Lundberg, Tactile vibration thresholds after acute poisonings with organophosphate insecticides, *Int. J. Occup. Environ. Health*, 2002, **8**, 212–19.
23. J. Miranda, I. Lundberg, R. McConnell, E. Delgado, R. Cuadra, E. Torres, C. Wesseling and M. Keifer, Onset of grip- and pinch-strength impairment after acute poisonings with organophosphate insecticides, *Int. J. Occup. Environ. Health*, 2002, **8**, 19–26.
24. J. Miranda, R. McConnell, C. Wesseling, R. Cuadra, E. Delgado, E. Torres, M. Keifer and I. Lundberg, Muscular strength and vibration thresholds during two years after acute poisoning with organophosphate insecticides, *Occup. Environ. Med.*, 2004, **61**, e4.
25. E. Delgado, R. McConnell, J. Miranda, M. Keifer, I. Lundberg, T. Partanen and C. Wesseling, Central nervous system effects of acute organophosphate poisoning in a two-year follow-up, *Scand. J. Work Environ. Health*, 2004, **30**, 362–70.
26. M. Lotti, Low-level exposures to organophosphorus esters and peripheral nerve function, *Muscle Nerve*, 2002, **25**, 492–504.
27. S. Darvesh, D. A. Hopkins and C. Geula, Neurobiology of butyrylcholinesterase, *Nat. Rev. Neurosci.*, 2003, **4**, 131–38.
28. B. H. Wilson, M. J. Hooper, M. E. Hansen and P. S. Nieberg, Reactivation of organophosphorus inhibited AChE with oximes, in *Organophosphates—Chemistry, Fate and Effects*, ed. J. E. Chambers and P. E. Levi, Academic Press, San Diego, CA, 1992, pp. 107–37.
29. Y. Li, M. L. Tse, I. Gawarammana, N. Buckley and M. Eddleston, Systematic review of controlled clinical trials of gastric lavage in acute organophosphorus pesticide poisoning, *Clin. Toxicol.*, 2009, **47**, 179–92.
30. M. Eddleston, E. Juszczak, N. A. Buckley, L. Senarathna, F. Mohamed, W. Dissanayake, A. Hittarage, S. Azher, K. Jeganathan, S. Jayamanne, M. H. R. Sheriff and D. A. Warrell, Ox-Col Poisoning Study collaborators, Multiple-dose activated charcoal in acute self-poisoning: a randomised controlled trial, *Lancet*, 2008, **371**, 583–39.
31. P. G. Blain, Organophosphorus poisoning (acute), *Clin. Evidence*, 2011, **2011**, pii: 2102.
32. J. Joubert, P. H. Joubert, M. van der Spuy and E. van Graan, Acute organophosphate poisoning presenting with choreoathetosis, *J. Toxicol. Clin. Toxicol.*, 1984, **22**, 187–91.
33. J. Joubert and P. H. Joubert, Chorea and psychiatric changes in organophosphate poisoning, *S. Afr. Med. J.*, 1988, **74**, 32–34.

34. D. M. Smith, Organophosphorus poisoning from emergency use of a handsprayer, *Practitioner*, 1977, **218**, 877–83.
35. G. P. Wedin, Comment: Anticholinesterase Insecticide Reactions, *Drug Intell. Clin. Pharm.*, 1988, **22**, 919–20.
36. M. Eddleston, N. A. Buckley, H. Checketts, L. Senarathna, F. Mohamed, M. H. R. Sheriff and A. Dawson, Speed of initial atropinisation in significant organophosphorus pesticide poisoning—a systematic comparison of recommended regimens, *J. Toxicol. Clin. Toxicol.*, 2004, **42**, 865–75.
37. C. Bismuth, R. H. Inns and T. C. Marrs, Efficacy, toxicity and clinical use of oximes in anticholinesterase poisoning, in *Clinical and Experimental Toxicology of Organophosphates and Carbamates*, ed. B. Ballantyne and T. C. Marrs, Butterworth-Heinemann, Oxford, 1992, pp. 555–77.
38. K. Sakurada, K. Matsubara, K. Shimizu, H. Shiono, Y. Seto, K. Tsuge, M. Yoshino, I. Sakai, H. Mukoyama and T. Takatori, Pralidoxime iodide (2-PAM) penetrates across the blood-brain barrier, *Neurochem. Res.*, 2003, **28**, 1401–7.
39. J. L. Willems, H. C. De Bisschop, A. G. Verstraete, C. Declerck, Y. Christiaens, P. Vanscheeuwyck, W. A. Buylaert, D. Vogelaers and F. Colardyn, Cholinesterase reactivation in organophosphorus poisoned patients depends on the plasma concentrations of the oxime pralidoxime methylsulphate and of the organophosphate, *Arch. Toxicol.*, 1993, **67**, 79–84.
40. F. Worek, M. Bäcker, H. Thiermann, L. Szinicz, U. Mast, R. Klimmek and P. Eyer, Reappraisal of indications and limitations of oxime therapy in organophosphate poisoning, *Hum. Exp. Toxicol.*, 1997, **16**, 466–72.
41. T. C. Marrs, Organophosphates: history, chemistry, pharmacology, in *Organophosphates and Health*, ed. L. Karalliedde, S. Feldman, J. Henry and T. Marrs, Imperial College Press, London, 2001, pp. 1–36.
42. H. J. De Silva, R. Wijewickrema and N. Senanayake, Does pralidoxime affect outcome of management in acute organophosphorus poisoning?, *Lancet*, 1992, **339**, 1136–38.
43. M. Eddleston, L. Szinicz, P. Eyer and N. Buckley, Oximes in acute organophosphorus pesticide poisoning: a systematic review of clinical trials, *Q. J. Med.*, 2002, **95**, 275–83.
44. M. Eddleston, P. Eyer, F. Worek, E. Juszczak, N. Alder, F. Mohamed, L. Senarathna, A. Hittarage, S. Azher, K. Jeganathan, S. Jayamanne, L. Von Meyer, A. H. Dawson, M. H. Rezvi Sheriff and N. A. Buckley, Pralidoxime in acute organophosphorus insecticide poisoning—a randomised controlled trial, *PLoS Med.*, 2009, **6**, e1000104.
45. P. B. Casey, L. Blakey, S. M. Bradberry and J. A. Vale, Late reactivation of erythrocyte cholinesterase activity by pralidoxime in a case of chlorpyrifos poisoning, *Przegl. Lek.*, 1995, **52**, 206.
46. P. B. Casey, E. Gosden, L. Blakey, J. P. Thompson and J. A. Vale, Plasma pralidoxime concentrations following bolus injection and continuous infusion, *Przegl. Lek.*, 1995, **52**, 203–4.

47. G. M. Tush and M. I. Anstead, Pralidoxime continuous infusion in the treatment of organophosphate poisoning, *Ann. Pharmacother.*, 1997, **31**, 441–44.
48. J. L. Willems and F. M. Belpaire, Anticholinergic poisoning: an overview of pharmacotherapy, in *Clinical and Experimental Toxicology of Organophosphates and Carbamates*, ed. B. Ballantyne and T. C. Marrs, Butterworth-Heinemann, Oxford, 1992, pp. 536–42.
49. H. Thiermann, U. Mast, R. Klimmek, P. Eyer, A. Hibler, R. Pfab, N. Felgenhauer and T. Zilker, Cholinesterase status, pharmacokinetics and laboratory findings during obidoxime therapy in organophosphate poisoned patients, *Hum. Exp. Toxicol.*, 1997, **16**, 473–80.
50. H. Thiermann, L. Szinicz, F. Eyer, F. Worek, P. Eyer, N. Felgenhauer and T. Zilker, Modern strategies in therapy of organophosphate poisoning, *Toxicol. Lett.*, 1999, **107**, 233–39.
51. T. Zilker, N. Felgenhauer, A. Hibler, R. Pfab, H. Thiermann, F. Worek and P. Eyer, Factors influencing the efficacy of obidoxime in organophosphate pesticides poisoning, *Przegl. Lek.*, 1997, **54**, 662–64.
52. R. Kusic, D. Jovanovic, S. Randjelovic, D. Joksovic, V. Todorovic, B. Boskovic, M. Jokanovic and V. Vojvodic, HI-6 in man: efficacy of the oxime in poisoning by organophosphorus insecticides, *Hum. Exp. Toxicol.*, 1991, **10**, 113–18.
53. I. J. Hayward, H. G. Wall, N. K. Jaax, J. V. Wade, D. D. Marlow and J. B. Nold, Decreased brain pathology in organophosphate-exposed rhesus monkeys following benzodiazepine therapy, *J. Neurol. Sci.*, 1990, **98**, 99–106.
54. D. R. Anderson, L. W. Harris, F.-C. T. Chang, W. B. Baze, B. R. Capacio, S. L. Byers and W. J. Lennox, Antagonism of soman-induced convulsions by midazolam, diazepam and scopolamine, *Drug Chem. Toxicol.*, 1997, **20**, 115–31.
55. D. Bokonjic, D. Jovanovic, M. Jokanovic and M. Maksimovic, Protective effects of oximes HI-6 and PAM-2 applied by osmotic minipumps in quinalphos-poisoned rats, *Arch. Int. Pharmacodyn. Ther.*, 1987, **288**, 309–18.
56. S. Rump and E. Grudzińska, Investigations on the effects of diazepam in acute experimental intoxication with fluostigmine, *Arch. Toxikol.*, 1974, **31**, 223–32.
57. S. Rump, J. Faff, T. Szymańska, W. Bak and G. Borkowska, Efficacy of repeated pharmacotherapy in experimental acute poisonings with fluostigmine, *Arch. Toxicol.*, 1976, **35**, 275–80.
58. Z. Kleinrok and E. Jagiełło-Wojtowicz, Wplyw wybranych leków przeciwpadaczkowych na toksycznosc ostra estru dwuizopropylowego kwasu fluorofosforowego u myszy, *Ann. Univ. Mariae Curie Sklodowska*, 1977, **32**, 155–60.
59. R. C. Gupta, Acute malathion toxicosis and related enzymatic alterations in bubalus bubalis: antidotal treatment with atropine, 2-PAM, and diazepam, *J. Toxicol. Environ. Health*, 1984, **14**, 291–303.
60. H. Krutak-Krol and E. F. Domino, Comparative effects of diazepam and midazolam on paraoxon toxicity in rats, *Toxicol. Appl. Pharmacol.*, 1985, **81**, 545–50.

61. J. Kassa and J. Bajgar, Treatment of the stressogenic effect of dichlorvos, *Sb. Ved. Pr. Lek. Fak. Karlovy Univerzity Hradci Kralove*, 1994, **37**, 29–32.
62. T. C. Marrs, Diazepam in the treatment of organophosphorus ester pesticide poisoning, *Toxicol. Rev.*, 2003, **22**, 75–81.
63. D. Barckow, G. Neuhaus and W. D. Erdmann, Zur Behandlung der schweren Parathion-(E605(R))-Vergiftung mit dem Cholinesterase-Reaktivator Obidoxim (Toxogonin(R)), *Arch. Toxikol.*, 1969, **24**, 133–46.
64. J. A. Vale and G. W. Scott, Organophosphorus poisoning, *Guy's Hosp. Rep.*, 1974, **123**, 13–25.
65. M. Yacoub, H. Skouri, M. Amamou and F. Besencenot, Intoxications aiguës par les insecticides organo-phosphorés. A propos de 908 cas observés au Centre Anti-Poisons de Tunis, *J. Toxicol. Med.*, 1981, **1**, 165–88.
66. D. G. Merrill and F. G. Mihm, Prolonged toxicity of organophosphate poisoning, *Crit. Care Med.*, 1982, **10**, 550–551.
67. J. Martí, C. Reverté, M. Roig, I. Pascual and J. Prats, Uso del diacepam en la intoxicación por insecticidas organofosforados, *An. Esp. Pediatr.*, 1985, **23**, 299–301.
68. F. N. LeBlanc, B. E. Benson and A. D. Gilg, A severe organophosphate poisoning requiring the use of an atropine drip, *J. Toxicol. Clin. Toxicol.*, 1986, **24**, 69–76.
69. W. L. A. M. de Kort, S. H. Kiestra and B. Sangster, The use of atropine and oximes in organophosphate intoxications: a modified approach, *J. Toxicol. Clin. Toxicol.*, 1988, **26**(3–4), 199–208.
70. D. Jovanovic, S. Randjelovic and D. Joksovic, A case of unusual suicidal poisoning by the organophosphorus insecticide dimethoate, *Hum. Exp. Toxicol.*, 1990, **9**, 49–51.
71. A. Weissmann-Brenner, A. David, A. Vidan and A. Hourvitz, Organophosphate poisoning: a multihospital survey, *Isr. Med. Assoc. J.*, 2002, **4**, 573–76.
72. M. K. Johnson and J. A. Vale, Clinical management of acute organophosphate poisoning: an overview, in *Clinical and Experimental Toxicology of Organophosphates and Carbamates*, ed. B. Ballantyne and T. C. Marrs, Butterworth-Heinemann, Oxford, 1992, pp. 528–35.
73. T. C. Y. Tsao, Y.-C. Juang, R.-S. Lan, W.-B. Shieh and C.-H. Lee, Respiratory failure of acute organophosphate and carbamate poisoning, *Chest*, 1990, **98**, 631–36.
74. R. J. Zwiener and C. M. Ginsburg, Organophosphate and carbamate poisoning in infants and children, *Pediatrics*, 1988, **81**, 121–26.
75. S. Y. Tay, D. Y. H. Tai, E. Seow and Y. T. Wang, Patients admitted to an intensive care unit for poisoning, *Ann. Acad. Med. Singapore*, 1998, **27**, 347–52.
76. K. Ameno, S.-K. Lee, S.-W. In, J.-Y. Yang, Y.-C. Yoo, S. Ameno, T. Kubota, H. Kinoshita and I. Ijiri, Blood carbofuran concentrations in suicidal ingestion cases, *Forensic Sci. Int.*, 2001, **116**, 59–61.
77. K. S. Pinakini and T. S. M. Kumar, Serial cholinesterase estimation in carbamate poisoning, *J. Clin. Forensic Med.*, 2006, **13**, 274–76.

78. S. Sarkar, A. Narang and S. Singh, Transplacentally acquired carbamate insecticide (Baygon) poisoning in a neonate, *Indian Pediatr.*, 1994, **31**, 343–46.
79. J. S. Lima and C. A. G. Reis, Poisoning due to illegal use of carbamates as a rodenticide in Rio de Janeiro, *J. Toxicol. Clin. Toxicol.*, 1995, **33**, 687–90.
80. J. A. Liddle, R. D. Kimbrough, L. L. Needham, R. E. Cline, A. L. Smrek, L. W. Yert, D. D. Bayse, A. C. Ellington and P. A. Dennis, A fatal episode of accidental methomyl poisoning, *J. Toxicol. Clin. Toxicol.*, 1979, **15**, 159–67.
81. K. E. Ferslew, A. N. Hagardorn and W. F. McCormick, Poisoning from oral ingestion of carbofuran (Furadan 4F), a cholinesterase-inhibiting carbamate insecticide, and its effects on cholinesterase activity in various biological fluids, *J. Forensic Sci.*, 1992, **37**, 337–44.
82. T. Miyazaki, M. Yashiki, T. Kojima and F. Chikasue, A. Ochiai and Y. Hidani, Fatal and non-fatal methomyl intoxication in an attempted double suicide, *Forensic Sci. Int.*, 1989, **42**, 263–70.
83. U. Hoffmann, U. Hecker and P. Abel, Acute poisoning by pirimicarb: clinical and toxicological features, *Clin. Toxicol.*, 2008, **46**, 694–96.
84. A. Tracqui, F. Flesch, P. Sauder, J. S. Raul, A. Géraut, B. Ludes and A. Jaeger, Repeated measurements of aldicarb in blood and urine in a case of nonfatal poisoning, *Hum. Exp. Toxicol.*, 2001, **20**(12), 657–60.
85. M. Lifshitz, E. Shahak, A. Bolotin and S. Sofer, Carbamate poisoning in early childhood and in adults, *J. Toxicol. Clin. Toxicol.*, 1997, **35**, 25–27.
86. P.-Y. Yang, T. C. Y. Tsao, J.-L. Lin, R.-K. Lyu and P.-C. Chiang, Carbofuran-induced delayed neuropathy, *J. Toxicol. Clin. Toxicol.*, 2000, **38**, 43–46.
87. R. Santinelli, C. Tolone, A. D'Avanzo, E. M. del Giudice, L. Perrone, M. D'Avanzo and Pontine myelinolysis in a child with carbamate poisoning, *Clin. Toxicol.*, 2006, **44**, 327–28.
88. N. Paul and T. J. Mannathukkaran, Intermediate syndrome following carbamate poisoning, *Clin. Toxicol.*, 2005, **43**, 867–68.
89. D. Siegal, M. A. Kotowycz, M. Methot and A. Baranchuk, Complete heart block following intentional carbamate ingestion, *Can. J. Cardiol.*, 2009, **25**, e288–e290.
90. A. Thierauf, H. Gnann, M. Bohnert, B. Vennemann, V. Auwärter and W. Weinmann, Suicidal poisoning with mercaptodimethur—morphological findings and toxicological analysis, *Int. J. Legal Med.*, 2009, **123**, 327–31.
91. A. M. Tsatsakis, A. K. Tsakalof, Y. Siatitsas and E. N. Michalodimitrakis, Acute poisoning with carbamate pesticides: the Cretan experience, *Sci. Justice*, 1996, **36**, 35–39.
92. M. S. J. Al-Samarraie, R. Karinen, T. Rognum, I. Hasvold, M. O. Stokke and A. S. Christophersen, Lethal poisoning with ethiofencarb and ethanol, *J. Anal. Toxicol.*, 2009, **33**, 389–92.

93. J. H. Garcia, G. R. Coelho, G. A. Marques, J. B. Gadelha, J. B. Vasconcelos, Jr. J. T. Valenca, R. M. Esmeraldo, J. A. Meija, C. A. Leite and E. R. Almeida, Successful organ transplantation from donors poisoned with a carbamate insecticide, *Am. J. Transplant.*, 2010, **10**, 1490–1492.
94. G. R. Simpson and S. Bermingham, Poisoning by carbamate pesticides, *Med. J. Aust.*, 1977, **2**, 148–49.
95. F. Testud, A. De Larquier and J. Descotes, Occupational acute poisonings with methomyl: role of dermal exposure and need for preventive measures, *J. Environ. Med.*, 1999, **1**, 137–39.
96. R. McConnell and A. J. Hruska, An epidemic of pesticide poisoning in Nicaragua: implications for prevention in developing countries, *Am. J. Public Health*, 1993, **83**, 1559–62.
97. A. M. E. Coleman, A. Smith and L. Watson, Occupational carbamate pesticide intoxication in three farm workers. implications and significance for occupational health in Jamaica, *West Indian Med. J.*, 1990, **39**, 109–13.
98. H. N. Saiyed, H. G. Sadhu, V. K. Bhatnagar, A. Dewan, K. Venkaiah and S. K. Kashyap, Cardiac toxicity following short-term exposure to methomyl in spraymen and rabbits, *Hum. Exp. Toxicol.*, 1992, **11**, 93–97.
99. J. H. Lim, H. S. Kim, H. O. Kim and Y. M. Park, Stevens–Johnson syndrome following occupational exposure to carbamate insecticide, *J. Dermatol.*, 2010, **37**, 182–84.
100. E. M. Richardson and R. I. Batteese, Jr., An incident of Zectran poisoning, *J. Maine Med. Assoc.*, 1973, **64**, 158–59.
101. W. J. Driskell, D. F. Groce, R. H. Hill, Jr. and M. M. Birky, Methomyl in the blood of a pilot who crashed during aerial spraying, *J. Anal. Toxicol.*, 1991, **15**, 339–40.
102. G. G. Cable and S. Doherty, Acute carbamate and organochlorine toxicity causing convulsions in an agricultural pilot: a case report, *Aviat. Space Environ. Med.*, 1999, **70**, 68–72.
103. D. J. Dickoff, O. Gerber and Z. Turovsky, Delayed neurotoxicity after ingestion of carbamate pesticide, *Neurology*, 1987, **37**, 1229–31.
104. F. Umehara, S. Izumo, K. Arimura and M. Osame, Polyneuropathy induced by *m*-tolyl methyl carbamate intoxication, *J. Neurol.*, 1991, **238**, 47–48.
105. M. Lotti and A. Moretto, Do carbamates cause polyneuropathy?, *Muscle Nerve*, 2006, **34**, 499–502.
106. L. Winnik, D. Pach, T. Gawlikowski, D. Targosz and P. Hydzik, Multiorgan damage in acute oral carbamates poisonings, *Przegl. Lek.*, 1997, **54**, 684–88.
107. D. J. Ecobichon, Carbamate insecticides, in *Handbook of Pesticide Toxicology. Principles*, ed. R. I. Krieger, Academic Press, San Diego, CA, 2001, pp. 1087–106.
108. N. K. Baban, D. L. Nunley, A. S. Borges, Jr., T. M. Roy and A. S. Borges Jr., Human sequelae of severe carbamate poisoning, *Tenn. Med.*, 1998, **91**, 103–6.

109. C. P. Carpenter, C. S. Weil, P. E. Palm, M. W. Woodside, J. H. Nair, III and H. F. Smyth Jr., Mammalian toxicity of 1-naphthyl-*N*-methylcarbamate (Sevin insecticide), *J. Agric. Food Chem.*, 1961, **9**, 30–39.
110. S. H. Sterri, B. Rognerud, S. E. Fiskum and S. Lyngaas, Effect of toxogonin and P2S on the toxicity of carbamates and organophosphorus compounds, *Acta Pharmacol. Toxicol.*, 1979, **45**, 9–15.
111. L. W. Harris, B. G. Talbot, W. J. Lennox and D. R. Anderson, The relationship between oxime-induced reactivation of carbamylated acetylcholinesterase and antidotal efficacy against carbamate intoxication, *Toxicol. Appl. Pharmacol.*, 1989, **98**, 128–33.
112. D. M. Sanderson, Treatment of poisoning by anticholinesterase insecticides in the rat, *J. Pharm. Pharmacol.*, 1961, **13**, 435–42.
113. I. L. Natoff and B. Reiff, Effect of oximes on the acute toxicity of anticholinesterase carbamates, *Toxicol. Appl. Pharmacol.*, 1973, **25**, 569–75.
114. M. Mercurio-Zappala and J. B. Hack, A. Salvador and R. S. Hoffman, Pralidoxime in carbaryl poisoning: an animal model, *Hum. Exp. Toxicol.*, 2007, **26**, 125–29.
115. B. R. Ekins and R. J. Geller, Methomyl-induced carbamate poisoning treated with pralidoxime chloride, *West. J. Med.*, 1994, **161**, 68–70.
116. J. L. Burgess, J. N. Bernstein and K. Hurlbut, Aldicarb poisoning: a case report with prolonged cholinesterase inhibition and improvement after pralidoxime therapy, *Arch. Intern. Med.*, 1994, **154**, 221–24.
117. L. S. Nelson, J. Perrone, F. DeRoos, C. Stork and R. S. Hoffman, Aldicarb poisoning by an illicit rodenticide imported into the United States: Tres Pasitos, *J. Toxicol. Clin. Toxicol.*, 2001, **39**, 447–52.
118. R. Goswamy, A. Chaudhuri and A. A. Mahashur, Study of respiratory failure in organophosphate and carbamate poisoning, *Heart Lung*, 1994, **23**, 466–72.
119. A. Farago, [Fatal suicidal case of sevin (1-naphthyl-*N*-methylcarbamate) poisoning], *Arch. Toxikol.*, 1969, **24**, 309–15.
120. D. H. Phua, C. C. Lin, M.-L. Wu, J.-F. Deng and C.-C. Yang, Neonicotinoid insecticides: an emerging cause of acute pesticide poisoning, *Clin. Toxicol.*, 2009, **47**, 336–41.
121. Y.-M. Hung and K. H. Meier, Acute ®Confidor (Imidacloprid-*N*-methyl pyrrolidone) insecticides intoxication with mimicking cholinergic syndrome, *Toxicol. Ind. Health*, 2005, **21**, 137–40.
122. Y. M. Hung, S. L. Lin, K. J. Chou and H. M. Chung, Imidacloprid-*N*-methyl pyrrolidone insecticides poisoning mimicking cholinergic syndrome, *Clin. Toxicol.*, 2006, **44**, 771–72.
123. N.-C. Huang, S.-L. Lin, C.-H. Chou, Y.-M. Hung, H.-M. Chung and S.-T. Huang, Fatal ventricular fibrillation in a patient with acute imidacloprid poisoning, *Am. J. Emerg. Med.*, 2006, **24**, 883–85.
124. F. Mohamed, I. Gawarammana, T. A. Robertson, M. S. Roberts, C. Palangasinghe, S. Zawahir, S. Jayamanne, J. Kandasamy, M. Eddleston, N. A. Buckley, A. H. Dawson and D. M. Roberts, acute human

self-poisoning with imidacloprid compound: a neonicotinoid insecticide, *PLoS ONE*, 2009, **4**, e5127.

125. T. Imamura, Y. Yanagawa, K. Nishikawa, N. Matsumoto and T. Sakamoto, Two cases of acute poisoning with acetamiprid in humans, *Clin. Toxicol.*, 2010, **48**, 851–53.
126. I.-W. Wu, J.-L. Lin and E.-T. Cheng, Acute poisoning with the neonicotinoid insecticide imidacloprid in *N*-methyl pyrrolidone, *J. Toxicol. Clin. Toxicol.*, 2001, **39**, 617–21.
127. I.-J. Yeh, T.-J. Lin and D.-Y. Hwang, Acute multiple organ failure with imidacloprid and alcohol ingestion, *Am. J. Emerg. Med.*, 2010, **28** (255.e1–255), e3.
128. D. David, I. A. George and J. V. Peter, Toxicology of the newer neonicotinoid insecticides: imidacloprid poisoning in a human, *Clin. Toxicol.*, 2007, **45**, 485–86.
129. R. Agarwal and R. Srinivas, Severe neuropsychiatric manifestations and rhabdomyolysis in a patient with imidacloprid poisoning, *Am. J. Emerg. Med.*, 2007, **25**, 844–45.
130. A. D. Karatas, Severe central nervous system depression in a patient with acute imidacloprid poisoning, *Am. J. Emerg. Med.*, 2009, **27**, 1171.
131. A. K. Panigrahi, D. K. S. Subrahmanyam and K. K. Mukku, Imidacloprid poisoning: a case report, *Am. J. Emerg. Med.*, 2009, **27**, (256.e5–256.e6).
132. R. Iyyadurai, I. A. George and J. V. Peter, Imidacloprid poisoning—newer insecticide and fatal toxicity, *J. Med. Toxicol.*, 2010, **6**, 77–78.
133. F. Mohamed, A. H. Dawson and D. Roberts, Factors influencing variability in clinical outcomes from imidacloprid self-poisoning, *Clin. Toxicol.*, 2009, **47**, 836–37.
134. N. Daneshvar, S. Aber, A. Khani and M. H. Rasoulifard, Investigation of adsorption kinetics and isotherms of imidacloprid as a pollutant from aqueous solution by adsorption onto industrial granular activated carbon, *J. Food Agric. Environ.*, 2007, **5**, 425–29.
135. W. Chefurka, P. Kashi and E. J. Bond, The effect of phosphine on electron transport in mitochondria, *Pestic. Biochem. Physiol.*, 1976, **6**, 65–84.
136. R. Dua and K. D. Gill, Effect of aluminium phosphide exposure on kinetic properties of cytochrome oxidase and mitochondrial energy metabolism in rat brain, *Biochim. Biophys. Acta*, 2004, **1674**, 4–11.
137. A. T. Proudfoot, Aluminium and zinc phosphide poisoning, *Clin. Toxicol.*, 2009, **47**, 89–100.
138. R. Kochhar, U. Dutta, K. S. Poornachandra, K. Vaiphei, S. Bhagat, B. Nagi and K. Singh, Clinical profile and outcome of aluminum phosphide-induced esophageal strictures, *J. Med. Toxicol.*, 2010, **6**, 301–6.
139. P. Alter, W. Grimm and B. Maisch, Lethal heart failure caused by aluminium phosphide poisoning, *Intensive Care Med.*, 2001, **27**, 327.
140. R. Katira, G. P. Elhence, M. L. Mehrotra, S. S. L. Srivastava, A. Mitra, R. Agarwala and A. Ram, A study of aluminium phosphide (AlP)

poisoning with special reference to electrocardiographic changes, *J. Assoc. Physicians India*, 1990, **38**, 471–73.

141. B. Arora, R. S. Punia, R. Kalra, S. N. Chugh and D. R. Arora, Histopathological changes in aluminium phosphide poisoning, *J. Indian Med. Assoc.*, 1995, **93**, 380–381.
142. A. Darbari, S. Tandon, S. Chaudhary, M. Bharadwaj, A. Kumar and G. P. Singh, Esophageal injuries due to aluminum phosphide tablet poisoning in India, *Asian Cardiovasc. Thorac. Ann.*, 2008, **16**, 298–300.
143. V. Gupta, J. Singh, S. S. Doodan and S. K. Bali, Multisystem organ failure (MSOF) in aluminium phosphide (AlP) poisoning, *JK Pract.*, 2000, **7**, 287–88.
144. S. N. Chugh, Dushyant, S. Ram, B. Arora and K. C. Malhotra, Incidence & outcome of aluminium phosphide poisoning in a hospital study, *Indian J. Med. Res.*, 1991, **94**, 232–35.
145. M. Akkaoui, S. Achour, K. Abidi, B. Himdi, A. Madani, A. Ali-Zeggwagh and R. Abouqal, reversible myocardial injury associated with aluminum phosphide poisoning, *Clin. Toxicol.*, 2007, **45**, 728–31.
146. R. Bajaj, H. S. Wasir, R. Agarwal, A. Malhotra, P. Chopra and M. L. Bhatia, Aluminium phosphide poisoning. Clinical toxicity and outcome in eleven intensively monitored patients, *Natl. Med. J. India*, 1988, **1**, 270–274.
147. G. S. Kalra, I. S. Anand, I. Jit, B. Bushnurmath and P. L. Wahi, Aluminium phosphide poisoning: haemodynamic observations, *Indian Heart J.*, 1991, **43**, 175–78.
148. S. B. Siwach, H. Singh, Jagdish, V. K. Katyal and G. Bhardwaj, Cardiac arrhythmias in aluminium phosphide poisoning studied by on continuous holter and cardioscopic monitoring, *J. Assoc. Physicians India*, 1998, **46**, 598–601.
149. G. P. Singh, Mortality in acute aluminium phosphide poisoning having hypomagnesaemia with and without ECG changes, *J. Indian Med. Assoc.*, 2000, **98**, 446.
150. V. Shah, S. Baxi and T. Vyas, Severe myocardial depression in a patient with aluminium phosphide poisoning: a clinical, electrocardiographical and histopathological correlation, *Indian J. Crit. Care Med.*, 2009, **13**, 41–43.
151. K. Soltaninejad, S. Shadnia, B. Ziyapour and J. Brent, Aluminum phosphide intoxication mimicking ischemic heart disease led to unjustified treatment with streptokinase, *Clin. Toxicol.*, 2009, **47**, 908–9.
152. S. Nayyar and M. Nair, Brugada pattern in toxic myocarditis due to severe aluminum phosphide poisoning, *Pacing Clin. Electrophysiol.*, 2009, **32**, e16–17.
153. S. M. Jain, A. Bharani, G. C. Sepaha, V. C. Sanghvi and P. G. Raman, Electrocardiographic changes in aluminium phosphide (AlP) poisoning, *J. Assoc. Physicians India*, 1985, **33**, 406–9.
154. G. Sasanian, S. Shadnia, M. Abdollahi, P. Allami, A. Hosseini, A. Ranjbar and N. Amini-Shirazi, A retrospective 7-years study of aluminum phosphide poisoning in Tehran, *Toxicol. Lett.*, 2010, **196**(Suppl 1), S80–81.

155. S. Singh, D. Singh, N. Wig, I. Jit and B. K. Sharma, Aluminum phosphide ingestion—a clinico-pathologic study, *J. Toxicol. Clin. Toxicol.*, 1996, **34**, 703–6.
156. S. N. Chugh, S. Ram, L. K. Mehta, B. B. Arora and K. C. Malhotra, Adult respiratory distress syndrome following aluminium phosphide ingestion. report of 4 cases, *J. Assoc. Physicians India*, 1989, **37**, 271–72.
157. R. B. Singh, R. G. Singh and U. Singh, Hypermagnesemia following aluminum phosphide poisoning, *Int. J. Clin. Pharmacol. Ther. Toxicol.*, 1991, **29**, 82–85.
158. M. S. Gupta, A. Malik and V. K. Sharma, Cardiovascular manifestations in aluminium phosphide poisoning with special reference to echocardiographic changes, *J. Assoc. Physicians India*, 1995, **43**(11), 773–74, 779–80.
159. N. Sharma, S. Mahi and S. Varma, Unusual complications of aluminium phosphide poisoning, *J. Med. Toxicol.*, 2008, **4**, 204.
160. M. K. Jain, S. K. Khanijo, N. Pathak, M. Sharma and P. Binayake, Electrocardiographic diagnosis of atrial infarction in aluminium phosphide poisoning, *J. Assoc. Physicians India*, 1992, **40**(10), 692–93.
161. R. M. Kaushik, R. Kaushik and S. K. Mahajan, Subendocardial infarction in a young survivor of aluminium phosphide poisoning, *Hum. Exp. Toxicol.*, 2007, **26**, 457–60.
162. G. S. Wander, S. Arora and S. B. Khurana, Acute Pericarditis in Aluminium Phosphide Poisoning, *J. Assoc. Physicians India*, 1990, **38**, 675.
163. S. N. Chugh and K. C. Malhotra, Acute pericarditis in aluminium phosphide poisoning, *J. Assoc. Physicians India*, 1992, **40**, 564.
164. R. L. Suman and M. Savani, Pleural effusion—a rare complication of aluminium phosphide poisoning, *Indian Pediatr.*, 1999, **36**, 1161–63.
165. A. K. Bayazit, A. Noyan and A. Anarat, A child with hepatic and renal failure caused by aluminum phosphide, *Nephron*, 2000, **86**, 517.
166. L. M. Siddaiah, S. M. Adhyapak, S. M. Jaydev, G. G. Shetty, K. Varghese, C. B. Patil and S. S. Iyengar, Intra-aortic balloon pump in toxic myocarditis due to aluminum phosphide poisoning, *J. Med. Toxicol.*, 2009, **5**, 80–83.
167. S. N. Khosla, N. Nand and P. Khosla, Aluminium phosphide poisoning, *J. Trop. Med. Hyg.*, 1988, **91**, 196–98.
168. H. Abder-Rahman, Effect of aluminium phosphide on blood glucose level, *Vet. Hum. Toxicol.*, 1999, **41**, 31–32.
169. G. Jain, S. K. Gupta, S. Tharwani and D. K. Singh, Rare case of hyperthermia in aluminum phosphide poisoning, *Indian J. Forensic Med. Toxicol.*, 2010, **4**, 26–27.
170. M. S. Gupta, L. Mehta, S. N. Chugh and K. C. Malhotra, Aluminium phosphide poisoning. Two cases with rare presentation, *J. Assoc. Physicians India*, 1990, **38**, 509–10.
171. S. K. Verma, S. Ahmad, N. Shirazi, S. P. Barthwal, D. Khurana, M. Chugh and H. S. Gambhir, Acute pancreatitis: a lesser-known complication of aluminum phosphide poisoning, *Hum. Exp. Toxicol.*, 2007, **26**(12), 979–81.

172. P. S. A. Sarma and J. Narula, Acute pancreatitis due to zinc phosphide ingestion, *Postgrad. Med. J.*, 1996, **72**, 237–38.
173. B. Mostafazadeh, A. Pajoumand, E. Farzaneh, A. Aghabiklooei and M. R. Rasouli, Blood levels of methemoglobin in patients with aluminum phosphide poisoning and its correlation with patient's outcome, *J. Med. Toxicol.*, 2011, **7**, 40–3.
174. W. T. Potter, S. Rong, J. Griffith, J. White and V. F. Garry, Phosphine-mediated Heinz body formation and hemoglobin oxidation in human erythrocytes, *Toxicol. Lett.*, 1991, **57**, 37–45.
175. K. L. Chin, X. Mai, J. Meaklim, G. R. Scollary and D. D. Leaver, The interaction of phosphine with haemoglobin and erythrocytes, *Xenobiotica*, 1992, **22**, 599–607.
176. S. B. Lall, S. S. Peshin and S. Mitra, Methemoglobinemia in aluminium phosphide poisoning in rats, *Indian J. Exp. Biol.*, 2000, **38**, 95–97.
177. C. Y. Frangides and I. A. Pneumatikos, Persistent severe hypoglycemia in acute zinc phosphide poisoning, *Intensive Care Med.*, 2002, **28**, 223.
178. D. Memis, D. Tokatlioglu, O. Koyuncu and S. Hekimoglu, Fatal aluminium phosphide poisoning, *Eur. J. Anaesthesiol.*, 2007, **24**, 292–93.
179. M. Mouhaoui, K. Khaleq, O. Rabbai, K. Yaqini and H. Louardi, Intoxication au phosphure d'aluminium et grossesse (à propos d'un cas), *Ann. Fr. Anesth. Réanim.*, 2008, **27**, 181–82.
180. O. Mehrpour, S. Alfred, S. Shadnia, D. E. Keyler, K. Soltaninejad, N. Chalaki and M. Sedaghat, Hyperglycemia in acute aluminum phosphide poisoning as a potential prognostic factor, *Hum. Exp. Toxicol.*, 2008, **27**, 591–95.
181. S. N. Chugh, K. Kishore, N. Aggarwal and S. Attri, Hypoglycaemia in acute aluminium phosphide poisoning, *J. Assoc. Physicians India*, 2000, **48**, 855–56.
182. S. K. Somani, Aluminium phosphide-induced esophageal stricture palliation with polyflex stent, *Dysphagia*, 2008, **23**, 411–12.
183. S. Kapoor, S. Naik, R. Kumar, S. Sharma, H. S. Pruthi and S. Varshney, Benign esophageal stricture following aluminium phosphide poisoning, *Indian J. Gastroenterol.*, 2005, **24**, 261–62.
184. S. Nijhawan, M. Rastogi, M. Tandon, A. Mathur and R. R. Rai, Aluminum phosphide-induced esophageal stricture: an unusual complication, *Endoscopy*, 2006, **38**(Suppl 5), E23.
185. R. Talukdar, D. K. Singal and R. K. Tandon, Aluminium phosphide-induced esophageal stricture, *Indian J. Gastroenterol.*, 2006, **25**, 98–99.
186. R. K. Verma, S. N. Gupta, D. V. Bahl and A. Gupta, Aluminium phosphide poisoning: late presentation as oesophageal stricture, *Jk Science*, 2006, **8**, 235–36.
187. K. Madan, S. B. Chalamalasetty, M. Sharma and G. Makharia, Corrosive-like strictures caused by ingestion of aluminium phosphide, *Natl. Med. J. India*, 2006, **19**, 313–14.
188. J. Tiwari, B. Lahoti, K. Dubey, P. Mishra and S. Verma, Tracheo-oesophageal fistula—an unusual complication following Celphos poisoning, *Indian J. Surg.*, 2003, **65**, 442–44.

189. A. Darbari, A. Kumar, G. Chandra and S. Tandon, Tracheo-oesophageal fistula with oesophageal stricture due to aluminium phosphide (Celphos tablet) poisoning, *J. Chest Dis. Allied Sci.*, 2007, **49**, 241–42.
190. S. P. Misra and M. Dwivedi, Aluminum phosphide-induced esophageal strictures: a new cause of benign esophageal strictures, *J. Clin. Gastroenterol.*, 2009, **43**, 405–9.
191. S. N. Chugh, S. Dabla, V. Pal, R. K. Yadav and A. Garg, Oesophageal stricture following aluminium phosphide (Celphos) poisoning: a rare complication, *Int. J. Med. Toxicol. Legal Med.*, 2007, **10**, 63–64.
192. D. Agrawal, A. Kumar, M. A. Ansari, V. Srivastava and T. K. Lahiri, Aluminium phosphide poisoning with esophageal stricture and tracheoesophageal fistula, *Indian J. Thoracic Cardiovasc. Surg.*, 2010, **26**, 198–203.
193. R. K. Jain, N. B. Gouda, V. K. Sharma, T. N. Dubey, A. Shende, R. Malik and G. Tiwari, Esophageal complications following aluminium phosphide ingestion: an emerging issue among survivors of poisoning, *Dysphagia*, 2009, **25**, 271–76.
194. S. Bhargava, R. Rastogi, A. Agarwal and G. Jindal, Esophagobronchial fistula—a rare complication of aluminum phosphide poisoning, *Ann. Thorac. Med*, 2011, **6**, 41–42.
195. V. Srivastava, M. Ansari, S. Basu, D. Agrawal, T. K. Lahiri and A. Kumar, Colonic conduit for esophageal bypass in Celphos-induced tracheoesophageal fistula: our experience of two cases, *Int. J. Colorectal Dis.*, 2009, **24**, 727–28.
196. S. Shadnia, O. Mehrpour and K. Soltaninejad, A simplified acute physiology score in the prediction of acute aluminum phosphide poisoning outcome, *Indian J. Med. Sci.*, 2010, **64**, 532–39.
197. M. Hajouji Idrissi, L. Oualili, K. Abidi, R. Abouqal, O. Kerkeb and A. Zeggwagh, Facteurs de gravité de l'intoxication aiguë au phosphure d'aluminium (Phostoxin®), *Ann. Fr. Anesth. Réanim.*, 2006, **25**, 382–85.
198. S. Jaiswal, R. K. Verma and N. Tewari, Aluminum phosphide poisoning: effect of correction of severe metabolic acidosis on patient outcome, *Indian J. Crit. Care Med.*, 2009, **13**, 21–24.
199. J. Chacko and C. Shivaprasad, Fatal aluminium phosphide poisoning due to myocardial depression refractory to high dose inotropic support and intra-aortic balloon counterpulsation, *Indian J. Crit. Care Med.*, 2008, **12**, 37–38.
200. B. Pepelko, J. Seckar, P. R. Harp, J. H. Kim, D. Gray and E. L. Anderson, Worker exposure standard for phosphine gas, *Risk Anal.*, 2004, **24**, 1201–13.
201. M. Q. Chaudhry, A review of the mechanisms involved in the action of phosphine as an insecticide and phosphine resistance in stored-product insects, *Pestic. Sci.*, 1997, **49**, 213–28.
202. IPCS, *Environmental Health Criteria 73. Phosphine and Selected Metal Phosphides*, International Programme on Chemical Safety, World Health Organization, Geneva, 1988.
203. T. Dumas and E. J. Bond, Separation of phosphine from odour-producing impurities, *J. Stored Prod. Res.*, 1974, **10**, 67–68.

204. D. L. Sudakin, Occupational exposure to aluminium phosphide and phosphine gas? A suspected case report and review of the literature, *Hum. Exp. Toxicol.*, 2005, **24**, 27–33.
205. H. A. Abder-Rahman, Aluminium phosphide fatalities at mild exertion in asymptomatic children: a clue to understand the variations of the autopsy findings, *J. Forensic Legal Med.*, 2009, **16**, 312–15.
206. A. T. Jones, R. C. Jones and E. O. Longley, Environmental and clinical aspects of bulk wheat fumigation with aluminum phosphide, *Am. Ind. Hyg. Assoc. J.*, 1964, **25**, 376–79.
207. R. Wilson, Jr., F. H. Lovejoy, R. J. Jaeger and P. L. Landrigan, Acute phosphine poisoning aboard a grain freighter: epidemiologic, clinical and pathological findings, *JAMA*, 1980, **244**, 148–50.
208. J. L. Pérez Navero, I. Ibarra de la Rosa, M. A. Frías Pérez, M. J. Arroyo Marín and P. Pérez Jorge, Intoxicación letal por inhalación accidental de fosfuro alumínico, *An. Pediatr.*, 2009, **71**, 427–31.
209. R. B. Vohra, K. A. Schwarz, S. R. Williams and R. F. Clark, Phosphine toxicity with echocardiographic signs in railcar stowaways, *Clin. Toxicol.*, 2006, **44**, 719–20.
210. D. Schoonbroodt, P. Guffens, P. Jousten, J. Ingels and J. Grodos, Acute phosphine poisoning? A case report and review, *Acta Clin. Belg.*, 1992, **47**, 280–284.
211. A. Dueñas, J. L. Pérez-Castrillon, M. A. Cobos and V. Herreros, Treatment of the cardiovascular manifestations of phosphine poisoning with trimetazidine, a new anti-ischemic drug, *Am. J. Emerg. Med.*, 1999, **17**, 219–20.
212. A. Ghose, H. Kabir, A. A. Sayeed, H. Ahmed, A. Basher, M. Rashid, M. R. Amin and A. M. Faiz, Aluminium phosphide poisoning in Bangladesh: inhalational and intentional, *J. Med. Toxicol.*, 2008, **4**, 203–4.
213. S. M. Bradberry, S. A. Cage, A. T. Proudfoot and J. A. Vale, Poisoning due to pyrethroids, *Toxicol. Rev.*, 2005, **24**, 93–106.
214. D. M. Llewellyn, A. Brazier, R. Brown, J. Cocker, M. L. Evans, J. Hampton, B. P. Nutley and J. White, Occupational exposure to permethrin during its use as a public hygiene insecticide, *Ann. Occup. Hyg.*, 1996, **40**, 499–509.
215. Z. Adamis, A. Antal, I. Füzesi, J. Molnár, L. Nagy and M. Susán, Occupational exposure to organophosphorus insecticides and synthetic pyrethroid, *Int. Arch. Occup. Environ. Health*, 1985, **56**, 299–305.
216. S. Chen, Z. Zhang, F. He, P. Yao, Y. Wu, J. Sun, L. Liu and Q. Li, An epidemiological study on occupational acute pyrethroid poisoning in cotton farmers, *Br. J. Ind. Med.*, 1991, **48**, 77–81.
217. Z. Zhang, J. Sun, S. Chen, Y. Wu and F. He, Levels of exposure and biological monitoring of pyrethroids in spraymen, *Br. J. Ind. Med.*, 1991, **48**, 82–86.
218. M. F. Wilkes, Pyrethroid-induced paresthesia—a central or local toxic effect?, *J. Toxicol. Clin. Toxicol.*, 2000, **38**, 103–5.

219. F. He, S. Wang, L. Liu, S. Chen, Z. Zhang and J. Sun, Clinical manifestations and diagnosis of acute pyrethroid poisoning, *Arch. Toxicol.*, 1989, **63**, 54–58.
220. W. N. Aldridge, An assessment of the toxicological properties of pyrethroids and their neurotoxicity, *Crit. Rev. Toxicol.*, 1990, **21**, 89–104.
221. S. B. Tucker, S. A. Flannigan and M. H. Smolensky, Comparison of therapeutic agents for synthetic pyrethroid exposure, *Contact Derm.*, 1983, **9**, 316.
222. J. M. Knox and S. B. Tucker, A new cutaneous sensation caused by synthetic pyrethroids, *Clin. Res.*, 1982, **30**, 915a.
223. J. M. Knox II, S. B. Tucker and S. A. Flannigan, Paresthesia from cutaneous exposure to a synthetic pyrethroid insecticide, *Arch. Dermatol.*, 1984, **120**, 744–46.
224. F. He, H. Deng, X. Ji, Z. Zhang, J. Sun and P. Yao, Changes of nerve excitability and urinary deltamethrin in sprayers, *Int. Arch. Occup. Environ. Health*, 1991, **62**, 587–90.
225. S. A. Flannigan and S. B. Tucker, Variation in cutaneous sensation between synthetic pyrethroid insecticides, *Contact Derm.*, 1985, **13**, 140–147.
226. B. Kolmodin-Hedman, Å. Swensson and M. Åkerblom, Occupational exposure to some synthetic pyrethroids (permethrin and fenvalerate), *Arch. Toxicol.*, 1982, **50**, 27–33.
227. K. Brandenburg, A. S. Deinard, J. DiNapoli, S. J. Englender, J. Orthoefer and D. Wagner, 1% permethrin cream rinse vs 1% lindane shampoo in treating pediculosis capitis, *Am. J. Dis. Child.*, 1986, **140**, 894–96.
228. D. C. Kalter, J. Sperber, T. Rosen and S. Matarasso, Treatment of pediculosis pubis. clinical comparison of efficacy and tolerance of 1% lindane shampoo vs 1% permethrin creme rinse, *Arch. Dermatol.*, 1987, **123**, 1315–19.
229. IPCS, *Environmental Health Criteria 94. Permethrin*, International Programme on Chemical Safety, World Health Organization, Geneva, 1990.
230. B. Kolmodin-Hedman, M. Åkerblom, S. Flato and G. Alex, Symptoms in forestry workers handling conifer plants treated with permethrin, *Bull. Environ. Contam. Toxicol.*, 1995, **55**, 487–93.
231. J. E. Lessenger, A rash and chemical burns in a cowboy exposed to permethrin, *J. Agromed.*, 1995, **2**, 25–28.
232. J. B. DiNapoli, R. D. Austin, S. J. Englender, M. P. Gomez and J. F. Barrett, Eradication of head lice with a single treatment, *Am. J. Public Health*, 1988, **78**, 978–80.
233. P. Lisi, Sensitization risk of pyrethroid insecticides, *Contact Derm.*, 1992, **26**, 349–50.
234. H. Müller-Mohnssen, Chronic sequelae and irreversible injuries following acute pyrethroid intoxication, *Toxicol. Lett.*, 1999, **107**(1–3), 161–76.
235. S. L. Wagner, Allergy from pyrethrin or pyrethroid insecticides, *J. Agromed.*, 1994, **1**, 39–45.

236. L. Elfman, C. Hogstedt, K. Engvall, E. Lampa and C. H. Lindh, Acute health effects on planters of conifer seedlings treated with insecticides, *Ann. Occup. Hyg.*, 2009, **53**, 383–90.
237. S. Choudhary, Burns due to anti-lice lotion, *Burns*, 1999, **25**, 184–85.
238. S. A. Box and M. R. Lee, A systemic reaction following exposure to a pyrethroid insecticide, *Hum. Exp. Toxicol.*, 1996, **15**, 389–90.
239. C. I. Coleman, E. L. Gillespie and C. M. White, Probable topical permethrin-induced neck dystonia, *Pharmacotherapy*, 2005, **25**, 448–50.
240. D. E. Ray, Toxicology of pyrethrins and synthetic pyrethroids, in *Pesticide Toxicology and International Regulation*, ed. T. C. Marrs and B. Ballantyne, John Wiley & Sons, Chichester, 2004, pp. 129–58.
241. A. Moretto, Indoor spraying with the pyrethroid insecticide lambda-cyhalothrin: effects on spraymen and inhabitants of sprayed houses, *Bull. World Health Organ.*, 1991, **69**, 591–94.
242. J. E. Lessenger, Five office workers inadvertently exposed to cypermethrin, *J. Toxicol. Environ. Health*, 1992, **35**, 261–67.
243. L. Fuortes, Urticaria due to airborne permethrin exposure, *Vet. Hum. Toxicol.*, 1999, **41**, 92–93.
244. J. Magdalan, M. Zawadzki and A. Merwid-Lad, Fatal intoxication with hydrocarbons in deltamethrin preparation, *Hum. Exp. Toxicol.*, 2009, **28**, 791–93.
245. P.-Y. Yang, J.-L. Lin, A. H. Hall, T. C. Y. Tsao and M.-S. Chern, Acute ingestion poisoning with insecticide formulations containing the pyrethroid permethrin, xylene, and surfactant: a review of 48 cases, *J. Toxicol. Clin. Toxicol.*, 2002, **40**, 107–13.
246. S. M. B. Grant, An unusual case of burning mouth sensation, *Br. Dent. J.*, 1993, **175**, 378–80.
247. R. N. Das and S. Parajuli, Cypermethrin poisoning and anti-cholinergic medication—a case report, *Internet J. Med. Update*, 2006, **1**, 1–3.
248. E. D. Caldas, F. M. Rebelo, V. O. Heliodoro, A. F. A. Magalhães and R. M. Rebelo, Poisonings with pesticides in the Federal District of Brazil, *Clin. Toxicol.*, 2008, **46**(10), 1058–63.
249. D. Hou, Pesticide 'Decis' induced massive hemorrhage of gastric intestinal tract, *Chin. J. Dig. Dis.*, 1986, **2**, 116.
250. Poisindex, Pyrethroids, in *POISINDEX® Managements*, MICROMEDEX, Greenwood Village, CO, 2005.
251. Y. Gotoh, M. Kawakami, N. Matsumoto and Y. Okada, Permethrin emulsion ingestion: clinical manifestations and clearance of isomers, *J. Toxicol. Clin. Toxicol.*, 1998, **36**, 57–61.
252. P. Garg and P. Garg, Mosquito coil (Allethrin) poisoning in two brothers, *Indian Pediatr.*, 2004, **41**(11), 1177–78.
253. F. Berthier, M. Orrillard, M. Blanchard and R. Parisot, Intoxication par un insecticide: la deltamethrine, *J. Eur. Urgences*, 1995, **8**, 25–28.
254. L. Poulos, S. Athanaselis and A. Coutselinis, Acute intoxication with cypermethrin (NRDC 149), *J. Toxicol. Clin. Toxicol.*, 1982, **19**, 519–20.

255. M. O'Malley, Clinical evaluation of pesticide exposure and poisonings, *Lancet*, 1997, **349**, 1161–66.
256. R. Aggarwal and S. Diddee, Organophosphate or organochlorines or something else..., *Indian J. Crit. Care Med.*, 2009, **13**, 31–33.
257. S. Ghosh, A. Ahlawat, K. Rai and A. Arora, An unusual cause of status epilepticus, *Indian J. Crit. Care Med.*, 2009, **13**, 106–7.
258. M. Tripathi, R. Pandey, S. P. Ambesh and M. Pandey, A mixture of organophosphate and pyrethroid intoxication requiring intensive care unit admission: a diagnostic dilemma and therapeutic approach, *Anesth. Analg.*, 2006, **103**, 410–12.
259. F. Lee and J.-L. Lin, Intermediate syndrome after organophosphate intoxication in patient with end-stage renal disease, *Ren. Fail.*, 2006, **28**, 197–200.
260. E. Shahar and J. Andraws, Extra-pyramidal parkinsonism complicating organophosphate insecticide poisoning, *Eur. J. Paediatr. Neurol.*, 2001, **5**, 261–64.
261. J. Joubert and P. H. Joubert, Chorea and psychiatric changes in organophosphate poisoning. a report of 2 further cases, *S. Afr. Med. J.*, 1988, **74**, 32–34.
262. N. Gunay, Z. Kekec, Y. Cete, C. Eken and A. T. Demiryurek, Oral deltamethrin ingestion due to a suicide attempt, *Bratisl. Lek. Listy*, 2010, **111**, 303–5.
263. N. Gunay, Z. Kekec, C. Yildirim, Y. Cete, C. Erken and A. Demiryurek, Oral deltamethrin ingestion due to a suicide attempt: a case report, *Toxicol. Lett.*, 2007, **172**(Suppl), S132.
264. C. I. Spencer and J. S. Sham, Mechanisms underlying the effects of the pyrethroid tefluthrin on action potential duration in isolated rat ventricular myocytes, *J. Pharmacol. Exp. Ther.*, 2005, **315**, 16–23.
265. Dz. Naumovski, N. Bekarovski, G. Pilovski, D. Capareska, L. Petkovska, A. Cibisev, D. Petrovski and L. Melovska, Increase of transaminase levels in patients intoxicated with deltametrin, *Arch. Toxicol. Kinet. Xenobiot. Metab.*, 2000, **8**, 169–70.
266. S. B. Tucker, S. A. Flannigan and C. E. Ross, Inhibition of cutaneous paresthesia resulting from synthetic pyrethroid exposure, *Int. J. Dermatol.*, 1984, **23**, 686–89.
267. S. A. Flannigan, S. B. Tucker, M. M. Key, C. E. Ross, E. J. Fairchild, B. A. Grimes and R. B. Harrist, Synthetic pyrethroid insecticides: a dermatological evaluation, *Br. J. Ind. Med.*, 1985, **42**, 363–72.
268. Advisory Committee on Pesticides, *Evaluation Number 55: Esfenvalerate*, Ministry of Agriculture Fisheries and Food, London, 1992.
269. J.-H. Song and T. Narahashi, Selective block of tetramethrin-modified sodium channels by (+/-)-α-tocopherol (vitamin E), *J. Pharmacol. Exp. Ther.*, 1995, **275**, 1402–11.

CHAPTER 12
Veterinary Pesticides

K N WOODWARD

TSGE, Concordia House, St James Business Park, Grimbald Crag Court, Knaresborough, North Yorkshire, UK. Email: knwoodward@talktalk.net

12.1 Background

Veterinary pesticides are used to treat a number of conditions in animals, usually ectoparasites. In fact 'veterinary pesticides' is a slightly misleading term. In most countries, including the European Union (EU), ectoparasiticides applied topically are regulated as veterinary medicinal products as they are used to treat diseases of animals. Hence the regulatory authorities that deal with pharmaceuticals and biological products such as vaccines usually control them. The notable exception is the United States where the Environmental Protection Agency (EPA) regulates pesticides, including those used as ectoparasiticides, whereas pharmaceuticals are controlled by the Food and Drug Administration's Center for Veterinary Medicine and vaccines and other biologicals by the United States Department of Agriculture. In contrast, products given orally or parenterally are controlled by drug regulatory authorities.[1–4]

Ectoparasitic diseases of animals are common and vary from cat and dog fleas to organisms that inflict major economic damage to farm animals and which threaten their welfare and health. They vary widely depending on geographical area and climactic conditions.

Cat and dog fleas and other external parasites are conveniently susceptible to external treatments such as with synthetic pyrethroids (excluding cats, which are susceptible to the toxic effects of these agents) and selamectin.[5–7] Nuisance and other flies and ticks of sheep and cattle can be controlled with a variety of agents including pour-on formulations and ear tags.[9–17] Sheep, cattle and other animals are susceptible to attack from blowfly strike, and myiasis, a condition

Issues in Toxicology No. 12
Mammalian Toxicology of Insecticides
Edited by Timothy C. Marrs

Published by the Royal Society of Chemistry, www.rsc.org

where the larvae of various flies infest living or necrotic tissues.[18–21] In addition, sheep are highly susceptible to sheep scab caused by the sheep scab mite *Psoroptes ovis*.[22,23] These diseases can be treated, and frequently prevented by a variety of chemotherapeutic agents including those containing organophosphorus (OP) compounds, synthetic pyrethroids, dicyclanil and cyromazine applied by spraying, plunge dipping or topical applications.[22,24–32] Of course, these are only a few, albeit important examples of ectoparasitic diseases and their treatments. Newer agents have been introduced in recent years including imidacloprid, spinosad, lufenuron and fipronil for the treatment of fleas on cats and dogs.[33–40] Much of this new development has been in the field of companion animals, and particularly with regard to flea control. The exception to this has been in the extension of the uses of the macrocyclic lactone endectocides for use in the treatment of external parasites in both companion and farm animals, including fish. These are discussed in Chapter 13.[41]

In aquaculture (fish farming), several agents are used for the treatment and control of sea lice infestations caused by *Lepeophtheirus salmonis* and *Caligus* spp. on Atlantic salmon. Azamethiphos, cypermethrin, deltamethrin, hydrogen peroxide, emamectin benzoate, diflubenzuron and teflubenzuron are currently used as chemotherapeutic agents, while in the past, dichlorvos and trichlorfon have been widely used.[41–46]

Some examples of the various common or novel agents used to control ectoparasites in companion and farmed animals are given in Table 12.1 and discussed in this chapter.

12.2 Toxicity of Individual Substances

12.2.1 Pyrethroids

Pyrethroids that do not contain an α-cyano group such as permethrin and allethrin give rise to pronounced repetitive activity in sense organs and sensory nerve fibres resulting in the so-called T syndrome. This is characterized by fine or coarse tremor, hypersensitivity or stimuli, and aggressive sparring. Pyrethroids in this group are classified as type I pyrethroids. The data available suggests that these compounds act directly on the axon through interference with sodium channel gating mechanisms (see Chapter 5).[47–50]

Pyrethroids which possess an α-cyano group such as deltamethrin, cyhalothrin, cyfluthrin and cypermethrin produce the so-called CS syndrome. This is characterized by marked choreoathetosis (sinuous writhing), salivation (hence, CS), coarse tremor and convulsions. Pyrethroids in this group are classified as type II pyrethroids. The data suggest that that α-cyano pyrethroids act on sodium channels in the nerve membrane and cause persistent prolongation of the transient increase in sodium permeability of the membrane during excitation. Type II pyrethroids tend to be more toxic than type I compounds, and more potent than the naturally occurring pyrethrins obtained from *Chrysanthemum cinerariaefolium*.[47,51–53]

Table 12.1 Some agents used to control ectoparasites in companion and farmed animals.

Substance	*Class*	*Presentations*	*Species*	*Indications*
Cyhalothrin	Synthetic pyrethroid	Topical	Cattle	Lice, biting flies
Deltamethrin	Synthetic pyrethroid	Impregnated collar Topical	Dogs, cattle, sheep, salmon	Ticks, sandflies, lice, biting and nuisance flies, sea lice
Cypermethrin	Synthetic pyrethroid	Ear tags, topical	Cattle, sheep	Nuisance and biting flies, blowfly, sheep scab
Alpha-cypermethrin	Synthetic pyrethroid	Plunge dip	Sheep	Blowfly strike
Permethrin	Synthetic pyrethroid	Spot-on	Horses, cattle, dogs, rabbits, guinea-pigs, hamsters, rats, mice, chinchillas	Fleas, ticks, lice, biting flies
Imidacloprid	Neonicotinoid	Spot-on	Dogs, cats, rabbits	Fleas
Azamethiphos	Organophosphorus	Topical (bath)	Salmon	Sea lice
Diazinon	Organophosphorus	Topical (plunge dip, shower)	Sheep, cattle	Sheep scab, blowfly, myiasis, other biting and nuisance flies
Metaflumizone	Semicarbazone; sodium channel blocker	Topical	Cats, dogs Dogs	Fleas Demodicosis (with amitraz)
Indoxacarb	Oxadiazine; sodium channel blocker	Topical	Cats, dogs	Fleas
Fipronil	Phenylpyrazole; $GABA_A$ chloride channel blocker	Topical	Cats, dogs	Fleas, ticks
Pyriprole	Phenylpyrazole; $GABA_A$ chloride channel blocker	Topical	Dogs	Fleas, ticks
Amitraz	Formamidine; α_2-adrenergic receptor agonist	Topical	Dogs, cattle, sheep, pigs, horses	Fleas, ticks
Dicyclanil	Pyrimidine; insect growth regulator	Topical	Sheep	Blowfly
Cyromazine	Triazine; insect growth regulator	Topical	Sheep, rabbits	Blowfly, myiasis
Lufenuron	Insect growth regulator	Subcutaneous, oral	Cats, dogs	Fleas, ticks
Diflubenzuron	Insect growth regulator	Topical	Salmon	Sea lice
Teflubenzuron	Insect growth regulator	Topical	Salmon	Sea lice
Spinosad		Topical	Dogs	Fleas

It has been postulated that type II pyrethroids exert some of their effects through binding to the GABA receptor. However, they are only moderately inhibitory to the GABA receptor, particularly when compared with the activities of the cyclodiene insecticides.[54]

The data suggests that type I pyrethroids act on peripheral nerves whereas type II act primarily on the central nervous system (CNS). Data indicate that for example, that deltamethrin concentrations in the brains of mice correlate with the severity of the type II response and type II agents injected intracerebrally are more potent than when given by the intraperitoneal route, and when compared with those causing type I responses.[55–57]

The main mode of action is thought to involve alteration of the sodium channels in the excitable membrane of nerve cells resulting in prolonged sodium permeability of the neuronal membrane.[49,50] Cyhalothrin (and permethrin) has been shown to be a potent inhibitor of the mitochondrial complex I *in vitro*. Using rat liver isolated mitochondria, there was a concentration-dependent inhibition of glutamate and succinate stage 3 respiration suggesting that an effect on cellular respiration may be a contributory factor to the effects of pyrethroids.[58]

Cyhalothrin is a member of a group of pyrethroids which are relatively more toxic and biologically more potent than the naturally occurring pyrethrin-I isolated from *Chrysanthemum cinerariaefolium*. Other members of the group include cyfluthrin, lambda-cyhalothrin, deltamethrin, fenvalerate and cypermethrin. They are chemically similar in that they possess the α-cyano-3-phenoxybenzyl alcohol group.[53]

The pyrethroids are discussed in Chapter 5.

12.2.1.1 Cyhalothrin

Introduction and Metabolism. Cyhalothrin is the ISO name for (*RS*)-α-cyano-3-phenoxybenzyl (Z)-(1*RS*,3*RS*)-2(chloro-3,3,3-trifluoropropenyl)-2,2-dimethyl-cyclopropanecarboxylate and it consists of 4 of a 16 possible isomers. These isomers comprise two pairs of enantiomers, A and B, in a ratio of 60:40, as shown in Figure 12.1. The related compound, lambda-cyhalothrin, contains only the B isomers but this substance is not used for veterinary purposes.

A pair of enantiomers Z (1R) *cis* (R)α-CN **and** Z (1S) *cis* (S)α-CN

B pair of enantiomers Z (1R) *cis* (S)α-CN **and** Z (1S) *cis* (R)α-CN

Hence, commercial products contain the following combinations of cyhalothrin isomers:

~30%	Z(1R) *cis* (R)α-CN	**A pair of enantiomers**
~30%	Z(1S) *cis* (S)α-CN	
~20%	Z(1R) *cis* (S)α-CN	**B pair of enantiomers**
~20%	Z(1S) *cis* (R)α-CN	

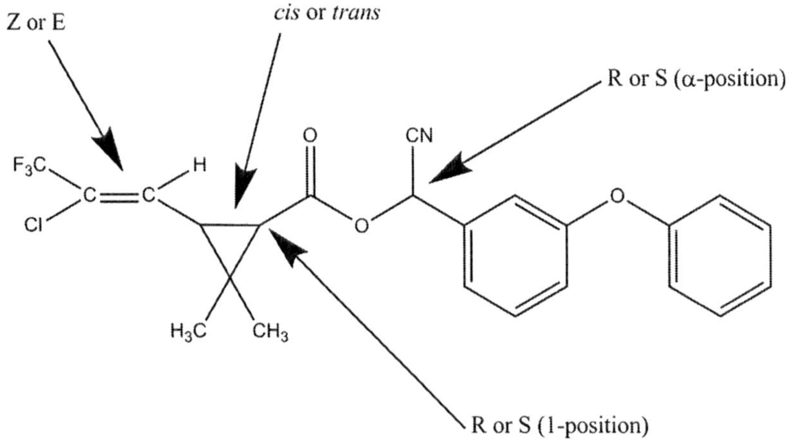

Figure 12.1 Structural formula of cyhalothrin.

Cyhalothrin is well absorbed after oral administration in corn oil to rats with around 40% being recovered in urine and the remainder in faeces. Around 4–10% was excreted in bile. In dogs and rats, metabolism is similar and yields the 3-phenoxybenzoic acid and the triflurochloropropenylcyclopropyl carboxylic acid moieties, through hydrolysis of the ester bond, a common pathway in pyrethroid metabolism.[59] The former is hydroxylated to 3-(4′-hydroxyphenoxy)benzoic acid in the rat and dog which is converted to the sulfate conjugate, and in the dog to give the glycine and glucuronide conjugates. The latter is conjugated to yield the glucuronide in both species. Similar biotransformation pathways have been reported in cattle and goats. The majority of absorbed cyhalothrin is found in the liver and adipose tissues.[60] In humans exposed to lambda-cyhalothrin during pesticide applications, the three main metabolites found in the dog, namely the trifluorochloropropenylcyclopropyl carboxylic acid, 3-phenoxybenzoic acid and 3-(4′-hydroxyphenoxy)benzoic acid were the major urinary metabolites.[61]

Toxicity. Cyhalothrin was moderately toxic after oral administration in corn oil to rats, with LD_{50} values being in the range 114–240 mg kg^{-1} bw although higher toxicity has been noted in some studies. It was more toxic to mice after oral administration in corn oil with LD_{50} values of 37 and 62 mg kg^{-1} bw being calculated. It was much less acutely toxic after dermal application, with LD_{50} values being in excess of 3500 mg kg^{-1} bw. Signs of toxicity were indicative of neurotoxicity with salivation, incontinence, ataxia, piloerection and abnormalities of gait being reported.[60]

In repeat dose studies, signs of neurotoxicity, occasionally severe, were noted in mice and rats in oral studies ranging from 5 days to 4 weeks for mice and 10 days to 3 months for rats. Like the acute toxicity studies, these signs included ataxia, piloerection and abnormal gait. Muscle trembling, collapse, convulsions and gait abnormalities have been reported in dogs given cyhalothrin for periods of 4 to 26 weeks. The no-observed effect levels (NOELs) for toxicity in mice, rats and dogs were 0.7, 0.5 and 2.5 mg kg^{-1} bw day^{-1}.[60]

Cyhalothrin has been tested in a range of studies for genotoxic potential including the reverse mutation test for point mutations, an *in vivo* cytogenetics test in the rat and a mouse lymphoma assay, but only negative results were obtained and where tested, similar effects were seen with lambda-cyhalothrin.[60,62] However, some studies with λ-cyhalothrin have yielded anomalous results. It induced micronuclei in rat bone marrow and intestinal epithelial cells and, in an unusual test using the fish *Cheirodon interruptus,* it also induced micronuclei in erythrocytes.[63,64] Nevertheless, cyhalothrin has been tested in combined chronic toxicity/carcinogenicity studies in mice and rats and there was no evidence for carcinogenic responses although signs of neurotoxicity were noted.[60]

The substance provided no evidence of reproductive toxicity in a three-generation study in mice and only a small decrease in litter size in a three-generation study in rats.[60,65] In a teratology study in rats given cyhalothrin orally in corn oil, there was no evidence of fetal abnormalities even at doses which produced maternal toxicity. The NOEL in this study was 10 mg kg^{-1} bw day^{-1}.[60] However, in other studies in rats, where cyhalothrin was given in an aqueous vehicle where low absorption would be expected, there was a delay in testicular descent in male pups. Delays in the development of fur, and ear and eye openings were also reported despite lower doses than in the previous study in rats.[66] There are no immediate explanations for the disparities in the results of these studies. No adverse effects were noted in teratology studies in the rabbit.[60] Together, the results suggest that cyhalothrin is not a mammalian teratogen.

As might be expected, cyhalothrin did not induce delayed neurotoxicity in the hen.[60] In a study of neurobehavioral effects cyhalothrin was tested in the inclined plane test in the rat. However, at the lowest dose used, 50 mg kg^{-1} bw day^{-1}, signs of neurotoxicity occurred including lethargy, writhing, ataxia, splayed gait, salivation, increased activity and vocalization. These effects possibly masked any subtle signs of neurobehavioral toxicity. To further investigate any possible effects, cyhalothrin was tested in the acute startle response test with auditory habituation after dosing. In the main part of this study, rats were given oral doses of between 5 and 75 mg kg^{-1} bw cyhalothrin. Clinical signs of toxicity, including ataxia, hypersensitivity to touch and a high-stepping gait occurred in high-dose animals but not in those given 5 or 15 mg kg^{-1} bw day^{-1}. There was however, no effect on time to maximum amplitude in the acute startle reflex (ASR) test itself performed 1 or 8 days after dosing.[60] Six functional domains have been suggested for the investigation of neurobehavioral effects of chemicals: sensorimotor, autonomic, neuromuscular, physiological, activity and excitability.[67–69] It can be argued that cyhalothrin has been subjected to standard toxicity tests, described above, where these endpoints have been addressed. Most effects were only seen at relatively high doses and even in the ASR test, signs of toxicity were only seen at the highest dose employed and a clear NOAEL of 15 mg kg^{-1} bw day^{-1} was identified. Moreover, in a functional observational battery in rats with supermethrin (a congener of cyhalothrin), lambda-cyhalothrin was used as a 'model compound'.[70] Some, but not all of the proposed functional domains mentioned above were affected by lambda-cyhalothrin, a substance that is generally more toxic than cyhalothrin, at a dose of 18 mg kg^{-1} bw. Taken together, these data confirm that the standard toxicity studies are sufficient in the case of cyhalothrin to investigate its neurobehavioral effects.

12.2.1.2 Deltamethrin

Introduction and Metabolism. Deltamethrin is the ISO name for (*S*)-α-cyano-3-phenoxybenzyl (1*R*3*R*)-3-(2,2-dibromovinyl)-2,2-dimethylcyclopropanecarboxylate is very similar structurally to cyhalothrin (Figure 12.2), the only difference being that two bromine atoms replace a chlorine atom and a trifluoromethyl group on the terminal vinyl group. As a result, there are eight possible stereoisomers for deltamethrin as the two bromine atoms remove the possibility for *cis–trans* isomerization about the vinyl group. Like cyhalothrin, deltamethrin is a type II pyrethroid and its toxicology has been reviewed by the International Programme on Chemical Safety and by the Joint FAO/WHO Meeting on Pesticide Residues (JMPR).[71,72]

Absorption of deltamethrin after oral administration is vehicle-dependent. It is well absorbed when given in polyethylene glycol 400 or glycerol formal, but less well absorbed when given in vegetable oil. Following oral administration to rats in sesame oil, deltamethrin was rapidly but incompletely absorbed. Soon after an oral dose of 26 mg kg^{-1} bw, plasma levels reached almost 1 μg mL^{-1}, as did concentrations of the major metabolite, 4′-hydroxydeltamethrin.[73] When given to rats in an aqueous vehicle it was virtually non-toxic, suggesting poor absorption. Around 70% of an orally administered dose was excreted in the urine. In rats and mice, deltamethrin and its metabolites are widely distributed to all parts of the body but concentrations were higher in lipid rich tissue including the fat and myelin.[74,75]

The biotransformation of deltamethrin is complex. In rats, the first metabolite is probably 4′-hydroxydeltamethrin followed by further hydroxylation and cleavage of the ester linkage. Many of the metabolites formed are converted to glucuronic acid, sulfate and glycine conjugates. Metabolism in other species has many similarities, but some notable differences. For example, in cattle, chickens and rats, a metabolic pathway gives rise to 4′-hydroxphenoxybenzoic acid. However, only cattle and chickens appear to form hydroxyl metabolites of the bromovinyl moiety, as well as the hydroxylactone derivative.[72,73,76–82]

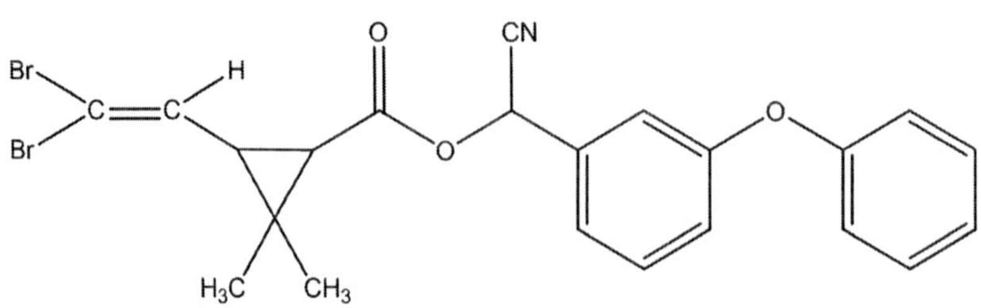

Figure 12.2 Structural formula of deltamethrin.

Toxicity. The acute oral toxicity of deltamethrin varies widely depending on the vehicle used and on the species. In rodents, the compound was very toxic to rats and mice when given in PEG 200 but marginally less so when given in sesame oil. Oral LD_{50} values were in the range 18–40 mg kg^{-1} bw. These correspond to values cited in published articles where oral LD_{50} values in the range of 25–140 mg kg^{-1} bw in rodents with arachis oil as vehicle have been reported,[83–85] and 25–63 mg kg^{-1} bw in rats administered the substance in glycerol formal.[55] The compound appears to be less toxic after oral administration to dogs with the LD_{50} value in excess of 300 mg kg^{-1} bw despite the use of PEG 200 which appeared to exacerbate the acute toxicity in rodents. When given to mice in aqueous methylcellulose, the acute toxicity was very low with the LD_{50} value being 6800 mg kg^{-1} bw.[86]

The studies also showed that deltamethrin was highly toxic to rodents and dogs after intraperitoneal and intravenous administration, with LD_{50} in the range 12–26 mg kg^{-1} bw (intraperitoneal; rodents) or 2–4 mg kg^{-1} bw (intravenous; rodents and dogs). Again, similar values have been reported after intravenous or intraperitoneal administration to rodents.[83,85,87]

Topical administration of deltamethrin resulted in low toxicity in rats and rabbits despite the use of occlusive dressing techniques. LD_{50} values were in the range 2–3 g kg^{-1} bw. This was almost certainly due to poor absorption due to the use of an aqueous vehicle. However, even when xylene, a good solvent for deltamethrin, was used, the dermal LD_{50} value exceeded the highest dose used: 800 mg kg^{-1} bw in the rat.[86]

Inhalation toxicity using acute exposures was investigated in rats, mice and guinea-pigs. However, these studies were conducted using a complex exposure media including isophorone, BHT and oil. The studies were poorly reported and overall, little interpretative value can be placed on the results although the studies suggested a low order of toxicity. A more conventional acute inhalation study in the rat using deltamethrin dust suggested a higher order of toxicity with a 6-h LC_{50} value of 600 mg m^{-3}. This is in keeping with published values of 940 and 785 mg m^{-3} in male and female Sprague-Dawley rats using approximately 2-h exposures.[85]

Signs of toxicity noted in these studies included muscular stiffening, clonic-tonic convulsions, ataxia, weakness, salivation and cyanosis in rodents with hyperexcitability, hind limb stiffness and vomiting in dogs. These signs are typical of those noted with type II pyrethroids and have been reported in published studies.[85,88,89]

When mice were administered deltamethrin at 5 or 25 mg kg^{-1} bw day^{-1} for 28 days, no major signs of overt toxicity were noted, in keeping with the majority of findings in the studies described above.[90] However, these co-workers also examined hepatic, haematopoietic and other effects in some detail and found that even the lower of the two doses caused liver, kidney and splenic degenerative changes, while deltamethrin at both doses caused stimulation of erythropoiesis, increased haematocrit, and increased numbers of leucocytes in males but not in females. Oral studies in rabbits using a 1-week dosing period failed to produce any major adverse effects although it was not clear what doses were used.[91] Others have reviewed a 21-day dermal toxicity studies in the rat with doses of up to 1000 mg kg^{-1} bw day^{-1} and PEG 400 as vehicle.[89] There were no signs of systemic toxicity, probably reflecting poor absorption by this route. Rats exposed to 6 or 12 mg m^{-3} deltamethrin as an aerosol 30 min each day for 45 days showed a variety of pulmonary effects including pneumonia, focal haemorrhage, foamy macrophage accumulation, emphysema and damage to alveolar lining cells.[92] However, it was unclear as to how the aerosol was generated or even what its components were. Moreover, the nature of the controls (no exposure or exposure to aerosol components only) was not stated. Consequently, it is difficult to derive any useful information from this study. Intraperitoneal administration to rats of deltamethrin at 7.2 mg kg^{-1} bw day^{-1} for 28 days produced no evidence of pyrethroid toxicity, although minor hepatic effects did occur (increase in mitochondria, change in their morphology).[93]

Although there are no standard reproductive studies available in the open literature with deltamethrin, other reports suggest that it may induce testicular degeneration, at least in the rat. Administration of doses of 1 or 2 mg kg^{-1} bw day^{-1} to male rats for 65 consecutive days in order to cover a complete spermatogenic cycle resulted in decreases in weights of the testes, seminal vesicles and prostates. There were also reductions in sperm counts and plasma testosterone concentrations. These findings were accompanied by a reduction in male fertility following mating with untreated female rats.[94] In a study where rats were given daily intraperitoneal injections of 1 mg kg^{-1} bw deltamethrin for 21 days, arrest of spermatogenesis occurred accompanied by degenerative changes in the testes and increases in the rate of apoptosis in basal germ cells and primary and secondary spermatocytes. Plasma levels of nitric oxide were increased over control values.[95] Deltamethrin produced slight changes in ejaculate volume and declines in sperm numbers in rabbits treated with deltamethrin although it was not entirely clear as to what doses had been used (cited as fractions of the LD_{50} value).[96]

When rats were exposed *in utero* to deltamethrin and its metabolites by treating pregnant female rats with the substance with oral gavage doses of 0, 1, 2 or 4 mg kg^{-1} bw day^{-1} from day 1 of gestation to day 21 of lactation, there were reductions in testicular weights of male offspring at the highest dose. The time to reach sexual maturity was not affected and there were no effects on sperm morphology or plasma testosterone levels.[97]

In vitro studies suggested that several pyrethroids had oestrogenic activity whereas a number of OP compounds were devoid of such activity.[98] However, studies in immature female rats and castrated males treated with up to 4 mg kg^{-1} bw day^{-1} deltamethrin for 3 days by oral gavage showed no oestrogenic or androgenic effects.[99] Similar negative results were obtained when rats were treated with a combination of deltamethrin and endosulfan.[100] However, deltamethrin resulted in testicular degenerative changes accompanied by reductions in testosterone and the arrest of spermatogenesis.[101]

The results of these studies suggest that deltamethrin may adversely affect the male reproductive system causing degenerative changes and reductions in sperm count, possibly involving an increase in apoptosis. However, it does not appear to have significant oestrogenic activity *in vivo*. Two multigeneration studies have been reviewed elsewhere. In one of these studies where there were three generations and two litters per generation, deltamethrin was given at 0, 20 or 50 ppm in the diet. The only effects seen were at the highest dietary concentration where there was reduced bodyweight in F_0 females and reduced food consumption in F_1 males. There were no effects on reproductive performance.[89] The NOEL was 20 ppm or 2 mg kg^{-1} bw day^{-1} using the WHO conversion factor.[102] In a second multigeneration study, rats were given dietary deltamethrin at 0, 5, 20, 80 or 320 ppm deltamethrin. The animals were doses for 12 weeks before a 3-week mating period, and then throughout gestation and lactation. Deltamethrin-related deaths occurred in the F_1 generation accompanied by ataxia, impaired righting reflex, urine stained fur and dark material in the stomachs of animals given the highest dietary concentration. Surviving animals also showed signs of pyrethroid-induced toxicity at this dietary level. The bodyweights of rats were also significantly reduced at 320 ppm. However, in this study there were no effects on reproductive performance. The NOEL for toxicity was 80 ppm or approximately 4.3 mg kg^{-1} bw day^{-1} as the calculated dose while the NOEL for reproductive effects was the highest concentration tested, 320 ppm or 18 mg kg^{-1} bw as the calculated dose.[103]

Deltamethrin has been tested in adequate developmental studies in the mouse, rat and rabbit at sensitive periods of gestation where it produced evidence of maternal toxicity in the rabbit for which the NOEL was 4 mg kg^{-1} bw day^{-1}. There were few signs of toxicity in mice and rats and no evidence for any teratogenic effects in any species tested.[103]

Similar findings have been reported in published studies where mice were given up to 38 mg kg^{-1} bw day^{-1} or rats up to 5 mg kg^{-1} bw day^{-1} during gestation. Although maternal signs of pyrethroid toxicity were noted, there was no evidence of teratogenic effects.[85,104,105] There is thus no evidence for teratogenic potential for deltamethrin.

There have been several publications dealing with the genotoxicity of deltamethrin. In these studies, negative results were obtained in *Salmonella* reversion assays, in the V79 Chinese hamster mutation test, in a dominant lethal assay in mice and in *in vivo* studies for clastogenicity in mice and the substance did not induce excision repair.[106–109]

However, some studies, largely for clastogenic activity, have reported positive or equivocal results with deltamethrin and other substances including other synthetic pyrethroids. Thus, positive results were seen in mouse micronucleus tests and clastogenicity studies using human lymphocytes. In the micronucleus tests, positive results were generally only seen at high doses, while in the human lymphocyte studies, the results appeared to depend on the donor rather than on the substance being tested.[110–117] For some substances, including deltamethrin, some of these effects may be due to contaminants and impurities.[118]

Overall, the available data suggest that deltamethrin is not a genotoxic material, in accordance with the findings of others and of regulatory bodies.[81,89,103,119]

Two carcinogenicity studies (mouse and rat) were commenced with deltamethrin. These were reported at 12 months as interim reports and they are discussed here under repeat dose toxicity. However, the final reports covering the entire 2-year period are not available, probably because agreement between the sponsor and the supplier of deltamethrin (and owner of the studies) was terminated.

However, the carcinogenicity studies were finalized and discussed at JMPR, European Medicines Agency (EMA) and by the European Commission.[81,103,119] In the mouse study, mice were fed diets containing 0, 1, 5, 25 and 100 ppm deltamethrin for 2 years. As described here, there were no significant findings at the 12-month interim report. Proliferative lesions were noted in some organs at 24 months but these were of similar incidence in treated and control animals and were considered to be spontaneous. There was no increased incidence of neoplasms in treated mice when compared to control values.[103]

In the rat study, animals were given diets containing 0, 5, 20 or 50 ppm deltamethrin for 2 years. Again, as described here, there were no untoward findings at the 12-month interim time point. No major signs of toxicity were noted during the 2-year study and at termination there was no increased incidence of tumours in treated animals when compared with controls.[103]

Other carcinogenicity studies in rats and mice have been conducted but the full results are not available. Thus, deltamethrin was given in the diets of mice at levels of up to 2000 ppm for 2 years. There was no increased incidence of tumours in treated animals. Similarly, rats were given diets containing up to 800 ppm deltamethrin for 2 years. Evidence of toxicity was noted in some animals but at termination there was no increased incidence of any tumour type.[103]

In studies conducted by the International Agency for Research on Cancer (IARC), mice were given gavage doses of deltamethrin in arachis oil at doses of 0, 1, 4 or 8 $mg\,kg^{-1}\,bw\,day^{-1}$ for 2 years. There was no compound-related increased incidence of any tumour type. Similarly, rats were given doses of 0, 3 or 6 $mg\,kg^{-1}\,bw\,day^{-1}$ deltamethrin for 2 years. Again, there was no evidence of compound-related carcinogenic effects and the authors concluded that deltamethrin does not appear to be carcinogenic in rats or mice.[120]

In mechanistic studies using topical application, deltamethrin was not carcinogenic in a mouse model. There was some evidence to suggest that it had some initiating activity when applied to mouse skin and the area treated with phorbol esters, but there were no data to suggest that it was a complete carcinogen.[121] It gave negative results in a rat model of hepatocarcinogenesis where animals were treated with diethylnitrosamine, followed by deltamethrin and then subject to partial hepatotectomy.[122]

Overall, there is overwhelming evidence that deltamethrin is not a carcinogen in rodent models.

As discussed earlier, type II pyrethroids are neurotoxic and many of their toxic effects can be attributed to this. All the pyrethroids affect motor function, but type II pyrethroids, including deltamethrin, are the most potent in this respect.[123] Deltamethrin has been shown to produce prolonged increases of excitability following nerve impulses in rats[124] and in the giant axons of the annelid worm *Myxicola infundibulum*,[125] while the substance reduced the amplitude of the action potential in isolated frog sciatic nerve.[126] When injected directly into the CNS, deltamethrin is extremely toxic with fatalities at doses of approximately 11 μg kg^{-1} bw in rats after intraventricular administration.[127,128] Intraperitoneal doses of 12.5 mg kg^{-1} bw to rats induce degenerative changes in the brain including apoptosis. Apoptosis also occurs in cultured cerebral cortex neurons *in vitro*.[129,130]

Deltamethrin has been tested in a range of neurobehavioral toxicity studies. The compound was found to increase spontaneous activity in both rats (7 mg kg^{-1} bw in corn as a commercial pesticide emulsion, orally) and mice (0.7 mg kg^{-1} bw in arachis oil–egg lecithin aqueous emulsion, orally). In the rat study, aggressive behaviour and the relearning index were also increased whereas there was no effect on learning and memory.[131,132] Oral doses of up to 8 mg kg^{-1} bw in corn oil resulted in decreases in maze behaviour in rats; the NOEL was 2 mg kg^{-1} bw day^{-1}. In this same work, deltamethrin resulted in a dose-dependent decrease in amplitude and an increase in latency of the ASR, again with an NOEL of 2 mg kg^{-1} bw.[133] When rats were studied in the ASR test with oral doses of up to 6 mg kg^{-1} bw deltamethrin in corn oil, the ASR was attenuated at the highest dose. The NOEL was 4 mg kg^{-1} bw.[134]

In a study of the scheduled-controlled response in mice given 0.3–3 mg kg^{-1} bw deltamethrin given by the intraperitoneal route, responses were reduced by doses greater than 0.1 mg kg^{-1} bw (the NOEL). The ED_{50} was approximately 1 mg kg^{-1} bw.[135] A similar study in rats used intraperitoneal doses of 2 mg kg^{-1} bw which reduced the operant response by 80%.[136] Unfortunately, no other doses were used and so an NOEL could not be identified.

Deltamethrin (15 and 150 mg kg^{-1} bw orally) was subjected to a number of tests in a battery of neuropharmacological studies. These comprised a test for motor coordination using a rotarod and the effects on pentobarbitone sleeping time and pentylenetetrazole-induced convulsions. Deltamethrin treatment significantly decreased sleeping time while it increased the duration of convulsions. In the rotarod test, significant ataxia occurred.[137] The latter findings are perhaps not surprising in view of the induction of ataxia and convulsions in standard toxicity studies.

Deltamethrin at 0.08 mg kg^{-1} bw led to alterations in swimming behaviour, motor activity and striatal dopamine concentrations when rats were exposed prenatally.[138] The data suggested a higher level of dopaminergic activity but the implications of these observations are unknown.

Rats treated with inducers of cytochrome P450 showed increases in some neurobehavioral parameters while those treated with cobalt chloride to deplete cytochrome P450 showed decreases, indicating that a metabolite of deltamethrin may be responsible for some of the effects seen in rodents.[139]

As noted in the discussion of cyhalothrin, it has been suggested that the neurobehavioral effects of substances should be assessed by reference to six functional domains: sensorimotor, autonomic, neuromuscular, physiological, activity and excitability.[67,69] These effects have not been studied systematically in the studies described here nor have dose–response relationships been established. However, in two GLP-compliant studies reviewed by JMPR, deltamethrin was administered to rats at oral doses of 0, 5, 15 or 50 mg kg^{-1} bw in corn oil. This battery included posture, convulsions, tremors, biting, eyelid closure, faecal consistency, ease of handling, lachrymation, salivation, respiration rate, muscular tone, motility, rearing, convulsions, tremors, bizarre or stereotypical behaviour, gait, startle, touch, tail pinch, grip strength, rotarod performance, body temperature and locomotor activity. At the highest dose, tremors, clonic and tonic convulsions and numerous behavioural changes occurred and all six domains described previously were affected.[69] At 15 mg kg^{-1} bw, slight salivation, slightly stained fur and impaired ability were noted in individual animals. The NOEL was 5 mg kg^{-1} bw. In a further study by the same authors, rats were given diets containing 0, 50, 200 or 800 ppm deltamethrin for 91 days, and subjected to the same battery of tests and observations as described immediately previously. Signs of toxicity and notably neurotoxicity were seen in animals given the highest dietary concentration and the lowest dietary concentration without adverse effect was 200 ppm, equivalent to 54 mg kg^{-1} bw day^{-1}.[103]

In conclusion therefore, deltamethrin induces neurotoxicity characterized by tremors, clonic and tonic convulsions and ataxia in rodents. It also produces more subtle neuropharmacological/toxicological effects. The oral NOELs for these effects is in the range of 2–5 mg kg^{-1} bw, although this was higher at 54 mg kg^{-1} bw in a 90-day neurobehavioral study.

Deltamethrin has been tested in a study in the hen. This test has been developed to demonstrate OP-induced delayed neuropathy (OPIDN), which occurs following an initial cholinergic crisis, and involves a selective degeneration of long and large fibres of the spinal cord and peripheral nervous system.[140] The domestic chicken has been shown to be a model, albeit with some limitations for this effect of OP compounds, and it is widely used to investigate or predict this specific toxicity.

Synthetic pyrethroids have not been shown to cause this effect and their mode of action would not suggest that it is likely and neuropathy target esterase (NTE) is almost exclusively a target for OP compounds.[141] Hence, this study can be considered to be superfluous. Unsurprisingly, the study was negative and there was no indication of the induction of a delayed neuropathy.

12.2.1.3 Cypermethrin and α-Cypermethrin

Introduction and Metabolism. Cypermethrin is the ISO name for (*RS*)-α-cyano-3-phenoxybenzyl (1*RS*, 3*RS*;1*RS*,3SR)-3-(2,2-dichlorovinyl)-2,2-dimethylcyclopropanecarboxylate. It is a type II pyrethroid similar in structure to deltamethrin (Figure 12.3) but here two chlorine atoms replace the two bromine atoms on the vinyl structure. Like deltamethrin, cypermethrin is a mixture of eight isomers. α-Cypermethrin (ISO) is a racemate composed of the 1*R*, *cis* and 1*S*, *cis* isomers.[142,143] Cypermethrin is well absorbed after oral

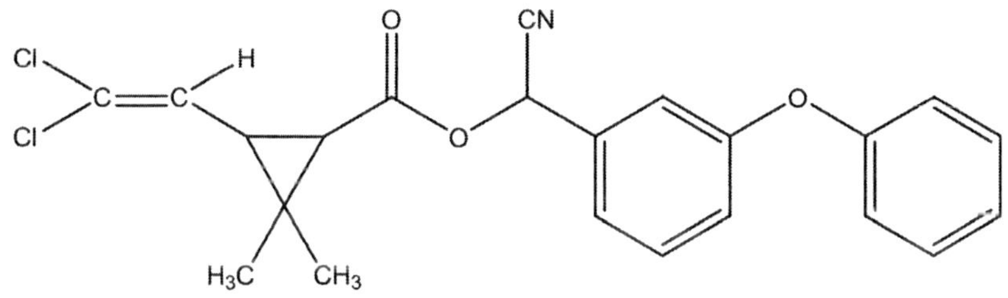

Figure 12.3 Structural formula of cypermethrin.

administration and rapidly excreted in urine and faeces. As with other pyrethroids, the ester bond is cleaved to yield the 3-phenoxybenzoyl moiety which is further oxidized and subject to sulfate conjugation, and the corresponding cyclopropane carboxylic acid. Minor metabolites include the glycine conjugate of 3-phenoxybenzoic acid and 3-(4-hydroxyphenoxy)benzoic acid.[144] Cypermethrin residues are found in most tissues in several animal species, but concentrations are significantly higher in adipose tissues.[142–144]

Toxicity. Like other pyrethroids, cypermethrin and α-cypermethrin are more toxic orally when administered in oil than when given in predominantly aqueous solvents. For example, the oral LD_{50} value for cypermethrin in rats was 251 mg kg^{-1} bw in corn oil but 4000 mg kg^{-1} bw when given in 40% aqueous dimethyl sulfoxide and 3423 mg kg^{-1} bw when given as a 50% aqueous suspension. Similar results were seen with α-cypermethrin but this was more acutely toxic than cypermethrin with LD_{50} values in the rat of 64 mg kg^{-1} bw when given in corn oil. However, as with cypermethrin, toxicity was reduced when given in aqueous dimethyl sulfoxide (LD_{50} 4000 mg kg^{-1} bw) or as an aqueous suspension ($LD_{50} > 5000$ mg kg^{-1} bw).

Signs of toxicity included ataxia, splayed gait, tip-toe walking, tremors and clonic convulsions.[142–144]

In repeat dose studies where mice were given a variety of dietary concentrations of α-cypermethrin for 29 days, animals developed signs of neurotoxicity at doses of 1200 or 800 mg kg^{-1} feed and above. Signs included ataxia, abnormal gait, overactivity and hunched posture. The NOEL was approximately 57 mg kg^{-1} bw day^{-1}. When groups of mice were given the substance at up to 1000 mg kg^{-1} feed for 13 weeks, the major adverse sign noted was hair loss at all dose levels. Other signs included ungroomed fur and encrustations of the dorsal body surfaces. There were a number of variations in organ weights but no frank evidence of neurotoxicity.[144] Effects on the bactericidal activity of neutrophils, and increase in the numbers of monocytes and lymphocytes seen in one study with α-cypermethrin in mice, are difficult to interpret in the absence of data from other pyrethroids and in other species.[145]

Cypermethrin and α-cypermethrin have been tested in a range of repeat dose studies in rats ranging from 35 to 90 or 95 days. With cypermethrin, the main effects were decreases in haemoglobin, mean corpuscular volume and eosinophil counts and increases in relative liver weights. The NOEL from these studies was 5 mg kg^{-1} bw day^{-1}. α-Cypermethrin decreases in mean corpuscular volume and haemoglobin were also noted. Animals given 800 and 1200 mg kg^{-1} feed showed signs of neurotoxicity including high stepping, splayed gait, abasia and hypersensitivity, with cachexia in severe cases. Similar effects were seen in a separate study with doses of 540 mg kg^{-1} feed. The NOEL from these studies was equivalent to 3 mg kg^{-1} bw day^{-1}. Signs of neurotoxicity have also been reported in dogs given cypermethrin or α-cypermethrin in repeat dose studies. These included tremors, ataxia, incoordination and hyperaesthesia. At 1500 mg kg^{-1} feed, 50% of animals in a dose group had to be killed due to the toxic effects of cypermethrin. The NOEL from this particular study was 12.5 mg kg^{-1} bw day^{-1}. Similar signs were reported with α-cypermethrin although this substance appeared to be more toxic than cypermethrin. The lowest NOEL with α-cypermethrin was 1.5 mg kg^{-1} bw day^{-1}.[144]

Cypermethrin had no adverse effects on reproductive performance in a three-generation study in rats. Both substances produced signs of maternal toxicity in teratogenicity studies in rats, while α-cypermethrin produced fetotoxicity at higher doses. Neither compound produced any evidence of teratogenicity. Similarly, both substances have been tested in teratogenicity studies in rabbits. Neither produced evidence of embryotoxicity or teratogenicity. The NOELs from these studies was 70 mg kg^{-1} bw day^{-1} for cypermethrin and 9 mg kg^{-1} bw day^{-1} for α-cypermethrin in rats and 120 mg kg^{-1} bw day^{-1} for cypermethrin and 30 mg kg^{-1} bw day^{-1} for α-cypermethrin in rabbits.[144]

Both substances have been investigated in a number of *in vitro* and *in vivo* studies of genotoxicity covering a variety of endpoints including the induction of mutations and clastogenicity. All of these studies yielded negative results. There was no evidence of carcinogenicity with cypermethrin in long-term studies in mice and rats,[144] and in view of this result, and the results from genotoxicity studies, there is no reason to believe that α-cypermethrin would be carcinogenic.

Cypermethrin produced a significant but transient functional impairment in the inclined plane test in rats that was maximal after 7 days of administration. However, there were no adverse effects in the ASR test.[146,147] α-Cypermethrin was investigated in a functional observational battery of tests for measurements of forelimb and hindlimb grip strength, hindlimb landing foot splay and motor activity. Signs of toxicity included abnormal and splayed gaits, prostration, vocalization, piloerection and hunched posture while gait abnormalities were observed during the functional observational battery phase of the study. At higher doses, slight degeneration of the fibres of the sciatic nerve occurred at higher doses. The NOEL for this study was 4 mg kg^{-1} bw day^{-1}.

12.2.1.4 Permethrin

Introduction and Metabolism. Permethrin is the ISO name for 3-phenoxybenzyl(1*RS*,3*RS*;1*RS*3*SR*)-3-(2,2-dichlorovinyl)-2,2-dimethylcyclopropanecarboxylate and, as mentioned above, it lacks the cyano group that characterizes cyhalothrin, deltamethrin, cypermethrin and other, similar synthetic

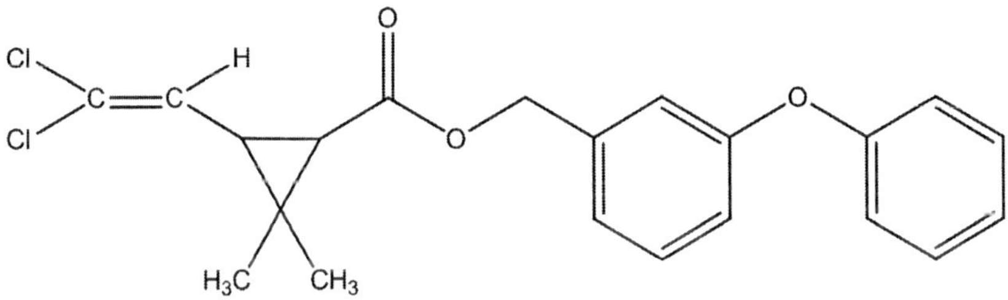

Figure 12.4 Structural formula of permethrin.

pyrethroids (Figure 12.4). Unlike these substances, permethrin, like allethrin and resmethrin, is a type I pyrethroid whose actions are thought to be due to its effects on ion channels in axons.[148,149] The metabolism of permethrin, like that of many other pyrethroids, is complex. In the rat, the major metabolites arise from cleavage of the ester bonds to yield the hydroxyphenoxybenzoic acid derivatives, which are conjugated with sulfate. Phenoxybenzoic acid, probably *via* phenoxybenzyl alcohol, is also formed and this is converted to the glucuronide and glycine conjugates.[148,149]

Toxicity. In the absence of an oily vehicle, the acute oral toxicity of permethrin is low with LD_{50} values in excess of 2000 mg kg^{-1} bw, or even 8000 mg kg^{-1} bw in rats. However, when administered in corn oil the toxicity increases markedly. LD_{50} values in the range 230–1700 mg kg^{-1} bw have been reported in mice, and 220–1600 mg kg^{-1} bw in rats. The toxicity also depends on the isomeric composition of the material used with permethrin with a *cis:trans* ratio of 80–100:20–0 being up to 24 times more toxic than that with a *cis:trans* ratio of 10–25:90:75. In rabbits, the oral LD_{50} value was greater than 2000 mg kg^{-1} bw for *cis:trans* 55:45 and 40:60 permethrin.[149]

In repeat dose studies in mice, no significant clinical findings occurred when animals were given permethrin (*cis:trans* 39:56) in the diet at concentrations equivalent to up to 560 mg kg^{-1} bw day^{-1} for 28 days. There were no major gross or microscopic findings except for increased liver weights at the two highest concentrations with some eosinophilia in centrilobular hepatocytes. The NOEL was equivalent to 140 mg kg^{-1} bw day^{-1}. In rats given a similar isomeric composition of permethrin in the diet for 28 days, 63% of those given the equivalent of 500 mg kg^{-1} bw day^{-1} and 100% of those given the equivalent of 1000 mg kg^{-1} bw day^{-1} died. Animals showed whole-body tremors, hyperactivity and piloerection. In a similar study where the highest dietary level was equivalent to 630–660 mg kg^{-1} bw day^{-1} for 30 days, all the high-dose animals died within 24 h to 1 week of the start of the study. There was very high mortality at the next dietary level, equivalent to around 250 mg kg^{-1} bw day^{-1}. The main clinical signs included slight to moderate tremors and staining of fur. The NOEL values from these two studies were 50 and 250 mg kg^{-1} bw day^{-1} respectively.[149]

When rats were given dietary permethrin at concentrations of 0, 20, 50, 100, 250, 500 and 1000 mg kg^{-1} bw day^{-1} for 28 days, all animals given the highest dietary intake died within 3 days of the start of the study. There was also high mortality at the 500 mg kg^{-1} bw day^{-1} dose. Prior to death, animals displayed whole-body tremors, hyperactivity and piloerection while surviving animals given 500 mg kg^{-1} bw day^{-1} showed urinary incontinence. At the end of the study, animals given $\geq$250 mg kg^{-1} bw day^{-1} had increased absolute and relative liver weights. The NOEL in this study was equal to 50 mg kg^{-1} bw day^{-1}. Similar qualitative findings were made in other repeat dose studies of similar durations and in other studies extending to 90 days or 6 months. Hypersensitivity was also seen in these studies.[140]

In dogs given repeat oral doses of permethrin of up to 500 mg kg^{-1} bw day^{-1} for 90 days, a range of clinical signs of toxicity has been reported including tremors, emesis, transient narcosis, nystagmus, ataxia and aggressive behaviour at the highest dose employed. However, when given at 250 mg kg^{-1} bw day^{-1} for 180 days, no major signs except emesis were observed. At 1000 mg kg^{-1} bw day^{-1} in a 52-week study, convulsions, muscle tremor and lack of coordination occurred. Lower doses produced reductions in bodyweight. The lowest NOEL from studies in dogs was 5 mg kg^{-1} bw day^{-1}.[140]

Permethrin was not a skin irritant in studies in rabbits and produced only signs of mild eye irritation in this species. It was tested in a number of maximization tests in the guinea-pig but produced no evidence of dermal sensitization.[140]

In a battery of tests for various endpoints of genotoxicity, permethrin gave negative results. These tests included those for gene mutations, the induction of unscheduled DNA synthesis and DNA repair. It has not been tested for clastogenic activity *in vivo* but there was evidence of clastogenicity, or at least equivocal results, in mammalian cells *in vitro*.[149–162]

In CD-1 mice, a slightly elevated incidence of alveolar adenoma and alveolar carcinoma were seen in an oral carcinogenicity study with permethrin. However, the incidence of the former was lower in test animals than in controls, while the incidence of the latter was only slightly increased in females (0% in all males and in female controls and in 1 each in 59 or 60 mice given the low, intermediate and high doses). A second study was conducted and here much higher incidences of both alveolar adenomas and carcinomas were noted than in the previous study. There was no apparent dose relationship in males, but the incidence of both tumour types was increased in females, 13, 24, 35 and 49% for adenomas and 8, 9, 15 and 20% for carcinomas in controls, low (3 mg kg^{-1} bw day^{-1}), intermediate (15 mg kg^{-1} bw day^{-1}) and high (75 mg kg^{-1} bw day^{-1}) respectively. There were also slight increases in hepatocellular adenoma as is often seen in long-term mouse studies, but the incidence of hepatocellular carcinoma was not increased. In a separate study using Alderley Park mice, and dietary concentrations equivalent to 0, 38, 150 or 380 mg kg^{-1} bw day^{-1}, there was no increased incidence of tumours of any type. Hence, there is conflicting evidence for the ability of permethrin to induce tumours in mice. However, in two long-term studies in rats (Alderley Park and Long Evans) there was no evidence of a carcinogenic effect. In the CD-1 mice studies where lung tumours were seen, there was a high background (*i.e.* incidence in controls) of alveolar adenomas in mice of both sexes in one study, and of adenomas and carcinomas in the other whereas a third study gave negative results. In view of this, the overwhelming number of negative results in tests for genotoxicity, and negative results in two rat carcinogenicity studies, permethrin can be considered not to be a genotoxic carcinogen in animal models.[149,163]

Permethrin has been tested in a number of studies of neurotoxicity. After intravenous administration to mice, permethrin induced hyperactivity, increased sensitivity to external stimuli and whole-body tremor. This eventually led to prostration and death with the ED_{50} being 20 mg kg^{-1} bw for the *cis* isomer, 36 mg kg^{-1} bw for the technical mixture and 93 mg kg^{-1} bw for the *trans* isomer. The ED_{50} was much lower after intracerebroventricular administration (0.09, 0.15 and 1.1 mg kg^{-1} bw respectively). The neurotoxicology of permethrin has been investigated in functional observational studies in rats. In one study where animals were given permethrin orally in corn oil, those given the highest dose had whole-body tremors, exaggeration of flexion of the hindlimbs, staggered gait, splayed hindlimbs, and abnormal posture with convulsions on the first day of the observational tests. The NOEL for this study was 150 mg kg^{-1} bw. In another observational battery study, rats were given diets containing permethrin for 13 weeks. Signs of toxicity appeared at dietary concentrations of $\geq$1500 ppm and included staggered gait, splayed hindlimbs, and tremors. The NOEL was equivalent to 15 mg kg^{-1} bw day.[149,164,165]

12.2.1.5 Overview of the Toxic Effects of the Pyrethroids

The synthetic pyrethroids are characterized by their neurotoxicity. They are not reproductive toxicants or teratogens and in general they yield negative results in genotoxicity and carcinogenicity studies. However, regardless of the type of study, if high enough doses are given, then neurotoxicity will ensue. As mentioned at the beginning of this section on synthetic pyrethroids, this toxicity is dependent on the presence or absence of the α-cyano group and is perhaps more advantageously characterized after intracerebroventricular administration.[48,57] Despite the advances made in research in recent years, the biochemical and molecular mechanisms of pyrethroid toxicity are still not fully understood and new information continues to become available. For example, although it is known that the pyrethroids cause decreases in the voltage-gated sodium channel inactivation rates, it is not known how these effects result in the characteristic effects of different synthetic pyrethroids and why the types I and II compounds differ in their activities. However, this could be related to differential effects on glutamergic and other neuronal networks.[166] Moreover, some sodium channels may be more sensitive to the effects of those pyrethroids containing the α-cyano group.[167] Deltamethrin may also affect the dopaminergic system, specifically in the hippocampus and striatum, and may even result in nerve cell damage through induction of mitochondrial-mediated apoptosis.[168–171] Deltamethrin is a potent inducer of brain-derived neurotropic factor in neurons suggesting its capacity to act as an inducer of neuronal hyperexcitation[172] However, none of these findings, or those discussed elsewhere in this chapter fully explain the range of signs of toxicity elicited in mammals by the synthetic pyrethroids.

12.2.1.6 Toxicity to Cats

Permethrin and other synthetic pyrethroids are widely used as ectoparasiticides for the treatment of dogs. These are usually applied as a spot-on formulations and are used largely to control fleas or, in combination with other active substances, to control other parasitic organisms. Problems occur when these pyrethroid products are used in cats. This can, and often does, result in severe toxicity and frequently in death. In fact, a recent survey of 286 cases found that 96.9% of cats exposed to permethrin developed clinical effects and over 10% died or were euthanized. Warnings have been issued, and continue to be issued, by regulatory authorities to try to minimize this problem. This extreme toxicity in cats is probably related to a number of factors including the animal's grooming habits and the inabilities of felines (large as well as small) to conjugate xenobiotics, thus ensuring that they are not adequately detoxified and excreted.[173–180]

12.2.1.7 Toxicity to Humans

Neurotoxicity occurs in animals exposed to relatively high doses of synthetic pyrethroids; doses higher than humans are likely to encounter during the normal use of veterinary medicinal products or indeed during the use of pyrethroid-containing pesticide products. In fact the major form of adverse effect associated with exposure to synthetic pyrethroid in humans is cutaneous paraesthesia.[182–184] This develops some time, usually hours, after exposure, and produces a burning or stinging sensation on the skin. It is associated with type II pyrethroids. However, when exposure to higher doses does occur, dizziness, burning and tingling sensations, epigastric pain, vomiting, anorexia, chest tightness, blurred vision, palpitations, muscular fasciculations, and disturbances of consciousness may occur. In severe poisoning cases, convulsions and loss of consciousness occur. All of these effects are reversible except when death occurs.[184–193]

12.2.2 Neonicotinoids

12.2.2.1 Imidacloprid

Introduction and Metabolism. Imidacloprid is the ISO name for 1-(6-chloro-3-pyridinylmethyl)-*N*-nitroimidazolidin-2-ylideneamine; Figure 12.5. It is a relatively new insecticide which is used as a pest control agent on crops and as a

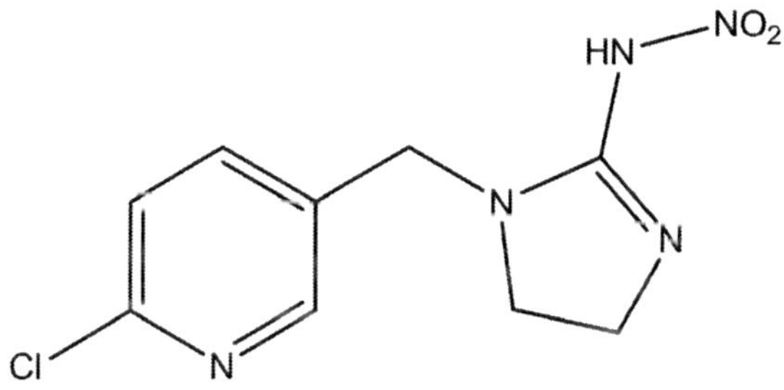

Figure 12.5 Structural formula of imidacloprid.

veterinary medicinal product. It is a member of the neonicotinoid group of agents. These are chemically related to natural nicotine and in addition to imidacloprid the class includes nitenpyram, thiacloprid, and acetamiprid. Imidacloprid, like many neonicotinoids, is water soluble (0.6 g L^{-1}) and most of an oral dose is excreted unchanged in the urine although a significant quantity may be excreted in the bile in rats.[192,193] The substance is widely distributed in the body. It is extensively metabolized to yield, *inter alia,* 6-hydroxynicotinic acid, 6-methylmercaptonicotinic acid and its glycine conjugate, and various metabolites derived from the nitro-2-imidazolidinimine moiety.[194,195]

The neonicotinoid agents act as agonists for the nicotinic acetylcholine receptors (nAChR) in insects and mammals, and particularly for the $\alpha4\beta2$ subtype.[192] For further discussion of the neonicotinoids see Chapter 6.

Toxicity. Imidacloprid is moderately acutely toxic in rodents with oral LD_{50} values in the range 380–650 mg kg^{-1} bw in rats and 130–170 mg kg^{-1} bw in mice. It was practically non-toxic ($LD_{50} > 5000$ mg kg^{-1} bw) following dermal application to rats. Where toxicity occurred, signs included transient trembling and spasms, motility abnormalities, respiratory signs, and behavioural effects.[194,195] In repeat dose studies in rodents, the main signs of toxicity were frequently limited to elevated enzymes and depressed protein, albumin, cholesterol and triglycerides. In rats, there was decreased activity of acetylcholine in serum and in the brain.[195,196] In dogs, ataxia, vomiting and tremor occurred.[193]

There was no evidence of carcinogenic effects in long-term studies in mice and rats although there was evidence of mineralization of the thalamus in mice and an increased incidence of mineralization of the thyroid in rats.[193] Imidacloprid has been tested in a range of studies for genotoxic potential covering a wide range of endpoints including those for mutations, DNA damage, sister chromatid exchange, and the *in vivo* micronucleus test. In the vast majority of those where imidacloprid was tested alone, negative results were obtained except in one test for sister chromatid exchange where a positive result was noted.[195] When tested in combination with the OP insecticide methamidophos, positive results were noted in a *Salmonella* reversion assay, in a rat bone marrow chromosome aberration assay and in a mouse micronucleus test; similar results were noted when the compounds were tested individually.[197]

The significance of these results remains unclear in the light of the results of the numerous (16) studies mentioned above, and overall the data strongly suggest that imidacloprid is not genotoxic.

In a multigeneration study in rats, the major effects were reduced bodyweights in parental and F_1 and F_2 animals. There was no evidence of teratological effects rats or rabbits but fetotoxicity was seen (wavy ribs, delayed ossification) in offspring from animals that showed signs of maternotoxicity.[195]

In studies of neurotoxicity, tremors, gait abnormalities, decreased activity and coolness to touch have been reported. The lowest NOEL from these studies was 9.3 mg kg^{-1} bw day^{-1}.[195]

Effects in Treated Animals. Imidacloprid has a good safety profile in treated animals. However, it may result in skin irritation and caustic burns in cats.[198] There has been a report of a exfoliative erythroderma in a cat treated with imidacloprid, but this animal also had a thymoma and the skin condition may have been related to this rather than to treatment.[199] Studies with the combination of imidacloprid and moxidectin in cats and dogs demonstrated excellent tolerance with no adverse reactions,[200,201] while ingestion from self-grooming may result in drooling and retching in cats.[202]

Effects in Humans. The major adverse effects in humans appear to be occasional mild cases of dermatitis in pet owners who had treated their animals with a commercial formulation of the product. However, these effects were probably due to other constituents of the formulation rather than to the active substance itself.[195] In a double-blind crossover study of tree planters who had handled trees treated with imidacloprid and cypermethrin, no adverse effects were found.[203] No signs of toxicity were seen in a 4-year-old child who ingested approximately 200 mg of imidacloprid, equivalent to about 10 mg kg^{-1} bw.[195] Ingestion of relatively large quantities of imidacloprid formulations which resulted in measurable plasma concentrations (mean 10.58 ng L^{-1}; range 0.02–51.25 ng L^{-1}) resulted in mild symptoms such as nausea, vomiting, headache and diarrhoea. One patient had respiratory failure and another prolonged sedation, but no deaths occurred.[204]

In summary, imidacloprid is a safe and effective agent for the treatment of external parasites of small animals. It has relatively low mammalian toxicity, but where signs of toxicity are seen these are indicative of neurotoxicity. In insects at least, imidacloprid appears to act on a number of nAChR subtypes and it almost certainly exerts its neurotoxic effects through these interactions in the insect targets and in mammals.[194,205]

12.2.3 Organophosphorus Compounds

The toxicity of the OP compounds in mammals, including humans, is well documented.[206,207] Their acute toxicity is characterized by inhibition of acetylcholine esterase (AChE) resulting in the accumulation of acetylcholine and overstimulation of nicotinic and muscarinic receptors.[207–209] However, there is also an intermediate phase characterized by weakness and paralysis of proximal limb and other muscles and delayed peripheral neuropathy, and these occur after the initial cholinergic phase, although the latter, rarely, may occur in its absence.[207–210] OPIDN, which only occurs with certain compounds, occurs around 5 weeks after acute poisoning and the molecular target is thought to be NTE.[207–210] However, in addition to these effects, OP compounds may cause a variety of other effects on a number of organ systems and many of them possess genotoxic properties.

In the past, a number of these substances have been used in veterinary medicine including phosmet, propetamphos, chlorfenvinphos and dichlorvos. However, the major uses are now restricted to diazinon and azamethiphos.

A comprehensive account of the toxicity of the OP compounds is given in Chapter 4.

12.2.3.1 Diazinon

Introduction and Metabolism. Diazinon (ISO); formerly dimpylate, *O,O*-diethyl *O*-[4-methyl-6-(propan-2-yl)pyrimidin-2yl] phosphorothioate; Figure 12.6; is an OP compound that is widely used in agriculture but whose use in veterinary medicine is now largely restricted to sheep-dipping for the treatment and control of ectoparasites. Its toxicity has been reviewed by JMPR.[211] Diazinon is well absorbed in rats after oral administration with up to 80% of the administered dose being excreted in urine. Similarly, after oral administration to mice, guinea-pigs and dogs, absorption was extensive. After topical application to rats, up to 80% of the applied dose was recovered in urine demonstrating facile dermal absorption.

The biotransformation of diazinon is complex, with both minor and major pathways being identified.[211] The major metabolites are the oxon (diazoxon), diethylphosphate and diethylthiophosphate along with 2-isopropyl-4-methyl-6-hydroxypyrimidine. The activation to the oxon has been shown to be due to the

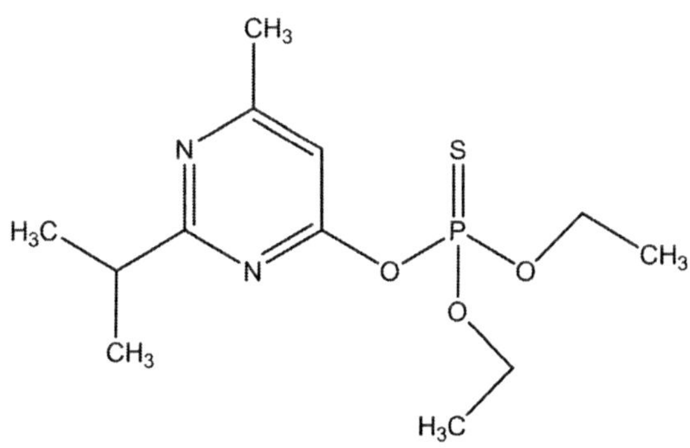

Figure 12.6 Structural formula of diazinon.

action of cytochrome P450 and in human liver *in vitro*, through CYP2C19.[211–215] Although diazinon itself is considered to be a weak inhibitor of AChE, the oxons are known to be potent inhibitors.

Toxicity. Diazinon is acutely toxic, with oral LD_{50} values in rats being in the range 200–1250 mg kg^{-1} bw in the rat. Following topical exposure, LD_{50} values were greater than 2150 mg kg^{-1} bw in the rat and greater than 2020 mg kg^{-1} bw in the rabbit.[211] Signs of toxicity were typical of OP compounds and those expected from AChE inhibitors, namely decrease in activity, sedation, dyspnoea, ataxia, tremors, convulsions, lacrimation and diarrhoea. These signs resolved in animals that survived. In repeat dose toxicity studies, the major signs were again those associated with AChE inhibition, and where measured, there were significant reductions in erythrocyte, and more importantly in brain AChE activities.[211] Diazinon has been tested for the induction of delayed neuropathy in the hen. There was no evidence of this effect and no adverse histopathological findings.[211]

Diazinon gave negative results in the majority of the studies for genotoxic activity, with the exception of the mouse lymphoma assay for forward mutations and an *in vitro* test for chromosome aberrations.[211] These results are perhaps not surprising because it is recognized that some OP compounds are genotoxic, although diazinon fell into the category of 'mainly negative results'.[212] Nevertheless, in some studies diazinon has given positive results including those for chromosomal aberrations and DNA damage.[216–224] This too is not surprising given that many OP compounds are alkylating agents.[223] However, although diazinon has been evaluated in carcinogenicity studies in mice and rats, there was no evidence for the induction of tumours.[211]

Diazinon has been tested in reproductive and in developmental studies. In reproduction studies, the main findings were reduced viability of pups and reduced weight gain of dams. The substance did not induce birth defects in the offspring of rats or rabbits treated during sensitive periods of gestation.[211] However, diazinon and other OP compounds may adversely affect the testes, sperm quality and motility, effects for which melatonin may be protective.

Many of the OP compounds are immunotoxic and immunosuppressive. The mechanisms by which diazinon and other OP compounds cause immunotoxicity are not fully understood but diazinon is cytotoxic and results in necrotic areas of the spleen and thymus in mice. It also leads to hyperplasia of the medulla of the thymus and lymph nodes, and hyperplasia of the white and red pulp in the spleen. Some of these changes may result in decreases in cytokines such as interleukins and interferons.[225–229]

Toxicity to Humans. There have been several incidents involving toxicity with diazinon. Some of these were deliberate and some accidental.[230–233] Of the accidental cases, these usually involved ingestion of stored liquid formulations.[234–236] Some cases involved topical exposure during applications for lice treatment while others involved children.[234–240] The majority of cases resulted in acute toxicity and signs were typical of AChE inhibition. Most cases responded well to atropine and oxime inhibition. In severe cases, death has resulted.[233]

There have been several cases of intermediate toxicity with diazinon. These resulted in paralysis of the proximal limb muscles, the neck flexor muscles and respiratory muscles. These signs are not responsive to atropine therapy.[231,241,242]

There have also been reports of occupational exposure to diazinon resulting in toxicity. Occasionally this was due to contamination with TEPP or monothio-TEPP and other impurities. In some of these cases, death have occurred and in addition, several cattle died in one incident, almost certainly as a result of exposure to monothio-TEPP present as a contaminant in the formulation.[243–245] In the period 1945–1989 pesticides were responsible for 1012 of the 87 385 deaths due to poisoning in the UK. Of these deaths, only one was due to diazinon.[246]

In the UK, reports began to appear of adverse reactions in farm workers and shepherds following exposure to sheep dips, particularly those containing OP compounds, and notably diazinon. These reactions included a transient influenza-like illness widely referred to as 'dipper's flu' and longer-term clinical signs and symptoms that were frequently vague in nature but included loss of memory, depression, pyrexia and headaches.[247–249] A considerable amount of research was commissioned to try to obtain data that might cast some light on this phenomenon, but none the efforts provided conclusive evidence to associate it with exposure to diazinon during sheep-dipping and although dipping of sheep continues, the numbers of reports of adverse reactions has fallen markedly.[247,250,251]

In addition to the anticholinergic effects of OP compounds, there are other, more subtle effects on the nervous system some of which may contribute to the long-term effects. Indeed, some of these effects occur at concentrations that are unlikely to inhibit AChE. These include effects on other enzymes and transcriptional changes in genes involved in neuronal cell differentiation, in cell signalling and enhancement of the toxicity of other xenobiotics. There may also be developmental effects arising from exposure to diazinon.[252–268] As an example, increases in sister chromatid exchange in peripheral lymphocytes have been reported in sheep-dip workers after dipping at exposures which did not result in an increase in diazinon urinary metabolites.[269] Moreover, there may be genetic components in the activity of certain enzymes as there appears to be with paraoxonase, an enzyme that can hydrolyse several oxons, including diazoxon.[270] Sheep farmers with lower paraoxonase activity had a greater risk of being affected by ill-health allegedly associated with sheep-dipping.[271]

12.2.3.2 Azamethiphos

Introduction and Metabolism. Azamethiphos (ISO); *S*-6-chloro-2,3-dihydro-2-oxo-1,3-oxazolo[4,5-*b*]pyridine-3-ylmethyl *O,O*-dimethyl phosphorothioate; Figure 12.7; is an OP compound that has been used to control flies and other

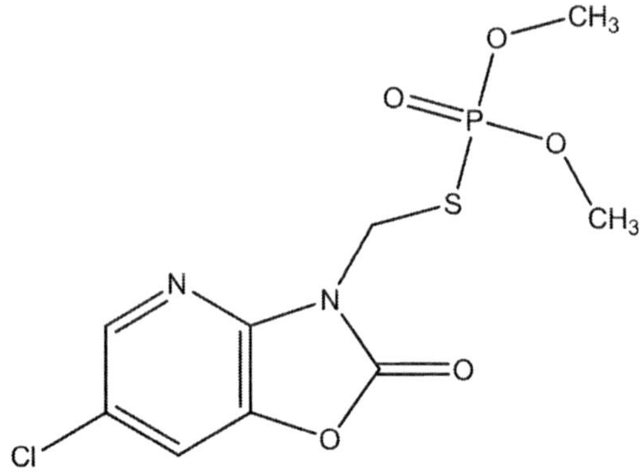

Figure 12.7 Structural formula of azamethiphos.

insects in animal houses, and to control mosquitoes, tsetse flies, cockroaches and other insects that pose a public hygiene problem. In veterinary medicine it is used for the treatment of sea lice on farmed Atlantic salmon and other fish (see earlier).[272–274]

Compared with many other pesticides, the metabolism and toxicity of azamethiphos are poorly documented. In fact there are more publications available on its environmental and ecotoxicological effects than there are on its toxicological effects. The only readily available document describing the toxicity of azamethiphos is that published by the European Medicines Agency as a summary of the data that supported its application for a European maximum residue limit (MRL).[275] This documents that azamethiphos is well absorbed after oral administration to rats, with most of the administered dose being found in urine. The substance was poorly absorbed after topical administration.[275] A number of metabolites are formed in rats and goats, including the sulfate and glucuronic acid conjugates of the chloroaminopyridine moiety, suggesting cleavage of the dimethyl phosphorothioate group and of the five-membered heterocyclic ring.[276]

Toxicity. Azamethiphos is of moderate toxicity to mammals with the acute oral LD_{50} value of 1180 mg kg^{-1} bw and the acute dermal LD_{50} value greater than 2150 mg kg^{-1} bw. The major signs of toxicity are those associated with cholinesterase inhibition. In a 13-week repeat dose study in rats, the major sign of toxicity was a reduction in blood cholinesterase activity which persisted for the duration of a 28-day recovery period. Unfortunately, brain cholinesterase was not monitored. In one study, azamethiphos decreased serum cholinesterase in rats while cholinesterase isozyme fractions gave conflicting results with some decreasing and some increasing.[277] In a three-generation study in rats, the major effects were on parental bodyweights. In developmental studies in rats and mice fetotoxicity was noticed at the highest doses employed but there was no evidence of teratogenic effects.[275]

Like diazinon and other OP compounds, azamethiphos gave equivocal results in genotoxicity tests. For example, it gave positive results in the *Salmonella* reverse mutation test, in a test for DNA damage in mammalian cells *in vitro*, in the *S. cerevisiae* D7 mutation assay and in the *Drosophila* wing spot test,[272,275] and it has been shown to be a proficient alkylating agent.[278] However, it gave negative results in a dominant lethal assay and in a test for unscheduled DNA repair. Despite the occurrence of positive results in these tests, azamethiphos gave negative results in two carcinogenicity studies in rats and in a study, unfortunately not conducted in accordance with Good Laboratory Practice, in mice. It did not induce delayed neuropathy.[275]

There are no data on the toxicity of azamethiphos to humans.

12.2.3.3 Adverse Effects of Diazinon and Azamethiphos in Animals

In view of the large number of sheep dipped in the UK each year, and the large numbers of sheep and cattle dipped or sprayed in other countries, it is perhaps surprising that there are virtually no reports of OP-related toxicity in treated animals. This may be related to the dilutions involved, to the poor absorption through skin and because in sheep at least, lanolin forms a protective barrier to absorption of the aqueous formulation. In one of the few cases reported, several sheep died after a 15-year-old sheep dip was used and this was found to have a high concentration of monothio-TEPP. In a separate incident in the same report, four yearling bulls were affected by cholinergic symptoms soon after treatment. One animal died but the others recovered.[279] As a formulated product, azamethiphos has been shown to be safe under normal circumstances of use in the European eel, sea bass and trout. Unlike the comparable medication containing dichlorvos, there have been no reports of OP toxicity in fish treated with azamethiphos-containing formulations.[280–284]

12.2.4 Semicarbazone

12.2.4.1 Metaflumizone

Introduction and Metabolism. Metaflumizone (ISO); (EZ)-2′-[2-(4-cyanophenyl)-1-(α,α,α-trifluoro-*m*-tolyl)ethylidene]-4-(trifluoromethoxy)carbanilhydrazide is a new semicarbazone insecticide (Figure 12.8). Its mode of action involves state-dependent blockage of sodium channels in insects leading to paralysis. This probably involves selective binding to the slow-inactivated state of the sodium channel, which is characteristic of the mode of action of the pyrazoline sodium channel insecticides developed during the 1970s and from which metaflumizone was derived.[285,286]

After oral administration to rats, the majority of the administered dose is excreted in faeces (~90%), with only small amounts in bile and urine. Maximum plasma concentrations were achieved after 10–48 h. Once absorbed, metaflumizone is metabolized by hydroxylation and hydrolysis and the metabolites

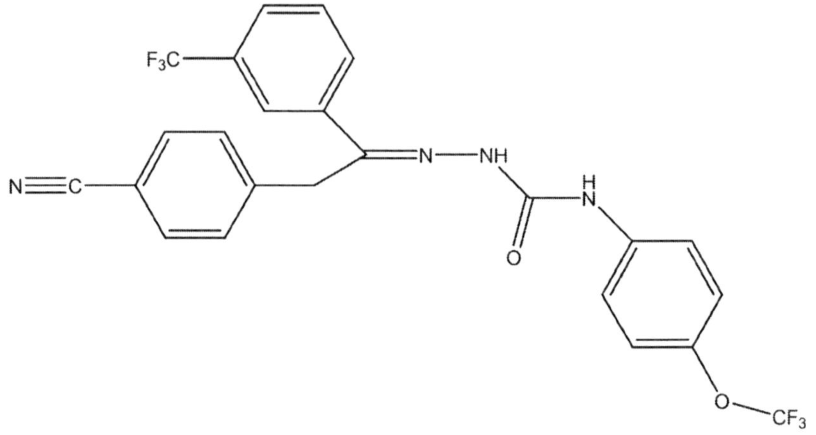

Figure 12.8 Structural formula of metaflumizone.

formed are subject to sulfate and glucuronic acid conjugation. After dermal administration to rats for 6 h, absorption was only 0.08–0.13%.[288] After topical application in cats and dogs, plasma levels were virtually undetectable soon after administration. Metaflumizone was well distributed throughout the hair of cats and dogs. These results suggest that metaflumizone acts locally and not systemically, at least in cats and dogs.[287,288]

Toxicity. As metaflumizone is a new insecticide in veterinary medicine, there are no reports of significant health effects either in treated animals or in exposed humans. The substance is of very low acute toxicity in the rat and mouse with the acute oral LD_{50} values being greater than 5000 mg kg^{-1} bw in both species. It had low toxicity after inhalation exposure (4 h, nose only) using the rat ($LC_{50} > 5.2$ mg L^{-1}). It was only a slight eye irritant in the rabbit and was not a skin irritant. There was no evidence of skin-sensitizing potential in the guinea-pig maximization test.[289]

Metaflumizone has been tested in a range of repeat dose studies including 28-day and 13-week oral gavage studies in the rat and 90-day and 1-year studies in the dog. There were no notable effects in the rat except for a slight liver lesion at the highest dose and the NOEL was established at 60 mg kg^{-1} bw day^{-1}. Toxicity was more marked in the dog where it was given in gelatine capsules. At doses of 30 mg kg^{-1} bw day^{-1} and above, reduced food intake was seen and animals had reduced bodyweight gain and bodyweight loss. The NOEL in the dogs was 12 mg kg^{-1} bw day^{-1}. Repeat dose studies using the inhalation route showed metaflumizone to be of low toxicity in rats. Similarly, it was of low toxicity in a 90-day dermal toxicity study, again in the rat.[289]

The substance was tested in a range of genotoxicity studies, both *in vitro* and *in vivo,* covering a range of endpoints. These included studies for gene mutation, for clastogenicity and for the induction of DNA damage. Negative results were obtained for the majority of these studies but a positive result occurred in an *in vitro* assay for chromosome aberrations. However, this only occurred in the absence of metabolic activation. Moreover, negative results were obtained in an *in vivo* study for clastogenic activity, the mouse micronucleus test, and the compound did not induce unscheduled DNA synthesis in an assay in rats. Overall, there are no data to suggest that metaflumizone is genotoxic. Moreover, in combined chronic toxicity and carcinogenicity studies in the mouse and rat, there was no increased incidence of any tumour type. In these studies, the only adverse effects were microscopic liver lesions at the two highest doses used (60 and 300 $mg\,kg^{-1}\,bw\,day^{-1}$) in rats, and brown pigment deposition in the spleen of mice at the highest dose (1000 $mg\,kg^{-1}\,bw\,day^{-1}$). The NOELs in these studies were 30 $mg\,kg^{-1}\,bw\,day^{-1}$ and 250 $mg\,kg^{-1}\,bw\,day^{-1}$ in rats and mice respectively.[289]

In a two-generation study in the rat, the highest dose (75 $mg\,kg^{-1}\,bw\,day^{-1}$) resulted in maternotoxicity characterized by poor general health and reduced bodyweights. This resulted in high pup mortality. The NOEL for maternotoxicity was determined to be 20 $mg\,kg^{-1}\,bw\,day^{-1}$. For successive generations, the maximum dose used was 50 $mg\,kg^{-1}\,bw\,day^{-1}$ which had no effects on reproductive performance. There were no indications of teratogenic potential in the rat and rabbit although maternal toxicity occurred in both species and this was accompanied by fetotoxicity.[289]

In specific studies for neurotoxicity in the rat, one an acute study and the other a 90-day oral study, the only signs seen were of general toxicity largely in the form of poor condition. However, there were no signs indicative of neurotoxicity.[289]

Safety in Animals and Humans. Metaflumizone is a relatively new compound to veterinary use and to date, there are no reports of adverse effects in humans. It has been evaluated for efficacy and safety in cats, kittens and dogs, in the latter species alone and in combination with amitraz. It has not evoked any adverse reactions in these trials.[290–294] However, a recent report has associated the therapeutic use of a metaflumizone–amitraz formulation in dogs with the development of a pemphigus foliaceus-like cutaneous reaction at the site of application. Some animals developed systemic disease and some required immunosuppressive treatment.[295]

12.2.5 Oxadiazone

12.2.5.1 Indoxacarb

Introduction and Metabolism. Indoxacarb (Figure 12.9), like metaflumizone, is a sodium channel-blocking insecticide that has been used against lepidopteran pests. However, it has recently been given a positive opinion in the EU by the Committee for Medicinal Products for Veterinary Use (CVMP) for use against fleas in dogs and cats by topical application. Indoxacarb is the ISO name for

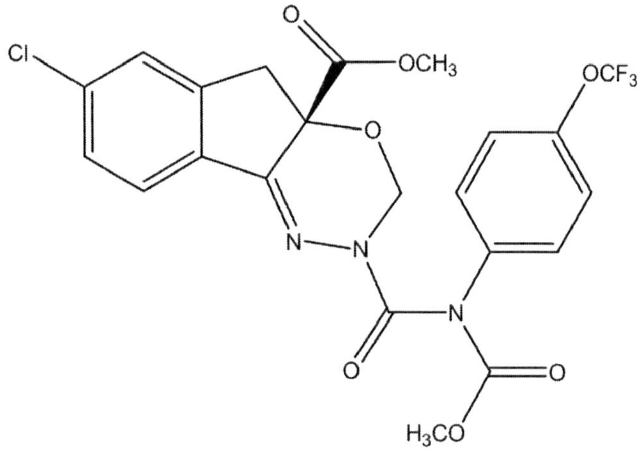

Figure 12.9 Structural formula of indoxacarb.

methyl (*S*)-*N*-[7-chloro-2,3,4a,5-tetrahydro-4a-(methoxycarbonyl)-indeno[1,2*e*]-[1,3,4]oxadiazin-2-ylcarbonyl]-4′-(trifluoromethoxy)carbanilate.[296–298] Indoxacarb exists as two enantiomers (*S* and *R*), also known as DPX-KN128 and DPX-KN127 respectively, but only the *S* enantiomer has insecticidal activity. There are no widely available data on indoxacarb, but it has been reviewed by the JMPR.[299]

After oral administration in aqueous vehicles, absorption in rats was saturable and was estimated to be 70–80% following a dose of 5 mg kg^{-1} bw but only 8–14% following 150 mg kg^{-1} bw. When the substance was radiolabelled in the indanone group, the half-life was 45–59 h. However, when radiolabelled in the trifluoromethoxy phenyl moiety, the half-life was 92 and 114 h for males and females respectively, suggesting some degree of biotransformation. Only low levels of indoxacarb-associated material were found in tissues. However, indoxacarb is metabolized both in insects and in rats to the more toxic IN-JT333 (methyl 7-chloro-2,5-dihydro-2-[[[4-(trifluoromethoxy)phenyl]amino]carbonyl] indeno-[1,2*e*] [1,3,4]oxadiazine-4a(3H) carboxylate) that was found in relatively high concentrations in fat.[299]

Metabolism of indoxacarb is complex; as well as IN-JT333, a range of other metabolites is formed in rats and these are further converted into sulfate and other conjugates.[299]

Toxicity. In acute oral toxicity studies in rats with DPX-MP062, an isomer blend containing 75% *S* and 25% *R* enantiomers, administered in corn oil, higher toxicity was noted in females where the LD_{50} value was 268 mg kg^{-1} bw compared with 1730 mg kg^{-1} bw in males. In this study, all rats given the two highest doses (3000 and 5000 mg kg^{-1} bw) died, with deaths occurring up to 20 days after dosing. Clinical signs included ataxia, hunched posture and ruffled fur. After day 5, signs included general spasms, immobility, lethargy, piloerection and tremors. After dermal administration of the same isomeric mix as an aqueous paste, the LD_{50} was greater than 5000 mg kg^{-1} bw in both sexes. An acute inhalation study was conducted with the racemic mixture of isomers, DPX-JW062. Toxicity was low, with the 4-h LC_{50} values being 5400 mg m^{-3} and 4200 mg m^{-3} in males and females respectively. Signs of toxicity included lethargy, abnormal gait and hunched posture.[299]

In mice given DPX-JW062 in the diet for 28 days, signs of toxicity included abnormal gait, head tilt and tremors. The NOEL based on reduced bodyweights in females was equivalent to 35.3 mg kg^{-1} bw day^{-1}. In a 90-day study in mice, similar clinical signs occurred. There were also haematological findings characterized by Heinz bodies within erythrocytes, increases in neutrophil and lymphocytes counts, and an increased incidence of haemosiderosis in the spleen and liver. These effects are typical of mild haemolysis. The NOEL in this study was 5.5 mg kg^{-1} bw based on haematological effects.[299]

When rats were given DPX-JW062 in the diet for 28 days, there were no major signs of toxicity in males at dietary levels of up to 235 ppm except for reductions in bodyweight when compared with control values. However, three out of five females given 400 ppm and two out of five females given 235 ppm died and signs of toxicity included abnormal gait, dehydration and ruffled fur. The NOELs based on effects on bodyweights were 8.9 and 2.6 mg kg^{-1} bw day^{-1} for male and female rats respectively. When administered in the diets of rats at dietary levels of up to 250 ppm for 90 days, there were no signs of toxicity except for decreased bodyweights and reduced food consumption in animals given the highest dietary concentrations. However, evidence of haemolysis occurred in treated animals with haemosiderosis in spleens, increased reticulocyte counts and erythrocytic hyperplasia. The NOELs based on haematological effects were 2.3 mg kg^{-1} bw day^{-1} for females and 3.9 mg kg^{-1} bw day^{-1} for males. In another 90-day dietary study with rats but using DPX-MP062, evidence of haemolysis was again the main finding. Haematological findings included reductions in erythrocyte numbers, haemoglobin and increases in mean corpuscular volume accompanied by haemosiderosis, increased splenic haematopoiesis and bone marrow hyperplasia. The overall NOEL in this study was 2.1 mg kg^{-1} bw day^{-1} in females based on haemolysis.[299]

Similar haemolytic effects were seen in dogs fed diets containing DPX-JW062 for periods of up to 1 year. In the 1-year dog study, the NOEL was 1.1 mg kg^{-1} bw day^{-1} based on haemolytic effects.[299]

These haemolytic effects are similar to those noted with aniline and its derivatives such as *p*-chloroaniline, the urea herbicides and other compounds which result in haemolysis and methaemoglobinaemia.[300–306] Some of the metabolites of indoxacarb are aniline derivatives or compounds closely related to these.

Indoxacarb gave negative results in a battery of short-term tests for genotoxicity and was not carcinogenic in animal bioassays in rats and mice. In multigeneration studies in rats, the only notable effects were parental toxicity and in developmental studies in rats and rabbits, there was no evidence of teratogenic effects. However, maternal toxicity was noted in these studies and this was accompanied by decreased fetal weights.[299]

Safety in Animals and Humans. As with metaflumizone, the veterinary use of indoxacarb is new and there are no reports of adverse effects in treated animals or in humans occupationally exposed to the substance. In overdose studies, three times and five times the intended therapeutic doses given topically to cats and dogs were well tolerated and produced only minor adverse effects including head shaking and squinting. Humans who have ingested indoxacarb formulations have developed methaemoglobinemia requiring methylene blue treatment and supportive care.[310–312] These effects are predictable from the animal studies (see above).

12.2.6 Phenylpyrazoles

12.2.6.1 Fipronil

Introduction and Metabolism. Fipronil (ISO); (±)-5-amino-1-(2,6-dichloro-α, α, α-trifluoro)-4-trifluoromethyl-sulfinylpyrazole-3-carbonitrile; Figure 12.10; has been used as an insecticide in agriculture for many years but has recently been introduced as an ectoparasiticide for use in companion animals. It acts by blocking chloride channels at the $GABA_A$ receptor and it is believed that it is more selective at this receptor through the β3 subunit in insects than it is in mammals, although this selectivity may be less pronounced with the sulfone metabolite and with the desulfinyl photodegradation product.[310–315] The toxicity of fipronil is also discussed in Chapter 9.

In rats, fipronil was well absorbed after oral administration with maximum blood concentrations being achieved after approximately 6 h. The elimination half-lives were 183 and 245 h in male and female rats respectively. However, after dermal application to rats, absorption was poor with less than 1% of the applied dose being recovered in blood, carcases, urine, faeces and cage washings.[316]

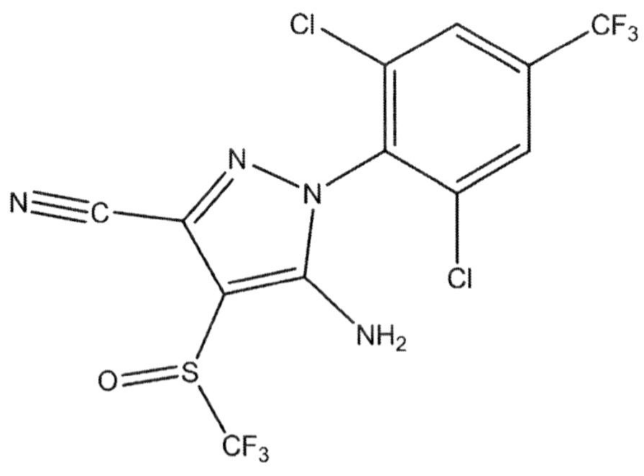

Figure 12.10 Structural formula of fipronil.

In mammals, fipronil is metabolized through reduction, oxidation and hydrolysis. The major product of reductive metabolism is the sulfinyl compound, found in rat faeces and, as a conjugate, in rat urine. The sulfoxide is formed through oxidative metabolism and this is also found in the faeces of rats and, again as a conjugate, in rat urine as well as in tissues and milk of goats. The product formed from the hydrolysis of the carbonitrile group is found in the faeces and urine of rats and in goat tissues. The sulfoxide may undergo further metabolism through cleavage of the pyrazole ring or by loss of the sulfoxide moiety.[316]

Toxicity. Fipronil has high acute toxicity in rats and mice, with oral LD_{50} values being in the range 91–103 mg kg^{-1} bw. The major signs seen were attributable to neurotoxicity and included convulsions, tremors, abnormal gait, and hunched posture. The substance was also toxic following inhalation exposure with 4-h nose-only LC_{50} values of 0.36–0.68 mg L^{-1} in rats. However, it was much less toxic after dermal application with LD_{50} values greater than 2000 mg kg^{-1} bw in the rat when given using an aqueous vehicle. However, it was more toxic in the rabbit by this route when given moistened with corn oil when the LD_{50} values were 445 and 354 mg kg^{-1} bw in males and females respectively, probably reflecting greater absorption with this lipid material.[316]

When mice were given fipronil in the diet for 13 weeks at concentrations of up to 25 ppm (equal to 3.2 and 4.5 mg kg^{-1} bw day^{-1}) there were no major signs of toxicity and the only notable effects were reduced bodyweight gains at the highest dietary level. Histopathological examination revealed the main effect to be a periacinar hypertrophy of the liver. Focal necrosis was observed in one male given the highest dietary level. As the hepatic effects were seen in mice given even the lowest dietary concentration, an NOEL could not be identified.[316]

In rats given diets containing up to 400 ppm fipronil for four weeks, there were no clinical signs of toxicity. At termination, males given 20 and 400 ppm fipronil, and females given all dietary concentrations, showed increased liver weights. There was thyroid follicular cell hypertrophy in almost all treated animals and because of this and the hepatic effects, an NOEL could not be identified. In a separate study, rats were given diets containing up to 300 ppm fipronil for 13 weeks. There was a clonic convulsion in one male given the highest dietary level of fipronil but there were no other signs attributable to compound intake. At necropsy, changes were again seen in the liver and thyroid. There were increases in relative liver weights in animals given the two highest dietary levels of 30 and 300 ppm. There was an increased incidence of thyroid follicular cell hyperplasia in animals given the highest dietary level. Based on these and other minor effects, some of which were probably related to effects on thyroid function, the NOEL was 5 ppm, equal to 0.333 mg kg^{-1} bw day^{-1}.[316]

In dogs given up to 10 mg kg^{-1} bw day^{-1} fipronil orally in gelatin capsules, significant signs of toxicity were seen in high-dose animals. These included emaciation, lack of activity, and hunched posture. Two dogs had to be euthanized. Other signs at the high dose included those suggestive of neurotoxicity including hypothermia, excessive salivation, convulsions, head nodding, tremors, limb jerk, ataxia and muscle twitching. Some animals also had an irregular heart rate. There were no histopathological findings that were considered to be compound related. The NOEL was 0.5 $mg\,kg^{-1}$ bw day^{-1}. In a 1-year study in dogs given up to 5 $mg\,kg^{-1}$ bw day^{-1} as gelatin capsules, signs of toxicity, some related to neurotoxicity, were seen at the high and intermediate (2 $mg\,kg^{-1}$ bw day^{-1}) doses. These included convulsions, tremors, twitches, ataxia, abnormal gait and aggression. One male given the intermediate dose and two males given the highest dose had to be euthanized. There were no consistent histopathological findings and the NOEL in this study was 0.2 $mg\,kg^{-1}$ bw day^{-1}. In a further study in dogs, fipronil was given in the diet at doses equivalent to 0, 0.075, 0.3, 1 or 3 $mg\,kg^{-1}$ bw day^{-1}. As well as routine observations, neurological examinations were also carried out and blood samples were taken and analysed for triiodothyronine (T3) and thyroxine (T4). One animal given the highest dietary level was euthanized on day 32 because of poor health and evidence of neurotoxicity. Neurological examination showed an absence of visual placing reactions, depressed menace and startle reactions, and abnormal gait. Other signs of toxicity included convulsions, head nodding, and twitching or tremors of muscles. There were no effects on T3 or T4.[316]

Fipronil has been tested in a battery of genotoxicity tests and, generally, negative results were obtained. However, in one study, a test for the induction of chromosome aberrations in a Chinese hamster lung cell line, a positive result was obtained. However, this only occurred under conditions of a 6-h pulse treatment and in the absence of metabolic activation and fipronil can be considered to be non-genotoxic.[316]

In a carcinogenicity study in mice, animals were given fipronil in the diet at concentrations of up to 60 ppm for 18 months. The only notable effects were an increased incidence of hepatocellular carcinoma in animals given the highest dietary level. This incidence was similar to that in historical controls for the testing facility concerned and was considered not to be compound related. In a study in rats, animals were given fipronil in the diet at concentrations of up to 300 ppm for 1 year to assess chronic toxicity while other groups were given the substance in the diet for 2 years. In this study evaluations were made of thyroid function by measuring T3 and T4 as well as thyroid stimulating hormone (TSH). At the highest dietary level, convulsions, sometimes fatal, were observed as well as other signs of neurotoxicity. Signs of neurotoxicity also occurred at the next dietary level, 30 ppm. T3 concentrations were not affected during treatment but were increased in females given 30 and 300 ppm from 4 weeks after treatment while T4 concentrations were depressed in male and female rats. TSH concentrations were increased in animals given 300 ppm and in males given 30 ppm fipronil in the diet. At necropsy, there was an increased incidence of thyroid follicular cell adenomas and carcinomas in animals given dietary fipronil.[316]

This increase in thyroid follicular tumours is not restricted to fipronil. Similar effects have been noted with a range of substances including several pesticides such as amitrole and fenbuconazole and with a number of pharmaceuticals including sulfadimidine (sulfamethazine), a sulfonamide antimicrobial agent. These agents are goitrogenic in rats resulting in constant stimulation of the thyroid by TSH and the eventual induction of thyroid tumours. Humans are insensitive to this mechanism of carcinogenesis and hence the results have no significance for human health.[317–328]

In a two-generation study with fipronil in rats, the main signs of toxicity were thyroid and liver hypertrophy in parental animals along with sings of neurotoxicity such as convulsions. It also resulted in decreased litter size and bodyweights and reduced the fertility index of parental animals. There were also reductions in post-implantation and postnatal survival. The NOEL for reproductive toxicity was 2.5 $mg\,kg^{-1}\,bw\,day^{-1}$.

Fipronil may lengthen the oestrous cycle in rats and alter concentrations of progesterone and oestradiol. In this study, where fipronil was applied topically to rats, the fertility index was also reduced.[327]

Fipronil was not teratogenic in rats and rabbits. However, in rats and rabbits, there were indications of some degree of maternal toxicity.[316]

Toxicity to Humans. Ingestion of or dermal contact with small quantities of fipronil are either asymptomatic or result in short-lived, non-specific effects.[329,330] In seven cases of fipronil ingestion in Sri Lanka, generalized tonic-clonic convulsions occurred in two patients who had peak fipronil plasma concentrations of 1600 and 3744 $\mu g\ L^{-1}$. These patients were managed with diazepam. Another patient with a peak plasma concentration of 1040 $\mu g\ L^{-1}$ was asymptomatic. In another case where fipronil ingestion had occurred, the patient required intubation and ventilation, and he developed seizures and pneumonia and died.[330] However, this patient may have ingested either a large dose of the agent or another substance such as endosulfan.[331,332] Surveillance data from 11 states in the USA for the period 2001–2007 for individuals exposed to fipronil identified 103 cases where exposure had occurred. The majority of these (86%) experienced exposure in a private dwelling and of these 37% had exposure to companion animal products and 26% had occupational exposures. Most of these had mild and temporary effects including headaches, dizziness and paraesthesia. Other effects included ocular, gastrointestinal, respiratory and dermal signs.[333]

Adverse Effects in Animals. Fipronil toxicity can occur in cats and dogs due to accidental ingestion or due to ingestion of applied product during grooming. Signs include tremors, convulsions, seizures and death.[334] Adult rabbits appear to be tolerant of fipronil but younger animals may be more susceptible and there have been reports of anorexia, lethargy, convulsions and death following treatment.[335]

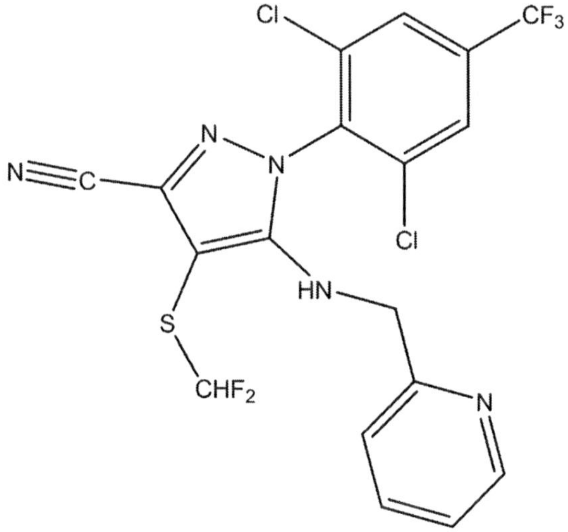

Figure 12.11 Structural formula of pyriprole.

12.2.6.2 Pyriprole

Pyriprole (ISO); 1-[2,6-dichloro-4-(trifluoromethyl)phenyl]-4-[(difluoromethyl)-sulfanyl]-5-[(pyridinylmethyl)amino]-1H-pyrazole-3-carbonitrile is structurally closely related to fipronil (Figure 12.11). It has been shown to be effective against cat and dog fleas and, alone or in combination with other ectoparasiticides, against other organisms.[336–341] Unfortunately, there are no readily available data on mammalian toxicity or on adverse effects in humans or animals. In view of its structural similarity with fipronil and its identical mode of action, it is likely that its safety profile will be qualitatively similar to that of fipronil.

12.2.7 Amitraz

12.2.7.1 Introduction and Metabolism

Amitraz (ISO); *N*-methylbis(2,4-xylyliminomethyl)amine; Figure 12.12; is a formamidine insecticide which has been in widespread use in agriculture since the early 1970s.[342] It is an α_2-adrenergic receptor agonist with some partial structural similarities with the therapeutic drugs clonidine, guanfacine and guanabenz,[343] and with the acaricide chlordimeform (*N′*-(4-chloro-2-methyl-phenyl)-*N,N*-dimethylimidoformamide).[342,344]

The substance prolongs gastric transit time and induces bradycardia in dogs. Through its actions on central and peripheral adrenoreceptors it may induce CNS depression and neurotoxicity, and some of its actions may be antagonized by the action of yohimbine.[345–350] Amitraz induces glucose intolerance in rats, possibly by the inhibition of insulin release through its action on α_2-adreno-ceptors.[351] Like other α_2-adrenoceptor agonists, amitraz can adversely affect

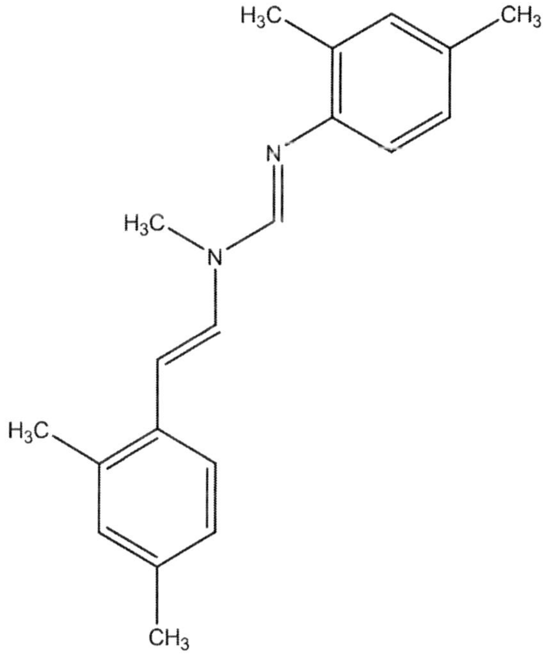

Figure 12.12 Structural formula of amitraz.

the mammalian reproductive system by binding to presynaptic α_2-adrenoceptors in the hypothalamus, thus inhibiting noradrenaline (norepinephrine) release and decreasing the secretion of gonadotropin-releasing hormone. This can result in a number of effects in mammals including decreased ovulation, decreases in litter size, and in males, reduced sperm production.[352–354] In human luteinized granulosa cells *in vitro*, amitraz inhibited basal and human chorionic gonadotropin-stimulated oestrogen production but had no effect on oestrogen.[355] Hence, amitraz might be expected to have adverse effects on reproduction in animal toxicology studies.

In mice given dietary amitraz, the majority of a radiolabelled dose was excreted in the urine. Similarly, in rats given amitraz orally, up to 85% was excreted in urine with the remainder in the faeces. In dogs, around 80% of an oral dose was excreted in urine, while in pigs and baboons the major part of the oral dose was subject to urinary excretion. Following topical administration to animals, most of the dose (~99%) was found at the application site and systemic absorption was very low. In mammals, amitraz is cleaved to yield the formamide and formamidine derivatives of 2,4-xylidine that may then be subject to conjugation. In addition, the former is metabolized to 4-*N*-methylformidoyl amino-*m*-toluic acid while the latter may be subject to further biotransformation to ultimately give 4-acetamido-*m*-toluic acid and 4-amino-*m*-toluic acid.[356,357]

12.2.7.2 Toxicity

The major signs of acute toxicity of amitraz in mice and rats were hyperexcitability, ataxia, tremor and ptosis. Rabbits developed CNS depression, decreased rectal temperatures and decreased pulse rates while dogs had CNS depression, ataxia, muscular weakness and spasms, micturition and decreased rectal temperatures and pulse rates. The oral LD_{50} values in the mouse and rat were greater than 1600 mg kg^{-1} bw and 600 mg kg^{-1} bw respectively, while in the dog the value was 100 mg kg^{-1} bw.[356]

In repeat dose studies in mice and rats, the main findings were increased aggression or irritability. Rats also had ataxia, increased nasal secretion, polyuria and bodyweight loss. In rabbits, decreased heart rates and rectal temperatures with a transient sedative effect were the main findings. Reduced rectal temperatures, decreased heart rates, vomiting, ataxia and CNS depression were the major findings in dogs. The lowest NOEL from these repeat dose studies was 0.25 mg kg^{-1} bw day^{-1} in the dog.[356]

Amitraz gave uniformly negative results in a battery of genotoxicity tests. In a mouse carcinogenicity study, there was an increased incidence of hepatocellular carcinomas in animals given in excess of the maximum tolerated dose, but in rats there was no increased incidence of any tumour type. The main findings in these studies were increases in aggressive behaviour of male mice and rats.[356]

In a rat multigeneration study, the most notable finding was a reduction in fertility and viability of offspring from the F_0 generation. The dose that produced this effect (16 mg kg^{-1} bw) was removed for the next generation, but nevertheless, the numbers of young alive at day 21 was reduced in all generations and the NOEL was 4.4 mg kg^{-1} bw.[356]

No major effects were reported in teratology studies in rats and rabbits at doses of up to 30 mg kg^{-1} bw day^{-1} in rats and up to 25 mg kg^{-1} bw day^{-1} in rabbits, administered at sensitive periods of gestation.[357] However, in a separate teratology study in rats using doses of up to 30 mg kg^{-1} bw day^{-1} in days 1–19 of pregnancy, there were increases in fetal death rates, decreases in litter size and a reduction in fetal bodyweight gain. There were also increases in the incidences of fetal external, visceral and skeletal anomalies at the maternotoxic dose (30 mg kg^{-1} bw day^{-1}) and the NOEL was 3 mg kg^{-1} bw day^{-1}.[358] In a cross-fostering study, control pups were nursed by control dams, control pups were nursed by treated dams, treated pups were nursed by treated dams and treated pups were nursed by control dams. Pups with prenatal exposure to amitraz showed decreases in the age for vaginal opening. Those from the group comprising treated pups and treated dams had higher locomotor activity and rearing frequency. There were no effects on the ASR or on open field behaviour.[359] Amitraz prolonged pro-oestrous in mice and resulted in lower concentrations of progesterone. Rats treated with amitraz had longer oestrous cycles resulting from prolonged oestrous or dioestrous.[356] Amitraz given orally to rats for 28 days resulted in increased relative adrenal weights; there were also decreases in mean corpuscular volume, in splenic plaque forming cells and in the delayed hypersensitivity reactions.[360]

12.2.7.3 Toxicity to Humans

The major signs of acute amitraz poisoning in humans are somnolence, coma, miosis, mydriasis, bradycardia, respiratory failure, hypo- and hyperthermia and increased blood glucose concentrations. The respiratory failure may require mechanical ventilation in severe cases.[361–365] A dose of around 10 g of amitraz, accompanied by around 35 g of xylene (a commercial amitraz formulation), proved almost fatal in a 72-year-old man who drank the product by mistake. He developed dizziness, coma, miosis, respiratory failure and hyperglycaemia. However, instead of bradycardia, atrial fibrillation developed which was successfully treated with digoxin. The man gradually recovered over a 3-day period.[366] From double-blind cross-over trials in humans and from volunteer studies, the NOEL for effects in humans is around 0.13–0.3 $mg\,kg^{-1}\,bw\,day^{-1}$.[356]

In children, the main sign of toxicity is altered consciousness along with nausea and vomiting. The other major signs were coma, convulsions, respiratory depression, hyperglycaemia, miosis, mydriasis, bradycardia, hypotension, hypothermia and polyuria.[367,368]

12.2.7.4 Adverse Reactions in Animals

In dogs and cats, the most common adverse reactions are anorexia, sedation and dry skin and may be associated with grooming areas treated with amitraz or with ingestion of medicated collars containing the agent. Sedation, bradycardia, polyuria, hypothermia and hyperglycaemia may also occur.[369–371] Prolongation of the QT interval has been reported in an English bulldog following topical treatment with an amitraz-containing medication.[372] Horses sprayed with an amitraz formulation showed signs of tranquilization, depression, ataxia, muscular incoordination and impaction colic. Following treatment with dexamethasone, intravenous fluids and enemas, all three affected animals returned to normal.[373]

In conclusion, therefore, the effects of amitraz in laboratory animal studies, in human toxicity cases and in animals exposed to amitraz during veterinary treatment are virtually all attributable to its pharmacological properties as an α_2-adrenoceptor agonist. The majority of these effects are reversible with supportive therapy.

12.2.8 Dicyclanil

12.2.8.1 Introduction and Metabolism

Dicyclanil (ISO); 4,6-diamino-2-(cyclopropylamino)pyrimidine-5-carbonitrile is an insect growth regulator used primarily to prevent blowfly strike on sheep although it can be used against other ectoparasites (Figure 12.13).[32,374–380] It exerts its actions through the inhibition of chitin synthesis.[381–383] The toxicity of insect growth regulators is further described in Chapter 7.

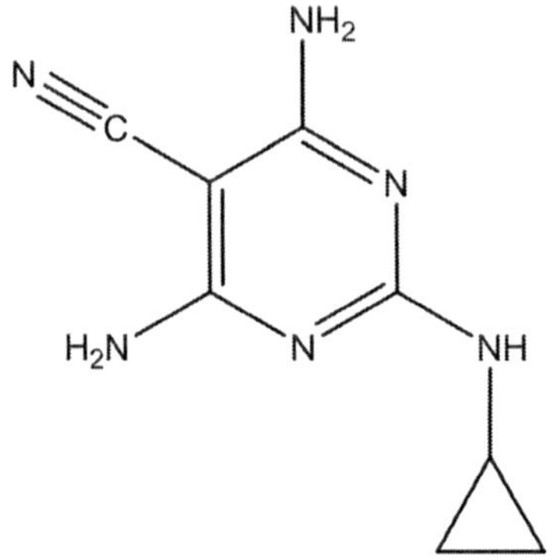

Figure 12.13 Structural formula of dicyclanil.

Absorption from the gastrointestinal tract of rats is extensive, with around 85% of the administered dose being excreted in urine. The highest concentrations of test compound were found in in liver, kidneys and carcase. In the rat, a number of metabolites are formed but the major fraction (~50% of the dose) is *N*-4, 6-diamino-5-cyanopyrimidin-2-yl)proprionamide. Other metabolites identified included 2,4,6-triaminopyrimidin-5-carbonitrile, 3-(4,6-diamino-5-cyanopyrimidin-2-ylamino)propionic acid and 2-(diamino-5-cyanopyrimidin-2-ylamino)-3-hydroxypropionic acid, as well as unchanged parent compound.[384]

12.2.8.2 Toxicity

In rats, the acute oral LD_{50} values were 560 mg kg^{-1} bw in males and 500 mg kg^{-1} bw in females. Signs of toxicity included piloerection, hunched posture and dyspnoea. There was reduced locomotor activity and ataxia in some males. After dermal application the LD_{50} value was greater than 2000 mg kg^{-1} bw. In a 4-h nose-only acute inhalation toxicity test, the LC_{50} value was 3400 and 3000 mg m^{-3} respectively and signs of toxicity included piloerection, hunched posture, dyspnoea and reduced locomotor activity.[384]

When groups of rats were given diets containing dicyclanil at concentrations of up to 2000 mg kg^{-1} of diet for 28 days in a range-finding study, the main sign of toxicity was piloerection. Dose related reductions in food consumption, bodyweight gain and final bodyweights occurred in all treatment groups. Reduced spermatogenesis was noted in animals given the high and intermediate (500 mg kg^{-1} of feed) dietary levels. In the corresponding females, polyovular ovaries occurred. In a 3-month dietary study in rats with dietary concentrations of up to 500 mg kg^{-1} of diet, there were no treatment related deaths and no signs of toxicity except for reductions in bodyweight at dietary concentrations of 125 or 500 mg kg^{-1} of feed. The NOEL in this study was 1.6 mg kg^{-1} bw day^{-1}. No significant signs of toxicity occurred in a 4-week study in rats where dicyclanil was applied topically at doses of up to 1000 mg kg^{-1} bw day^{-1}. Microscopic examination showed hepatic hypertrophy at 300 and 1000 mg kg^{-1} bw day^{-1}

in females and at 1000 mg kg^{-1} bw day^{-1} in males. The NOEL in this study was 30 mg kg^{-1} bw day^{-1}.[384]

Dietary concentrations of up to 2500 mg kg^{-1} of diet given to beagle dogs for 4 weeks resulted in those given the highest dietary concentration developing tremors, vomiting and dyspnoea. There was some evidence of testicular atrophy and focal hepatic necrosis in this group. When groups of beagles were given diets containing dicyclanil at concentrations of up to 1500 mg kg^{-1} of feed for 3 months, one high-dose male was found dead at the end of week 11 of the study. Animals in this group began to show signs of toxicity from week 11 and these included slight ataxia, raised tails and frequent shaking. At necropsy, microscopic examination revealed focal or multifocal subcapsular inflammation of the liver with fibrosis in some male and female dogs given the highest dietary concentration. High-dose males had a degree of tubular atrophy of the testes with a marked reduction in spermatogenesis. There was also an increase in inflammatory changes of the urinary bladder associated with epithelial hyperplasia in females given the 100, 500 and 1500 mg kg^{-1} of diet concentrations. The NOEL in this dog study was 0.6 mg kg^{-1} bw day^{-1} based on a number of findings including the bladder changes, increased plasma cholesterol and changes in the prostate. In a 1-year beagle dog study animals were given diets containing up to 750 mg kg^{-1} of food. One high-dose female was found dead on day 13 of the study and one high-dose male was euthanized *in extremis* on day 32. Females given the highest dietary concentration vomited, and food consumption was reduced in this group. Throughout the study, plasma cholesterol levels were increased in animals given the highest dietary level and in males given the next lower level of 150 mg kg^{-1} food. This persisted during the 4-week recovery period. Microscopic findings were limited to the two animals which died or were euthanized before the end of the study and these were considered to be incidental findings and not related to compound treatment. The NOEL in this study was 0.7 mg.kg bw day^{-1} based on increased cholesterol concentrations in plasma.[384]

Dicyclanil gave only negative results in a battery of genotoxicity tests.[384,385] In a mouse carcinogenicity study conducted over 18 months, animals were given diets containing up to 1500 mg kg^{-1} of feed. Males and females given the highest dietary concentration showed injuries obtained through excessive scratching and this group was terminated during weeks 58–59. At termination, the major findings were in the liver where Kupffer cell pigmentation and hepatocellular necrosis occurred in males given $\geq$ 100 mg kg^{-1} of dicyclanil in feed. Increases in the numbers of hepatocytes with mitotic figures and multinucleated hepatocytes occurred in high-dose males. There was an increased incidence of hepatocellular adenomas in females given the two highest doses and of hepatocellular carcinomas in females given the highest dose. The increased incidence of liver tumours occurred in animals where the maximum tolerated dose was given.[384]

Despite liver tumours in mice being common in mouse carcinogenicity studies, some workers have investigated the association between the occurrence of these neoplasms and exposure to dicyclanil. It has been suggested that they may arise from DNA damage caused by oxidative stress in the absence of direct genotoxicity and initiation activity, in combination with inhibition of apoptosis and failure to repair oxidative DNA damage.[385–389] Thus, dicyclanil-induced oxidative damage may mediate liver tumour promotion in mice.[390]

Dicyclanil had no notable effects in a two-generation study in rats at doses of up to 500 mg kg^{-1} bw day^{-1} and it was not teratogenic in rats or rabbits.[381]

12.2.8.3 Adverse Effects in Animals and Humans

There are no reports of significant effects in treated animals or exposed humans.

12.2.9 Cyromazine

12.2.9.1 Introduction and Metabolism

Cyromazine (ISO); *N*2-cyclopropyl-1,3,5-triazine-2,4,6-triamine is a triazine derivative which nonetheless, is similar in structure to dicyclanil (Figure 12.14) and has the same mode of action (see also Chapter 7).[27,30,376,378,379,382,391,392]

After oral administration to rats, the major route of excretion is urinary with approximately 94% of the administered dose being recovered. Only small amounts (3.8%) were recovered in faeces. In a separate experiment where rats were given single oral or intravenous doses of cyromazine, similar results were obtained irrespective of the route of administration which suggested that around 70–80% of the administered dose was excreted in the urine with up to 7.5% in faeces. Plasma concentrations reached a maximum approximately 30 min after oral administration and rapidly declined over the next 24 h. Extremely low concentrations were found in the tissues and carcases of treated animals. Similar findings were made in monkeys (*Macaca fascicularis*) given single oral doses of cyromazine, with up to 83% of the administered dose being recovered in urine and less than 2% in faeces. After dermal application to rats, only around 5% of the applied dose was found in urine and faeces with the

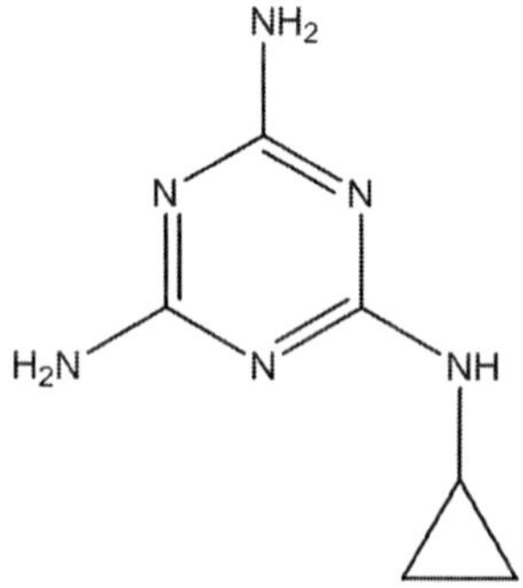

Figure 12.14 Structural formula of cyromazine.

remainder being recovered from the application site suggesting poor absorption.[393]

The major component of urine in rats was unchanged parent compound. However, small quantities of melamine (1,3,5-triazine-2,4,6-triamine) were present along with methylcyromazine and hydroxycyromazine. A slightly higher quantity of melamine was produced in monkeys than in rats.[393]

12.2.9.2 Toxicity

The acute toxicity of cyromazine in rats is low, with the acute and dermal LD_{50} values being 3387 and greater than 3170 mg kg^{-1} bw respectively. Similarly, acute toxicity following inhalation exposure is also low (4-h LC_{50} 3.6 mg L^{-1}). The major signs of toxicity seen in the acute toxicity studies were sedation, dyspnoea and a curved body posture, and animals recovered within 9–12 days. After inhalation exposure the main signs were decreased activity, piloerection and nasal discharge, and animals recovered by the second day following exposure.[393]

Cyromazine was of low toxicity in repeat dose toxicity studies where rats were given the substance for up to 90 days in the diet and rabbits were given topical administrations for 6 h each day for 3 weeks. The NOEL values were 232 mg kg^{-1} bw day^{-1} and 2000 mg kg^{-1} bw day^{-1} respectively. In dogs given diets containing up to 3000 ppm cyromazine for 90 days, the main findings were reductions in bodyweight gain at the highest dietary level. Treated animals also increased incidences of slightly relaxed nictitating membranes when compared with controls. There were no macroscopic or microscopic findings associated with compound intake. The lowest NOEL was 300 ppm equal to 12 mg kg^{-1} bw day^{-1} based on reductions in bodyweight gain.[393]

In a 1-year dog study, animals were given diets containing up to 3500 ppm cyromazine. One female in the high-dose group was found dead during week 3 and one male in a group given 200 ppm cyromazine in the diet had to be euthanized during week 29 due to aggressiveness. There were no other signs of toxicity in treated animals but females given the highest dietary level had reduced bodyweights in the first weeks of treatment. Haematological changes included a slight hypochromic and microcytic anaemia with lower haemoglobin concentrations, erythrocyte volume fraction, mean corpuscular volume and mean corpuscular haemoglobin in high-dose males and females. The absolute and relative weights of the heart and liver were increased in high-dose males and females and in females given 800 ppm dietary cyromazine. Relative kidney weights were also increased in high-dose females. The main macroscopic and microscopic findings were in high-dose animals with a hard myocardium and severe chronic myocarditis in the right atria. There was also chronic epithelial regeneration of the kidney tubules and hypercellularity of bone marrow. The NOEL in this study was 5.7 mg kg^{-1} bw day^{-1} based on haematological findings.[393]

Cyromazine was tested in an extensive battery of tests for genotoxicity but gave uniformly negative results except in the mouse spot test where the result was inconclusive.[393]

In a carcinogenicity study in mice, animals were given diets containing up to 3000 ppm cyromazine for 2 years. There were no effects on survival. Males, but not females showed reductions in bodyweight at 1000 and 3000 ppm. At termination, there was no significant increase in the incidence of tumours when compared with controls except for a slight increase in the incidences of hepatocellular adenoma and carcinomas, notably in male mice. These were not considered to be treatment related as there was no evident dose response and the effect was limited to males. In a carcinogenicity study in rats, animals were given diets containing up to 3000 ppm cyromazine for 2 years. There were no effects on condition, behaviour or survival but there were decreases in bodyweights in rats given the highest dietary level. There was a higher incidence of mammary tumours in females given the highest dietary level but the values were within those of historical controls.[393]

The metabolite melamine is of low acute oral toxicity. In repeat dose toxicity studies in mice and rats, the main findings were urinary tract calculi, with bladder ulceration associated with inflammation. Calculi were occasionally found in the kidneys.[391] In a mouse carcinogenicity study the major findings were in the urinary tract and consisted of acute and chronic inflammation and epithelial hyperplasia and urinary bladder calculi. In rats, chronic inflammation of the kidney was observed in some females given diets contacting melamine, and calculi were also found. Examination of the calculi showed them to be melamine.[391] In a separate study in rats, administration of melamine was associated with an increased incidence of papillomas and carcinomas of the urinary bladder and ureters.[392] Similar findings were made in a further study with melamine in rats, but at low doses of melamine with sodium chloride the incidence of urinary tract lesions diminished, suggesting that the associated increased urinary outputs decreased the incidence of inflammation and associated pathology.[396]

Melamine has been shown to be non-genotoxic.[393] Other chemicals that induce calculi in the urinary tract, such as saccharin, *o*-phenylphenol and uracil, have also been shown to induce tumours in the bladder and associated organs. These calculi induce inflammation which may in turn lead to carcinogenesis.[397–405] Most of these substances probably cause calculi because of their poor solubility in water or, more specifically, in urine. It seems likely that the small quantities of melamine formed from the biotransformation of cyromazine are insufficient to result in concentrated solutions in urine and therefore in the deposition of calculi.

In a multigeneration study, dietary concentrations of 1000 and 3000 ppm cyromazine resulted in reduced parental bodyweights. Male fertility was reduced at 3000 ppm but reproductive performance was not affected. The lowest NOEL, for parental toxicity and minor effects on offspring, was 51 mg kg^{-1} bw day^{-1}.[393]

Cyromazine was not teratogenic in rats and rabbits, but maternotoxicity was observed in both species. In rabbits, this resulted in maternal deaths, abortions and decreases in bodyweight with embryotoxicity. The lowest NOEL from these studies was 10 mg kg^{-1} bw day^{-1} based on maternotoxicity.[393]

12.2.9.3 Adverse Effects in Humans and Animals

Despite extensive use of cyromazine in veterinary medicine and in crop protection, there are no reports of adverse effects in humans or in animals. There have been reports of suicide attempts with substances of the triazine class but these were generally asymptomatic, suggesting low toxicity to humans, as might be expected from their low acute toxicity to animals.[406]

12.2.10 Benzoylureas—Diflubenzuron/Lufenuron/Teflubenzuron

12.2.10.1 Introduction and Metabolism

The benzoylureas, or at least those referred to here, are derivatives of 2,6-difluorobenzoylurea and differ only in the substitution pattern of the phenyl ring attached to the terminal nitrogen of the urea group (Figure 12.15). They are insect growth regulators that act through the inhibition of chitin synthesis.[407–411] Lufenuron is used to control fleas and ticks on companion animals and, unlike many agents used to control external parasites, it is effective after oral administration; diflubenzuron and teflubenzuron are used for the control of sea lice, largely on Atlantic salmon.[42,44–46,412–416]

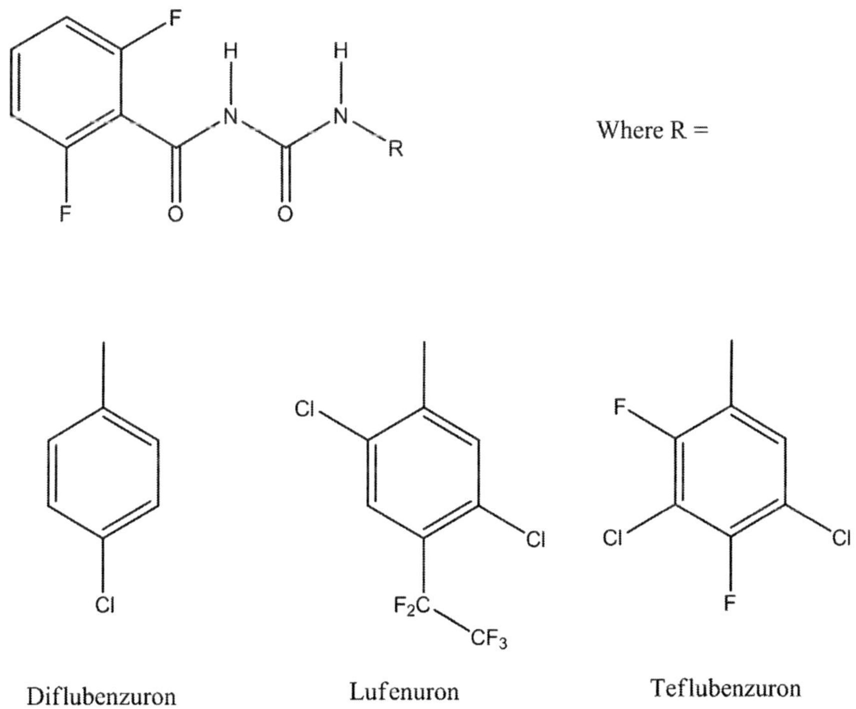

Figure 12.15 Structural formula of benzoylureas (diflubenzuron, lufenuron and teflubenzuron).

Few data are available on lufenuron and teflubenzuron but diflubenzuron has been reviewed by both the JMPR and by the International Programme on Chemical Safety.[417,418]

Diflubenzuron (ISO); 1-(4-chloropheny)-3-(2,6-difluorobenzoyl)urea is well absorbed in rats and mice after oral administration with up to about 30% of the administered dose being recovered from urine. The remainder was voided in faeces and only a small amount is accounted for by biliary excretion. The degree of absorption decreases with increasing dose. Thus, at 4 mg kg^{-1} bw in rats, around 42% of the dose was excreted in urine but at 1000 mg kg^{-1} bw this was reduced to only 1%. When applied topically to rats, less than 0.5% of the applied dose was absorbed. Similar results were obtained when diflubenzuron, as an aqueous suspension, was applied topically to rabbits.[418]

A number of metabolites are produced in rats but the major metabolic pathways give rise to 2,5-diflurobenzoic acid and 3-chloro-5-hydroxyaniline. The aniline moiety of diflubenzuron itself is hydroxylated and this and the 3-chloro-5-hydroxyaniline are subject to conjugation with sulfate and glutamate. The main metabolites formed in rats are 2-hydroxydiflubenzuron (7–10%), 4-chlorophenylurea (5–6%), and 2,6-difluorobenzamide (2–4%).[418]

12.2.10.2 Toxicity

Diflubenzuron has low acute toxicity with oral LD_{50} values in the mouse and rat being in excess of 4600 mg kg^{-1} bw. In rats, the dermal LD_{50} was greater than 10 000 mg kg^{-1} bw while with whole-body exposure, the LC_{50} with an unspecified exposure time was in excess of 2.9 mg L^{-1} air. No clinical signs were noted in these studies.[418]

In a 13-week feeding study in rats with dietary concentrations of up to 50 000 ppm, the main findings were chronic hepatitis, haemosiderosis, congestion of the spleen and erythroid hyperplasia of the bone marrow. These lesions increased in severity with increasing dose. There was also a degree of methaemoglobinaemia. As adverse effects, notably methaemoglobinemia, occurred at all dose levels, including the lowest dose employed (160 ppm, equal to 8 mg kg^{-1} bw day^{-1}), an NOEL could not be identified.[418]

In a study using dermal application, rats were given diflubenzuron in 0.25% aqueous gum tragacanth using 6-h semi-occlusive dressings and doses of up to 1000 mg kg^{-1} bw day^{-1}. The only death in this study was incidental and there were no signs of toxicity including effects on bodyweights and food consumption. There were reductions in erythrocyte parameters in females given 500 mg kg^{-1} bw day^{-1} and in animals of both sexes given 1000 mg kg^{-1} bw day^{-1}. Those given 500 mg kg^{-1} bw day^{-1} also showed polychromasia, hyperchromasia and anisocytosis. The NOEL in this study was 20 mg kg^{-1} bw day^{-1}.[418]

Reductions in haemoglobin and in erythrocyte volume occurred in rats exposed to diflubenzuron by inhalation for 4 weeks. This only occurred at the highest exposure concentration (110 mg/m^3) and the NOEL was 34 mg m^{-3} of air.[418]

In dogs given up to 160 ppm in food for 13 weeks, there were no mortalities and no signs of toxicity. The main findings in this study were haematological and consisted of reductions in erythrocytes and increases in haemoglobin and methaemoglobin. The NOEL was 40 ppm, equal to 1.6 mg kg^{-1} bw day^{-1}. When diflubenzuron was given to beagle dogs at doses of up to 250 mg kg^{-1} bw day^{-1} for 52 weeks, there were two deaths thought not to be related to compound treatment but no other consistent differences between controls and treated animals. As with the 13-week study, the major findings in this study were haematological, with increases in sulfhaemoglobin and methaemoglobin at weeks 4, 13, 26 and 52. In addition, there were decreases in haemoglobin and increases in Heinz bodies and pigmented Kupffer cells. These were seen at all dose levels (10, 50 and 250 mg kg^{-1} bw day^{-1}) but were marginal at the lowest dose of 2 mg kg^{-1} bw day^{-1}. The NOEL in this study was 2 mg kg^{-1} bw day^{-1}.[418]

Diflubenzuron has been tested in an extensive battery of tests for genotoxicity and uniformly negative results were obtained.[418,419]

In a mouse carcinogenicity study where animals were given diets containing up to 10 000 ppm diflubenzuron for 91 weeks, the major findings were again haematological and included methaemoglobinaemia. However, there was no evidence of any carcinogenic effects. Rats were given diflubenzuron in the diet at concentrations of up to 160 ppm for 2 years. Once again, the major findings were haematological and included methaemoglobinaemia. There was no increased incidence of any tumour type.[418] It can be concluded that diflubenzuron is not a genotoxic carcinogen.

There were no major effects in a multigeneration study in rats except for a reduction in pup bodyweight gain during lactation, with significant decreases in F_1 pup weights at days 4, 8 and 21 of lactation. The NOEL for effects on offspring was 430 mg kg^{-1} bw day^{-1}. There was no evidence that diflubenzuron was teratogenic in rats or rabbits and the NOEL in both studies was the highest dose tested, 1000 mg kg^{-1} bw day^{-1}.

In summary, therefore, diflubenzuron has low mammalian toxicity and is not genotoxic or carcinogenic. However, it does cause haematotoxicity, largely through the induction of methaemoglobinemia. As with indoxacarb (see earlier) some of the metabolites of diflubenzuron are aniline derivatives, compounds that are known to cause methaemoglobinaemia.[300–306]

The toxicities of lufenuron (ISO); (*RS*)-1-[2,5-dichloro-4-(1,1,2,3,3,3-hexafluoropropoxy)phenyl]-3-(2,6-difluorobenzoyl)urea and teflubenzuron (ISO); 1-(3,5-dichloro-2,4-difluorophenyl)-3-(2,6-difluorobenzoyl)urea might be expected to be at least qualitatively similar to that of diflubenzuron.

No toxicity data are available on lufenuron. However, there is a brief summary report available for teflubenzuron published by the European Medicines Agency as part of the consideration for an MRL.[420] This shows that teflubenzuron is metabolized in rats to give similar products as those noted with diflubenzuron but in extremely low quantities; most of the orally administered dose was excreted unchanged. Like diflubenzuron, the acute toxicity was low after oral administration to mice and rats with the LD_{50} value in excess of 5000 mg kg^{-1} bw. However, in repeat dose toxicity studies in rats, mice and dogs, there was no apparent haematotoxicity, including methaemoglobinaemia, perhaps reflecting the small quantities of aniline-related metabolites formed.

Like diflubenzuron, teflubenzuron was not genotoxic and gave negative results in carcinogenicity studies in mice and rats. It had no appreciable effects in a two-generation study in rats and was not teratogenic in rats or rabbits.

It seems likely therefore that for the benzoylureas to produce haematotoxicity, and specifically to produce methaemoglobinaemia, significant oral absorption is required accompanied by conversion to appreciable quantities of aniline-related metabolites. In a study of haematological effects of five benzoylphenylurea compounds, only diflubenzuron and triflumuron resulted in haematological effects in rats.[421]

12.2.10.3 Adverse Effects in Humans and Animals

There are no data on the safety of teflubenzuron or diflubenzuron to fish, the target species, or to humans. However, the toxicity is expected to be low, based on the effects in laboratory animals. Teflubenzuron was well tolerated by Atlantic salmon at single doses of up to 1500 mg kg^{-1} bw, and after repeated doses of 5 times the recommended therapeutic dose. After administration of 130 mg kg^{-1} bw day^{-1} (13 times the recommended dose) for 10 days, swimming activity decreased and the fish had difficulties maintaining a fixed position in the water, but the effects resolved on cessation of treatment.[420] Lufenuron has a high margin of safety in cats and dogs and there are no reports of ill-effects following use.[202,422]

The only report of exposure in humans concerns a lactating and breast-feeding woman who accidentally ingested lufenuron. The infant was exposed to an average lufenuron dose of 0.032 mg kg^{-1} day^{-1} but no adverse effects were noted during a 7-month follow-up period.[423]

12.2.11 Spinosad

12.2.11.1 Introduction and Metabolism

Spinosad (ISO); 50–95% (*2R*,3a*S*,5a*R*,5b*S*,9*S*,13*S*,14*R*,16a*S*,16b*R*)-2-(6-deoxy-2,3,4-tri-*O*-methyl-α-L-mannopyranosyloxy)-13-(4-dimethylamino-2,3,4,6-tetradeoxy-β-D-erythropyranosyloxy)-9-ethyl-2,3,3a,5a,5b,6,7,9,10,11,12,13,14, 15,16a,16b-hexadecahydro-14-methyl1*H-as*-indaceno[3,2-*d*]oxacyclododecine-7,15-dione [spinosyn A] and 50–5% (*2S*,3a*R*,5a*S*,5b*S*,9*S*,13*S*,14*R*,16a*S*,16b*R*)-2-(6-deoxy-2,3,4-tri-*O*-methyl-α-L-mannopyranosyloxy)-13-(4-dimethylamino-2, 3,4,6-tetradeoxy-β-D-erythropyranosyloxy)-9-ethyl-2,3,3a,5a,5b,6,7,9,10,11,12,13, 14,15,16a,16b-hexadecahydro-4,14-dimethyl1*H-as*-indaceno[3,2-*d*]oxacyclododecine-7,15-dione [spinosyn D] is a novel insecticide (Figure 12.16).[424–426] It is a biological product derived from the soil actinomycete *Saccharopolyspora spinosa* and it is active against a number of insect pests.[424]

Spinosad appears to act on the CNS causing involuntary muscle contractions and excitation of the nervous system.[427,428] However, it seems to act at a receptor site which as yet remains unidentified.[429] It is extremely effective against fleas on dogs,[430] and against head lice in humans.[431]

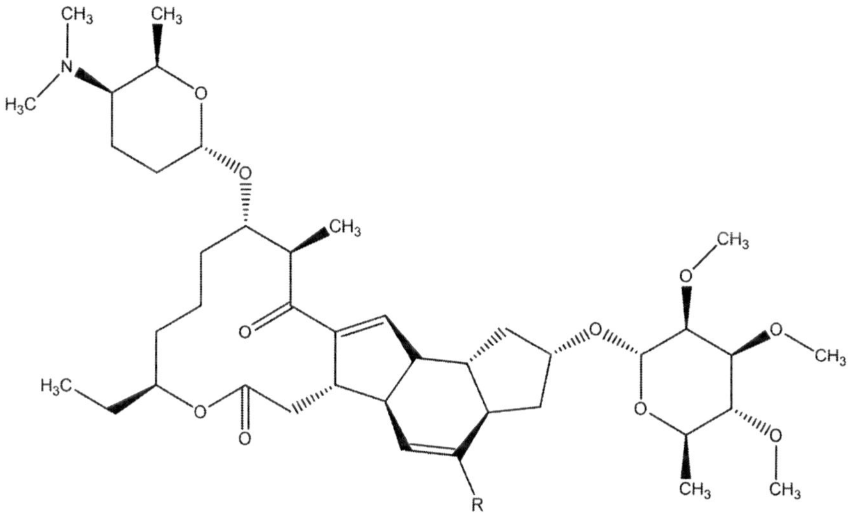

Spinosyn A, R = H; Spinosyn D, R = CH_3

Figure 12.16 Structural formula of spinosad.

After oral administration to rats, the main route of excretion was faecal with up to 88% of the administered dose being voided by this route and 6–10% being excreted in the urine. Faecal elimination was biphasic, suggesting biliary excretion, and in a study with bile duct cannulated rats 28–40% of the orally administered dose was excreted in the bile. The highest concentrations of substance were found in the kidneys, lymph nodes, fat and thyroid and, in males, in the liver.[418]

The major routes of biotransformation in rats were *N*-demethylation and *O*-demethylation of spinosyns A and D. The demethylation products and parent compounds were subject to conjugation with glutathione.[418]

12.2.11.2 Toxicity

Spinosad had low acute toxicity in the rat after oral administration in aqueous vehicles, with LD_{50} values being in excess of 5000 mg kg^{-1} bw. Similar findings were made for spinosyns A and D. There were no signs of toxicity. After dermal application to rabbits the LD_{50} was in excess of 5000 mg kg^{-1}.[418]

In repeat dose studies in mice where animals were given diets containing spinosad for 13 weeks, the main findings were intracellular vacuolation of histiocytic and epithelial cells in numerous tissues and organs including the kidneys, liver, spleen, thymus, pancreas, ovaries, cervix, vagina and epididymis, probably as a result of phospholipidosis. The NOEL for these effects was equal to 6 mg kg^{-1} bw day^{-1}.[418,432] Similar effects were seen in repeat dose studies in rats with dosing periods of up to 1 year and the lowest NOEL was 7.7 mg kg^{-1} bw day^{-1}.[418,433]

In a range-finding study with one animal of each sex per dose level, dogs were given diets containing up to 4000 ppm spinosad for 4 weeks. The two dogs given 4000 ppm (equal to 120 mg kg^{-1} bw day^{-1} for the male and 92 mg kg^{-1} bw day^{-1} for the female) were euthanized on day 23 because of extremely poor condition. These animals had loose or watery faeces with blood and mucus, and vomiting. There was occult blood in the urine of these animals. At 2000 ppm, there was vacuolation of several organs and tissues similar in appearance to those noted in rodent studies, while at 200 ppm microgranulomas of the liver and spleen with focal haemorrhage of the caecum occurred. A further study used dietary concentrations of up to 900 ppm for females and 1350 ppm for males, and this was given to dogs for 13 weeks. The 1350 ppm dietary level for males had to be reduced to 900 ppm at day 38 as one animal was euthanized *in extremis*. Loose and black faeces occurred in high-dose females and at week 13, the bodyweights of animals of both sexes given the highest concentrations were markedly reduced compared to control values. Dogs given spinosad had numerous haematological and blood biochemical abnormalities including reductions in erythrocytes, haemoglobin, lymphocytes and platelets. The haematological findings, along with bone marrow necrosis and hypercellularity seen on microscopic examination, are suggestive of the early phases of aplastic anaemia. At termination, there was vacuolation of several tissue and organ systems and the NOEL from this study was 150 ppm, equal to 4.9 mg kg^{-1} bw day^{-1}. In a 12-month study in dogs where animals were given diets containing up to 300 ppm spinosad, only slight effects were noted. There were some abnormalities in clinical chemistry and the vacuolation of various tissues but only at the 300 ppm level. The NOEL in this study was 100 ppm, equal to 2.7 mg kg^{-1} bw day^{-1}.[418]

Spinosad has been tested in a battery of tests for genotoxicity and only negative results were observed. It was tested in carcinogenicity studies in rats and mice with dietary concentrations of up to 360 ppm for 18 months in mice and up to 1000 ppm for up to 2 years in rats. There was no evidence of any carcinogenic effects in these studies but the vacuolation effects noted in repeat dose studies were seen in both the mouse and the rat. An NOEL of 80 ppm equal to 11 mg kg^{-1} bw day^{-1} was identified in mice and 50 ppm equal to 2.4 mg kg^{-1} bw day^{-1} in rats.[418,432,433]

The main finding in a two-generation study with rats and dietary doses equal to 0, 3, 10 or 100 mg kg^{-1} bw day^{-1} was parental toxicity with effects on the offspring. There were increases in relative and absolute liver, kidney, heart, spleen and kidney weights and vacuolation of various tissues and organs at the highest dietary concentration used. Litter sizes were reduced at this concentration and survival of pups was reduced. The NOEL for this study was 10 mg kg^{-1} bw day^{-1}.[418,434]

Spinosad was not teratogenic in developmental toxicity studies in rats and rabbits at doses of up to 200 mg kg^{-1} bw day^{-1} and 50 mg kg^{-1} bw day^{-1} respectively. However, significant maternal toxicity was noted in both species, with some embryo- and fetotoxicity. The NOEL values for maternal toxicity were 50 and 10 mg kg^{-1} bw day^{-1} and for embryo- and fetotoxicity 200 and 50 mg kg^{-1} bw day^{-1}, in rats and rabbits respectively.[418,435]

Hence, spinosad is of low acute toxicity but is more toxic in repeated dose studies where the major effects are vacuolation of various organs and tissues due to phospholipidosis. Phospholipidosis usually arises from inhibition of phospholipase activity and is caused by cationic amphiphilic drugs and chemicals including amiodarone, tamoxifen, gentamicin, chlorphentermine and chloroquine.[436–439]

12.2.11.3 Adverse Effects in Humans and Animals

There are no reports of adverse effects of spinosad in treated animals. Treatment of human patients suffering from head lice with a 0.9% spinosad topical medication resulted only in occasional and mild ocular hyperaemia and application site erythema and irritation.[431]

12.3 Conclusions

A number of agents, indeed classes of agents, are available to treat animals suffering from ectoparasitic disease, or indeed to prevent such disease. Their pharmacological effects are exerted through a number of mechanisms, although the majority are neurotoxic. They can exert a variety of toxic effects and quite frequently these effects are related to the pharmacodynamic activity of the agent. However, these compounds are important components of the armoury against parasitic disease in animals (and occasionally, in humans), and when used properly and in accordance with label recommendations, are effective and safe for both users and treated animals.

References

1. K. N. Woodward, Veterinary pharmacovigilance in the European Union, in *Veterinary Pharmacovigilance: Adverse Reactions to Veterinary Medicinal Products*, ed. K. N. Woodward, Wiley-Blackwell, Chichester, 2009, pp. 19–46.
2. K. N. Woodward, Regulation of veterinary drugs, in *General and Applied Toxicology*, ed. B. Ballantyne, T. Marrs and T. Syversen, MacMillan Reference, Basingstoke, 2nd edn, 2000, pp. 1633–1652.
3. K. N. Woodward, Veterinary Pharmacovigilance, Part 1. The legal basis in the European Union, *Vet. Pharmacol. Ther.*, 2005, **28**, 131–147.
4. P. Frank and J. H. Schafer, Animal health products, in *Regulatory Toxicology*, ed. S. C. Gad, Taylor & Francis, London, 2nd edn, 2001, pp. 70–84.
5. R. G. Endris, V. E. Reuter, J. D. Nelson and J. O. Nelson, Efficacy of 65% permethrin applied as a topical spot-on against walking dandruff caused by the mite *Cheyletiella yasguri* in dogs, *Vet. Ther.*, 2000, **1**, 273–279.

6. R. G. Endris, R. Everett, J. Cunningham, T. L. Katz and K. Thompson, *Efficacy of two 65% permethrin spot-on formulations against canine infestations of Ctenocephalides felis* and *Rhipicephalus sanguineus, Vet. Ther.*, 2002, **3**, 326–332.
7. R. G. Endris, D. Cooke, D. Amodie, D. L. Sweeney and T. L. Katz, Repellency and efficacy of 65% permethrin and selamectin spot-on formulations against *Ixodes ricinus* ticks on dogs, *Vet. Ther.*, 2002, **3**, 64–71.
8. R. Molina, J.-M. Lohse and J. Nico, Evaluation of a topical solution containing 65% permethrin against the Sandfly (*Phlebotomus perniciosus*) in dogs, *Vet Ther.*, 2001, **2**, 261–267.
9. J. E. Hillerton, A. M. Bramley and N. H. Yarrow, Control of flies (Diptera: *Muscidae*) on dairy heifers by Flectron ear-tags, *Br. Vet. J.*, 1985, **141**, 160–167.
10. N. French, A. J. Wright, W. R. Wilson and D. B. R. Nichols, Control of headfly on sheep, *Vet. Rec.*, 1977, **100**, 40–43.
11. H. D. Bailie and D. W. T. Morgan, Field trials to assess the efficacy of permethrin for the control of flies on cattle, *Vet. Rec.*, 1980, **106**, 124–127.
12. J. A. Harris, J. E. Hillerton and S. V. Morant, Effect on milk production of controlling muscid flies, and reducing fly-avoidance behaviour, by the use of fenvalerate ear tags during the dry period, *J. Dairy Res.*, 1987, **54**, 165–171.
13. J. E. Hillerton, S. V. Morant and J. A. Harris, Control of Muscidae on cattle by flucythrinate ear-tags, the behaviour of these flies on cattle and the effects on fly-dislodging behaviour, *Entomol. Exp. Appl.*, 1986, **41**, 213–218.
14. R. N. Titchener, Insecticidal ear tags control cattle ectoparasiticides, *Parasitol. Today*, 1986, **2**, 26–27.
15. J. A. Shemanechuk, Repellent action of permethrin, cypermethrin and resmethrin against black flies (*Simulium* spp.) attacking cattle, *Pestic. Sci.*, 1981, **12**, 412–416.
16. M. H. Muma, B. E. Hill, E. Hixon and L. Harries, Control of sheep-ticks on feeder lambs, *J. Econ. Entomol.*, 1952, **45**, 833–838.
17. D. W. Tarry, The sheep head fly and the problem of pasture flies, *ADAS Q. Rev.*, 1976, **23**, 283–293.
18. M. J. Hal, Traumatic myiasis of sheep in Europe: a review, *Parasitologia*, 1997, **39**, 409–413.
19. J. M. Brougham and R. Wall, Fly abundance and climate as determinants of sheep blowfly strike incidence in southwest England, *Med. Vet. Entomol.*, 2007, **21**, 231–238.
20. B. Bisdorf and R. Wall, Sheep blowfly strike risk and management in Great Britain: a survey of current practice, *Med. Vet. Entomol.*, 2008, **22**, 303–308.
21. S. E. Aiello and A. Mays (ed.). Flies, in *The Merck Veterinary Manual*, Merck & Co., Whitehouse Station, NJ, 8th edn, 1998, pp. 637–658.
22. P. Bates, Sheep scab (*Psoroptes ovis*) in *Diseases of Sheep*, ed. I. D. Aitken, Blackwell Publishing, Oxford, 2007, pp. 321–325.

23. A. H. van den Broek and J. F. Huntley, Sheep scab: the disease, pathogenesis and control, *J. Comp. Pathol.*, 2003, **128**, 79–81.
24. P. Bates, M. Rankin, D. Clifford and L. Stubbings, Shower dipping in diazinon or cypermethrin dipwash to control ovine psoroptic mange (sheep scab), *Vet. Rec.*, 2005, **156**, 655.
25. R. Wall and P. Bates, Sheep scab control using trans-cinnamic acid, *Vet. Parasitol.*, 2011, **175**, 129–134.
26. B. Harvey, M. Bakewell, T. Felton, K. Stafford, G. C. Coles and R. Wall, Comparison of traps for the control of sheep blowfly in the U.K., *Med. Vet. Entomol.*, 2010, **24**, 210–13.
27. B. Lonsdale, D. W. Tarry, F. L. Bowen and D. G. Stansfield, Cyromazine pour-on for the prevention of cutaneous myiasis of sheep, *Vet. Rec.*, 1990, **126**, 207–210.
28. W. N. Beesley, Sheep dipping, with special reference to the UK, *Pest. Outlook*, 1994, **5**, 16–21.
29. R. Wall, N. P. French and A. Fenton, Sheep blowfly strike: a model approach, *Res. Vet. Sci.*, 2000, **69**, 1–9.
30. D. J. O'Brien and G. Fahey, Control of blowfly strike in sheep by means of a pour-on formulation of cyromazine, *Vet. Rec.*, 1991, **129**, 351–353.
31. F. L. Bowen, P. Fisara, P. Junquera, D. T. Keevers, R. H. Mahoney and H. R. Schmid, Long-lasting prevention against blowfly strike using the insect growth regulator dicyclanil, *Aust. Vet. J.*, 1999, **77**, 454–460.
32. B. Lonsdale, H. R. Schmid and P. Junquera, Prevention of blowfly strike on lambs with the insect growth regulator dicyclanil, *Vet. Rec.*, 2000, **147**, 540–544.
33. B. L. Blagburn, D. R. Young, J. A. Meyer, A. Leigh-Effron, T. Paarlberg, A. G. Zimmermann, D. Mowrey, S. Wiseman and D. E. Snyder, Effects of orally administered spinosad (Comfortis) in dogs on adult and immature stages of the cat flea (*Ctenocephalides felis*), *Vet. Parasitol.*, 2010, **168**, 312–327.
34. D. E. Snyder, A. G. Zimmermann, M. Qiao, S. J. Gissendanner, L. R. Cruthers, R. L. Slone and D. R. Young, Preliminary studies on the effectiveness of the novel pulicide, spinosad, for the treatment and control of fleas on dogs, *Vet. Parasitol.*, 2007, **150**, 345–351.
35. M. K. Rust, Advances in the control of *Ctenophalides felis* (cat flea) on cats and dogs, *Trends Parasitol.*, 2005, **21**, 232–236.
36. R. Marsella, Advances in flea control, *Vet. Clin. North Am. Small Anim. Pract.*, 1999, **29**, 1407–24.
37. M. W. Dryden, Flea and tick control in the 21st century: challenges and opportunities, *Vet. Dermatol.*, 2009, **20**, 435–440.
38. T. Schneider, S. Wolken and N. Mencke, Comparative efficacy of imidacloprid, selamectin, fipronil-(S)-methoprene, and metaflumizone against cats experimentally infested with, *Ctenocephalides felis, Vet. Ther.*, 2008, **9**, 176–183.
39. M. Franc and E. Bouhsira, Efficacy of a combination of fipronil-(S)-methoprene spot-on formulation and a deltamethrin-impregnated collar in controlling fleas and sandflies on dogs, *Vet. Ther.*, 2009, **10**, 71–77.

40. B. L. Blagburn and M. W. Dryden, Biology, treatment and control of flea and tick infestations, *Vet. Clin. North Am. Small Anim. Pract.*, 2009, **39**, 1173–1120.
41. K. N. Woodward, Macrocyclic lactone endectocides, This volume, Chapter 13, pp. 427–467.
42. M. Roth, R. H. Richards and C. Sommerville, Current practices in the chemotherapeutic control of sea lice infestations in aquaculture: a review, *J. Fish Dis.*, 1993, **16**, 1–26.
43. J. Stone, I. H. Sutherland, C. S. Sommerville, R. H. Richards and K. J. Varma, The efficacy of emamectin benzoate as an oral treatment of sea lice, *Lepeophtheirus salmonis* (Krøyer) infestations in Atlantic salmon, *Salmo salar* L, *J. Fish Dis.*, 1999, **22**, 261–270.
44. B. M. MacKinnon, Sea lice: a review, *World Aquac.*, 1997, **28**, 5–10.
45. M. Roth, The availability and use of chemotherapeutant sea lice control products, *Contrib. Zool.*, 2000, **69**, http//:dpc.uva.nl/ctz/vol69/nr01/art12
46. M. J. Costello, Review of methods to control sea lice (Caligidae: Crustacea) infestations on salmon (*Salmo salar*) farms in *Pathogens of Wild and Farmed Fish: Sea Lice* ed. G. A. Boxall and D. Defaye, Ellis Horwood, London, 1993, pp. 219–252.
47. W. N. Aldridge, An assessment of the toxicological properties of pyrethroids and their neurotoxicity, *Crit. Rev. Toxicol.*, 1990, **21**, 89–103.
48. D. W. Gammon, M. A. Brown and J. E. Casida, Two classes of pyrethroid action in the cockroach, *Pest. Biochem. Physiol.*, 1981, **15**, 181–191.
49. H. P. M. Vijverberg and J. van den Bercken, Neurotoxicological effects and the mode of action of pyrethroid insecticides, *Crit. Rev. Toxicol.*, 1990, **21**, 105–126.
50. H. P. M. Vijverberg, J. M. van den Zalm and J. van den Bercken, Similar mode of action of pyrethroids and DDT on sodium channel gating in myelinated nerves, *Nature*, 1982, **295**, 601–603.
51. J. R. Coats, Mechanisms of toxic action and structure relationships for organochlorine and synthetic pyrethroid insecticides, *Environ. Health Persp.*, 1990, **87**, 255–262.
52. T. Narahashi, Nerve membrane ionic channels as the primary target of pyrethroids, *Neurotoxicology*, 1985, **2**, 3–22.
53. L. Pap, D. Bajomi and I. Székely, I. The pyrethroids, An overview, *Int. Pest Control*, 1996, **38**, 15–19.
54. J. R. Bloomquist, P. M. Adams and D. M. Soderlund, Inhibition of gamma-amino butyric acid-stimulated chloride flux in mouse brain vesicles by polychlorocycloalkane and pyrethroid insecticides, *Neurotoxicology*, 1986, **7**, 11–20.
55. J. M. Barnes and R. D. Verschoyle, Toxicity of new pyrethroid insecticides, *Nature*, 1974, **248**, 711.
56. L. O. Ruzo, J. L. Engel and J. E. Casida, Decamethrin metabolites from oxidative, hydrolytic and conjugative reactions in mice, *J. Agric. Food Chem.*, 1979, **27**, 725–731.

57. L. J. Lawrence and J. E. Casida, Pyrethroid toxicology: mouse intracerebral structure-toxicity relationships, *Pest. Biochem. Physiol.*, 1982, **18**, 9–14.
58. B. Gassner, A. Wüthrich A, G. Scholtysik and M. Solioz, The pyrethroids permethrin and cyhalothrin are potent inhibitors of the mitochondrial complex I, *J. Pharmacol. Exp. Ther.*, 1997, **281**, 855–860.
59. J. Miyamoto, Degradation, metabolism and toxicity of synthetic pyrethroids, *Environ. Health Perspect.*, 1976, **14**, 15–28.
60. JECFA, *Toxicological Evaluation of Certain Veterinary Drug Residues in Food*, 54th Meeting of the Joint FAO/WHO Expert Committee on Food Additives, WHO Food Additive Series 45, World Health Organization, Geneva, 2000, pp. 41–74.
61. G. Chester, N. N. Sabathy and B. H. Woollon, Exposure and health assessment during application of lambda-cyhalothrin for malaria vector control in Pakistan, *Bull. World Health Organ.*, 1992, **70**, 615–619.
62. IPCS, Environmental Health Criteria 99, *Cyhalothrin*, International Programme on Chemical Safety, World Health Organization, Geneva, 1990.
63. M. A. Campana, A. M. Panzeri, V. J. Moreno and F. N. Dulout, Genotoxic evaluation of the pyrethroid lambda-cyhalothrin using the micronucleus test in erythrocytes of the fish *Cheirodon interruptus interruptus*. *Mutat. Res.*, 1999, **438**, 155–161.
64. A. Çelik, B. Mazmanci, Y. Çamlica, A. Aşkin and U. Çömelekoglu, Induction of micronuclei by lambda-cyhalothrin in Wistar rat bone marrow and gut epithelial cells, *Mutagenesis*, 2005, **20**, 125–129.
65. P. B. Desmukh, Three-generation study of a synthetic pyrethroid—cyhalothrin, *Toxicol. Lett.*, 1992, **64/65**, 770–781.
66. M. de Silva Gomes, M. M. Bernardi and H. de Souza Spinosa, Effects of prenatal pyrethroid insecticide exposure on the sexual development of rats, *Vet. Human Toxicol.*, 1991, **33**, 427–428.
67. V. C. Moser, Screening approaches to neurotoxicity: A functional observational battery, *J. Am. Coll. Toxicol.*, 1989, **8**, 85–93.
68. V. C. Moser, J. P. McCormick, J. P. Creason and R. C. MacPhail, Comparison of chlordimeform and carbaryl using a functional observational battery, *Fund. Appl. Toxicol.*, 1988, **11**, 189–206.
69. V. C. Moser and R. C. MacPhail, Comparative sensitivity of neurobehavioral tests for chemical screening, *Neurotoxicol.*, 1990, **11**, 335–354.
70. M. Hornychová, E. Frantik, J. Kubát and J. Formánek, Neurotoxicity profile of supermethrin, a new pyrethroid insecticide, *Centr. Eur. J. Publ. Health*, 1995, **3**, 210–218.
71. IPCS, *Environmental Health Criteria 97, Deltamethrin*, International Programme on Chemical Safety, World Health Organization, Geneva, 1990.
72. JMPR, *Pesticide Residues in Food—2000, Evaluations 2000. Part II—Toxicological*, Joint FAO/WHO Meeting on Pesticide Residues, World Health Organization, Geneva, 2001.

73. A. Anadón, M. R. Larrañaga, M. Fernandez-Cruz, M. J. Diaz, M. C. Fernandez and M. A. Martinez, Toxicokinetics of deltamethrin and its 4′-hydroxy metabolite in rat, *Toxicol. Appl. Pharmacol.*, 1996, **141**, 8–16.
74. L. M. Cole, L. O. Ruzo, E. J. Wood and J. E. Casida, Decamethrin metabolites from oxidative, hydrolytic, and conjugative reactions in mice, *J. Agric. Food Chem.*, 1979, **27**, 725–731.
75. L. O. Ruzo, T. Unai and J. E. Casida, Pyrethroid metabolism. Comparative fates in rats of tralomethrin, tralocythrin, deltamethrin and (1R, alpha S)-cis-cypermethrin, *J. Agric. Food Chem.*, 1982, **30**, 631–636.
76. K. E. Appel and S. Gericke, Neurotoxicity and toxicokinetics of pyrethroids, *Bundesgesundheitblatt*, 1993, **6**, 219–228.
77. S. S. Anand, K.-B. Kim, J. W. Fisher and J. V. Bruckner, Ontogeny of hepatic and plasma metabolism of deltamethrin in vitro. Role in age-dependent acute neurotoxicity, *Drug. Met Rev.*, 2006, **34**, 389–397.
78. M. H. Akhtar, Metabolism of deltamethrin by cow and chicken liver enzyme preparations, *J. Agric. Food Chem.*, 1984, **32**, 258–262.
79. M. H. Akhtar, R. M. G. Hamilton and H. L. Trenholm, Metabolism, distribution and excretion of deltamethrin by leghorn hens, *J. Agric. Food Chem.*, 1985, **33**, 753–756.
80. M. H. Akhtar, K. E. Hartin and H. L. Trenholm, Fate of [^{14}C] deltamethrin in the lactating dairy cow, *J. Agric. Food Chem.*, 1986, **34**, 753–756.
81. EMA/CVMP, *Deltamethrin, MRL Summary Report (3), EMEA/MRL/779/01-FINAL*, European Medicines Agency/Committee for Veterinary Medicinal Products, 2001. www.ema.europa.eu/ (accessed January 2011).
82. M. A. Saleh, N. A. Ibrahim, N. Z. Soliman and M. K. El Sheimy, Persistence and distribution of cypermethrin, deltamethrin and fenvalerate in laying chickens, *J. Agric. Food Chem.*, 1986, **34**, 895–898.
83. P. H. Chanh, C. Navarro-Delmasure, A. Chanh, P. Clavel, G. Van Haverbeke and S. L. Cheav, Toxicological studies of deltamethrin, *Int. J. Tiss. Reac.*, 1984, **VI**, 127–133.
84. T. B. Gaines and R. E. Linder, Acute toxicity of pesticides in adult and weanling rats, *Fundam. Appl. Toxicol.*, 1986, **7**, 299–308.
85. R. Kavlock, N. Chernoff, R. Baron, R. Linder, E. Rogers, B. Carver, J. Dilley and V. Simmon, Toxicity studies with decamethrin, a synthetic pyrethroid insecticide, *J. Environ. Pathol. Toxicol.*, 1979, **2**, 751–765.
86. S. H. Kowalczyk-Bronisz, J. Gieldanowski and B. Bubak, Immunological profile of animals exposed to pesticide—deltamethrin, *Arch. Immunol. Therap. Exper.*, 1990, **38**, 229–238.
87. M. Bleys, J. Cotonat and P. Foulhoux, Letter to the editor, *J. Toxicol. Clin. Exper.*, 1986, **6**, 211–212.
88. P. H. Chanh, C. Navarro-Delmasure, A. Chanh, S. L. Cheav, F. Ziade and F. Samaha, Pharmacological studies of deltamethrin on the central nervous system, *Arzneim. Forsch.*, 1984, **34**, 175–181.
89. S. M. Barlow, F. M. Sullivan and J. Lines, Risk assessment of the use of deltamethrin on bednets for the prevention of malaria, *Food. Chem. Toxicol.*, 2001, **39**, 407–22.

90. S. Tós-Luty, A. Haratym-Maj, J. Latuszyńska, D. Obuchowska-Przebirowska and M. Tokarska-Rodak, Oral toxicity of deltamethrin and fenvalerate in Swiss mice, *Ann. Agric. Environ. Med.*, 2001, **8**, 245–254.
91. N. Shaker, G. A. Hassan, F. D. El-Nouty, Z. Abo-Elezz and G. A. Abd-Allah, In vivo chronic effect of dimethoate and deltamethrin on rabbits, *J. Environ. Sci. Health*, 1988, **B23**, 387–399.
92. S. Erdogan, E. H. Zeren, M. Emre, O. Aydin and D. Gumurdulu, Pulmonary effects of deltamethrin inhalation: an experimental study in rats, *Ecotoxicol. Environ. Safety*, 2006, **63**, 318–323.
93. J. Cotonat, M. Bleys and P. Foulhoux, Antagonistic effect of phenprobamate and mephenesin carbamate on deltamethrin toxicity, *J. Toxicol. Clin. Exper.*, 1987, **7**, 5–19.
94. M. I. Abd-El-Aziz, A. M. Sahlab and M. Adb El-Khalik, Influence of diazinon and deltamethrine on reproductive organs and fertility of male rats, *Dtsch. Tierärztl. Wschr.*, 1994, **101**, 213–248.
95. M. El-Gohary, W. M. Awara, S. Nassar and S. Hawas, Deltamethrin-induced testicular apoptosis in rats: the protective effect of nitric oxide synthase inhibitor, *Toxicology*, 1999, **132**, 1–8.
96. M. H. Salem, Z. Abo-Elezz, G. A. Abd-Allah, G. A. Hassan and N. Shaker, Effects of organophosphorus (dimethoate) and pyrethroid (deltamethrin) pesticides on semen characteristics in rabbits, *J. Environ. Sci. Health*, 1988, **B23**, 279–290.
97. A. J. M. Andrade, S. Araújo, G. M. Santana, M. Ohi and P. R. Dalsenter, Screening for in vivo (anti)estrogenic and (anti)androgenic activities of technical and formulated deltamethrin, *Regul. Toxicol. Pharmacol.*, 2002, **35**, 379–382.
98. H. Chen, J. Xiao, G. Hu, J. Zhou, H. Xiao and X. Wang, Estrogenicity of organophosphorus and pyrethroid pesticides, *J. Toxicol. Environ. Health*, 2002, **65**, 1419–1435.
99. A. J. M. Andrade, S. Araújo, G. M. Santana, M. Ohi and P. R. Dalsenter, Reproductive effects on deltamethrin on male offspring of rats exposed during pregnancy and lactation, *Regul. Toxicol. Pharmacol.*, 2002, **36**, 310–317.
100. K. M. Presibella, D. H. Kita, C. B. Carneiro, A. J. M. Andrade and P. R. Dalsenter, Reproductive evaluation of two pesticides combined (deltamethrin and endosulfan) in female rats, *Reprod. Toxicol.*, 2005, **20**, 95–101.
101. C. Issam, H. Samir, H. Zohra, Z. Monia and B. Cheikh, Hassen, Toxic responses to deltamethrin on gonads, sex hormones and lipoperoxidation in male rats following subcutaneous implants, *J. Toxicol. Sci.*, 2009, **34**, 663–670.
102. IPCS, *Environmental Health Criteria 70, Principles for the Safety Assessment of Food Additives and Contaminants in Food,* World Health Organization, Geneva, 1987.
103. JMPR, *Evaluations 2000, Part II. Toxicological, Deltamethrin*, Joint FAO/WHO Meeting on Pesticide Residues, World Health Organization, Geneva, 2000, pp. 79–110.

104. M. M. Abdel-Khalik, M. S. M. Hanafy and M. I. Abdel Aziz, Studies on the teratogenic effects of deltamethrin in rats, *Dtsch. Tierärtztl. Wschr.*, 1993, **100**, 142–143.
105. R. J. Kavlock, N. Chernoff and E. H. Rogers, The effect of acute maternal toxicity on fetal development in the mouse, *Teratogen. Carcinogen. Mutagen.*, 1985, **5**, 3–13.
106. H. Poláková and M. Vargová, Evaluation of the mutagenic effects of decamethrin: cytogenetic analysis of bone marrow, *Mutat. Res.*, 1983, **120**, 167–171.
107. M. Pluijmen, M. Drevon, R. Montesano, A. Malaveille, A. Hautefeuille and H. Bartsch, Lack of mutagenicity of synthetic pyrethroids in *Salmonella typhimurium* strains and in V79 hamster cells, *Mutat. Res.*, 1984, **137**, 7–15.
108. Y. Shukla and P. Taneja, Mutagenic evaluation of deltamethrin using rodent dominant lethal assay, *Mutat. Res.*, 2000, **467**, 119–127.
109. J. Surrallés, N. Xamena, A. Creus and R. Marcos, The suitability of the micronucleus assay in human lymphocytes as a new biomarker of excision repair, *Mutat. Res.*, 1995, **342**, 43–59.
110. D. K. Agarwal, L. K. S. Chauhan, S. K. Gupta and V. Sundararaman, Cytogenetic effects of deltamethrin on rat bone marrow, *Mutat. Res.*, 1994, **311**, 133–139.
111. S. P. Bhunya and P. C. Pati, Effect of deltamethrin, a synthetic pyrethroid, on the induction of chromosome aberrations, micronuclei and sperm abnormalities in mice, *Mutagenesis*, 1990, **5**, 229–232.
112. L. K. S. Chauhan, D. K. Agarwal and V. Sundararaman, In vivo induction of sister chromatid exchange in mouse bone marrow following oral exposure to commercial formulations of alpha-cyano pyrethroids, *Toxicol. Lett.*, 1997, **93**, 153–157.
113. P. Dolara, M. Salvadori, T. Capobianco and F. Torricelli, Sister-chromatid exchanges inn human lymphocytes induced by dimethoate, omethoate, deltamethrin, benomyl and their mixture, *Mutat. Res.*, 1992, **283**, 113–118.
114. G. Gandhi, J. B. Chowdhury, P. K. Sareen and V. P. S. Dhillon, Genotoxic effects of deltamethrin in the mouse bone marrow micronucleus assay, *Mutat. Res.*, 1995, **346**, 203–206.
115. C. K. Grisolia, A comparison between mouse and fish micronucleus test using cyclophosphamide, mitomycin C and various pesticides, *Mutat. Res.*, 2002, **518**, 145–150.
116. J. Surrallés, N. Xamena, A. Creus, J. Catalán, H. Norppa and R. Marcos, Induction of micronuclei by five pyrethroid insecticides in whole-blood and isolated human lymphocytes cultures, *Mutat. Res.*, 1995, **341**, 169–184.
117. M. Villarini, M. Moretti, P. Pasquini, G. Scassellati-Sforzolini, C. Fatigoni, M. Marcarelli, S. Monarca and A. V. Rodríguez, In vitro genotoxic effects of the insecticide deltamethrin in human peripheral blood leukocytes: DNA damage ('comet' assay) in relation to the induction of sister-chromatid exchanges and micronuclei, *Toxicology*, 1998, **130**, 129–139.

118. W. Hadnagy, N. H. Seemayer, K.-H. Kühn, G. Leng and H. Idel, Induction of mitotic cell division disturbances and mitotic arrest by pyrethroids in V79 cell cultures, *Toxicol. Lett.*, 1999, **107**, 81–87.
119. European Commission, *Review Report for the Active Substance Deltamethrin*, 6504/VI/99-Final, 2002.
120. J. R. P. Cabral, D. Galendo, M. Laval and N. Lyandrat, Carcinogenicity studies with deltamethrin in mice and rats, *Cancer Lett.*, 1990, **49**, 147–152.
121. Y. Shukla, A. Arora and A. Singh, Tumourigenic studies on deltamethrin in Swiss albino mice, *Toxicology*, 2001, **163**, 1–9.
122. K. Hakoi, R. Cabral, T. Hoshiya, R. Hasegawa, T. Shirai and N. Ito, Analysis of carcinogenic activity of some pesticides in a medium-term liver bioassay in the rat, *Teratogen. Carcinogen. Mutagen.*, 1992, **12**, 269–276.
123. M. J. Wolansky, C. Gennings and K. M. Crofton, Relative potencies for acute effects of pyrethroids on motor function in rats, *Toxicol. Sci.*, 2006, **89**, 271–277.
124. M. Takahashi and P. M. Le Quesne, The effects of the pyrethroids deltamethrin and cismethrin on nerve excitability in rats, *J. Neurol. Neurosurg. Psychiatr.*, 1982, **45**, 1005–1011.
125. H. Duclohier and D. Georgescauld, The effects of the insecticide decamethrin on action potential and voltage-clamp currents of *Myxicola* giant axon, *Comp. Biochem. Physiol.*, 1979, **62C**, 217–2223.
126. G. Theophilidis, M. Benaki and E. Papadopoulou-Mourkidou, Neurotoxic action of six pyrethroid insecticides on the isolated sciatic nerve of a frog (*Rana ridibunda)*), *Comp. Biochem. Physiol.*, 1997, **118C**, 97–103.
127. M. E. Brodie, Deltamethrin infusion into different sites in the neuraxis of freely-moving rats, *Neurobehav. Toxicol. Teratol.*, 1985, **7**, 51–55.
128. A. J. Gray and J. Rickard, Toxicity of pyrethroids to rats after direct injection into the central nervous system, *Neurotoxicology*, 1982, **3**, 25–35.
129. A. Wu and Y. Liu, Apoptotic death in rat brain following deltamethrin treatment, *Neurosci. Letters*, 2000, **279**, 85–88.
130. A. Wu, L. Li and Y. Liu, Deltamethrin induces apoptotic cell death in cultured cerebral cortical neurons, *Toxicol. Appl. Pharmacol.*, 2003, **187**, 50–57.
131. P. Eriksson and A. Fredriksson, Neurotoxic effects of two different pyrethroids, bioallethrin and deltamethrin, on immature and adult mice: changes in behavioural and muscarinic receptor variables, *Toxicol. Appl. Pharmacol.*, 1991, **108**, 78–85.
132. R. Husain, R. Husain, V. M. Adhami and P. K. Seth, Neurobehavioral, neurochemical, and neuromorphological effects of deltamethrin in adult rats, *J. Toxicol. Environ. Health*, 1996, **48**, 515–526.
133. K. M. Crofton and L. W. Reiter, Effects of two pyrethroid insecticides on motor activity and the acoustic startle response in the rat, *Toxicol. Appl. Pharmacol.*, 1984, **75**, 318–328.

134. T. H. Hijzen and J. L. Slanger, JL. Effects of Type I and Type II pyrethroids on the startle response in rats, *Toxicol. Lett.*, 1988, **40**, 141–152.
135. J. R. Glowa, Acute and sub-acute effects of deltamethrin and chlordimeform on schedule-controlled responding in the mouse, *Neurobehav. Toxicol. Teratol.*, 1986, **8**, 97–102.
136. A. S. Bloom, G. Staatz and T. Dieringer, Pyrethroid effects on operant responding and feeding, *Neurobehav. Toxicol. Teratol.*, 1983, **5**, 321–324.
137. S. Manna, D. Bhattacharyya, T. K. Mandal and S. Dey, Neuropharmacological effects of deltamethrin in rats, *J. Vet. Sci.*, 2006, **7**, 133–136.
138. C. A. Lazarini, J. C. Florio, I. P. Lemonica and M. M. Bernardi, Effects of prenatal exposure to deltamethrin on forced swimming behaviour, motor activity, and striatal dopamine levels in male and female rats, *Neurotoxicol. Teratol.*, 2006, **23**, 665–673.
139. M. Dayal, D. Parmar, A. Dhawan, M. Ali, U. N. Dwivedi and S. K. Seth, Effect of pre-treatment of cytochrome P450 (P450) modifiers on neurobehavioral toxicity induced by deltamethrin, *Food Chem. Toxicol.*, 2003, **41**, 431–437.
140. B. Veronesi, Validation of a rodent model of organophosphorus-induced delayed neuropathy, in *Clinical and Experimental Toxicology of Organophosphates and Carbamates*, ed. B. Ballantyne and T. C. Marrs, Butterworth Heinemann, Oxford, 1992, pp. 114–125.
141. M. J. Wolansky and J. A. Harril, Neurobehavioral toxicology of pyrethroid insecticides in adult animals: a critical review, *Neurotoxicol. Teratol.*, 2008, **30**, 55–78.
142. IPCS, *Environmental Health Criteria 82, Cypermethrin*, International Programme on Chemical Safety, World Health Organization, Geneva, 1989.
143. IPCS, *Environmental Health Criteria 142, Alpha-cypermethrin*, International Programme on Chemical Safety, World Health Organization, Geneva, 1992.
144. JECFA, *Toxicological Evaluation of Certain Veterinary Drug Residues in Food*, WHO Food Additive Series 38, 47th Meeting of the Joint FAO/WHO Expert Committee on Food Additives, World Health Organization, Geneva, 1996, pp. 133–170.
145. S. Luty, J. Latuszyńska, D. Obuchowska-Przebirowska, M. Tokarska and A. Haratym-Maj, Subacute toxicity of orally applied alpha-cypermethrin in Swiss mice, *Ann. Agric. Environ. Med.*, 2000, **7**, 33–41.
146. G. P. Rose and A. J. Dewar, Intoxication with four synthetic pyrethroids fails to show any correlation between neuromuscular dysfunction and neurobiochemical abnormalities in rats, *Arch. Toxicol.*, 1983, **53**, 297–316.
147. T. H. Hijzen, R. de Beun and J. L. Slanger, Effects of pyrethroids on acoustic startle reflex in the rat, *Toxicology*, 1988, **49**, 271–276.
148. IPCS, *Environmental Health Criteria 94, Permethrin*, International Programme on Chemical Safety, World Health Organization, Geneva, 1990.

149. JMPR, *Evaluations 1999. Part II—Toxicological,* Joint FAO/WHO Meeting on Pesticide Residues, World Health Organization, Geneva, 2000.
150. H. Bartsch, C. Malaveille, A. M. Camus, G. Martel-Planche, G. Brun, A. Huatefeuille, N. Sabadie, A. Barbin, T. Kuroki, C. Drevon, C. Piccoli and R. Montesano, Validation and comparative studies on 180 chemicals with *S. typhimurium* strains and V79 hamster cells in the presence of various metabolizing systems, *Mutat. Res.*, 1980, **76**, 1–50.
151. M. Moriya, T. Ohta, K. Watanabe, T. Miyazawa, K. Kato and Y. Shirasu, Further mutagenicity studies on pesticides in bacterial reversion assay systems, *Mutat. Res.*, 1983, **116**, 185–216.
152. N. Kornuta, E. Bagley and N. Nedopitanskaya, Genotoxic effects of pesticides, *J. Environ. Pathol. Toxicol. Oncol.*, 1996, **15**, 75–78.
153. M. D. Pednekar, S. R. Gandhi and M. S. Netrawali, Evaluation of mutagenic activities of endosulfan, phosalone, malathion and permethrin, before and after metabolic activation, in the Ames *Salmonella* test, *Bull. Environ. Contam. Toxicol.*, 1987, **38**, 925–933.
154. A. Herrera and E. Laborda, Mutagenic activity in synthetic pyrethroids in *Salmonella typhimurium, Mutagenesis*, 1988, **3**, 509–514.
155. A. Herrera, C. Barrueco, C. Caballo and E. de la Peña, Effect of permethrin on the induction of sister chromatid exchanges and micronuclei in cultured human lymphocytes, *Environ. Mol. Mutagen.*, 1992, **20**, 218–222.
156. R. C. Woodruff, J. P. Phillips and D. Irwin, Pesticide-induced complete and partial chromosome loss in screens with repair-defective females of *Drosophila melanogaster. Environ. Mutagen.*, 1983, **5**, 835–846.
157. R. K. Gupta, Z. A. Mehr, D. W. Korte and L. C. Routledge, Mutagenic potential of permethrin in *Drosophila melanogaster* (Diptera: Drosophilidae) sex-linked recessive lethal test, *J. Econ. Entomol.*, 1990, **83**, 721–724.
158. C. Barrueco, A Herrera, C. Caballo and E. de la Peña, Cytogenetic effects of permethrin in cultured human lymphocytes, *Mutagenesis*, 1992, **7**, 433–437.
159. C. Barrueco, A Herrera, C. Caballo and E. de la Peña, Induction of structural chromosome aberrations in human lymphocyte cultures and CHO cells by permethrin, *Teratogen. Mutagen. Carcinogen.*, 1994, **14**, 31–38.
160. J. Surrallès, N. Xamena, A. Creus, J. Catalan, H. Norppa and R. Marcos, Induction of micronuclei by five pyrethroid insecticides in whole-blood and isolated human lymphocytes cultures, *Mutat. Res.*, 1995, **341**, 169–164.
161. J. Surrallès, N. Xamena, A. Creus and R. Marcos, The suitability of the micronucleus assay in human lymphocytes as a new biomarker of excision repair, *Mutat. Res.*, 1995, **342**, 43–59.
162. J. Flodström, L. Warngard, S. Lundquist and U. G. Ahlborg, Inhibition of metabolic cooperation in vitro and enhancement of enzyme foci incidence in rat liver by the pyrethroid insecticide fenvalerate, *Arch. Toxicol.*, 1988, **61**, 218–223.

163. J. Ishmael and M. H. Litchfield, Chronic toxicity and carcinogenic evaluation of permethrin in rats and mice, *Fundam. Appl. Toxicol.*, 1988, **11**, 308–322.
164. C. G. Staatz, A. S. Bloom and J. J. Lech, A pharmacological study of pyrethroid toxicity in mice, *Pest. Biochem. Physiol.*, 1982, **17**, 287–292.
165. D. W. Gammon, L. J. Lawrence and J. E. Casida, Pyrethroid toxicology: protective effects of diazepam and phenobarbital in the mouse and cockroach, *Toxicol. Appl. Pharmacol.*, 1982, **66**, 290–296.
166. T. J. Shafer, S. O. Rijal and G. W. Gros, Complete inhibition of spontaneous activity in neuronal networks in vitro by deltamethrin and permethrin, *NeuroToxicol.*, 2008, **29**, 203–212.
167. C. A. Meacham, P. D. Brodfuehrer, J. A. Watkins and T. Shafer, Developmentally-regulated sodium channel subunits are differentially sensitive to α-cyano containing pyrethroids, *Toxicol. Appl. Pharmacol.*, 2008, **231**, 273–281.
168. S. K. Tayebati, M. A. Di Tullio, A. Ricci and F. Amenta, Influence of dermal exposure to the pyrethroid insecticide deltamethrin on rat brain microanatomy and cholinergic/dopaminergic neurochemistry, *Brain Res.*, 2009, **1301**, 180–188.
169. D. Chen, X. Huang, L. Lin and N. Shi, Deltamethrin induces mitochondrial membrane permeability an altered expression of cytochrome c in rat brain, *J. Appl. Toxicol.*, 2007, **27**, 368–372.
170. T. Morikawa and K. Furuhama, Effects of NMDA receptor antagonists on deltamethrin-induced striatal dopamine release in conscious unrestrained rats, *J. Vet. Med. Sci.*, 2009, **71**, 1129–1132.
171. D. E. Ray, S. A. Burr and T. Lister, The effects of combined exposure to the pyrethroids deltamethrin and *S*-bioallethrin on hippocampal inhibition and skeletal muscle hyperexcitability in rats, *Toxicol. Appl. Pharmacol.*, 2006, **216**, 354–362.
172. L. Isamura, M. Yasuda, K. Kuramitsu, D. Hara, A. Tabuchi and M. Tsuda, Deltamethrin, a pyrethroid insecticide, is a potent inducer for the activity-dependent gene expression of brain-derived neurotrophic factor in neurons, *J. Pharmacol. Exp. Ther.*, 2006, **316**, 136–143.
173. K. N. Woodward, Adverse effects of veterinary pharmaceutical products in animals, in *Veterinary Pharmacovigilance: Adverse Reactions to Veterinary Medicinal Products*, ed. K. N. Woodward, Wiley-Blackwell, Chichester, 2009, pp. 393–421.
174. W. M. Valentine, Pyrethrin and pyrethroids insecticides, *Vet. Clin. North Am: Small Anim. Pract.*, 1990, **20**, 375–382.
175. P. A. Volmer, S. A. Kahn, M. W. Knight and S. R. Hansen, Warning against use of some permethrin products in cats, *J. Am. Vet. Med. Assoc.*, 1998, **213**, 800.
176. E. K. Meyer, Toxicosis in cats erroneously treated with 45 to 65% permethrin products, *J. Am. Vet. Med.*, 1999, **215**, 198–203.

177. N. Sutton, N. Bates and A. Campbell, Seasonal rise in permethrin 'spot-on' poisoning in cats, *Vet. Rec.*, 2007, **161**, 244.
178. N. M. Sutton, N. Bates and A. Campbell, Clinical effects and outcome of feline permethrin spot-on poisonings reported to the Veterinary Poisons Information Service (VPIS), London, *J. Feline Med. Surg.*, 2007, **9**, 335–339.
179. P. J. Linnett, Permethrin toxicosis in cats, *Austral. Vet. J.*, 2008, **86**, 32–35.
180. R. Malik, M. P. Ward, A. Seavers, A. Fawcett, E. Bell, M. Govendir and S. Page, Permethrin spot-on intoxication of cats. Literature review and survey of veterinary practitioners in Australia, *J. Feline Med. Surg.*, 2010, **12**, 5–14.
181. S. A. Box and M. R. Lee, A systemic reaction following exposure to a pyrethroid insecticide, *Hum. Exp. Toxicol.*, 1996, **15**, 389–390.
182. P. M. LeQuesne, I. C. Maxwell and S. T. Butterworth, Transient facial sensory symptoms following exposure to synthetic pyrethroids: a clinical and electrophysiological assessment, *Neurotoxicol.*, 1980, **2**, 1–11.
183. S. B. Tucker and S. A. Flannigan, Cutaneous effects from occupational exposure to fenvalerate, *Arch. Toxicol.*, 1983, **54**, 195–202.
184. F. He, L. Liu, S. Chen, Z. Zhang and J. Sun, Clinical manifestations and diagnosis of acute pyrethroid poisoning, *Arch. Toxicol.*, 1989, **63**, 54–58.
185. A. T. Proudfoot, Poisoning due to pyrethrins, *Toxicol. Rev.*, 2005, **24**, 107–113.
186. D. E. Ray and J. R. Fry, A reassessment of the neurotoxicity of pyrethroid insecticides, *Pharmacol. Ther.*, 2006, **111**, 174–193.
187. L. G. Costa, G. Giordano, M. Guizetti and A. Vitalone, Neurotoxicity of pesticides: a brief review, *Front. Biosci.*, 2008, **13**, 1240–1249.
188. K. Hammond and J. B. Leikin, Topical pyrethrin toxicity leading to acute-onset stuttering in a toddler, *Am. J. Ther.*, 2008, **15**, 323–324.
189. J. Magdalen, M. Zawadzki and A. Merwid-Lad, Fatal intoxication with hydrocarbons in deltamethrin preparation, *Hum. Exp. Toxicol.*, 2009, **28**, 791–793.
190. N. Gunay, Z. Kekec, Y. Cete, C. Eken and A. T. Demiryurek, Oral deltamethrin ingestion due in a suicide attempt, *Bratisl. Lek. Listy.*, 2010, **111**, 303–305.
191. R. Ahdab, S. S. Ayache, F. Maltoni, P. Brugières and J. P. Lefaucheur, Motor neuron disorder with tongue spasms due to pyrethroid insecticide toxicity, *Neurology*, 2011, **76**, 196–197.
192. S. M. Bradberry, S. A. Cage, A. T. Proudfoot and J. A. Vale, Poisoning due to pyrethroids, *Toxicol. Rev.*, 2005, **24**, 93–106.
193. S. A. Flannigan, S. B. Tucker, M. M. Key, C. E. Ross, E. J. Fairchild, B. A. Grimes and R. B. Harris, Synthetic pyrethroid insecticides: a dermatological evaluation, *Br. J. Ind. Med.*, 1985, **42**, 363–372.
194. M. Tomizawa and J. E. Casida, Neonicotinoid insecticide toxicology: mechanisms of selective action, *Annu. Rev. Pharmacol.*, 2005, **45**, 247–268.

195. JMPR, *Pesticide Residues in Food—2001, Evaluations 2001. Part II—Toxicological*, Joint FAO/WHO Meeting on Pesticide Residues, World Health Organization, Geneva, 2002.
196. S. Bhardwaj, M. K. Srivastava, U. Kapoor and L. P. Srivastava, A 90 days oral toxicity of imidacloprid in female rats: morphological, biochemical and histopathological examination, *Food Chem. Toxicol.*, 2010, **48**, 1185–1190.
197. N. U. Karabay and M. G. Oguz, Cytogenetic and genotoxic effects of the insecticides, imidacloprid and methamidophos, *Genet. Mol. Res.*, 2005, **4**, 653–662.
198. D. M. Boothe, Dermatologic therapy, in *Small Animal Clinical Pharmacology and Therapeutics*. W. B. Saunders, London, 2001, pp. 654–675.
199. D. R. Godfrey, Dermatosis and associated systemic signs in a cat with thymoma and recently treated with an imidacloprid preparation, *J. Small Anim. Pract.*, 1999, **40**, 333–337.
200. D. Traversa, A. Di Cesare, P. Milillo, B. Lohr, R. Iorio, F. Pampurini, R. Schaper, B. Paoletti and J. Heine, Efficacy and safety of imidacloprid 10%/moxidectin 1% spot-on formulation in the treatment of feline aelurostrongylosis, *Parasitol. Res.*, 2009, **105**, S55–62.
201. J. J. Fourie, P. C. Delport, L. J. Fourie, J. Heine, I. G. Horak and K. J. Krieger, Comparative efficacy and safety of two treatment regimens with a topically applied combination of imidacloprid and moxidectin (Advocate®) against generalised demodicosis in dogs, *Parasitol. Res.*, 2009, **105**, S115–S124.
202. L. R. Hovda and S. B. Hooser, Toxicity of newer pesticides for use in dogs and cats, *Vet. Clin. Small. Anim.*, 2002, **32**, 455–467.
203. L. Elfman, C. Hogstedt, K. Engvall, E. Lampa and C. H. Lindh, Acute health effects on planters of conifer seedlings treated with insecticides, *Ann. Occup. Hyg.*, 2009, **53**, 383–390.
204. F. Mohamed, I. Gawarammana, T. A. Robertson, M. S. Roberts, C. Palangasinghe, S. Zawahir, S. Jayamanne, J. Kandasamy, M. Eddleston, N. A. Buckley, A. H. Dawson and D. M. Roberts, Acute human self-poisoning with imidacloprid compound: a neonicotinoid insecticide, *PLoS One*, 2009, **4**, e5127.
205. S. D. Buckingham, B. Lapied, H. Le Corronc, F. Grolleau and D. B. Battelle, Imidacloprid actions on insect neuronal acetylcholine receptors, *J. Exp. Biol.*, 1997, **200**, 2685–2692.
206. B. Ballantyne and T. C. Marrs, Overview of the biological and clinical aspects of organophosphates and carbamates, in *Clinical and Experimental Toxicology of Organophosphates and Carbamates*, ed. B. Ballantyne and T. C. Marrs, Butterworth Heinemann, London, 1992, pp. 3–14.
207. M. Lotti, Central neurotoxicity and behavioural effects of anticholinesterases, in *Clinical and Experimental Toxicology of Organophosphates and Carbamates*, ed. B. Ballantyne and T. C. Marrs, Butterworth Heinemann, London, 1992, pp. 75–83.

208. M. K. Johnson and P. Glyn, Neuropathy target esterase (NTE) and organophosphorus-induced delayed polyneuropathy (OPIDP): recent advances, *Toxicol. Lett.*, 1995, **82/83**, 459–463.
209. M. B. Abou-Donia, Organophosphorus ester-induced chronic neurotoxicity, *Arch. Environ. Health*, 2003, **58**, 484–497.
210. A. Moretto, Experimental and clinical toxicology of anticholinesterase agents, *Toxicol. Lett.*, 1998, **102/103**, 509–513.
211. JMPR, *Pesticide Residues in Food—1993, Evaluations. Part II—Toxicological*, Joint FAO/WHO Meeting on Pesticide Residues, World Health Organization, Geneva, 1994.
212. E. Mutch and F. Williams, Diazinon, chlorpyrifos and parathion are metabolised by multiple cytochromes P450 in human liver, *Toxicology*, 2006, **224**, 22–32.
213. T. S. Poet, H. Wu, A. A. Kousba and C. Timchalk, In vitro rat hepatic and intestinal metabolism of the organophosphate pesticides chlorpyrifos and diazinon, *Toxicol. Sci.*, 2003, **72**, 193–200.
214. W. A. Kappers, R. J. Edwards, S. Murray and A. R. Boobis, Diazinon is activated by CYP2C19 in human liver, *Toxicol. Appl. Pharmacol.*, 2001, **177**, 68–76.
215. M. Jokanović, Biotransformation of organophosphorus compounds, *Toxicology*, 2001, **166**, 139–160.
216. K. M. Atherton, F. M. Williams, S. Jameson and E. Mutch, DNA damage by single doses of five organophosphate pesticides, *Toxicology*, 2008, **253**, 8–9.
217. E. Salazar-Arredondo, M. de Jesús, Solis-Heredia, E. Rojas-García, I. Hernández-Ochoa and B. Quintanilla-Vega, Sperm chromatin alteration and DNA damage by methyl-parathion, chlorpyrifos and diazinon and their oxon metabolites in human spermatozoa, *Reprod. Toxicol.*, 2008, **25**, 455–460.
218. M. Tisch, P. Schmezer, M. Faulde, A. Groh and H. Maier, Genotoxicity studies on permethrin, DEET and diazinon in primary human nasal mucosal cells, *Eur. Arch. Otorhinolaryngol.*, 2002, **259**, 150–153.
219. K. Kuroda, Y. Yamaguchi and G. Endo, Mitotic activity, sister chromatid exchange, and rec assay of pesticides, *Arch. Environ. Contam. Toxicol.*, 1992, **23**, 13–18.
220. D. Lopez, C. Aleixandre, M. Merchan and E. Carrascal, In vitro induction of alterations in peripheral blood lymphocytes by different doses of diazinon, *Bull. Environ. Contam. Toxicol.*, 1986, **37**, 517–522.
221. D. Lopez and E. Carrascal, Sensitivity of human lymphocyte chromosome to diazinon at different times during cell culture, *Bull. Environ. Contam. Toxicol.*, 1987, **38**, 125–130.
222. Ş. Çakir and R. Sarikaya, Genotoxicity testing of some organophosphate insecticides in the *Drosophila* wing spot test, *Food Chem. Toxicol.*, 2005, **43**, 443–450.
223. N. E. Garrett, H. F. Stack, M. A. Jackson and M. D. Waters, Genotoxic and carcinogenic potential of anticholinesterases, in *Clinical and*

Experimental Toxicology of Organophosphates and Carbamates, ed. B. Ballantyne and T. C. Marrs, Butterworth Heinemann, London, 1992, pp. 223–240.

224. D. B. MacGregor, A. Brown, P. Cattenach, I. Edwards, D. McBride, C. Riach and W. J. Caspary, Responses of the L5178 TK^+/TK^- mouse lymphoma cell forward mutation assay: III. 72 coded chemicals, *Environ. Mol. Mutagen.*, 1988, **12**, 85–154.
225. E. Zabihi Neishabouri, Z. M. Hassa, E. Azizi and S. N. Ostad, Evaluation of immunotoxicity induced by diazinon in C57bl/6 mice, *Toxicology*, 2004, **196**, 173–179.
226. G. J. Oostingh, G. Wichmann, M. Schmittner, I. Lehmann and A. Duschl, The cytotoxic effects of the organophosphates chlorpyriphos and diazinon differ from their immunomodulating effects, *J. Immunotoxicol.*, 2009, **6**, 136–145.
227. R. D. Handy, H. A. Abd-El Samei, M. F. F. Bayomy, A. M. Mahran, A. M. Abdeen and E. A. El-Elaimy, Chronic diazinon exposure: pathologies of spleen, thymus, blood cells and lymph nodes are modulated by dietary protein or lipid in the mouse, *Toxicology*, 2002, **172**, 13–34.
228. T. Galloway and R. Handy, Immunotoxicity of organophosphorus pesticides, *Ecotoxicology*, 2003, **12**, 345–363.
229. A. M. Alluwaimi and Y. Hussein, Diazinon immunotoxicity in mice: modulation of cytokines level and their gene expression, *Toxicology*, 2007, **236**, 123–131.
230. V. N. Kabrawala, R. M. Shah and G. G. Oza, Diazinon poisoning (a study of 25 cases), *Indian Pract.*, 1965, **18**, 711–717.
231. R. S. Wadia, C. Sadagopan, R. B. Amin and H. V. Sardesai, Neurological manifestations of organophosphorus poisoning, *J. Neurol. Neurosurg. Psychiatry*, 1974, **37**, 841–847.
232. H. W. Klemmer, E. R. Reichert and W. L. Yauger, Five cases of intentional ingestion of 25 per cent diazinon with treatment and recovery, *Clin. Toxicol.*, 1978, **12**, 435–444.
233. A. Poklis, F. W. Kutz, J. F. Sperling and D. P. Morgan, A fatal diazinon poisoning, *Forensic Sci. Int.*, 1980, **15**, 135–140.
234. R. J. Zwiener and C. M. Ginsberg, Organophosphate and carbamate poisoning in infants and children, *Pediatrics*, 1988, **81**, 121–126.
235. Z. Weizman and S. Sofer, Acute pancreatitis in children with anticholinesterase insecticide intoxication, *Pediatrics*, 1992, **90**, 204–206.
236. IPCS, *Environmental Health Criteria 198, Diazinon.* International Programme on Chemical Safety, World Health Organization, Geneva, 1998.
237. F. Muratore, H. Faiolo and G. Ponzetta, On four cases of poisoning by organophosphorus acid esters (antiparasitic) with recovery, with special reference to hepatic involvement, *Minerva Med.*, 1960, **51**, 3342–3345.
238. A. Halle and D. D. Sloas, Percutaneous organophosphate poisoning, *South. Med. J.*, 1987, **80**, 1179–1181.

239. A. M. Mizra, N. Schaub, L. Samler and T. E. Reuchelderfer, Organophosphate insecticide diazinon poisoning in children, *Med. Ann. DC.*, 1972, **41**, 559–560.
240. S. L. Wagner and D. L. Orwick, Chronic organophosphate exposure associated with transient hypertonia in an infant, *Pediatrics*, 1994, **94**, 94–97.
241. S. W. Hall and B. B. Baker, Intermediate syndrome from organophosphate poisoning, *Vet. Hum. Toxicol.*, 1989, **31**, 355.
242. K. K. Samal and C. S. Sahu, Organophosphorus poisoning and intermediate neurotoxic syndrome, *J. Assoc. Phys. India*, 1990, **38**, 181–182.
243. S. A, Soliman, G. W. Sovocol, A. Curley, N. S. Ahmed and A. K. El-Fiki El-Sabae, Two acute human poisoning cases resulting from exposure to diazinon transformation products in Egypt, *Arch. Environ. Health*, 1982, **37**, 207–212.
244. D. Mello, F. Rodrigues Puga and R. Benintendi, Intoxications produced by degradation products of diazinon in its use as a tick killer, *Biologico*, 1972, **38**, 136–139.
245. L. Wecker, R. E. Mrak and W. D. Dettbran, Evidence of necrosis in human intercostal muscle following inhalation of an organophosphorus insecticide, *J. Environ. Pathol. Toxicol. Oncol.*, 1985, **6**, 171–175.
246. P. Casey and J. A. Vale, Deaths from pesticide poisoning in England and Wales: 1945–1989, *Human Exp. Toxicol.*, 1994, **13**, 95–101.
247. T. C. Marrs and P. Edwards, Medicines used to control and treat external parasites of sheep—toxicology and the phenomenon of reported adverse human responses to organophosphorus sheep dips, in *Veterinary Pharmacovigilance: Adverse Reactions to Veterinary Medicinal Products*, ed. K. N. Woodward, Wiley-Blackwell, Chichester, 2009, pp. 517–527.
248. D. Swanston and I. Shaw, Organophosphorus compounds, *Sheep Farmer*, 1990, **10**, 31–33.
249. G. A. Jamal, Long term neurotoxic effects of organophosphate compounds, *Adverse Drug React. Toxicol. Rev.*, 1995, **14**, 85–99.
250. K. N. Woodward, Adverse reactions in humans following exposure to veterinary drugs, in *Veterinary Pharmacovigilance: Adverse Reactions to Veterinary Medicinal Products*, ed. K. N. Woodward, Wiley-Blackwell, Chichester, 2009, pp. 475–515.
251. K. N. Woodward, Veterinary pharmacovigilance, Part 4. Adverse reactions in humans to veterinary medicinal products, *J. Vet. Pharmacol. Ther.*, 2005, **28**, 185–201.
252. J. C. Axelrad, C. V. Howard and W. G. McLean, The effects of acute pesticide exposure on neuroblastoma cells chronically exposed to diazinon, *Toxicology*, 2003, **185**, 67–78.
253. W. G. Carter, M. Tarhoni, A. J. Rathbone and D. E. Ray, Differential protein adduction by seven organophosphorus pesticides in both brain and liver, *Human Exp. Toxicol.*, 2007, **26**, 347–353.
254. L. Sarabia, I. Maurer and E. Bustos-Obregón, Melatonin prevents damage by the organophosphorus pesticide diazinon on the mouse testis, *Ecotoxicol. Environ. Safety*, 2009, **72**, 938–942.

255. S. H. Swan, R. L. Kruse, F. Liu, D. B. Barr, E. Z. Drobnis, J. B. Redmon, C. Wang, C. Brazil and J. W. Overstreet, and the Study for Future Families Research Group, Semen quality in relation to biomarkers of pesticide exposure, *Environ. Health Perspect.*, 2003, **111**, 1478–1484.
256. A. Okamura, M. Kamijima, K. Ohtani, O. Yamanoshita, D. Nakamura, Y. Ito, J. Ueyama, T. Suzuki, R. Imai, K. Takagi and T. Nakajima, Broken sperm, cytoplasmic droplets and reduced sperm motility are principal markers of decreased sperm quality due to organophosphorus pesticides in rats, *J. Occup. Health*, 2009, **51**, 478–487.
257. T. Rush, X. Q. Liu, J. Hjelmhaug and D. Lobner, Mechanisms of chlorpyrifos and diazinon induced neurotoxicity in cortical culture, *Neuroscience*, 2010, **166**, 899–906.
258. T. A. Slotkin, I. T. Ryde, E. D. Levin and F. J. Seidler, Developmental neurotoxicity of low-dose diazinon exposure of neonatal rats: effects of serotonin systems in adolescence and adulthood, *Brain Res. Bull.*, 2008, **75**, 640–647.
259. E. Sidiropoulou, M. Sachana, J. Flaskos, W. Harris, A. J. Hargreaves and Z. Woldehiwet, Diazinon oxon interferes with differentiation of rat C6 glioma cells, *Toxicol. In Vitro*, 2009, **23**, 1548–1552.
260. C. S. Roegge, O. Timofeeva, F. J. Seidler, T. A. Slotkin and E. D. Levin, Developmental neurotoxicity in rats: later effects on emotional response, *Brain Res. Bull.*, 2008, **75**, 166–172.
261. T. A. Slotkin, B. E. Bodwell, E. D. Levin and F. J. Seidler, Neonatal exposure to low doses of diazinon: long-term effects on neural cell development and acetylcholine systems, *Environ. Health Perspect.*, 2008, **116**, 340–348.
262. T. A. Slotkin and F. J. Seidler, Comparative developmental neurotoxicity of organophosphates *in vivo*: transcriptional responses of pathways for brain cell development, cell signalling, cytotoxicity and neurotransmitter systems, *Brain Res. Bull.*, 2007, **72**, 232–274.
263. T. A. Slotkin, F. J. Seidler and F. Fumagalli, Exposure to organophosphates reduces the expression of neurotrophic factors in neonatal rat brain regions: similarities and differences in the effects of chlorpyriphos and diazinon on the fibroblast growth factor superfamily, *Environ. Health Perspect.*, 2007, **115**, 909–916.
264. T. A. Slotkin, E. D. Levin and F. J. Seidler, Comparative developmental neurotoxicity of organophosphate insecticides: effects on brain development are separable from systemic toxicity, *Environ. Health Perspect.*, 2006, **114**, 746–751.
265. T. A. Slotkin, C. A. Tate, I. T. Ryde, E. D. Levin and F. J. Seidler, Organophosphate insecticides target the serotonergic system in developing rat brain regions: disparate effects of diazinon and parathion at doses spanning the threshold for cholinesterase inhibition, *Environ. Health Perspect.*, 2006, **114**, 1542–1546.
266. O. A. Timofeeva, C. S. Roegge, F. J. Seidler, T. A. Slotkin and E. D. Levin, Persistent cognitive alterations in rats after early postnatal

exposure to low doses of the organophosphate pesticide, diazinon, *Neurotoxicol. Teratol.*, 2008, **30**, 38–45.
267. M. Guizzetti, S. Pathak, G. Gennaro and L. G. Costa, Effect of organophosphorus insecticides and their metabolites on astroglial cell proliferation, *Toxicology*, 2005, **215**, 182–190.
268. D. E. Ray and P. G. Richards, The potential for toxic effects of chronic, low-dose exposure to organophosphates, *Toxicol. Lett.*, **120**, 343–351.
269. K. O'Leary, R. J. Edwards, M. M. Town and A. R. Boobis, Genetic and other sources of variation in the activity of serum paraoxonase/diazoxonase in humans: consequences for risk from exposure to diazinon, *Pharmacogenet. Genomics*, 2005, **15**, 51–60.
270. B. A. Hatjian, E. Mutch, F. M. Williams, P. G. Blain and J. W. Edwards, Cytogenetic response without changes in peripheral cholinesterase enzymes following exposure to a sheep dip containing diazinon in vivo and in vitro, *Mutat. Res.*, 2000, **472**, 85–92.
271. B. Mackness, P. Durrington, A. Povey, S. Thompson, M. Dippnall, M. Mackness, T. Smith and N. Cherry, Paraoxonase and susceptibility to organophosphorus poisoning in farmers dipping sheep, *Pharmacogenetics*, 2003, **13**, 81–88.
272. M. Roth, The availability and use of chemotherapeutant sea lice control products, *Contrib. Zool.*, 2000, **69**. http://dpc.uva.nl/ctz/vol69/nr01/art12
273. A. N. Grant, Medicines for sea lice, *Pest. Manag. Sci.*, 2002, **58**, 521–527.
274. L. Burridge, J. S. Weis, F. Cabello, J. Pizarro and K. Bostock, Chemical use in salmon aquaculture: A review of current practices and possible environmental effects, *Aquaculture*, 2010, **306**, 7–23.
275. CVMP, *Azamethiphos, Summary Report (1)*. Committee for Veterinary Medicinal Products, EMEA/MRL/001/95-FINAL.
276. T. Roberts and D. Hutson (ed.). Azamethiphos, in *Metabolic Pathways of Agrochemicals. Part two. Insecticides and Fungicides*, Royal Society of Chemistry, Cambridge, 1999, pp. 205–206.
277. K. Kohzaki, T. Masaoka, M. Nagayama, F. Akahori, K. Kakaguchi and R. Kohzaki, Effects of azamethiphos, an organophosphorus insecticide, on serum cholinesterase activity and isoenzymes in the rats, *Vet. Human Toxicol.*, 1991, **33**, 575–578.
278. V. Zitco, Alkylating potency of azamethiphos, *Bull. Environ. Contam. Toxicol.*, 2001, **66**, 283–286.
279. R. T. Sharpe, C. T. Livesey, I. H. Davies, J. R. Jones and A. Jones, Diazinon toxicity in sheep and cattle arising from misuse of unlicensed and out-of-date products, *Vet. Rec.*, 2006, **159**, 16–19.
280. C. Pretti, G. Soldani, A. M. Cognetti-Varriale, G. Monni, V. Meucci and L. Intorre, Efficacy and safety of azamethiphos for the treatment of pseudodactylogyrosis in the European eel, *J. Vet. Pharmacol. Ther.*, 2002, **25**, 155–157.
281. L. Intorre, G. Soldani, A. M. Cognetti-Varriale, G. Monni, V, Meucci and C. Pretti, Safety of azamethiphos in eel, seabass and trout, *Pharmacol. Res.*, 2004, **49**, 171–176.

282. T. E. Horsberg, T. Høy and I. Nafstad, Organophosphate poisoning of Atlantic salmon in connection with treatment against sea lice, *Acta. Vet. Scand.*, 1989, **30**, 385–390.
283. J. F. Burka, K. L. Hammell, T. E. Horsberg, G. R. Johnson, D. J. Rainnie and D. J. Speare, Drugs in salmonid aquaculture—a review, *J. Vet Pharmacol. Ther.*, 1997, **20**, 333–349.
284. J. E. Bron, C. Sommerville, R. Wooten and G. H. Rae, Influence of treatment with dichlorvos on the epidemiology of *Lepeophtheirus salmonis* (Krøyer, 1837) and *Caligus elongates* Nordmann, 1832 on Scottish salmon farms, in *Pathogens of Wild and Farmed Fish: Sea Lice*, ed. G. A. Boxall and D. Defaye, Ellis Horwood, London, 1993, pp. 263–274.
285. V. L. Salgado and J. H. Hayashi, Metaflumizone is a novel sodium channel blocker insecticide, *Vet. Parasitol.*, 2007, **150**, 182–189.
286. K. S. Silver, Y. Nomura, V. L. Salgado and K. Dong, Role of the sixth transmembrane segment of domain IV of the cockroach sodium channel in the action of sodium channel-blocker insecticides, *Neurotoxicol.*, 2009, **30**, 613–621.
287. R. L. DeLay, E. Lacoste, S. Delprat and F. Blond-Riou, Pharmacokinetics of metaflumizone in the plasma and hair of cats following topical application, *Vet. Parasitol.*, 2007, **150**, 258–262.
288. R. L. DeLay, E. Lacoste, T. Mezzasalma and F. Blond-Riou, Pharmacokinetics of metaflumizone and amitraz in the plasma and hair of dogs following topical application, *Vet. Parasitol.*, 2007, **150**, 251–257.
289. K. Hempel, F. G. Hess, C. Bögi, E. Fabian, J. Hellwig and I. Fegert, Toxicological properties of metaflumizone, *Vet. Parasitol.*, 2007, **150**, 190–195.
290. K. Heaney and R. G. Lindahl, Safety of a topically applied spot-on formulation of metaflumizone plus amitraz for flea and tick control in dogs, *Vet. Parasitol.*, 2007, **150**, 225–232.
291. K. Heaney and R. G. Lindahl, Safety of a topically applied spot-on formulation of metaflumizone for flea and tick control in cats and kittens, *Vet. Parasitol.*, 2007, **150**, 233–238.
292. K. Hellmann, K. Adler, L. Parker, K. Pfister, R. L. DeLay and D. Rugg, Evaluation of the efficacy and safety of a novel formulation of metaflumizone plus amitraz in dogs naturally infested with fleas and ticks in Europe, *Vet. Parasitol.*, 2007, **150**, 239–245.
293. K. Hellmann, K. Adler, L. Parker, K. Pfister, R. L. DeLay and D. Rugg, Evaluation of the efficacy and safety of a novel formulation of metaflumizone in cats naturally infested with fleas in Europe, *Vet. Parasitol.*, 2007, **150**, 246–250.
294. V. D. Tarallo, R. P. Lia, M. Sasanelli, C. Carfachia and D. Otranto, Efficacy of amitraz plus metaflumizone for the treatment of canine demodicosis associated with *Malssezia pachydermatis, Parasit. Vectors*, 2009, **2**, 13.
295. U. Oberkirchner, K. E. Linder, S. Dunston, P. Bizikova and T. Olivry, Metaflumizone-amitraz (Promeris)-associated pustular acantholytic

dermatitis in 22 dogs: evidence suggests contact drug-triggered pemphigus folicaeus, *Vet. Dermatol.*, 2011, **22**, 436–438.
296. B. Lapied, F. Grolleau and D. B. Battelle, Indoxacarb, an oxadiazine insecticide, blocks neuronal sodium channels, *Br. J. Pharmacol.*, 2001, **132**, 587–595.
297. T. Narahashi, Nerve membrane ion channels as the target site for insecticides, *Mini Rev. Med. Chem.*, 2002, **2**, 419–432.
298. EMA, *European Public Assessment Report, Activyl, EMA/529995/2010*, European Medicines Agency, February 2011.
299. JMPR, *Pesticide Residues in Food—1993, Evaluations 2005. Part II—Toxicological*, Joint FAO/WHO Meeting on Pesticide Residues, World Health Organization, Geneva, 2006.
300. H. Singh, E. Purnell and C. Smith, Mechanistic study on aniline-induced erythrocyte toxicity, *Arh. Hig. Rada Toksikol.*, 2007, **58**, 275–285.
301. R. S. Chhabra, M. Thompson, M. R. Elwell and D. K. Gerken, Toxicity of *p*-chloroaniline in rats and mice, *Food Chem. Toxicol.*, 1990, **28**, 717–722.
302. B. E. Watt, A. T. Proudfoot, S. M. Bradberry and J. A. Vale, Poisoning due to urea herbicides, *Toxicol. Rev.*, 2005, **24**, 161–166.
303. J. Umbreit, Methemoglobin—it's not just blue: a concise review, *Am. J. Hematol.*, 2007, **82**, 134–144.
304. S. M. Bradberry, Occupational methaemoglobinaemia. Mechanisms of production, features, diagnosis and management including the use of methylene blue, *Toxicol. Rev.*, 2003, **22**, 13–27.
305. G. Wang, X. Zhang, C. Yao and M. Tian, Four-week oral toxicity of three metabolites of nitrobenzene in rats, *Drug Chem. Toxicol.*, 2010, **33**, 238–243.
306. H. Singh and E. T. Purnell, Aniline derivative-induced methemoglobin in rats, *J. Environ. Pathol. Toxicol. Oncol.*, 2005, **24**, 57–65.
307. Y. J. Wu, Y. L. Lin, H. Y. Huang and B. G. Hsu, Methemoglobinemia induced by indoxacarb intoxication, *Clin. Toxicol.*, 2010, **48**, 766–767.
308. R. Chhabra, I. Singh, M. Tandon and R. Babu, Indoxacarb poisoning: a rare presentation as methemoglobinaemia, *Indian J. Anaesth.*, 2010, **54**, 239–241.
309. L. Prasan, M. S. Rao, V. Singh, R. Kujur and Gowrishankar, Indoxacarb poisoning: a rare presentation as methemoglobinemia, *Ind. J. Crit. Care*, 2008, **12**, 198–200.
310. D. Hainzl, L. M. Cole and J. E. Casida, Mechanisms for selective toxicity of fipronil insecticide and its sulfone metabolite and desulfinyl photoproduct, *Chem. Res. Toxicol.*, 1998, **11**, 1529–1535.
311. T. Narahashi, X. Zhao, T. Ikeda, K. Nagata and J. Z. Yeh, Differential actions of insecticides on target sites: basis for selective toxicity, *Human Exp. Toxicol.*, 2007, **26**, 361–366.
312. P. Varro, J. Győri and I. Világi, In vitro effects of fipronil on neuronal excitability in mammalian and molluscan nervous systems, *Ann. Agric. Environ. Med.*, 2009, **16**, 71–77.

313. G. S. Ratra, G. Kamitra and J. E. Casida, Role of human $GABA_A$ receptor β3 subunit in insecticide toxicity, *Toxicol. Appl. Pharmacol.*, 2001, **172**, 233–240.
314. D. Hainzl and J. E. Casida, Fipronil insecticide: novel photochemical desulfinylation with retention of neurotoxicity, *Proc. Natl. Acad. Sci. USA*, 1996, **93**, 12764–12767.
315. M. Eddleston, Phenylpyrazole insecticides, *Clin. Toxicol.*, 2008, **46**, 405.
316. JMPR, *Pesticide Residues in Food—1997, Evaluations 2005. Part II—Toxicological*, Joint FAO/WHO Meeting on Pesticide Residues, World Health Organization, Geneva, 1998.
317. P. M. Hurley, R. N. Hill and R. J. Whiting, Mode of carcinogenic action of pesticides inducing thyroid follicular cell tumors in rodents, *Environ. Health Perspect.*, 1998, **106**, 437–445.
318. K. N. Woodward, The toxicity of particular veterinary drug residues, in *Pesticide, Veterinary and Other Residues in Food*, ed., D. H. Watson, CRC Press/Woodhead Publishing, Boca Raton/Cambridge, 2000, pp. 175–223.
319. K. N. Woodward, Veterinary pharmacovigilance, Part 6. Predictability of adverse reactions in animals from laboratory toxicology studies, *J. Vet. Pharmacol. Ther.*, 2005, **28**, 213–231.
320. K. N. Woodward, Carcinogenicity of sulfadimidine, *Human Exp. Toxicol.*, 1992, **11**, 60–61.
321. J. E. Heath and N. A. Littlefield, Effects of subchronic oral sulfamethazine administration on Fischer 344 rats and B6C3F1 mice, *J. Environ. Pathol. Toxicol. Oncol.*, 1984, **5**, 201–214.
322. J. E. Heath and N. A. Littlefield, Morphological effects of subchronic oral sulfamethazine on Fischer 344 rats and B6C3F1 mice, *Toxicol. Pathol.*, 1984, **12**, 3–9.
323. N. A. Littlefield, W. G. Sheldon, R. Allen and D. W. Gaylor, Chronic toxicity/carcinogenicity studies of sulfamethazine in Fischer 344/N rats: two-generation exposure, *Food Chem. Toxicol.*, 1990, **28**, 156–167.
324. F. R. Fullerton, R. J. Kushmaul, R. L. Suber and N. A. Littlefield, Influence of oral administration of sulfamethazine on thyroid hormone levels in Fischer F344 rats, *J. Toxicol. Environ. Health*, 1987, **22**, 175–185.
325. R. N. Hill, T. M. Crisp, P. M. Hurley, S. L. Rosenthal and D. V. Singh, Risk assessment of thyroid follicular cell tumors, *Environ. Health Perspect.*, 1998, **106**, 447–457.
326. R. M. McClain, Mechanistic considerations for the relevance of animal data on thyroid neoplasia to human risk assessment, *Mutat. Res.*, 1995, **333**, 131–142.
327. L. A. Poirier, D. R. Doerge, D. W. Gaylor, M. A. Miller, R. J. Lorentzen, D. A. Casciano, F. F. Kadlubar and B. A. Schwetz, An FDA review of sulfamethazine toxicity, *Regul. Toxicol. Pharmacol.*, 1999, **30**, 217–222.
328. M. Ohi, R. R. Dalsenter, A. J. M. Andrade and A. J. Nascimento, Reproductive adverse effects of fipronil in Wistar rats, *Toxicol. Lett.*, 2004, **146**, 121–127.

329. H. T. Fung, K. K. Chang, W. M. Ching and C. W. Kam, A case of ingestion of ant bait containing fipronil, *J. Toxicol. Clin. Toxicol.*, 2003, **41**, 245–248.
330. Z. Chodorowski and J. S. Anand, Accidental dermal and inhalation exposure with fipronil – a case report, *J. Toxicol. Clin. Toxicol.*, 2004, **42**, 189–190.
331. F. Mohamed, L. Senarathna, A. Percy, M. Abeyewardene, G. Eaglesham, R. Cheng, S. Azher, A. Hittarage, W. Dissanayake, M. H. Rezvi Sheriff, W. Davies, N. Buckley and M. Eddleston, Acute human self-poisoning with the *N*-phenylpyrazole insecticide fipronil – a $GABA_A$-gated chloride channel blocker, *J. Toxicol. Clin. Toxicol.*, 2004, **42**, 955–963.
332. M. Eddleston, M. H. Rezvi Sheriff and K. Hawton, Deliberate self harm in Sri Lanka: an overlooked tragedy in the developing world, *BMJ*, 1998, **317**, 133–135.
333. S. J. Lee, P. Mulay, B. Diebolt-Brown, M. J. Lackovic, L. N. Mehler, J. Beckman, J. Waltz, J. B. Prado, Y. A. Mitchell, S. A. Higgins, A. Schwartz and G. M. Calvert, Acute illnesses associated with exposure to fipronil—surveillance data from 11 states in the United States, 2001–2007, *Clin. Toxicol.*, 2010, **48**, 737–734.
334. R. C. Gupta, Fipronil, in *Veterinary Toxicology. Basic and Clinical Principles*, Elsevier Academic Press, London, 2007, pp. 502–504.
335. M. Webster, Product warning: Frontline, *Aust. Vet. J.*, 1999, **77**, 202.
336. G. Schuele, S. Barnett, B. Bapst, T. Cavaliero, G. Strehlau, D. R. Young, C. Moran and P. Junquera, Efficacy of a single application of a 12.5% pyriprole topical solution against laboratory infestations with ticks (*Ixodes ricinus, Dermacentor reticulatus*, and *Rhipicephalus sanguineus*) in dogs, *Vet. Parasitol.*, 2008, **154**, 311–317.
337. S. Barnett, L. Luempert, G. Schuele, A. Quezada, G. Strehlau and P. Doherty, Efficacy of pyriprole topical solution against the cat flea, *Ctenocephalides felis*, on dogs, *Vet. Ther.*, 2008, **9**, 4–14.
338. G. Schuele, S. Barnett, B. Bapst, T. Cavaliero, L. Luempert, G. Strehlau, D. R. Young, C. Moran and P. Jonquera, The effect of water and shampooing on the efficacy of a pyriprole 12.5% topical solution against the brown dog tick (*Rhipicephalus sanguineus*) and cat flea (*Ctenocephalides felis*) infestations on dogs, *Vet. Parasitol.*, 2008, **151**, 300–311.
339. C. Thomas, M. Roques and M. Franc, The effectiveness of a pyriprole (125 mg/ml) and a metaflumizone (150 mg/ml) combined with amitraz (150 mg/ml) spot-on treatment in preventing *Phlebotomus perniciosus* from feeding on dogs, *Parasite*, 2008, **15**, 93–96.
340. E. Bouhsira, A. Fysikopoulos and M. Franc, Efficacy of fipronil-(S)-methoprene, metaflumizone, combined with amitraz, and pyriprole commercial spot-on products in preventing *Culex pipiens pipiens* from feeding on dogs, *Vet. Rec.*, 2009, **165**, 135–137.
341. J. J. Fourie, I. G. Horak and V. de la Puente Redondo, Efficacy of a formulation of pyriprole on dogs infested with *Sarcoptes scabei*, *Vet. Rec.*, 2010, **167**, 442–445.

342. R. M. Hollingworth, Chemistry, biological activity, and uses of formamidine pesticides, *Environ. Health Perspect.*, 1976, **14**, 57–69.
343. B. B. Hoffman and R. J. Lefkowitz, Catecholamines, sympathomimetic drugs, and adrenergic receptor antagonists, in *Goodman & Gilman's The Pharmacological Basis of Therapeutics*, ed. J. G. Hardman and L. E. Limbird, McGraw-Hill, London, 9th edn, 1996, pp. 199–248.
344. F. Matsumura and R. W. Beeman, Biochemical and physiological effects of chlordimeform, *Environ. Health Perspect.*, 1976, **14**, 71–82.
345. L. G. Costa, D. S. Wu, G. Olibet and S. D. Murphy, Formamidine pesticides and alpha 2-adrenoceptors: studies with amitraz and chlordimeform in rats and development of a radioceptor binding assay, *Neurotoxicol. Teratol.*, 1989, **11**, 405–411.
346. W. H. Hsu and D. L. Hopper, Effect of yohimbine on amitraz-induced CNS depression and bradycardia in dogs, *J. Toxicol. Environ. Health*, 1986, **18**, 423–429.
347. L. K. Curren and J. A. Reynolson, Central and peripheral-adrenoceptor actions of amitraz in the dog, *J. Vet. Pharmacol. Ther.*, 1990, **13**, 86–92.
348. L. G. Costa, G. Olibet and S. D. Murphy, Alpha 2-adrenoceptors as a target for formamidine pesticides: in vitro and in vivo studies in mice, *Toxicol. Appl. Pharmacol.*, 1988, **93**, 319–328.
349. W. H. Hsu, R. N. Shaw, D. D. Schaffer, M. H. Crump and M. H. Greer, Further evidence to support the alpha 2-adrenergic nature of amitraz-induced decrease in intestinal motility, *Arch. Int. Pharmacodyn. Ther.*, 1987, **286**, 145–151.
350. W. H. Hsu and S. V. McNeel, Amitraz-induced prolongation of gastrointestinal transit and bradycardia in dogs and their antagonism by yohimbine: preliminary study, *Drug Chem. Toxicol.*, 1985, **8**, 239–253.
351. B. E. Smith, W. H. Hsu and P. C. Yang, Amitraz-induced glucose intolerance in rats: antagonism by yohimbine but not by prazosin, *Arch. Toxicol.*, 1990, **64**, 680–683.
352. D. Altobelli, M. Matire, S. Maurizi and P. Preziosi, Interaction of formamidine pesticides with the presynaptic (alpha(2)-adrenoceptor regulating [^{3}H] noradrenaline release from rat hypothalamic synaptosomes, *Toxicol. Appl. Pharmacol.*, 2001, **172**, 179–185.
353. J. M. Goldman and R. L. Cooper, Assessment of toxicant-induced alterations in the luteinising hormone control of ovulation in the rat, in *Female Reproductive Toxicology*, ed. H. Chapin, Academic Press, San Diego, CA, 1993, pp. 79–91.
354. A. S. Al-Thani, A. Al-Thani, A. Elbetieha and H. Darmani, Assessment of reproductive and fertility effects of amitraz pesticides in male mice, *Toxicol. Lett.*, 2003, **138**, 235–260.
355. F. M. Young, M. F. Menadue and T. C. Lavranos, Effects of the insecticide amitraz, an α_2-adrenergic receptor agonist, on human luteinized granulosa cells, *Hum. Reprod.*, 2005, **20**, 3018–3025.

356. JMPR, *Pesticide Residues in Food—1993, Evaluations 1998. Part II—Toxicological*, Joint FAO/WHO Meeting on Pesticide Residues, World Health Organization, Geneva, 1999.
357. T. Roberts and D. Hutson (ed.). Amitraz, in *Metabolic Pathways of Agrochemicals. Part two. Insecticides and Fungicides*, Royal Society of Chemistry, Cambridge, 1999, pp. 729–733.
358. J. C. Kim, J. Y. Shin, Y. S. Yang, D. H. Shin, C. J. Moon, S. H. Kim, S. C. Park, Y. B. Kim, H. C. Kim and M. K. Chung, Evaluation of developmental toxicity of amitraz in Sprague-Dawley rats, *Arch. Environ. Contam. Toxicol.*, 2007, **52**, 137–144.
359. J. Palermo-Neto, J. C. Flório and M. Sakate, Developmental and behavioural effects of prenatal amitraz exposure in rats, *Neurotoxicol. Teratol.*, 1994, **16**, 65–70.
360. L. Institoris, H. Banfi, Z. Lengyel, A. Papp and L. Nagymajtenyi, A study on the immunotoxicological effects of subacute amitraz exposure in rats, *Hum. Exp. Toxicol.*, 2007, **26**, 441–445.
361. S. Ulukaya, K. Demirağ and A. R. Moral, Acute amitraz intoxication in human, *Intensive Care Med.*, 2001, **27**, 930–933.
362. P. G. Jorens, E. Zandijk, L. Belmans, P. J. C. Schepens and L. L. Bossaert, An unusual poisoning case with the unusual pesticide amitraz, *Human Exp. Toxicol.*, 1997, **16**, 600–601.
363. J. J. W. Ros and J. van Aken, A case of poisoning with amitraz, an agricultural pesticide, *Ned. Tidschr. Geneeskd.*, 1994, **138**, 776–778.
364. R. Garnier, D. Chataigner and D. Djebbar, Six human cases of amitraz poisoning, *Hum. Exp. Toxicol.*, 1998, **17**, 294.
365. O. Kennel, C. Prince and R. Garnier, Four cases of amitraz poisoning in humans, *Vet. Hum. Toxicol.*, 1996, **38**, 28–30.
366. S. Vucinic, D. Jovanovic, Z. Vucinic, D. Joksovic, Z. Segrt, M. Zlatovic and M. Jovanovic, A near-fatal case of acute poisoning by amitraz/xylene showing atrial fibrillation, *Forensic Toxicol.*, 2007, **25**, 41–44.
367. K. Aydin, S. Kurtoglu, M. H. Poyrazoglu, K. Üzüm, H. B. Üstünbas and I. K. Hallaç, Amitraz poisoning in children: clinical and laboratory findings of eight cases, *Hum. Exp. Toxicol.*, 1997, **16**, 680–682.
368. K. Aydin, H. Per, S. Kurtoglu, M. H. Poyrazoglu, N. Narin and D. Aslan, Amitraz poisoning in children, *Eur. J. Pediatr.*, 2002, **161**, 349–350.
369. M. R. Grossman, Amitraz toxicosis associated with ingestion of an acaricide collar in a dog, *J. Am. Vet. Med. Assoc.*, 1993, **203**, 55–57.
370. C. Hugnet, F. Buronfusse, X. Pineua, J.-L. Lorgue and P. J. Berney, Toxicity and kinetics of amitraz in dogs, *Am. J. Vet. Res.*, 1996, **57**, 1506–1510.
371. P. Gunaratnam, G. T. Wilkinson and A. A. Seawright, A study of amitraz toxicity in cats, *Aust. Vet. J.*, 1993, **60**, 55–57.
372. A. Malmasi and M. Selk Ghaffari, Electrocardiographic abnormalities in an English bulldog with amitraz toxicity, *Comp. Clin. Pathol.*, 2010, **19**, 103–105.

373. D. E. Auer, A. A. Seawright, C. C. Pollitt and G. Williams, Illness in horses following spraying with amitraz, *Aust. Vet. J.*, 1984, **61**, 257–259.
374. H. R. Schmid, G. van Tulder and P. Junquera, Field efficacy of the insect growth regulator dicyclanil for flystrike prevention on lambs, *Vet. Parasitol.*, 1999, **86**, 147–151.
375. R. M. Nottingham, B. C. Hosking, H. R. Schmid, G. Strehlau and P. Junquera, Prevention of blowfly strike on coarse and fine woolled sheep with the insect growth regulator dicyclanil, *Aust. Vet. J.*, 2001, **79**, 51–57.
376. S. Jess, C. Kearns and D. I. Matthews, A survey of annual pesticide usage during the control of sheep ectoparasites in Northern Ireland, 2005, *J. Agric. Sci.*, 2007, **145**, 517–528.
377. P. J. James, A. P. Cramp, J. Winkleman, R. Mophie and G. W. Brown, Strategic use of crutching and dicyclanil to protect unmulesed sheep against breech strike, *Aust. Vet. J.*, 2009, **87**, 138–141.
378. P. Bates, Therapies for ectoparasiticism in sheep, *In Practice*, 2004, **26**, 538–547.
379. M. A. Taylor, Recent developments in ectoparasiticides, *Vet. J.*, 2001, **161**, 253–268.
380. S. Sotiraki, A. Stefanakis, M. J. Hall, R. Frakas and J. F. Graf, Wohlfartiosis in sheep and the role of dicyclanil in its prevention, *Vet. Parasitol.*, 2005, **131**, 107–117.
381. H. Tunaz and N. Uygun, Insect growth regulators for insect pest control, *Turk. J. Agric. For.*, 2004, **28**, 377–387.
382. E. Cohen, Chitin synthesis and inhibition: a revisit, *Pest. Manag. Sci.*, 2001, **57**, 946–950.
383. H. Oberlander and D. L. Silhacek, Mode of action of insect growth regulators in lepidopteran tissue culture, *Pest. Sci.*, 1998, **54**, 300–302.
384. JECFA, *Toxicological Evaluation of Certain Veterinary Drug Residues in Food, WHO Food Additive Series 45*, 54th Meeting of the Joint FAO/WHO Expert Committee on Food Additives. World Health Organization, Geneva, 2000, pp. 75–99.
385. M. Moto, Y. F. Sasaki, M. Okamura, M. Fujita, Y. Kashida, N. Machida and K. Mitsumori, Absence of in vivo genotoxicity and liver initiation activity of dicyclanil, *J. Toxicol. Sci.*, 2003, **28**, 173–179.
386. M. Moto, M. Okamura, M. Mugumura, T. Ito, M. Jin, Y. Kashida and K. Mitsumori, Gene expression analysis of dicyclanil-induced hepatocellular tumors in mice, *Toxicol. Pathol.*, 2006, **34**, 744–751.
387. M. Moto, M. Okamura, T. Muto, Y. Kashida, N. Machida and K. Mitsumori, Molecular pathological analysis on the mechanism of liver carcinogenesis in dicyclanil-treated mice, *Toxicology*, 2005, **207**, 419–436.
388. T. Umemura, Y. Kurolwa, M. Tasaki, T. Okamura, Y. Ishii, Y. Kodama, T. Nohmi, K. Mitumori, A. Nishikawa and M. Hirose, Detection of oxidative DNA damage, cell proliferation and in vivo mutagenicity induced by dicyclanil, a non-genotoxic carcinogen, using *gpt* delta mice, *Mutat. Res.*, 2007, **633**, 46–54.

389. M. Moto, T. Umemura, M. Okamura, M. Muguruma, T. Ito, Y. Kashida and M. Mitsumori, Possible involvement of oxidative stress in dicyclanil-induced hepatocarcinogenesis in mice, *Arch. Toxicol.*, 2006, **80**, 694–702.
390. M. Jin, Y. Dewa, M. Kawai, J. Nishimura, Y. Saegusa, S. Kemmochi, T. Harada, M. Shibutani and K. Mitsumori, The threshold dose for liver tumor promoting effects of dicyclanil in ICR mice, *J. Toxicol. Sci.*, 2010, **35**, 69–78.
391. E. E. Grafton-Cardwell, L. D. Godfrey, W. E. Chaney and W. J. Bentley, Various novel pesticides are less toxic to humans, more specific to key pests, *Calif. Agric.*, 2005, **59**, 29–34.
392. G. W. Levot, Resistance and the control of sheep ectoparasites, *Int. J. Parasitol.*, 1995, **25**, 1355–1362.
393. JMPR, *Pesticide Residues in Food—2000, Evaluations 2006. Part II—Toxicological*, Joint FAO/WHO Meeting on Pesticide Residues, World Health Organization, Geneva, 2008.
394. NTP, *Carcinogenesis Bioassay of Melamine (CAS No 108-78-1) in F344/N Rats and B6C3F1 Mice*, Technical Report Series No 245, National Toxicology Program, Research Triangle Park, NC, 1983.
395. M. Okumura, R. Hasegawa, T. Shirai, M. Ito, S. Yamada and S. Fukushima, Relationship between calculus formation and carcinogenesis in the urinary bladder of rats administered non-genotoxic agents thymine or melamine, *Carcinogenesis*, 1992, **13**, 1043–1045.
396. M. Ogasawara, K. Imaida, H. Ishiwata, K. Toyoda, T. Kawanishi, C. Uneyama, S. Hayashi, M. Takahashi and Y. Hayashi, Urinary bladder carcinogenesis induced by melamine in F344 male rats: correlation between carcinogenicity and urolith formation, *Carcinogenesis*, 1995, **16**, 2773–2777.
397. S. Fukushima and T. Mirai, Calculi, precipitates and microcrystalluria associated with irritation and cell proliferation as a mechanism of urinary bladder carcinogenesis in rats and mice, *IARC Sci. Publ.*, 1999, **147**, 159–174.
398. K. Otori, Y. Yano, N. Takad, C. C. Lee, S. Hayashi, S. Otani and S. Fukushima, Reversibility and apoptosis in rat urinary bladder papillomatosis induced by uracil, *Carcinogenesis*, 1997, **18**, 1485–1489.
399. S. M. Cohen, Urinary bladder carcinogenesis, *Toxicol. Pathol.*, 1988, **26**, 121–127.
400. T. Sakata, T. Masui, M. St John and S. M. Cohen, Uracil-induced calculi and proliferative lesions of the mouse urinary bladder, *Carcinogenesis*, 1988, **9**, 1271–1276.
401. T. Fujii, K. Nakamura and K. Hiraga, Effects of pH on the carcinogenicity of *o*-phenylphenol and sodium *o*-phenylphenate in the rat urinary bladder, *Food Chem. Toxicol.*, 1987, **25**, 359–362.
402. N. Niho, M. Shibutani, K. Toyoda, H. Sato, A Hirose, K. Imaida, M. Takahashi, Y. Hayashi and M. Hirose, Dose- and time-response studies of sodium *o*-phenylphenate urinary bladder carcinogenicity in rats, *Food Chem. Toxicol.*, 2002, **40**, 715–722.

403. K. N. Woodward, Origins of injection-site sarcomas in cats: the possible role of chronic inflammation—a review, *ISRN Vet. Sci.*, 2011, **2011**(Art ID 210982), 16pp.
404. P. A. Oliveira, A. Colaco, P. L. F. De la Cruz and C. Lopes, Experimental bladder carcinogenesis in rats, *Exp. Oncol.*, 2006, **28**, 2–11.
405. H. A. Milman, Possible contribution of indomethacin to the carcinogenicity of nongenotoxic bladder carcinogens that cause bladder calculi, *Drug Chem. Toxicol.*, 2007, **30**, 161–166.
406. R. Loosli, Triazines, *Toxicology*, 1994, **91**, 59–62.
407. R. Tilak, A. K. Verma and U. B. Wankhade, Effectiveness of diflubenzuron in the control of houseflies, *J. Vector Borne Dis.*, 2010, **47**, 97–102.
408. N. P. Hajjar and J. E. Casida, Insecticidal benzoylphenyl ureas: structure activity relationships as chitin synthesis inhibitors, *Science*, 1978, **30**, 1499–1500.
409. F. Matsumura, Studies on the action mechanism of benzoylurea insecticides to inhibit the process of chitin synthesis in insects: a review on the status of research activities in the past, the present and future prospects, *Pest. Biochem. Physiol.*, 2010, **97**, 133–139.
410. B. D. Hammock and G. B. Quistad, Metabolism and mode of action of juvenile hormone, juvenoids, and other insect growth regulators, In *Progress in Pesticide Biochemistry*, vol. 1, ed. D. H. Hutson and T. R. Roberts, John Wiley & Sons, Chichester, 1981, pp. 1–83.
411. P. Cabras, Pesticides: toxicology and residues in food, In *Food Safety. Contaminants and Toxins*, ed. J. P. F. D'mello, CABI International, Wallingford, 2003, pp. 91–124.
412. G. Ritchie, S. S. Rønsberg, K. A. Hoff and E. J. Bransom, Clinical efficacy of teflubenzuron (Calicide) for the treatment of *Lepeophtheirus salmonis* infestations of farmed Atlantic salmon *Salmo salar* at low water temperature, *Dis. Aquat. Organ.*, 2002, **51**, 101–106.
413. M. Franc and M. C. Cadiergues, Use of injectable Lufenuron for treatment of infestations of *Ctenocephalides felis* in cats, *Am. J. Vet. Res.*, 1997, **58**, 140–142.
414. R. D. Smith, A. J. Paul, U. D. Kitron, J. R. Philip, S. Barnett, M. J. Piel, R. W. Ness and M. Evilsizer, Impact of orally administered insect growth regulator (Lufenuron) on flea infestations of dogs in a controlled simulated home environment, *Am. J. Vet. Res.*, 1996, **57**, 502–504.
415. Y. Nishida, C. Haga, K. Oda and T. Hayama, Disinfestation of experimentally infested cat fleas, *Ctenocephalides felis*, on cats and dogs by oral Lufenuron, *J. Vet. Med. Sci.*, 1995, **57**, 655–658.
416. S. R. Dean, R. W. Meola, S. M. Meola, H. Sittertz-Bhatkhar and R. Schenker, Mode of action of Lufenuron in adult *Ctenocephalides felis* (Siphonaptera: Pulicidae), *J. Med. Entomol.*, 1999, **36**, 486–492.
417. IPCS, *Environmental Health Criteria 194, Diflubenzuron*, International Programme on Chemical Safety, World Health Organization, Geneva, 1996.

418. JMPR, *Pesticide Residues in Food—2001, Evaluations, Part II—Toxicological*, Joint FAO/WHO Meeting on Pesticide Residues, World Health Organization, Geneva, 2002.
419. J. T. MacGregor, D. H. Gould, A. D. Mitchell and G. P. Sterling, Mutagenicity tests of diflubenzuron in the micronucleus test in E4 mice, the L5178Y mouse lymphoma forward mutation assay and the Ames Salmonella reverse mutation assay, *Mutat. Res.*, 1979, **66**, 45–53.
420. EMA/CVMP, *Teflubenzuron, MRL Summary Report (3)*, EMEA/MRL/221/97-FINAL, European Medicines Agency/Committee for Veterinary Medicinal Products, 1997. www.ema.europa.eu/
421. M. Tasheva and V. Hristeva, Comparative study of five benzoylphenylurea insecticides on haematological parameters in rats, *J. Appl. Toxicol.*, 1993, **13**, 67–68.
422. D. G. Stansfield, A review of the safety and efficacy of lufenuron in dogs and cats, *Canine Pract.*, 1997, **22**, 34–38.
423. B. Bar-Oz, S. Ito, V. Parks, M. P. Maurer and G. Koren, Estimation of neonatal exposure after accidental ingestion of lufenuron in a breast-feeding mother, *J. Hum. Lact.*, 2000, **16**, 229–230.
424. G. D. Thompson and R. Dutton, Spinosad—a case study: an example from a natural product discovery, *Pest. Manag. Sci.*, 2000, **56**, 696–702.
425. G. Thompson and S. Hutchins, Spinosad, *Pesticide Outlook*, 1999, **10**, 78–81.
426. M. Sarfraz, L. M. Dodsall and B. A. Keddie, Spinosad: a promising new tool for integrated pest management, *Outlooks Pest Manag.*, 2005, **16**, 78–84.
427. V. L. Salgado, Studies on the mode of action of spinosad: insect symptoms and physiological correlates, *Pest. Biochem. Physiol.*, 1998, **60**, 91–102.
428. V. L. Salgado, J. J. Sheets, G. B. Watson and A. L. Schmidt, Studies on the mode of action of spinosad, The internal effective concentration and the concentration dependence of neural excitation, *Pest. Biochem. Physiol.*, 1998, **60**, 103–110.
429. N. Orr, A. J. Shaffner, K. Richey and G. D. Crousse, Novel mode of action of spinosad: Receptor binding studies demonstrating lack of interaction with known insecticidal target sites, *Pest. Biochem, Physiol.*, 2009, **95**, 1–5.
430. J. Cisneros, D. Goulson, L. C. Derwent, D. I. Penagos, O. Hernández and T. Williams, Toxic effects of spinosad on predatory insects, *Biol. Control*, 2002, **23**, 156–163.
431. D. Stough, S. Shellabarger, J. Quiring and A. Gabrielsen, Efficacy and safety of spinosad and permethrin crème rinses for pediculosis capitis (head lice), *Pediatrics*, 2009, **124**, e389–395.
432. K. E. Stebbins, D. M. Bond, M. N. Novilla and M. J. Reasor, Spinosad insecticide: subchronic and chronic toxicity and lack of carcinogenicity in CD-1 mice, *Toxicol. Sci.*, 2002, **65**, 276–287.

433. B. L. Yano, D. M. Bond, M. N. Novilla, L. G. McFadden and M. J. Reasor, Spinosad insecticide: subchronic and chronic toxicity and lack of carcinogenicity in Fischer 344 rats, *Toxicol. Sci.*, 2002, **65**, 288–298.
434. T. R. Hanley, W. J. Breslin, J. F. Quast and E. W. Carney, Evaluation of spinosad in a two-generation dietary reproductive study using Sprague-Dawley rats, *Toxicol. Sci.*, 2002, **67**, 144–152.
435. W. J. Breslin, M. S. Marty, U. Vedula, A. B. Liberacki and B. L. Yano, Developmental toxicity of spinosad administered by gavage to CD rats and New Zealand white rabbits, *Food Chem. Toxicol.*, 2000, **38**, 1103–1112.
436. W. H. Halliwel, Cationic amphiphilic drug-induced phospholipidosis, *Toxicol. Pathol.*, 1997, **25**, 53–60.
437. M. J. Reasor and S. Kacew, Drug-induced phospholipidosis: are there functional consequences? *Exp. Biol. Med.*, 2001, **226**, 825–830.
438. L. A. Chatman, D. Morton, T. O. Johnson and S. D. Anway, A strategy for risk management of drug-induced phospholipidosis, *Toxicol. Pathol.*, 2009, **37**, 997–1005.
439. N. Anderson and J. Borlak, Drug-induced phospholipidosis, *FEBS Lett.*, 2009, **580**, 5533–5540.

CHAPTER 13

Macrocyclic Lactone Endectocides

K N WOODWARD

TSGE, Concordia House, St James Business Park, Grimbald Crag Court, Knaresborough, North Yorkshire, UK. Email: knwoodward@talktalk.net

13.1 Introduction

The macrocyclic lactone endectocides (drugs active against internal and external parasites), typified by abamectin and ivermectin, were first introduced as anthelmintic agents and are active against a wide range of nematodes including *Ostertagia* spp., *Strongyloides papillosus*, *Cooperia* spp., and *Dictyocaulus viviparous*. However, they are also active against a number of arthropod parasites of cattle and sheep, among them cattlegrubs and warbles (*e.g. Hypoderma bovis*), screwworm fly larvae (*e.g. Chrysoma bezziana*) sucking and biting lice, mange mites including the organism responsible for sheep scab, *Psoroptes ovis*, and various ticks including *Boophilus* species.[1–6] They are used in veterinary medicine to treat and control internal and external parasites and may be given orally, by subcutaneous injection or as pour-on formulations for topical application.

Ivermectin, a member of the avermectin group, was the first member of that group to be used in veterinary medicine and was introduced in many countries in the early 1980s.[7] Since then, a number of related compounds have been registered. These include abamectin, eprinomectin, doramectin, selamectin and emamectin benzoate. They are 16-membered macrocyclic lactones with a spiroketal component and are derivatives of naturally occurring avermectins

Issues in Toxicology No. 12
Mammalian Toxicology of Insecticides
Edited by Timothy C. Marrs

Published by the Royal Society of Chemistry, www.rsc.org

formed as fermentation products of *Streptomyces avermitilis*.[8,9] Ivermectin is 22, 23-dihydroavermectin (Figure 13.1) while abamectin is a mixture of avermectin B_{1a} and avermectin B_{1b}, and emamectin and eprinomectin are substituted derivatives.

Doramectin is closely related to abamectin and ivermectin but has a cyclohexyl substituent at the 25 position while selamectin is a derivative of doramectin. Moxidectin is a member of the milbemycin group. These are closely related structurally to the avermectins but lack the bisoleandrosyl substituent at the C-13 position (Figure 13.2).

Eprinomectin has been developed as a broad-spectrum endectocide for use topically in cattle,[10,11] while selamectin is specifically used topically for the treatment of external parasites of cats and dogs such as fleas and ear mites, and for the treatment of internal parasites including heartworm and *Toxocara* species.[12–14]

Emamectin benzoate, derived from emamectin, a compound used in crop protection,[15] is given orally in feed for the treatment of external parasites on

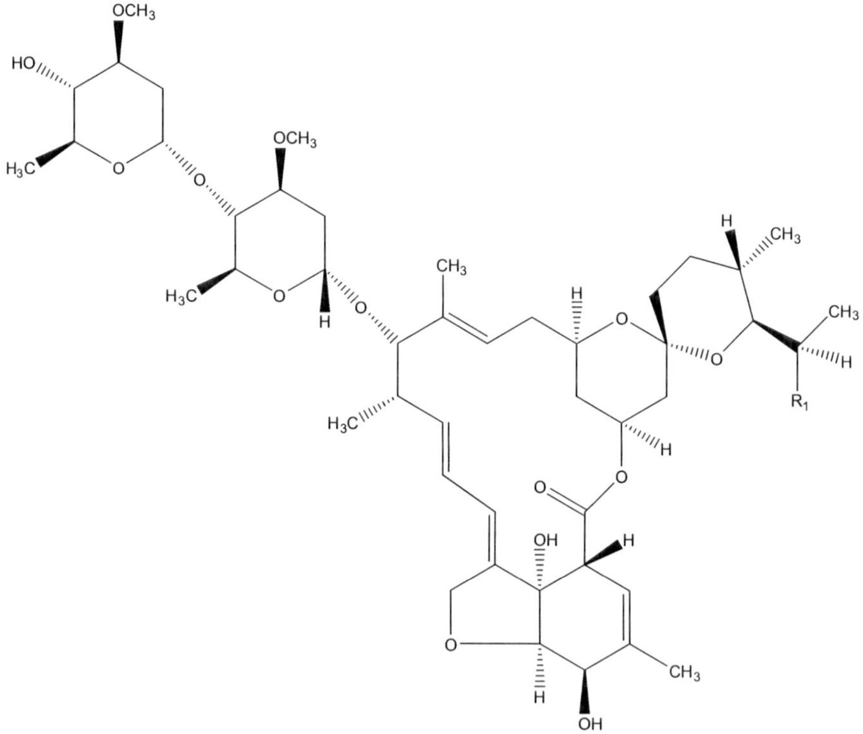

Ivermectin B_{1a}, $R_1 = C_2H_5$
Ivermectin B_{1b}, $R_1 = CH_3$

Figure 13.1 Structural formula of ivermectin.

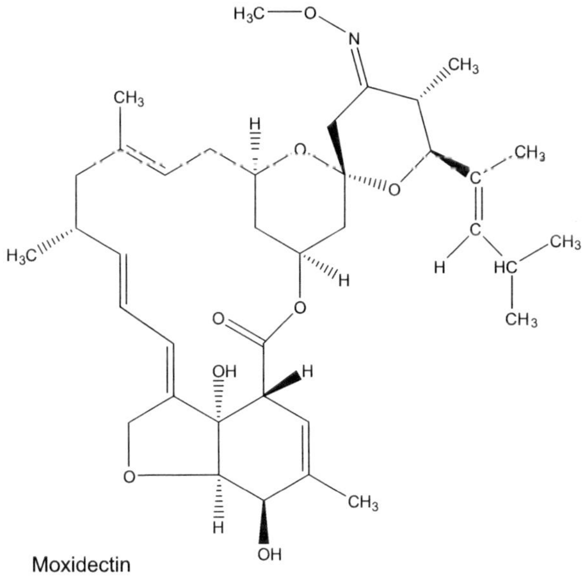

Figure 13.2 Structural formula of moxidectin.

farmed salmon and other aquaculture species.[16–23] The major ectoparasites of farmed and wild salmon are the copepods *Caligus* spp. and *Lepeophtheirus salmonis*, and emamectin and ivermectin are active against these organisms.[16–24] The use of emamectin benzoate has successfully supplemented the few other available treatments (*e.g.* hydrogen peroxide, organophosphorus compounds and cypermethrin) and replaced the off-label or illegal use of other chemotherapeutics, including that of ivermectin.[25,26]

13.2 Metabolism

Virtually all of the metabolism and toxicology data on the avermectins and milbemycins has been generated to support marketing authorization/approval dossiers, or dossiers for European Union maximum residue limit (MRL) applications.[26–30] As a consequence, these data are not publically available. However, the metabolism and toxicological studies for several avermectins and for moxidectin have been reviewed by the Joint FAO/WHO Expert Committee on Food Additive (JECFA) which proposes MRL values for veterinary drugs for use in the Codex Alimentarius process, while abamectin has been reviewed by the Joint Meeting on Pesticide Residues (JMPR) which conducts a similar function for pesticides.[31] The toxicological monographs prepared for the respective meetings of the committees are reviewed here. Emamectin benzoate has not been formally reviewed by JECFA but it is discussed in the eprinomectin monograph. There is also a summary of the MRL evaluation prepared by the European Medicines Agency and this will be referred to here. The metabolism of these compounds has been studied in a number of species.[32–42]

After oral administration of ivermectin to rats in sesame oil, the majority of the administered dose (up to 90%) was excreted in the faeces, suggesting poor absorption from the gastrointestinal tract, but absorbed ivermectin is subject to intestinal secretion.[43] Ivermectin is extremely lipophilic, and the highest concentrations were found in fat, followed by kidney, liver and muscle. Absorption is slow and peak plasma levels were reached 10 days after administration.

However, after subcutaneous administration, bioavailability is much higher, at least in cattle, with 40–55% of the administered dose being absorbed, depending on the formulation. In pigs, plasma ivermectin concentrations were significantly higher after subcutaneous administration when compared with oral dosing.

The major compound found *in vivo* is unchanged ivermectin. However, hepatic metabolism does occur, and the major liver metabolite is 24-hydroxymethyl-H_2B_{1a} in cattle, sheep and rats. In pigs, the 3′-*O*-desmethyl-H_2B_{1a} and 3′-*O*-desmethyl-H_2B_{1b} compounds are the major metabolites. Similar patterns of desmethylation and hydroxylation occur with doramectin, while eprinomectin is subject to *N*-deacetylation. Moxidectin is excreted largely unchanged although some minor metabolites are formed in the liver.

13.3 Toxicology

These compounds have been tested in a range of toxicology studies. As already mentioned, much of this data is not publically available and so the sources described earlier have provided much of the content for this discussion.

13.3.1 Acute Toxicity

Abamectin was moderately toxic to mice, rats, dogs and monkeys when given orally with LD_{50} values in the range of 11–41 mg kg^{-1} bodyweight (bw) in mice, 8.7–12.8 mg kg^{-1} bw in rats, approximately 8 mg kg^{-1} bw in dogs and more than 24 mg kg^{-1} bw in monkeys. In all cases, the major clinical signs were ataxia and tremors. The substance was less toxic when applied dermally with LD_{50} values of in excess of 330 mg kg^{-1} bw in the rat and in excess of 1600 mg kg^{-1} in the rabbit. In monkeys, the no-observed adverse effect level (NOAEL) was 1 mg kg^{-1} bw following a single dose.[32,33]

When beagles were given oral doses of ivermectin in sesame oil of up to 10 mg kg^{-1} bw, signs of toxicity including emesis and salivation were noted at the two highest doses employed (5 and 10 mg kg^{-1} bw). Tremors also occurred and one dog given the highest dose became ataxic and comatose. It later recovered. In another study in beagles, mortalities occurred with doses of 40 and

80 mg kg^{-1} bw and clinical signs included emesis and salivation. No deaths occurred at doses of 20 mg kg^{-1} bw. Following a single subcutaneous dose of 4.7, 9.4, 18.8, 37.5 and 75 mg kg^{-1} bw ivermectin, 50% of dogs given 9.4 mg kg^{-1} bw and all of those given higher doses died. Mydriasis and negative papillary responses were seen in all treated animals along with tremors, ataxia, salivation and decreased activity.[34]

The acute toxicity of ivermectin appeared to be much higher in the collie dog. Reports from in-use treatment of collies had already suggested that ivermectin was more toxic in some animals in this breed than in other breeds. In this study, groups of four collies were given oral doses of 50, 200 or 600 μg kg^{-1} bw ivermectin in fractionated coconut oil with 2% benzyl alcohol. One dog given 200 and one given 600 μg kg^{-1} bw ivermectin developed severe signs of toxicity including ataxia, depression, tremors, recumbency and mydriasis. The affected dog given 600 μg kg^{-1} bw was euthanized while that from the 200 μg kg^{-1} bw group died after approximately 50 h of administration. The affected dogs had higher concentrations of ivermectin in the central nervous system (CNS) than did unaffected animals.[35]

With ivermectin, LD_{50} values were in the range 43–53 mg kg^{-1} bw in the rat. However, in the specific strain of mouse used in the studies, the CF-1 mouse, the LD_{50} values were in the range 12–57 mg kg^{-1} bw. Again, the major signs noted were tremor and ataxia with paresis, paralysis and death. Neonatal rats were more susceptible to the acute toxicity of ivermectin than adult animals with an oral LD_{50} value of 2.3 mg kg^{-1} bw. In the dog, oral LD_{50} values varied from over 10 to around 80 mg kg^{-1} bw. After subcutaneous administration to this species the LD_{50} values were in the 8 to 10.5 mg kg^{-1} bw range.[34] Signs of systemic toxicity including tremor, bradypnoea and anorexia were seen in rabbits treated topically with 165, 330 or 660 mg kg^{-1} bw ivermectin and the percutaneous LD_{50} value was estimated to be 406 mg kg^{-1} bw.

Signs of neurotoxicity including depression and ataxia have been noted in cattle given 4 mg kg^{-1} bw ivermectin—four times the therapeutic dose. When they were given 8 mg kg^{-1} bw ivermectin, increased respiratory rate, muscular tremors, rigidity of the extremities and deaths occurred. Sheep given 4 mg kg^{-1} bw ivermectin showed signs of depression including recumbency and lack of coordination. Animals given 8 mg kg^{-1} bw ivermectin became ataxic and depressed. All animals were mildly depressed and lacked coordination 24 h after dosing. Horses given 2 mg kg^{-1} bw ivermectin showed signs of depression and impairment of vision. At 12 mg kg^{-1} bw, horses displayed ataxia and depression. One horse found in a state of lateral recumbency was euthanized. Signs of toxicity have been noted in pigs given single subcutaneous doses of 30 but not at 0.3, 3 or 15 mg kg^{-1} bw ivermectin. These signs included ataxia, tremor, mydriasis, laboured breathing and recumbency.[34,36]

Eprinomectin appeared to be less toxic than either abamectin or ivermectin. Oral LD_{50} values were in the range of 70 and 55 mg kg^{-1} bw in the mouse and rat respectively. The intraperitoneal LD_{50} value was 35 mg kg^{-1} bw in both species.[37]

With emamectin benzoate as the solvate and as the hydrate, the oral LD_{50} values in CD-1 mice were 120 and 107 mg kg^{-1} bw respectively. However, the substance was more toxic to CF-1 mice with LD_{50} values for the hydrochloride salt of 22 and 31 mg kg^{-1} bw for male and female animals respectively. In rats, the acute oral LD_{50} values were 88 and 76 mg kg^{-1} bw respectively. Signs of toxicity included tremors, ataxia, ptosis, bradypnoea and loss of righting reflex.[37,38]

Doramectin appeared to be slightly less toxic to mice and rats than some of the other compounds. When given in oil to male and female CD-1 mice, the LD_{50} values were between 75 and 250 mg kg^{-1} bw. It was even less toxic when an aqueous vehicle was used with LD_{50} values being more than 2000 mg kg^{-1} bw in mice. Similarly, when given in oil to rats the LD_{50} value was 50 and 100 mg kg^{-1} bw for male and female animals respectively, but when given as an aqueous formulation the LD_{50} values were 500 and 1000 mg kg^{-1} bw for male and female animals. These vehicle differences were also seen following intraperitoneal administration. In male mice the intraperitoneal LD_{50} values were 700 and 100 mg kg^{-1} bw with the aqueous and oil formulations respectively, and in male rats over 300 and 50 mg kg^{-1} bw with the aqueous and oil formulations respectively. The most frequent signs of toxicity were decreased activity, weakness, tremors and ataxia.[39]

When given orally in sesame oil to groups of mice, doses of 30, 100 and 300 mg kg^{-1} bw selamectin produced no mortalities. The major signs of toxicity were ptosis and mild piloerection at 100 and 300 mg kg^{-1} bw and increased respiration rates and moderate muscle weakness. No adverse effects occurred at 30 mg kg^{-1} bw.[12]

Moxidectin was moderately toxic to mice after oral administration with LD_{50} values of 84 (male and female), 42–50 (female) mg kg^{-1} bw. In male and female rats, the oral LD_{50} value was 106 mg kg^{-1} bw. It was slightly less toxic after subcutaneous administration in mice and even less so in rats (LD_{50} values of 86 and 394 mg kg^{-1} bw respectively) and even less toxic after subcutaneous administration (LD_{50} values in mice and rats of 263 and over 640 mg kg^{-1} bw respectively). In mice, the main signs of toxicity were decreased activity. In rats, decreased activity, tremors, prostration, decreased respiration, hypersensitivity to touch and sound and epistaxis occurred.[40]

The acute toxicity studies with these compounds demonstrate a number of points:

- The avermectins and moxidectin are moderately toxic after oral administration.
- The main signs of toxicity are those of neurotoxicity.
- The CF-1 mouse, where used in these studies, appears to be more sensitive to the acute toxicity of these compounds.
- Ivermectin is highly toxic to some collies and certainly more toxic in some individuals of this breed than it is to laboratory beagles.
- On the limited data available, the toxicity is affected by vehicle effects with higher toxicity being seen with oil and less with aqueous vehicles. This is undoubtedly due to the greater solubility of these substances in lipid solvents.

13.3.2 Repeat Dose Toxicity

A number of repeat dose toxicity studies have been performed with these compounds, but there are no data in the public domain for selamectin. The data generated emphasize the findings from acute toxicity studies, mainly that these substances are relatively toxic and that neurotoxicity is the main form of toxicity noted. The results of these studies are briefly presented in Table 13.1.

In these studies, typical signs of neurotoxicity included decreased activity, lethargy, tremors, mydriasis, ataxia and salivation. When tested in the CF-1 mouse, emamectin, but not moxidectin, was more toxic than in the CD-1 strain. Beagles were the only breed tested in these repeat dose studies, so no comparisons can be made with sensitive individuals within the collie population.

Further evidence for the neurotoxic effects of these drugs was provided by histopathological examination in some of these studies. Dogs given 2 mg kg^{-1} bw eprinomectin for a year in an aqueous vehicle showed degeneration of some neurons in the pons area or in the cerebellar nuclei.[34] Neuronal vacuolation or degeneration in the spinal cord and sciatic nerve occurred in rats given emamectin hydrochloride in the diet for 14 weeks or longer, while emamectin hydrochloride and emamectin benzoate produced neuronal degeneration in dogs in both the peripheral and central nervous systems.[38] However, in all these studies, clear NOAEL values were obtained, and these degenerative effects tended to occur at the highest doses used in these studies.

13.3.3 Carcinogenicity Studies

The available studies are reviewed in Table 13.2.

Ivermectin and emamectin were not carcinogenic in mice or rats in the studies reviewed here, nor was moxidectin. There are no publicly available data for selamectin and carcinogenicity studies are not available for ivermectin, eprinomectin or doramectin. However, ivermectin is similar in structure to abamectin while eprinomectin is similar to emamectin. Doramectin shares close similarities with the other avermectins. Moreover, all of these substances have been tested in batteries of tests for genotoxic potential including tests for point mutations in bacterial systems, tests for forward mutations in mammalian cells *in vitro*, and in *in vivo* studies, *e.g.* cytogenetic assay in rat bone marrow, tests for unscheduled DNA synthesis, and in the mouse bone marrow micronucleus test.[32–34,36–40,44–52] These tests overall gave convincingly negative results. Hence, taken together, the data strongly suggests that these compounds are not genotoxic and are not mammalian carcinogens.

However, the studies are more striking from the point of view that they emphasize the neurotoxic potential of these compounds even to the point of dosing rates having to be adjusted to ensure sufficient survival for the adequacy of the study.

Table 13.1 Results of repeat dose oral studies.[40–42,44–48]

Compound	*Species*	*Vehicle*	*Duration*	*Doses/route ($mg\ kg^{-1}$ bw day^{-1} except where shown)*	*Major Findings*	*No observed effect level ($mg\ kg^{-1}$ bw day^{-1})*	*Reference*
Abamectin	Mouse (CD-1)	Diet	12 weeks	0–20 ppm 40 or 60 ppm	Decrease in bodyweight gain at 60 ppm	40 ppm ≅ 8	31, 32
	Rat	Diet	8 weeks	0–60 ppm	Mortality, tremors, and decreased activity at 40 and 60 ppm; all animals were sacrificed at days 15 and 5 respectively	N/A – toxicity in all groups	
	Dog (beagle)	Diet	12 weeks	0–4	Signs of neurotoxicity, including tremors, were noted at highest dose which was reduced to 2 mg kg^{-1} bw day^{-1} in week 4. Mydriasis in dogs given 1 mg kg^{-1} bw day^{-1}. Depression of bodyweight gain at 2 and 4 mg kg^{-1} bw day^{-1}	0.5	
		Diet	53 weeks	0–1	Mydriasis at 0.5 and 1 mg kg^{-1} bw. One death and 2 dogs euthanized at 1 mg kg^{-1} bw day^{-1}	≤0.25	
Avermectin B_{1a}	Dog (beagle)	Not specified	18 weeks	0–8	Mortalities at 1/30, 3/30 and 3/30 at 0.5, 2 and 8 mg kg^{-1} bw day^{-1}. Signs of neurotoxicity at 0.5 mg kg^{-1} bw day^{-1}. These included convulsions at 2 and 8 mg kg^{-1} bw day^{-1}.	0.25	31, 32
Ivermectin	Rat	Unspecified	14 weeks	0.4–1.6/po	Splenic enlargement, reactive bone marrow hyperplasia	0.4	33, 35
	Dog	Sesame oil	14 weeks	0.5–2.0/po	Ptyalism, mydriasis, tremors, ataxia, and recumbency at highest dose	1.0	
	Rhesus monkey	Sesame oil	At least 14 days	0.3, 0.6, 1.2	None	1.2	
	Rhesus monkey (neonatal)	Sesame oil	14 days	0.04, 0.1	None	0.1	
Eprinomectin	Rat	Diet	23 days	0–10	Lowest dose of 0.5 mg kg^{-1} bw day^{-1} increased to 20 mg kg^{-1} bw day^{-1} at day 15 due to absence of toxicity. Decreases in bodyweight gain were seen at this dose	N/A (exploratory study only)	36

	Diet	26 days	0–60	Severe toxicity (ataxia, piloerection, and whole-body tremors) was seen at two highest doses, 40 and 60 mg kg^{-1} bw day^{-1} and these groups were terminated after week 1 and a new group (30 mg kg^{-1} bw day^{-1}) was introduced. In this group, milder signs were noted, and in females at next lower dose of 20 mg kg^{-1} bw day^{-1}	N/A—toxicity in all groups
	Diet	90 days	0–30	High dose was reduced to 20 mg kg^{-1} bw day^{-1} after week 4 due to high incidence of whole-body tremors but these still occurred at 20 mg kg^{-1} bw day^{-1}	5.0
Dog	Diet for first 13 days, then in aqueous vehicle by gavage for remainder of study due to unpalatability	6 weeks	0–4	Treatment at 4 mg kg^{-1} bw day^{-1} was discontinued after first gavage dose due to severe toxicity including mydriasis, ataxia, salivation, decreased activity and 1 mortality. Mydriasis was seen occasionally in animals at next lowest dose, 2 mg kg^{-1} bw day^{-1}	1.0
	Aqueous vehicle	90 days	0–3	Highest dose was lowered to 2 mg kg^{-1} bw day^{-1} from week 2 due to death of 50% of males and emesis, ataxia, salivation, lateral recumbency and weight loss. No other clinical signs were noted. Axonal degeneration of sciatic nerve seen in 2/4 high-dose females	0.8
	Aqueous vehicle	52 weeks	0–2	Only clinical sign was mydriasis in high dose dogs. Degeneration of small numbers of neurons in the pons area and/or cerebellar nuclei in 3/8 high-dose dogs	1.0

Table 13.1 (*Continued*)

Compound	*Species*	*Vehicle*	*Duration*	*Doses/route ($mg\ kg^{-1}$ bw day^{-1} except where shown)*	*Major Findings*	*No observed effect level ($mg\ kg^{-1}$ bw day^{-1})*	*Reference*
Emamectin (hydro-chloride)	Mouse (CD-1)	Diet	13 days	0–12.5	None	N/A (exploratory study only)	37
Emamectin (benzoate)	Mouse (CF-1)	Diet	15 days	0–0.9	Tremors, decreased activity, slow righting reflex and reduced bodyweight were seen at 0.3 and 0.9 mg kg^{-1} bw day^{-1}—several mice from these group were euthanized in moribund condition	0.1	
Emamectin (hydro-chloride)	Rat	Diet	14 weeks	0–12.5	Males given the highest dose were euthanized due to toxicity namely, whole-body tremors. Neuronal vacuolation was noted in brains of 2/20 rats given 2.5 mg kg^{-1} bw day^{-1} and in majority of high-dose animals	0.5	
Emamectin (benzoate)	Rat	Diet	14 weeks	0–5.0	This included functional observational battery and motor activity test component. Tremors observed at 5 mg kg^{-1} bw day^{-1} with effects on posture, gait, rearing, grip strength and righting reflex	1.0	
	Rat	Diet	53 weeks	0–2.5	Females at high dose began study on 5 mg kg^{-1} bw day^{-1} but this was reduced to 2.5 mg kg^{-1} bw day^{-1} after week 18 due to toxicity notably tremors. These animals improved following this dose reduction. Reduced abilities were noted in the functional observational battery in high-dose animals	1.0	
Emamectin (hydro-chloride)	Dog (beagle)	Not specified	5 weeks	0–1.5	Neuronal damage in dogs given 1.5 mg kg^{-1} bw day^{-1} including axonal degeneration and myelin degeneration	0.5	

	Dog (beagle)	Not specified	14 weeks	0, 0.5, 1.0 or 1.5 reduced to 0, 0.25, 0.5 and 1 at week 3	Tremors, mydriasis, lethargy and anorexia were seen at 1.0 or 1.5 mg kg^{-1} bw day^{-1} and 3 dogs had to be euthanized. Neuronal degeneration was seen in dogs given 1.5/1 and 1.0/0.5 mg kg^{-1} bw day^{-1}	0.25	
Emamectin (benzoate)	Dog (beagle)	Not specified	53 weeks	0–1	All dogs given 1 mg kg^{-1} bw day^{-1} and males given the next highest dose of 0.75 mg kg^{-1} bw day^{-1} were euthanized due to toxicity (body tremors, mydriasis, and decreased motor activity). One of 8 dogs given 0.8 mg kg^{-1} bw day^{-1} also had tremors. Axonal degeneration of the nerves in the peripheral and central nervous system was noted at doses of 0.5 mg kg^{-1} bw day^{-1} and above. Degeneration of the retinal ganglionic cells and axonal degeneration in the optic nerve occurred at 0.75 and 1 mg kg^{-1} bw day^{-1}	0.25	
Doramectin	Mouse (CD-1)	Diet	43 days	Target doses of 0–60 on days 1–14, 0–100 on days 15–28 and 0–600 on days 29–43	Signs of toxicity (lethargy, tremors and and hunched posture) were noted at 600 and next highest dose, 400 mg kg^{-1} bw day^{-1}	N/A (exploratory study only)	38
		Diet	92 days	0–300	Tremors, hunched postures and lethargy were noted at 300 and mid-dose, 200 mg kg^{-1} bw day^{-1} with mortalities in both dose groups	N/A—toxicity was seen in all treatment groups	
	Rat	Diet	14 days	0–10	No adverse effects noted	N/A	
		Diet	30 days	0–20	No clinical signs. Liver weights slightly increased in males given highest dose	10	
		Diet	38 days	0–80	Chromorhinorrhea, chromodacryorrhea and urogenital staining were noted at two highest doses of 40 and 80 mg kg^{-1} bw day^{-1} as were whole-body tremors	40	

Table 13.1 (*Continued*)

Compound	*Species*	*Vehicle*	*Duration*	*Doses/route (mg kg^{-1} bw day^{-1} except where shown)*	*Major Findings*	*No observed effect level (mg kg^{-1} bw day^{-1})*	*Reference*
		Sesame oil	38 days	0–10	Ataxia, tremors, and hunched appearance were seen in rats given highest dose	5	
		Diet	92	0–50	Tremors, chromorhinorrhea, chromodacryorrhea and urogenital staining were noted in all treated animals; 35% of rats at 30 mg kg^{-1} bw day^{-1}, 100% at 40 and 95% at 50 mg kg^{-1} bw day^{-1} died or were euthanized moribund	N/A—toxicity was seen in all treatment groups	
		Sesame oil	90 days	0–8	These animals were selected from an F_1 generation where parental animals had been exposed to doramectin *in utero*. There were no overt signs of toxicity but liver and kidney weights were increased at 8 mg kg^{-1} bw day^{-1}	2	
	Dog (beagle)	Diet	14 days	0–2	No signs of toxicity or effects on bodyweight, haematology or urinalysis	2	
		Sesame oil	29 days	0–4	Depression of bodyweight in 1/1 female and 1/1 male given the highest dose. Male had tremors, salivation, mydriasis, ataxia and decreased heart rate. Mydriasis was seen in all treated dogs	N/A—toxicity was seen in all treatment groups	
		Diet	36 days	0–4	Mydriasis was seen in 2/6 high dose dogs	2	
		Sesame oil	91 days	0–2	Mydriasis was seen in all treatment groups. One dog given 2 mg kg^{-1} bw day^{-1} showed ataxia, anorexia and tremors	N/A—toxicity was seen in all treatment groups	
		Sesame oil	92 days	0–0.3	Mydriasis in 1/3 females given highest dose was the only notable clinical sign	0.1	
Moxidectin	Mouse (CF-1)	Diet	28 days	0–32	Signs of toxicity included tremors, hypersensitivity to touch and urine stained fur at 18, 23 and 24 mg kg^{-1} bw day^{-1}. Mortality was high in the 3 highest dose groups (80–100%); one animal died at 18 mg kg^{-1} bw day^{-1}	6.9	39

Rat	Diet	28 days	0–31	Mortality was 100% in the two highest dose groups and 2/5 females given the next lower dose of 23 mg kg^{-1} bw day^{-1} died. Signs of toxicity included ataxia, tremors, salivation, piloerection and diuresis. At the two highest dose levels, there was diffuse atrophy of the liver, kidneys, heart, spleen, adrenals, thyroid, testes, ovaries and epididymis at the two highest dose levels and in the 2 females which died at 23 mg kg^{-1} bw day^{-1}. Hypersensitivity to touch was noted in all treatment groups	N/A—toxicity was seen in all treatment groups
	Diet	13 weeks	0–12	At the highest dose, signs of toxicity included hypersensitivity to touch, lethargy, and aggressive behaviour; hypersensitivity to touch was also noted at the next lower dose, 7.9 mg kg^{-1} bw day^{-1} but this resolved by day 14	3.9
Dog (beagle)	Diet	28 days	0–4	Anorexia, tremors, ataxia, prostration, hypersensitivity to touch and diarrhoea noted at highest doses.	0.5
	Diet	90 days	0–1.6	Dose-dependent reductions in food consumption and absolute bodyweight gain noted at two highest doses	0.3
	Diet	52 weeks	0–1.15	None	1.15

Table 13.2 Carcinogenicity/long-term feeding studies.

Compound	*Species*	*Vehicle*	*Duration*	*Doses (mg kg^{-1} bw day^{-1})*	*Major findings*
Abamectin	Mouse (CD-1)	Diet	94 weeks	0, 2, 4, 8	Treatment-related tremors in females in all dose groups with mortalities at 4 and 8 mg kg^{-1} bw day^{-1}. However, when repeated with further groups of females, tremors occurred only at 8 mg kg^{-1} bw day^{-1}. No increased tumour incidence
	Rat	Diet	2 years	0, 0.75, 1.5, or ~2	As no signs were noted at the highest dose, this was increased from 2 to 2.5 mg kg^{-1} bw day^{-1}. However, this resulted in tremors and severe CNS toxicity and the dose was reduced back to 2 mg kg^{-1} bw day^{-1}. Several of the high-dose animals died during the period they were given 2.5 mg kg^{-1} bw day^{-1} abamectin. No increased tumour incidence
Ivermectin	–	–	–	–	No studies available
Eprinomectin	–	–	–	–	No studies available
Emamectin (benzoate)	Mouse (CD-1)	Diet	79 weeks	0, 0.5, 2.5, 12.5	Animals given 12.5 mg kg^{-1} bw day^{-1} showed signs of toxicity and the dietary intake was adjusted to give a dose of 7.5 mg kg^{-1} bw day^{-1} during the study, and later to 5 mg kg^{-1} bw day^{-1}. Mortality was increased at this higher dose level. Animals in this group displayed tremors and vocalization. Several mice in the group given 5 mg kg^{-1} bw day^{-1} showed fasciculating forelimb tremors. Sciatic nerve degeneration was observed in 2 males given 12.5 mg kg^{-1} bw day^{-1}. No increased tumour incidence

Emamectin (benzoate)	Rat	Diet	2 years	0, 0.25, 1, 5	Bodyweight gain and food consumption reduced during first year of study at 1 and 5 mg kg^{-1} bw day^{-1}. At necropsy, neuronal vacuolation was noted in the brain and spinal cord in both sexes given 5 mg kg^{-1} bw day^{-1}. No increased tumour incidence
Doramectin	–	–	–	–	No studies available
Moxidectin	Mouse (CD-1)	Diet	2 years	0, 2.5, 5.1, 12	After 9 weeks of treatment, the highest dose was reduced to 7.9 mg kg^{-1} bw day^{-1} because of mortalities. Animals in the high dose group showed hunched postures, tremors, decreased activities, laboured breathing and coldness to touch. No other clinical signs were noted. No increased tumour incidence
Moxidectin	Rat	Diet	2 years	0, 0.8, 3.2, 9.8	After 8 weeks of treatment, the highest dose was reduced to 5.1 mg kg^{-1} bw day^{-1} because of mortalities. Signs of toxicity in these animals included hunched postures, tremors, hyperactivity, hypersensitivity to external stimuli and rough hair. The signs resolved when the dose was reduced to 5.1 mg kg^{-1} bw day^{-1}. No increased tumour incidence

13.3.4 Reproductive Studies

13.3.4.1 Studies of Reproductive Performance

A number of reproductive studies have been carried out with these compounds with dosing usually occurring prior to conception, throughout gestation and into lactation.[32–34,36–40] The doses used in these studies were limited by neonatal or maternal toxicity. In some of these studies, there was again evidence of neurotoxicity in maternal animals and in neonates. In one study with eprinomectin, there was evidence of neurotoxicity in pups resulting from excretion of the drug into the milk during the lactation period. Doses of 0.4 mg kg^{-1} bw day^{-1} ivermectin were toxic to neonatal rats and this was likely to be due to excretion in maternal milk rather than as a result of neonatal exposure. Doses of 0.05, 0.1 or 0.2 mg kg^{-1} bw day^{-1} were without effects on reproductive performance and neonatal toxicity.[44]

13.3.4.2 Teratology Studies

As with the studies of reproductive performance, neurotoxicity was frequently the limiting factor in teratology studies where the compounds where administered to experimental animals in gestation, including through periods of critical organogenesis.[32–34,36–40] There was no evidence for frank teratogenic effects in these studies. Where anomalies did occur they tended to be associated with maternal toxicity or with embryotoxicity, and they were generally the types of effects associated with these, *e.g.* wavy ribs, cleft palate and delayed ossification. Signs of neurotoxicity were occasionally seen in these studies.

As with acute and repeat dose studies, the CF-1 mouse, where used in the studies, appeared to be uniquely susceptible to the toxic effects of these compounds (Table 13.3).

13.3.5 Summary of Animal Toxicity Studies

When tested in toxicology studies, many drugs, pesticides and other chemicals display a spectrum of effects, *e.g.* on the liver, on kidneys, on blood elements, on reproductive outcomes and in carcinogenicity studies. However, the compounds discussed here together display one main feature, that of neurotoxicity. This is typified by mydriasis, tremors and convulsions, ataxia and abnormalities of locomotion and, where examined, neuronal degeneration at higher doses. Moreover, there is evidence that ivermectin may induce behavioural effects in rats, including following perinatal exposure,[45,46] while histopathological findings reveal degenerative lesions in the CNS and peripheral nervous system in some animals in longer-term studies. From the studies available, the CF-1 mouse and the collie dog appear to be more susceptible than other species or breeds to these effects. Of course it could well be that the neurotoxicity is so dose limiting that other toxic effects and signs of toxicity, which might appear at higher doses, are thus being prevented.

Table 13.3 Teratology.[40–42,44–48]

Compound	*Species*	*Vehicle*	*Doses/route ($mg\ kg^{-1}\ bw\ day^{-1}$)*	*Major findings*	*NOEL ($mg\ kg^{-1}\ bw\ day^{-1}$)*
Abamectin	Mouse (CF_1)	Not specified (gavage)	0.1, 0.2, 0.4, 0.8	Deaths occurred at all dose levels except 0.2, usually preceded by tremors. An increased incidence of cleft palate was noted at 0.4 and 0.8 mg kg^{-1} bw day^{-1}	<0.1 (maternotoxicity); 0.2 (fetotoxicity and teratogenic effects)
		Not specified (gavage)	0.025, 0.05, 0.075, 0.1	One mouse died at 0.1 mg kg^{-1} bw day^{-1} and tremors were noted in other animals at this dose level. No effects were noted in fetuses	0.05 (maternotoxicity)
Avermectin B_{1b}		Not specified (gavage)	0.025, 0.05, 0.075, 0.1	Two deaths occurred at 0.075 preceded by weight loss and tremors. Single fetuses at all dose levels had anomalies including cleft palates and exencephaly with incidences being ~1%. However, historical controls had incidences 1.6% for exencephaly and 1.3% for cleft palate	0.05 (maternotoxicity)
Abamectin	Rat	Not specified (gavage)	0.25, 0.5, 1, 2	No compound related effects except for reductions in bodyweight gain at highest dose	1
		Not specified (gavage)	0.4, 0.6, 1.6	No evidence of maternotoxicity. Increased incidence of exencephaly, cleft palate, distended ureters and gastroschisis, but all of these were within historical control values. No clear dose relationships for these effects	1.6
	Rabbit	Not specified	0.5, 1.2, 2, 3	Animals at highest dose appeared to be in a stupor after 4th and further doses and had yellow or green discharges from nose and mouth. No other effects noted	2 (maternotoxicity)
		Not specified	0.5, 1, 2	Single deaths occurred in all treatment groups. Fetuses from high-dose animals had cleft palate, omphaloceles and clubbed feet. Higher incidence of skeletal malformations (vertebral, branched and fused ribs)	1

Table 13.3 (*Continued*)

Compound	*Species*	*Vehicle*	*Doses/route ($mg\ kg^{-1}\ bw\ day^{-1}$)*	*Major findings*	*NOEL ($mg\ kg^{-1}\ bw\ day^{-1}$)*
Ivermectin (Avermectin B_{1a})	Mouse (CF_1)	Not specified	0.2, 0.4, 0.8, 1.6	2 maternal animals euthanized or died at all but the lowest dose level. Cleft palate noted at highest dose level	0.2
Ivermectin (Avermectin B_{1b})	Mouse (CF_1)	Not specified	0.4, 0.8, 1.6	1 animal euthanized at 0.8 and 1.6 mg kg^{-1} bw day^{-1}. Cleft palate occurred in fetuses at these dose levels	0.2
	Mouse (CRCF)	Not specified	0.1, 0.2, 0.4, 0.8	1 and 3 maternal animals euthanized at the 0.2 and 0.8 mg kg^{-1} bw day^{-1} dose levels; 3 mice euthanized or died at the 0.4 mg kg^{-1} bw day^{-1} level. Higher incidence of cleft palate noted at two highest doses	0.1
	Rat	Not specified	2.5, 5, 10	No maternotoxicity evident at two lower dose levels; 3 animals euthanized at highest dose level. Cleft palate seen at highest dose	5
	Rabbit	Not specified	1.5, 3	No maternotoxicity. Fetuses at higher dose had cleft palate and reduced litter weights	1.5
	Dog (beagle)	Not specified	0.5—every 10 days	No adverse effects noted	
Eprinomectin	Rat	Aqueous	0.5, 1.5, 5, 10, 15	1 maternal animal euthanized at the high-dose level. This rat had tremors, ptosis, severe weight loss, decreased activity and abnormal posture. Other dams at this dose had fine tremors and abnormal postures. Maternal bodyweights were increased at 5 and 10 mg kg^{-1} bw day^{-1} but decreased at 15 mg kg^{-1} bw day^{-1}. Fetal bodyweights were decreased at 1.5, 5, 10 and 15 mg kg^{-1} bw day^{-1}. No fetal anomalies	0.5

	Aqueous	0.5, 1, 3, 12	No evidence of maternotoxicity. At the two higher doses, maternal weight was higher than controls initially but this decreased later. No fetal anomalies	1
Rabbit	Aqueous	1.5, 4, 10, 25	Maternal animals at the two higher doses were euthanized due to poor condition and weight loss. Dilated pupils and slowed papillary responses occurred at $\geq 4\,mg\,kg^{-1}$ bw day^{-1}, while mild tremors and decreased food consumption were observed at $\geq 10\,mg\,kg^{-1}$ bw day^{-1}. No other adverse effects noted	4
	Aqueous	2, 4, 8	Mydriasis and slowed papillary responses occurred in all treatment groups. 2 high-dose animals euthanized due to weight loss and failure to eat. No effects noted on any index of reproduction and no fetal anomalies noted	N/A—tox.city noted in all groups
	Aqueous	0.5, 2, 8	Mydriasis and slowed papillary responses seen at 2 and $8\,mg\,kg^{-1}$ bw day^{-1}. The numbers of implants and numbers of live fetuses were reduced at 2 and $8\,mg\,kg^{-1}$ bw day^{-1} but these values were within historical control values and were probably due to reduced numbers of corpora lutea and not related to treatment. No effects on fetal weight and no evidence of any anomalies	0.5 based on maternotoxicity
	Aqueous	1.2, 2, 8	No treatment-related deaths and no abortions. High-dose animals showed reduced papillary responses and mydriasis. No effects on embryonic or fetal survival and no effects on the numbers of corpora lutea, implants, resorptions and live fetuses	2 based on materr.otoxicity

Table 13.3 (*Continued*)

Compound	*Species*	*Vehicle*	*Doses/route ($mg\ kg^{-1}\ bw\ day^{-1}$)*	*Major findings*	*NOEL ($mg\ kg^{-1}\ bw\ day^{-1}$)*
Emamectin (benzoate)	Rat	Not specified (gavage)	2, 4, 8	Signs of neurotoxicity including tremors and convulsions were observed in dams given highest dose. Dose-related reductions in bodyweights at 4 and 8 mg kg^{-1} bw day^{-1}. No evidence of teratogenic effects but high-dose fetuses had supernumerary ribs and delayed ossification	2 based on maternotoxicity
	Rabbit	Not specified (gavage)	1.5, 3, 6	Maternal bodyweights and food consumption were reduced at highest dose. These dams also had mydriasis and decreased papillary reactions. No evidence of fetotoxicity or teratogenicity	3 based on maternotoxicity
Doramectin	Rat	Sesame oil	0.1, 0.2, 0.4, 0.8, 1.6	No signs of toxicity in maternal animals. No embryotoxicity including embryolethality. No effects on numbers of live fetuses or fetal weights. Fetuses not examined for anomalies	1.6
		Not specified (Gavage)	1.5, 3, 6	No signs of toxicity in maternal animals. No embryotoxicity including embryolethality. No effects on numbers of live fetuses or fetal weights. Fetuses not examined for anomalies	6
		Sesame oil	1.5, 3, 6	No signs of toxicity in maternal animals. Slight increase in embryolethality at 6 mg kg^{-1} bw day^{-1}. Slight increases in rudimentary ribs, wavy ribs, unossified hyoid and 5th metatarsals and dilatation of the ureters and renal pelvis but all within historical control limits and not dose-related. 5 h after the 6 mg kg^{-1} bw day^{-1} dose, amniotic fluid concentrations were 0.014 μg ml^{-1} and fetal concentrations were 0.27–1.1 μg g^{-1}	3

	Rabbit	Sesame oil	0.2, 0.8, 1.6	Bodyweights decreased in does given 1.6 mg kg^{-1}bw day^{-1}. Embryolethality noted at this dose. Fetuses were not examined	0.8
		Sesame oil	1.5, 3, 6	Bodyweights decreased in does at 6 mg kg^{-1} bw day^{-1}. Embryolethality and depressed fetal weights occurred at 6 mg kg^{-1} bw day^{-1}. Fetuses were not examined	3
		Sesame oil	0.75, 1.5, 3	Food intake reduced at 1.5 and 3 mg kg^{-1} bw day^{-1} and bodyweights lower at the high dose. Embryolethality and fetal weights similar in all groups including untreated controls. At 3 mg kg^{-1} bw day^{-1} cleft palate was noted in 1 litter, and phocomelia, syndactyly and coelosomia were noted in another. Incidence of cleft palate was within the normal historical control range but the other anomalies had not been encountered before in the test facility. Ossification of the pubic bones was observed at 1.5 and 3 mg kg^{-1} bw day^{-1}.	0.75
Moxidectin	Rat	Corn oil	2.5, 5, 10, 12	No maternal deaths but 12 mg kg^{-1} bw day^{-1} led to urine-stained fur and chromodacryorrhoea. At 10 and 12 mg kg^{-1} bw day^{-1} there were significant decreases in maternal weight. At 10 and 12 mg kg^{-1} bw day^{-1} there were increases in the numbers of fetuses with anomalies including cleft palate, and wavy or incompletely ossified ribs which were probably due to embryotoxic effects. No evidence of any frank teratogenic effects	5
	Rabbit	Corn oil	1, 5, 10	Dose-related decreases in bodyweights at 5 and 10 mg kg^{-1} bw day^{-1}. No effects on numbers of corpora lutea, implantations or resorptions. No increases in the incidences of any anomalies	1

NOEL, no observed effect level.

13.3.6 Factors Relating to Toxicity

The avermectins and milbemycins are highly lipophilic drugs whose mode of toxicity in mammals remains unclear. In target organisms, there is a high-affinity binding site and the physiological response to binding is an increase in permeability to chloride ions through γ-aminobutyric acid (GABA)-gated chloride ion channels.[47,48] GABA is a major neurotransmitter in target parasites and in mammals and it is likely that the avermectins and milbemycins are agonists resulting in interference with the transmission of nerve impulses.[49–57] Ivermectin and other analogues bind to GABA receptors in rat brain.[50] In mammals, GABA receptors are found largely in the CNS and any disruption of nerve impulses could plausibly explain the toxicity noted in both laboratory animal studies, and as will soon become apparent, in target animal adverse reactions. It is likely that many of these adverse reactions are due to overdosing. However, the lessons from the CF-1 mouse, discussed above, cannot be ignored.

Subsequent studies have shown that a subpopulation of CF-1 mice is deficient in P-glycoprotein in the intestinal epithelium and brain endothelium and concentrations of radiolabelled ivermectin were found to be higher in brains of sensitive mice when compared to non-sensitive animals.[58] This deficiency allows greater intestinal absorption of orally administered avermectins, and higher brain penetration.[59] This sensitive population of CF-1 mice is also more susceptible to the reproductive effects of the avermectins. In a susceptible population of CF-1 mice, 100% of animals were affected by cleft palate while in the non-sensitive population there was almost a zero incidence at the same doses.[60] P-glycoprotein is a protein pump encoded for by a gene known as *MDR1* or *ABCB1*. It is one of a number of proteins belonging to an ATP-binding class and where present it can lead to drug resistance (or lowered susceptibility to toxicity).[61] It was so named because of the overexpression noted in multidrug resistant human tumour cells.[62] Deficiency or disruption of this gene or inhibition of P-glycoprotein leads to enhanced absorption or organ exposure, including enhanced brain penetration, to a number of drugs including ivermectin.[63–69] P-glycoprotein deficiency also leads to higher fetal exposure to some drugs.[70] Similar effects can be achieved by pharmacological blocking of P-glycoprotein.[3,71]

In the CF-1 mouse, this sensitive population constitutes around 25% of the animals. These animals have been shown to have low or absent P-glycoprotein in brain endothelial cells and these animals are those with the higher sensitivity to the effects of the macrocyclic lactones.[58–60]

There may be significant differences in the disposition of these macrocyclic lactones. For example, in P-glycoprotein deficient mice, ivermectin and eprinomectin, but to a much lesser extent moxidectin, were excreted by the intestine through a P-glycoprotein-dependent pathway whereas moxidectin excretion was P-glycoprotein independent. All three drugs accumulated in the brains of these mice but eprinomectin concentrations were lower, possibly because eprinomectin disposition is controlled by P-glycoprotein efflux.[72]

13.4 Human Toxicity

Other species or breeds known to be susceptible to the toxicity of the avermectins and other drugs, such as the collie mentioned earlier and the white Swiss shepherd dog, have been shown to have individuals with mutations of the *MDR1* gene[73–75] and there are concerns that P-glycoprotein polymorphisms in humans could result in increased susceptibility to the toxicity of ivermectin.[76–80] Ivermectin (as Mectizan) has been widely used in humans for the treatment and prophylaxis of filariasis due to *Wucheria bancrofti,* onchocerciasis caused by *Onchocerca volvulus*, loiasis, scabies and strongyloidiasis.[29,77–89] Treatments usually consist of 150 μg kg^{-1} bw (or approximately 10 mg for a 70-kg adult) and deaths are very rare as are adverse reactions in general which are usually mild.[89–95] A high oral dose of avermectin (414 mg kg^{-1} bw) resulted in coma, myoclonus and polyneuropathy.[96] The most common adverse reaction seen with ivermectin in humans is also a marker of its success. This is the Mazzotti reaction, which is an immune response to dying parasites characterized by urticaria, fever, swollen lymph nodes, arthralgias, hypotension, tachycardia, oedema and abdominal pain.[97–100] Under some circumstances, particularly when the initial parasite burden is high, this may be severe,[97–102] and has been observed with other drugs including diethylcarbamazine and praziquantel.[103–105] This may be due to neutrophil granule activation,[106,107] possibly induced by a myosin-like antigen released from the muscle of the parasite.[108] An encephalopathy associated with ivermectin treatment of onchocerciasis-infected patients in *Loa loa* endemic areas remains of uncertain origin, although this too is probably associated with parasite burden.[109,110] However, treatment with ivermectin or with ivermectin plus diethylcarbamazine is generally safe.[111]

Oral ingestion of small amounts of abamectin was generally asymptomatic. However, oral doses of around 23 mg kg^{-1} bw produced minor symptoms. More severe effects were produced by doses of around 15 mg kg^{-1} bw ivermectin and 115 mg kg^{-1} bw abamectin, including coma, aspiration with respiratory failure and hypotension. Of seven affected patients, six recovered after intensive supportive care but one later died with multiple organ failure.[112]

13.5 Adverse Effects in Target Species

Data on adverse effects in target animals are difficult to access. Much of this information is reported to regulatory authorities by veterinarians, by manufacturers and the public, under regulatory pharmacovigilance schemes, and remains confidential.[113–115] However, some data can be accessed from the website of the US Food and Drug Administration's Center for Veterinary Medicine (CVM) where there are cumulative summaries from 1987 to 2009.[116] Some caution must be applied to these data. There is a numerator in these lists, the numbers of animals with adverse events, but no denominator, *i.e.* the total numbers of animals treated. Hence no measure of incidence can be applied to the numbers. Similarly, for all drugs cited in the list there is no way of relating the number of animals affected with each clinical sign. There are also no data on doses given and so it is not possible to relate adverse drug reactions and clinical toxicity to overdosing. However, with these considerations in mind, there is useful information in the CVM data, some of which is reviewed in Table 13.4.

13.5.1 Cats

In the period 1987–2009, CVM reports 126 cases of cats with adverse drug reactions to oral ivermectin. Not surprisingly, of the numerous signs reported, many are indicative of neurotoxicity including ataxia, depression and lethargy, mydriasis, recumbency, hypersalivation, trembling, confusion, blindness, convulsions, hyperesthesia, staggering, collapse, difficulty rising and face trembling.[116] The reports have similarities with case reports where recumbency, lethargy, weakness, blindness, ataxia, tachycardia and coma have been noted.[117–120] The general findings of ataxia, emesis, disorientation, body tremors, blindness, bradycardia, hypothermia, coma and death are typical of toxicity resulting from overdose.[121,122] Doses of 200–1330 $\mu g\ kg^{-1}$ bw are said to be generally tolerated in cats,[121] but the oral therapeutic dose is, 24–48 $\mu g\ kg^{-1}$ bw day^{-1},[123] and higher doses may be toxic in individual cats.

Selamectin has been developed specifically for use in cats and dogs and has been shown to be effective against internal parasites and fleas in both species.[12–14,124] Its safety has been studied extensively in cats at the therapeutic topical dose range of 6–12 mg kg^{-1} bw and no adverse effects were seen.[125]

The CVM cumulative report lists 44 cats with adverse events following oral exposure to selamectin.[116] Clinical signs include emesis, hypersalivation,

Table 13.4 Adverse events with macrocyclic lactones in CVM cumulative database 1987–2009—miscellaneous species.[107]

Species	*Route*	*Substance* *Ivermectin*	*Doramectin*	*Selamectin*	*Moxidectin*
Fish	Oral	–	–	–	7
Goat	Oral	9	–	–	81
	Topical	–	–	–	8
	Parenteral	1	–	–	–
Sheep	Oral	–	–	–	137
	Parenteral	71	–	–	–
Pig	Parenteral	916	21 334	–	–
Guinea-pig	Oral	4	–	–	–
	Parenteral	1	–	–	–
Alpaca	Topical	–	–	1	–
Bobcat	Topical	–	–	1	–
Ferret	Topical	–	–	10	–
Rabbit	Topical	–	–	122	–
	Parenteral	4	–	–	–
Llama	Parenteral	1	–	–	–
Turtle	Parenteral	2	–	–	–

hyperactivity, tachycardia and trembling. After topical application, 4394 cats are listed. Many of these reports probably refer to lack of expected efficacy but signs of neurotoxicity including ataxia, depression and lethargy, trembling, convulsions, hypersalivation, recumbency, stupor, confusion, hyperesthesia, blindness, paresis, paralysis and vision disorders also appear.

13.5.2 Dogs

In dogs, the therapeutic oral dose of ivermectin is between 6 and 12 $\mu g\ kg^{-1}$ bw[123] and doses of 50–500 $\mu g\ kg^{-1}$ bw may be tolerated.[121,126] Adverse reactions in 1165 dogs given ivermectin orally included the familiar signs of neurotoxicity.[116] Similar signs were noted in 5 dogs after topical treatment and in 76 after parenteral treatments. After exposure to eprinomectin, dogs (2) showed mydriasis and blindness, while after oral exposure to doramectin (12 animals) signs of neurotoxicity were observed. Depression and lethargy, hypersalivation and death were among signs in three dogs treated topically with doramectin and ataxia, depression, lethargy, temporary blindness, convulsions, paresis, tachycardia and trembling occurred in three animals treated parentally. Blindness has been reported in dogs suffering from ivermectin toxicosis.[127] The electroretinogram was completely disrupted in one animal and abnormal in the other and there were areas of multifocal retinal oedema. Both animals made complete clinical recoveries.

Moxidectin is authorized in some countries for the treatment of heartworm in dogs, with oral therapeutic doses of 3 $\mu g\ kg^{-1}$ bw.[123,126] Recognized signs of neurotoxicity with moxidectin in dogs are similar to those noted with the avermectins.[122] These effects were among those observed and reported among 369 dogs exposed orally to moxidectin and reported to the CVM, and in 6 dogs exposed topically.[116]

One of the major concerns with the use of ivermectin or other macrocyclic lactones in dogs is the increased sensitivity to these compounds shown by some animals in the collie dog population as a result of the homozygous occurrence of the nt230 (del 4) deletion in the *MDR1* gene and defective P-glycoprotein, allowing greater intestinal and brain permeability and hence higher local exposure to these drugs. As discussed earlier and as noted elsewhere, ivermectin and the related compounds are more toxic to susceptible individuals in this breed,[35,121] and this has been noted with ivermectin, doramectin and moxidectin[128–136] although moxidectin appears to be less toxic than ivermectin[134–139] while ivermectin was less toxic when administered in a beef formulation.[139] The young of other breeds may be more sensitive.[140] Other breeds including the white Swiss shepherd dog, Australian shepherds and long-haired whippets also have this mutation and may be susceptible not only to the macrocyclic lactones but also to other drugs dependent on P-glycoprotein.[73,141–148] It should be noted that as with the CF-1 mouse and collie dog, not all animals are susceptible and only a proportion of animals in these breeds are likely to be exquisitely sensitive to the effects of these drugs.[121,130] In a small population (40) of collie dogs, for example, 22% were homozygous for the normal gene, 42% were heterozygous and 35% were homozygous for the mutant.[148]

Reactions including anorexia, lethargy, pyrexia and emesis and a shock-like syndrome have been seen following ivermectin treatment of dogs infected with microfilariae.[149–152] These reactions may be attributable to toxicity but they may also be due to the effects of dying parasites and thus may be analogous to the Mazzotti reaction noted in humans. They have been reported with other drugs used in dogs for this condition, including moxidectin.[153–158]

Consumption of moxidectin-containing equine products has resulted in toxicity in dogs, which in turn has given rise to regulatory concerns.[159–161]

Selamectin, as already described, has been developed as an endectocide specifically for use in cats and dogs.[12,13,122,162] It has been extensively studied for safety in the dog at its topical target dose range of 6–12 mg kg^{-1} bw.[163] Examination of the CVM cumulative list reveals 98 dogs orally exposed to selamectin displaying a range of clinical signs including depression and lethargy, hypersalivation, convulsions, trembling, weakness, ataxia and collapse.[116] For topical application there are a total of 11 427 animals which have been evaluated with a spectrum of clinical signs including lack of expected efficacy and signs which are unrelated to neurotoxicity. However, reactions suggestive of neurotoxicity have been noted.

13.5.3 Horses and Other Equines

The CVM cumulative report includes 387 horses with adverse effects following oral ivermectin. Some of these reports concern lack of expected efficacy but signs suggestive of neurotoxicity are also apparent.[116] Blindness, ataxia and convulsions were reported after donkeys (3) were exposed to oral ivermectin, while ataxia and staggering were reported in horses (3) after topical application. Cases of equine adverse events following ivermectin have been reported in the literature. These included a death, injection site swellings, mydriasis, depression, incoordination and disorientation, fore and hind limb ataxia, muscle fasciculations[164–166] and neurological effects and blindness occurring in a mule foal.[167] Depression, ataxia and blindness occurred in a Grant's zebra accidentally overdosed with ivermectin.[168]

Oral eprinomectin has been associated with abdominal pain in a horse, while oral doramectin was associated with abdominal pain; this latter animal was euthanized.[116] The CVM cumulative reports include 1262 animals associated with adverse events, including signs of neurotoxicity, following oral moxidectin exposures. Toxicity in horses, with depression, ataxia, tremors, seizures and coma, has been reported in the scientific literature.[165–171]

13.5.4 Cattle

In the CVM's database, 41 cattle are listed as undergoing adverse events following oral exposure to ivermectin with signs including anorexia, depression and lethargy and collapse.[116] Sign of toxicity have also been reported after topical administration.

For eprinomectin, only three cattle are listed following oral exposure and the only sign provided is death. However, after topical treatments, 540 animals are included with ataxia, recumbency, collapse, hypersalivation, blindness, depression and lethargy, and confusion. Doramectin was associated with 1001 affected animals after oral exposure where signs included recumbency, ataxia, blindness and hypersalivation. Toxicity in cattle following treatment with avermectins has been reported in the literature.[172–174] In one report of three separate incidents, calves overdosed with abamectin displayed signs of neurotoxicity and a number of deaths occurred.[173] Neurological signs have been reported in Murray Grey cattle given therapeutic doses of abamectin.[174] Protruding tongues resulting from lingual paralysis were noted in some animals. Some animals died. In a clinical investigation under controlled conditions, 208 Murray Grey cattle were given the drug strictly in accordance with label requirements and one animal developed neurological signs. These findings suggest that this breed may be more sensitive to the effects of avermectins than others. The authors suggested that the drug may be able to reach the CNS in this breed more easily than in other breeds. This is reminiscent of the P-glycoprotein phenomenon discussed in relation to other animal strains and breeds. However, there is no firm evidence to support this in Murray Grey cattle although, not surprisingly, P-glycoprotein plays a role in ivermectin transport in bovine blood vessels.[175] However, this breed also has other mutations which result in neurological signs including those that lead to spinal myelinopathy and mannosidosis[176–178] and such conditions might confound the effects of drugs.

13.5.5 Other Species

Adverse reactions, including signs of neurotoxicity with ivermectin, doramectin, selamectin and moxidectin, have been reported in several species including goats, sheep and pigs, as set out in Table 13.4.

13.6 Conclusions

The macrocyclic endectocides are exceptionally effective veterinary drugs which have revolutionized the treatment of several diseases, including those caused by internal and external parasites. Prior to their introduction there were only two classes of drug available for the treatment of internal parasites, levamisole/tetramisole and the benzimidazoles. The introduction of the macrocyclic lactones not only provided a third class of drugs but also offered greater scope for the rotation of drug class as an aid in avoiding the build-up of parasite resistance. Over the years since their individual introductions, countless millions of doses have been administered safely and effectively to animals. These are very important pieces of information which cannot be overemphasized—these drugs are highly effective, offer alternatives to existing drug classes and, when used according to label recommendations, are extremely safe.

However, these compounds do exert toxic effects and neurotoxicity is the predominant finding. This is particularly true in subpopulations of laboratory and other animals with mutations in the *MDR1* gene which predisposes to nervous system exposure.

In clinical use, adverse drug reactions to these drugs do occur. Neurological effects, typical of those noted in preclinical toxicity studies, are frequently observed. Although the data do not allow for further analysis it is likely that the majority of these arise from overdosing, from miscalculation of doses during off-label use and in animals with sensitivity to the toxic effects of this group of drugs.[49,179] In fact there are several reports of ivermectin toxicity in animals following off-label use, including mice, rhesus macaque, fruit bats, a chameleon, chelonians and chickens.[180–185] Off-label use of ivermectin in farmed salmon may result in toxicity with listlessness, inappetance and death.[186–188] Younger animals with more a permeable blood–brain barrier may be more susceptible than adults to the toxic effects of these substances.

References

1. I. H. Sutherland and W. C. Campbell, Development, pharmacokinetics and mode of action of ivermectin, *Acta Leiden.*, 1990, **59**, 161–168.
2. Q. A. McKellar, Ecotoxicology and residues of anthelmintic agents, *Vet. Parasitol.*, 1997, **72**, 413–435.
3. Q. A. McKellar and H. A. Benchaoui, Avermectins and milbemycins, *J. Vet. Pharmacol. Ther.*, 1996, **19**, 331–351.
4. J. C. Williams, Anthelmintic treatment strategies: current status and future, *Vet. Parasitol.*, 1997, **72**, 461–470.
5. S. Omura, Ivermectin: 25 years and still going strong, *Int. J. Antimicrob. Agents*, 2008, **31**, 91–98.
6. W. L. Shoop, H. Mrozik and M. H. Fisher, Structure and activity of avermectins and milbemycins in animal health, *Vet. Parasitol.*, 1995, **59**, 139–156.
7. J. Di Netta, List of registrations, in *Ivermectin and Abamectin*, ed. W. C. Campbell, Springer-Verlag, New York, 1989, pp. 344–346.
8. M. H. Fisher and H. Mrozik, Chemistry, in *Ivermectin and Abamectin*, ed. W. C. Campbell, Springer-Verlag, New York, 1989, pp. 1–23.
9. R. W. Burg and E. O. Stapley, Isolation and characterization of the producing organism, in *Ivermectin and Abamectin*, ed. W. C. Campbell, Springer-Verlag, New York, 1989, pp. 24–32.
10. W. L. Shoop, P. Dermontigny, D. W. Fink, J. B. Williams, J. R. Egerton, H. Mrozik, M. H. Fisher, B. J. Skelly and M. J. Turner, Efficacy in sheep and pharmacokinetics in cattle that led to the selection of eprinomectin as a topical endectocide for cattle, *Int. J. Parasitol.*, 1996, **26**, 1227–1235.
11. W. L. Shoop, J. R. Egerton, C. H. Eary, H. W. Haines, B. F. Michael, H. Mrozik, P. Eskola, M. H. Fisher, L. Slayton, D. A. Ostlind, B. J. Skelly,

R. K. Fulton, D. Barth, S. Costa, L. M. Gregory, W. C. Campbell, R. L. Seward and M. J. Turner, Eprinomectin: A novel avermectin for use as a topical endectocide for cattle, *Int. J. Parasitol.*, 1996, **26**, 1237–1242.
12. B. F. Bishop, C. I. Bruce, N. A. Evans, A. C. Goudie, K. A. F. Gration, S. P. Gibson, M. S. Pacey, D. A. Perry, N. D. A. Walshe and M. J. Witty, Selamectin: a novel broad spectrum endectocide for dogs and cats, *Vet. Parasitol.*, 2000, **91**, 163–176.
13. M. G. Boy, R. H. Six, C. A. Thomas, M. J. Novotny, C. D. Smothers, T. G. Rowan and A. D. Jernigan, Efficacy and safety of selamectin against fleas and heartworms in dogs and cats presented as veterinary patients in North America, *Vet. Parasitol.*, 2000, **91**, 233–250.
14. R. H. Six, G. H. Sture, C. A. Thomas, R. G. Clemence, H. A. Benchaoui, M. G. Boy, P. Watson, D. G. Smith, A. D. Jernigan and T. G. Rowan, Efficacy and safety of selamectin against gastrointestinal nematodes in cats presented as veterinary patients in North America, *Vet. Parasitol.*, 2000, **91**, 321–331.
15. I. Ishaaya, S. Kontsedalov and A. R. Horowitz, Emamectin, a novel insecticide for controlling field crop pests, *Pest. Manag. Sci.*, 2002, **58**, 1091–1095.
16. J. Brocklebank, Assessing fish health during sea lice infestations, *North. Aquac.*, 1995, **1**, 8–9.
17. J. Stone, I. H. Sutherland, C. S. Sommerville, R. H. Richards and K. J. Varma, The efficacy of emamectin benzoate as an oral treatment of sea lice, *Lepeophtheirus salmonis* (Krøyer) infestations in Atlantic salmon, *Salmo salar L. J. Fish Dis.*, 1999, **22**, 261–270.
18. B. M. MacKinnon, Sea lice: a review, *World Aquac.*, 1997, **28**, 5–10.
19. R. Armstrong, D. MacPhee, T. Katz and R. Endris, A field efficacy evaluation of emamectin benzoate for the control of sea lice on Atlantic salmon, *Can. Vet. J.*, 2000, **41**, 607–612.
20. L. E. Burridge, N. Hamilton, S. E. Waddy, K. Haya, S. M. Mercer, R. Greenhalgh, R. Tauber, S. V. Radecki, L. S. Crouch, P. G. Wislocki and R. G. Endris, Acute toxicity of emamectin benzoate (SLICE^{TM}) in fish feed to American lobster, *Homarus americanus. Aquac. Res.*, 2004, **35**, 713–722.
21. K. Willis and N. Ling, The toxicity of emamectin benzoate, an aquaculture pesticide, to planktonic marine copepods, *Aquaculture*, 2003, **221**, 289–297.
22. K. J. Willis, P. A. Gillibrand, C. J. Cromey and K. D. Black, Sea lice treatments on salmon farms have no adverse effects on zooplankton communities: a case study, *Mar. Pollut. Bull.*, 2005, **50**, 806–816.
23. T. C. Telfer, D. J. Baird, J. G. McHenery, J. Stone, I. Sutherland and P. Wislocki, Environmental effects of the anti-sea lice (Copepoda: Caligidae) therapeutant emamectin benzoate under commercial use conditions in the marine environment, *Aquaculture*, 2006, **260**, 163–180.
24. P. R. Smith, M. Maloney, A. McElligott, R. Palmer, J. O'Kelly and F. O'Brien, The efficiency of oral ivermectin in the control of sea lice

infestations of farmed Atlantic salmon, in *Pathogens of Wild and Farmed Fish: Sea Lice*, ed. G. A. Boxall and D. Defaye, Ellis Horwood, London, 1993, pp. 296–307.
25. M. Roth, R. H. Richards and C. Sommerville, Current practices in the chemotherapeutic control of sea lice infestations in aquaculture: a review, *J. Fish Dis.*, 1993, **16**, 1–26.
26. K. N. Woodward, Maximum residue limits, in *Veterinary Pharmacovigilance. Adverse Reactions to Veterinary Medicinal Products*, ed. K. N. Woodward, Wiley-Blackwell, Chichester, 2009, pp. 547–567.
27. K. N. Woodward, The assessment of user safety, exposure and risk to veterinary medicinal products in the European Union, *Regul. Pharmacol. Toxicol.*, 2008, **50**, 114–128.
28. K. N. Woodward, The evolution of safety assessments for veterinary medicinal products in the European Union, *Vet. Human Toxicol.*, 2004, **46**, 199–205.
29. K. N. Woodward, Regulation of veterinary drugs, in *General and Applied Toxicology*; ed. B. Ballantyne, T. C. Marrs and T. Syversen, Macmillan, Basingstoke, 2nd edn, 2000, pp. 1633–1652.
30. K. N. Woodward, Veterinary pharmacovigilance. Part 6. Predictability of adverse reactions in animals from laboratory toxicology studies, *J. Vet. Pharmacol. Ther.*, 2005, **28**, 213–231.
31. K. N. Woodward, The toxicity of particular veterinary drug residues, in *Pesticide, Veterinary and Other Residues in Food*, ed. D. H. Watson, Woodhead Publishing/CRC Press, Cambridge, 2004, pp. 175–223.
32. JMPR, *Pesticide Residues in Food—1992. Joint FAO/WHO Meeting on Pesticide Residues. Evaluations 1992. Part II—Toxicology*. International Programme on Chemical Safety, World Health Organization, Geneva, 1993, pp. 3–31.
33. JMPR, *Pesticide Residues in Food—1994. Joint FAO/WHO Meeting on Pesticide Residues. Evaluations 1994. Part II—Toxicology*. International Programme on Chemical Safety, World Health Organisation, Geneva, 1995, pp. 3–13.
34. JECFA, *Toxicological Evaluation of Certain Veterinary Drug Residues in Food. WHO Food Additive Series 27*. 36th Meeting of the Joint FAO/WHO Expert Committee on Food Additives. International Programme on Chemical Safety, World Health Organization, Geneva, 1991, pp. 27–73.
35. J. D. Pulliam, R. L. Seward, R. T. Henry and S. A. Steinberg, Investigating ivermectin toxicity in collies, *Vet. Med.*, 1985, **80**, 33–40.
36. JECFA, *Toxicological Evaluation of Certain Veterinary Drug Residues in Food. WHO Food Additive Series 31*. 40th Meeting of the Joint FAO/WHO Expert Committee on Food Additives. International Programme on Chemical Safety, World Health Organization, Geneva, 1993, pp. 23–36.
37. JECFA, *Toxicological Evaluation of Certain Veterinary Drug Residues in Food. WHO Food Additive Series 41*. 50th Meeting of the Joint FAO/WHO Expert Committee on Food Additives. International Programme

on Chemical Safety, World Health Organization, Geneva, 1998, pp. 3–22.

38. Committee for Veterinary Medicinal Products, *Emamectin. Summary Report (1)*, European Medicines Agency. EMEA/MRL/546/99-FINAL, 1999 http://www.cma.europa.eu/ (accessed 24 May, 2010).
39. JECFA, *Toxicological Evaluation of Certain Veterinary Drug Residues in Food. WHO Food Additive Series 36*. 45th Meeting of the Joint FAO/WHO Expert Committee on Food Additives. International Programme on Chemical Safety, World Health Organization, Geneva, 1996, pp. 3–25.
40. JECFA, *Toxicological Evaluation of Certain Veterinary Drug Residues in Food. WHO Food Additive Series: 36*. 45th Meeting of the Joint FAO/WHO Expert Committee on Food Additives. International Programme on Chemical Safety, World Health Organization, Geneva, 1996, pp. 27–50.
41. D. W. Fink and A. G. Porras, Pharmacokinetics of ivermectin in animals and humans, in *Ivermectin and Abamectin*, ed. W. C. Campbell, Springer-Verlag, New York, 1989, pp. 113–130.
42. S.-H. L. Chiu and A. Y. H. Lu, Metabolism and tissue residues, in *Ivermectin and Abamectin*, ed. W. C. Campbell, Springer-Verlag, New York, 1989, pp. 131–143.
43. C. M. Laffont, P.-L. Toutain, M. Alvinerie and A. Bousquet-Mélou, Intestinal secretion is a major route for parent ivermectin elimination in the rat, *Drug Metab. Dispos.*, 2002, **30**, 626–630.
44. G. R. Lankas, D. H. Minsker and R. T. Robertson, Effects of ivermectin on reproduction and neonatal toxicity in rats, *Food Chem. Toxicol.*, 1989, **27**, 523–529.
45. J. M. Poul, Effects of perinatal ivermectin exposure on behavioural development in rats, *Neurotoxicol. Teratol.*, 1988, **10**, 267–272.
46. I. Nafstad, E. Sannes, A. Hem, P. A. Engen and T. Sagvolden, Behavioural method to detect marginal neurotoxic effects of ivermectin in DA/Orl rats, *J. Exp. Anim. Sci.*, 1991, **34**, 81–86.
47. G. N. Burkhart, Ivermectin: an assessment of its pharmacology, microbiology and safety, *Vet. Human Toxicol.*, 2000, **42**, 30–35.
48. M. J. Turner and J. M. Schaeffer, Mode of action of ivermectin, in *Ivermectin and Abamectin*, ed. W. C. Campbell, Springer-Verlag, New York, 1989, pp. 73–88.
49. J. D. Roder and E. L. Stair, An overview of ivermectin toxicosis, *Vet. Human Toxicol.*, 1998, **40**, 369–370.
50. G. R. Dawson, K. A. Wafford, A. Smith, G. R. Marshall, P. J. Bayley, J. M. Schaeffer, P. T. Meinke and R. M. McKernan, Anticonvulsant and adverse effects of avermectins are mediated through the γ-aminobutyric $acid_A$ receptor, *J. Pharmacol. Exp. Ther.*, 2000, **295**, 1051–1060.
51. R. J. Martin, An electrophysiological preparation of *Ascaris suum* pharyngeal muscle reveals a glutamate-gated chloride channel sensitive to the avermectin analogue, Milbemycin D, *Parasitology*, 1996, **112**, 247–252.

52. J. M. Schaeffer and H. W. Haines, Avermectin binding to *Caenorhabditis elegans*. A two-state model for ivermectin binding site, *Biochem. Pharmacol.*, 1989, **38**, 2329–2338.
53. J. Yamazaki, K. Matsumoto, H. Ono and H. Fukuda, Macrolide compounds ivermectin and milbemycin D stimulate chloride channels through GABAergic drugs in cultured chick spinal neurons, *Comp. Biochem. Physiol. C*, 1989, **93**, 97–104.
54. K. L. Mealey, Ivermectin: macrolide parasitic agents, in *Small Animal Toxicology*, ed. M. E. Peterson and P. A. Talbot, Elsevier, St. Louis, 2nd edn, 2006, pp. 785–794.
55. D. F. Cully, P. S. Paress, K. K. Liu, J. M. Schaeffer and J. F. Arena, Identification of a *Drosophila melanogaster* glutamate-gated chloride channel sensitive to the antiparasitic agent avermectin, *J. Biol. Chem.*, 1996, **271**, 20187–20191.
56. B. Robertson, Actions of anaesthetics and avermectin on $GABA_A$ chloride channels in mammalian dorsal root ganglion neurones, *Br. J. Pharmacol.*, 1989, **98**, 167–176.
57. A. T. Eldefrawi and M. E. Eldefrawi, Receptors for gamma-aminobutyric acid and voltage-dependent chloride channels as targets for drugs and toxicants, *FASEB J.*, 1987, **1**, 262–271.
58. G. R. Lankas, M. E. Cartwright and D. Umbenhauer, P-glycoprotein deficiency in a subpopulation of CF-1 mice enhances avermectin-induced neurotoxicity, *Toxicol. Appl. Pharmacol.*, 1997, **143**, 357–365.
59. D. R. Umbenhauer, G. R. Lankas, T. R. Pippert, D. Wise, M. E. Cartwright, S. J. Hall and C. M. Beare, Identification of a P-glycoprotein-deficient subpopulation of the CF-1 mouse strain using a restriction fragment length polymorphism, *Toxicol. Appl. Pharmacol.*, 1997, **146**, 88–94.
60. G. R. Lankas, D. Wise, M. E. Cartwright, T. Pippert and D. Umbenhauer, Placental P-glycoprotein deficiency enhances susceptibility to chemically induced birth defects in mice, *Reprod. Toxicol.*, 1998, **12**, 457–463.
61. K. L. Mealey, Therapeutic implications of the MDR-1 gene, *J. Vet. Pharmacol. Ther.*, 2004, **27**, 257–264.
62. K. Ueda, C. Cardarelli, M. M. Gottesman and I. Pastan, Expression of a full-length cDNA for the human 'MDR-1' gene confers resistance to colchicine, doxorubicin and vinblastine, *Proc. Natl. Acad. Sci. USA*, 1987, **84**, 3004–3008.
63. M. Uhr, T. Steikler, A. Yassouridis and F. Holsboer, Penetration of amitriptyline, but not fluoxetine, into brain is enhanced in mice with blood-brain barrier deficiency due to mdr1a P-glycoprotein gene disruption, *Neuropsychopharmacology*, 2000, **22**, 380–387.
64. M. Uhr, F. Holsboer and M. B. Muller, Penetration of endogenous steroid hormones cortisone, cortisol, aldosterone and progesterone into the brain in mice deficient for both mdr1a and mdr1b P-glycoprotein, *J. Neuroendocrinol.*, 2002, **14**, 753–759.

65. A. H. Schinkel, J. J. Smit, O. van Tellingen, J. H. Beijnen, E. Wagenaar, L. van Deemter, C. A. A. M. Mol, M. A. van der Valk, E. C. Robanus-Maan dag, H. P. J. te Riele, A. J. M. Berns and P. Borat, Disruption of the mouse mdr1a P-glycoprotein gene leads to a deficiency in blood-brain barrier and to increased sensitivity to drugs, *Cell*, 1994, **77**, 491–502.
66. A. H. Schinkel, E. Wagenaar, C. A. A. M. Mol and L. van Deempter, P-glycoprotein in the blood-brain barrier of mice influences the brain penetration and pharmacological activity of many drugs, *J. Clin. Invest.*, 1996, **97**, 2517–2524.
67. G. Y. Kwei, R. F. Alvaro, Q. Chen, C. Jenkins, E. A. C. Hop, C. A. Keohane, V. T. Ly, J. R. Straus, R. W. Wang, Z. Wang, T. R. Pippert and D. R. Umbenhauer, Disposition of ivermectin and cyclosporin A in CF-1 mice deficient in mdr1a P-glycoprotein, *Drug Metab. Dispos.*, 1999, **27**, 581–587.
68. J. F. Pouliot, F. L'Heureux, Z. Liu, R. K. Prichard and E. Georges, Reversal of p-glycoprotein-associated multidrug resistance by ivermectin, *Biochem. Pharmacol.*, 1997, **53**, 17–25.
69. L. F. Marques-Santos, R. R. Bernardo, E. F. de Paula and V. M. Rumjanek, Cyclosporin A and trifluoperazine, two resistance-modulating agents, increase ivermectin neurotoxicity in mice, *Pharmacol. Toxicol.*, 1999, **84**, 125–129.
70. J. W. Smit, M. T. Huisman, O. van Tellingen, H. R. Wiltshire and A. H. Schinkel, Absence or pharmacological blocking of placental P-glycoprotein profoundly increases fetal drug exposure, *J. Clin. Invest.*, 1999, **104**, 1441–1447.
71. A. D. Didier and F. Loor, Decreased biotolerability for ivermectin and cyclosporine A in mice exposed to potent p-glycoprotein inhibitors, *Int. J. Cancer*, 1995, **63**, 263–267.
72. S. Kiki-Mvouaka, C. Ménez, C. Borin, F. Lyazrhi, M. Foucaud-Vignault, J. Dupuy, X. Collet, M. Alvinerie and A. Lespine, Role of p-glycoprotein in the disposition of macrocyclic lactones: A comparison between ivermectin, eprinomectin and moxidectin in mice, *Drug Metab. Dispos.*, 2010, **38**, 73–80.
73. J. Geyer, S. Klintzsch, K. Meerkamp, A. Wöhle, O. Distl, A. Moritz and E. Petzinger, Detection of the nt230(del4) MDR 1 mutation in White Swiss Shepherd dogs: case reports of doramectin toxicoses, breed predisposition and microsatellite analysis, *J. Vet. Pharmacol. Ther.*, 2007, **30**, 482–485.
74. J. L. Barbet, T. Snook, J. M. Gay and K. L. Mealey, ABCB1-1Δ (MDR1-1Δ) genotype is associated with adverse reactions in dogs treated with milbemycin oxime for generalized demodicosis, *Vet. Dermatol.*, 2009, **20**, 111–114.
75. L. L. Sartor, S. A. Bentjen, L. Trepanier and K. L. Mealey, Loperamide toxicity in a collie with the MDR1 mutation associated with ivermectin sensitivity, *J. Vet. Intern. Med.*, 2004, **18**, 117–118.
76. N. MacDonald and A. Gledhill, Potential impact of the ABCB1 (p-glycoprotein) polymorphisms on avermectin toxicity in humans, *Arch. Toxicol.*, 2007, **81**, 553–563.

77. G. Edwards, Ivermectin: does P-glycoprotein play a role in neurotoxicity? *Filaria J.*, 2003, **2**, (Suppl. 1), 58–63.
78. C. N. Burkhart and C. G. Burkhart, Another look at ivermectin in the treatment of scabies and head lice, *Int. J. Dermatol.*, 1999, **36**, 235.
79. M. L. Elgart, A risk-benefit assessment of agents used in the treatment of scabies, *Drug Saf.*, 1996, **14**, 386–393.
80. K. M. Burkhart, C. N. Burkhart and C. G. Burkhart, Comparing topical scabietic treatments will soon be extinct, *Arch. Dermatol.*, 1997, **133**, 1314.
81. M. A. Aziz, S. Diallo, I. M. Diop, M. Larivière and M. Porta, Efficacy and tolerance of ivermectin in human onchocerciasis, *Lancet*, 1982, **2**, 171–173.
82. J.-L. Cartel, N. L. Nguyen, J.-P. Moulia-Pelat, R. Plichart, P. M. V. Martin and A. Spiegel, Mass chemoprophylaxis of lymphatic filariasis with a single dose of ivermectin in a polynesian community with a high *Wucheria bancrofti* infection rate, *Trans. R. Soc. Trop. Med. Hyg.*, 1992, **86**, 537–540.
83. S. Diallo, M. Larivière, L. Diop-Mar, R. N'Diaye, S. Badiane, M. Porta and M. Aziz, Conduct in Senegal of first studies on efficacy and safety of ivermectin (MK 993) in human onchocerciasis, *Bull. Soc. Pathol. Exot. Filiales*, 1984, **77**, 196–205.
84. V. Kumaraswami, E. A. Ottesen, V. Vijayasekaran, U. Deli, M. Swaminathan, M. A. Aziz, G. R. Sarma, R. Prabhakar and S. P. Tripathy, Ivermectin for the treatment of *Wucheria bancrofti* filariasis, *JAMA*, 1988, **259**, 3105–3153.
85. C. Naquira, G. Jiminez, J. G. Guerra, R. Bernal, D. R. Nalin, D. Neu and M. Aziz, Ivermectin for human strongyloidosis and other intestinal helminths, *Am. J. Trop. Med. Hyg.*, 1989, **40**, 304–309.
86. D. Richard-Lenoble, M. Kombila, E. A. Rupp, E. S. Pappayliou, P. Gaxotte, C. Nguiri and M. A. Aziz, Ivermectin in loiasis and concomitant *O. vovlulus* and *M. perstans* infections, *Am. J. Trop. Med. Hyg.*, 1988, **39**, 480–483.
87. I. A. Conti Diaz and J. Amaro, Treatment of human scabies with oral ivermectin, *Rev. Inst. Trop. São Paulo*, 1999, **41**, 259–261.
88. J. Remme, G. De Sole, K. Y. Dadzie, R. H. Baker, J. D. Hatbema, A. P. Plaisier, G. J. Oortmarssen and E. M. Samba, Large scale ivermectin distribution and its epidemiological consequences, *Acta Leiden.*, 1990, **59**, 177–191.
89. M. C. Pacque, Z. Dukuly, B. M. Greene, E. Munoz, E. Keyvan-Larijani, P. N. Williams and H. R. Taylor, Community-based treatment of onchocerciasis with ivermectin: acceptability and early adverse reactions, *Bull. World Health Organ.*, 1989, **67**, 721–730.
90. J. P. Coulaud, M. Larivière, M. C. Gervias, P. Gaxotte, A. Aziz, A. M. Deluol and J. Cenac, Treatment of human onchocerciasis with ivermectin, *Bull. Soc. Pathol. Exot. Filiales*, 1983, **76**, 681–688.

91. D. A. Hilmarsdottir, R. Mayorga-Sagastume, M. Lyagoubi, P. Gaxotte, S. Biligui, J. Chodakewitz, D. Neu and M. Gentilini, Treatment of *Strongyloides stercoralis* infection with ivermectin compared to albendazole: results of an open study with 60 cases, *Trans. R. Soc. Trop. Med. Hyg.*, 1994, **88**, 344–345.
92. A. González-Canga, A. M. Sahagún Prieto, M. J. Diez Liébana, N. F. Martinez, M. S. Vega and J. J. García Vieitez, The pharmacokinetics and interactions of ivermectin in humans—a mini-review, *AAPS J.*, 2008, **10**, 42–46.
93. G. M. Burnham, Adverse reactions to ivermectin treatment for onchocerciasis. Results of a placebo-controlled, double blind-trial in Malawi, *Trans. R. Soc. Trop. Med. Hyg.*, 1993, **87**, 313–317.
94. J. P. Chippaux, N. Gardon-Wendel, J. Gardon and J. C. Ernould, Absence of any adverse effect of inadvertent ivermectin treatment during pregnancy, *Trans. R. Soc. Trop. Med. Hyg.*, 1993, **87**, 318.
95. C. A. Guzzo, C. L. Furtek, A. G. Porras, C. Chen, R. Tipping, C. M. Clineschmidt, D. G. Scibberas, J. Y. Hsieh and K. C. Lasseter, Safety, tolerability and pharmacokinetics of escalating high doses of ivermectin in healthy adult subjects, *J. Clin. Pharmacol.*, 2002, **42**, 1122–1133.
96. Y. F. Sung, C. T. Huang, C. K. Fan, C. H. Lin and S. P. Lin, Avermectin intoxication with coma, myoclonus, and polyneuropathy, *Clin. Toxicol.*, 2009, **47**, 686–688.
97. B. G. Olson and J. B. Domachowske, Mazotti reaction after presumptive treatment for schistosomiasis and strongyloidiasis in a Liberian refugee, *Pediatr. Infect. Dis. J.*, 2006, **25**, 466–468.
98. J. Gardon, N. Gardon-Wendel, Demanga-Ngangue, J. Kamgno, J. P. Chippaux and M. Boussinesq, Serious reactions after mass treatment of onchocerciasis with ivermectin in an area endemic for *Loa loa* infection, *Lancet*, 1997, **350**, 18–22.
99. J. P. Chippaux, M. Boussinesq, J. Gardon, N. Gardon-Wendel and J. C. Emould, Severe adverse reaction risks during mass treatment with ivermectin in loiasis-endemic areas, *Parasitol. Today*, 1996, **12**, 448–450.
100. G. De Sole, J. Remme, K. Awadzi, S. Accorsi, E. S. Alley, O. Ba, K. Y. Dadzie, J. Giese, M. Karam and F. M. Keita, Adverse reactions after large-scale treatment of onchocerciasis with ivermectin: combined results from eight community trials, *Bull. World Health Org.*, 1989, **67**, 707–719.
101. C. D. Mackenzie, T. G. Geary and J. A. Gerlach, Possible pathogenic pathways in the adverse clinical events seen following ivermectin administration to onchocerciasis patients, *Filaria J.*, 2003, **2**, (Suppl. 1), S5.
102. Scientific Working Group on Serious Adverse Events in Loa Loa Endemic Areas, Report of a Scientific Group on Serious Adverse Events following Mectizan® treatment of onchocerciasis in *Loa loa* endemic areas, *Filaria J.*, 2003, **2**, (Suppl. 1), S2.

103. D. Shorter, K. Hale and E. Elliot, Mazzotti-like reaction after treatment with praziquantel for schistosomiasis, *Pediatr. Infect. Dis. J.*, 2006, **25**, 1087–1088.
104. H. Francis, K. Awadzi and E. A. Ottesen, The Mazzotti reaction following treatment of onchocerciasis with diethylcarbamazine: clinical severity as a function of infection intensity, *Am. J. Trop. Med. Hyg.*, 1985, **34**, 529–536.
105. P. Stingl and M. Stingl, Leprosy, onchocerciasis, diethylcarbamazine and the Mazzotti reaction, *Lepr. Rev.*, 1982, **53**(4), 317–318.
106. G. M. Ackerman, H. Kephart, K. Francis, G. J. Awadzi, G. J. Gleich and E. A. Ottesten, Eosinophil deregulation. An immunologic determinant in the pathogenesis of the Mazotti reaction in human onchocerciasis, *J. Immunol.*, 1990, **144**, 3961–3969.
107. F. L. Njoo, C. E. Flack, J. Oosting, J. S. Stilma and A. Kijlstra, Neutrophil activation in ivermectin-treated onchocerciasis patients, *Clin. Exp. Immunol.*, 1993, **94**, 330–333.
108. N. E. Erondu and J. E. Donelson, Characterization of a myosin-like antigen from *Onchocera volvulus*. *Mol. Biochem. Parasitol.*, 1990, **40**, 213–224.
109. N. A. Y. Twum-Danso, Serious adverse events following treatment with ivermectin for onchocerciasis control: a review of reported cases, *Filaria J.*, 2003, **2**, (Suppl. 1), S3.
110. N. A. Y. Twum-Danso, *Loa loa* encephalopathy temporally related to ivermectin administration reported from onchocerciasis mass treatment programs from 1989 to 2001: implications for the future, *Filaria J.*, 2003, **2**, (Suppl. 1), S7.
111. N. D. E. Alexander, M. J. Bockarie, W. A. Kastens, J. W. Kazura and M. P. Alpers, Absence of ivermectin-associated excess deaths, *Trans. R. Soc. Trop. Med. Hyg.*, 1998, **92**, 342.
112. K. Chung, C.-C. Yang, M.-L. Wu, J.-F. Deng and W.-J. Tsai, Agricultural avermectins: an uncommon but potentially fatal cause of pesticide poisoning, *Ann. Emerg. Med.*, 1999, **34**, 51–57.
113. K. N. Woodward, Veterinary pharmacovigilance. Part 2. Veterinary pharmacovigilance in practice—the operation of a spontaneous reporting scheme in a European Union country—the UK, and schemes in other countries, *J. Vet. Pharmacol. Ther.*, 2005, **28**, 149–170.
114. K. N. Woodward, Veterinary pharmacovigilance in the European Union, in *Veterinary Pharmacovigilance. Adverse Reactions to Veterinary Medicinal Products*, ed. K. N. Woodward, Wiley-Blackwell, Chichester, 2009, pp. 19–46.
115. T. M. Hodge, Pharmacovigilance in the US—an industry perspective, in *Veterinary Pharmacovigilance. Adverse Reactions to Veterinary Medicinal Products*, ed. K. N. Woodward, Wiley-Blackwell, Chichester, 2009, pp. 231–285.
116. U.S. Food and Drug Administration, Center for Veterinary Medicine. http://www.fda.gov/AnimalVeterinary/SafetyHealth/ProductSafetyInformation/ (accessed 5 January, 2011).

117. D. T. Lewis, S. R. Merchant and T. M. Neer, Ivermectin toxicosis in a kitten, *J. Am. Vet. Med. Assoc.*, 1994, **205**, 584–585.
118. H. Frische and L. Hunt, Suspected ivermectin toxicity in kittens, *Can. Vet. J.*, 1991, **32**, 245.
119. J. Pritchard, Treating ivermectin toxicity in cats, *Vet. Rec.*, 2010, **166**, 766.
120. G. Muhammad, A. Jabbar, M. Z. Khan and M. Saqib, Use of neostigmine in massive ivermectin toxicity in cats, *Vet. Human Toxicol.*, 2004, **46**, 28–29.
121. R. A. Lovell, Ivermectin and piperazine toxicoses in dogs and cats, *Vet. Clin. North Am. Small Anim. Pract.*, 1990, **20**, 453–468.
122. J. D. Roder, Antiparasiticals, in *Clinical Veterinary Toxicology*; ed. K. H. Plumlee, Mosby, St. Louis, 2004, pp. 302–305.
123. C. R. Reinemeyer and C. H. Courtney, Chemotherapy of parasitic diseases, in *Veterinary Pharmacology and Therapeutics*, ed. H. R. Adams, Iowa State Press, Ames, 2001, pp. 947–991.
124. H. A. Benchaoui, R. G. Clemence, P. J. M. Clements, R. L. Jones, P. Watson, D. J. Shanks, D. G. Smith, G. H. Sture, A. D. Jernigan and T. G. Rowan, Efficacy and safety of selamectin against fleas on dogs and cats presented as veterinary patients in Europe, *Vet. Parasitol.*, 2000, **91**, 223–232.
125. M. J. Krautmann, M. J. Novotny, K. Keulenaer, C. S. Godin, E. I. Evans, J. W. McCall, C. Wang, T. G. Rowan and A. D. Jermigan, Safety of selamectin in cats, *Vet. Parasitol.*, 2000, **91**, 393–403.
126. A. Campbell and M. Chapman. *Handbook of Poisoning in Dogs and Cats*, Blackwell Science, Oxford, 2000, pp. 167–173.
127. P. J. Kenny, K. M. Vernau, B. Puscher and D. J. Maggs, Retinopathy associated with ivermectin toxicosis in two dogs, *J. Am. Vet. Med. Assoc.*, 2008, **233**, 279–284.
128. R. C. Lynn, Drugs for the treatment of heartworm infections, in *Small Animal Clinical Pharmacology and Therapeutics*, ed. D. M. Boothe, W. B. Saunders, London, 2001, pp. 267–279.
129. K. Hopper, J. Aldrich and S. C. Haskins, Ivermectin toxicity in 17 collies, *J. Vet. Intern. Med.*, 2002, **16**, 89–94.
130. A. J. Paul, W. J. Tranquilli, R. L. Seward, K. S. Todd and J. A. DiPietro, Clinical observations in collies given ivermectin orally, *Am. J. Vet. Res.*, 1987, **48**, 684–685.
131. M. W. Beal, R. H. Poppenga, W. J. Birdsall and D. Hughes, Respiratory failure attributable to moxidectin intoxication in a dog, *J. Am. Vet. Med. Assoc.*, 1999, **215**, 1813–1817.
132. W. J. Tranquilli, A. J. Paul, R. L. Seward, K. S. Todd and J. A. DiPietro, Response to physostigmine administration on collie dogs exhibiting ivermectin toxicosis, *J. Vet. Pharmacol. Ther.*, 1987, **10**, 96–100.
133. E. Yas-Natan, M. Shamir, S. Kleinbart and I. Aroch, Doramectin toxicity in a collie, *Vet. Rec.*, 2003, **153**, 718–720.
134. A. J. Paul, W. J. Tranquilli and D. E. Hutchens, Safety of moxidectin in avermectin-sensitive collies, *Am. J. Vet. Res.*, 2000, **61**, 482–483.

135. J. G. Sherman, A. J. Paul and L. D. Firkins, Evaluation of the safety of spinosad and milbemycin 5-oxime orally to collies with the MDR 1 gene mutation, *Am. J. Vet. Res.*, 2010, **71**, 115–119.
136. A. J. Paul, D. E. Hutchens, L. D. Firkins and M. Borgstrom, Dermal safety study with imidacloprid/moxidectin topical solution in the ivermectin-sensitive collie, *Vet. Parasitol.*, 2004, **121**, 285–291.
137. W. J. Tranquilli, A. J. Paul and K. S. Todd, Assessment of toxicosis induced by high-dose administration of milbemycin oxime in collies, *Am. J. Vet Res.*, 1991, **52**, 1170–1172.
138. N. J. Snowden, C. V. Helyar, S. R. Platt and J. Penderis, Clinical presentation and management of moxidectin toxicity in two dogs, *J. Small Anim. Pract.*, 2006, **47**, 620–624.
139. P. E. Fassler, W. J. Tranquilli, A. J. Paul, M. D. Soll, J. A. DiPietro and K. S. Todd, Evaluation of the safety of ivermectin administered in a beef-based formulation to ivermectin-sensitive collies, *J. Am. Vet. Med. Assoc.*, 1991, **199**, 457–460.
140. D. E. Crandell and G. L. Weinberg, Moxidectin toxicosis in a puppy successfully treated with intravenous lipids, *J. Vet. Emerg. Crit. Care*, 2009, **19**, 181–186.
141. L. L. Sartor, S. A. Bentjen, L. Trepanier and K. L. Mealey, Loperamide toxicity in a collie with the MDR1 mutation associated with ivermectin sensitivity, *J. Vet. Intern. Med.*, 2004, **18**, 117–118.
142. J. Geyer, O. Gavrilova and E. Petzinger, Brain penetration of ivermectin and selamectin in *mdr 1 a, b* P-glycoprotein- and *bcrp*-deficient knockout mice, *J. Vet. Pharmacol. Ther.*, 2008, **32**, 87–96.
143. V. A. Merola, S. Khan and S. Gwaltney-Brant, Ivermectin toxicosis in dogs: a retrospective study, *J. Am. Anim. Hosp. Assoc.*, 2009, **45**, 106–111.
144. P. Dowling, Pharmacogenetics: it's not just about ivermectin in collies, *Can. Vet. J.*, 2006, **47**, 1165–1168.
145. D. M. Houston, J. Parent and K. J. Matushek, Ivermectin toxicosis in a dog, *J. Am. Vet. Med. Assoc.*, 1987, **191**, 78–79.
146. M. K. Hadrick, S. E. Bunch and J. N. Kornegay, Ivermectin toxicosis in two Australian shepherds, *J. Am. Vet. Med. Assoc.*, 1995, **206**, 1147–1150.
147. M. Cotman and J. Zabavnik, Mutation of *MDR 1* gene associated with multidrug sensitivity in Australian shepherds in Slovenia, *Slov. Vet. Res.*, 2007, **44**, 19–24.
148. K. L. Mealey, S. A. Bentjen and D. K. Waiting, Frequency of the mutant MDR1 allele associated with ivermectin sensitivity in a sample population of collies from northwestern United States, *Am. J. Vet. Res.*, 2002, **63**, 479–481.
149. L. S. Blair, P. F. Malatesta and D. W. Ewanciw, Dose-response study of ivermectin against *Dirofilaria immitis* in dogs with naturally acquired infections, *Am. J. Vet. Res.*, 1983, **44**, 475–477.

150. E. C. McManus and J. D. Pulliam, Histopathologic features of canine heartworm microfilarial infection after treatment with ivermectin, *Am. J. Vet, Res.*, 1984, **45**, 91–97.
151. M. T. Suderman and T. M. Craig, Efficacy of ivermectin against *Dirofilaria immitis* microfilariae in naturally infected dogs, *Am. J. Vet. Res.*, 1984, **45**, 1031–1032.
152. P. F. Boreham and R. B. Atwell, Absence of shock-like reactions to ivermectin in dogs infected with *Dirofilaria immitis, J. Helminthol*, 1983, **57**, 279–281.
153. P. F. Boreham, R. B. Atwell and J. M. Euclid, Studies on the mechanism of the DEC reaction in dogs infected with *Dirofilaria immitis, Int. J. Parasitol.*, 1985, **15**, 543–549.
154. C. H. Carlisle, C. W. Prescott, P. J. McCosker and A. A. Seawright, The toxic effects of thiacetarsemide sodium in normal dogs and in dogs infested with *Dirofilaria immitis, Aust. Vet. J.*, 1974, **50**, 204–208.
155. Y. Sasaki, H. Kitagawa and K. Ishihara, Clinical application of milbemycin D as a prophylactic agent against *Dirofilaria immitis* infection in dogs: clinical findings in dogs with shock-like reaction, *Nippon Juigaku Zasshi.*, 1986, **48**, 1207–1214.
156. N. E. Palumbo, R. S. Desowitz and S. F. Perri, Observations on the adverse reaction to diethylcarbamazine in *Dirofilaria immitis*-infected dogs, *Tropenmed. Parasitol.*, 1981, **32**, 115–118.
157. P. F. Boreham and R. B. Atwell, Adverse drug reactions in the treatment of filarial parasites: haematological, biochemical, immunological and pharmacological changes in *Dirofilaria immitis* infected dogs treated with diethylcarbamazine, *Int. J. Parasitol.*, 1983, **13**, 547–556.
158. K. N. Woodward, Adverse effects of veterinary pharmaceutical products in animals, in *Veterinary Pharmacovigilance. Adverse Reactions to Veterinary Medicinal Products*, ed. K. N. Woodward, Wiley-Blackwell, Chichester, 2009, pp. 393–421.
159. D. Brown, Ingestion of equine moxidectin by dogs, *Vet. Rec.*, 2000, **147**, 340.
160. A. K. Gray, Avermectin toxicity in the dog, *Vet. Rec.*, 1997, **140**, 563.
161. S. Eaton, Ingestion of Equest oral gel by dogs, *Vet. Rec.*, 1999, **145**, 236.
162. L. R. Hovda and S. B. Hooser, Toxicology of the newer pesticides for use in dogs and cats, *Vet. Clin. Small Anim. Pract.*, 2002, **32**, 455–467.
163. M. J. Novotny, M. J. Krautmann, J. C. Ehrhart, C. S. Godin, E. I. Evans, J. W. McCall, F. Sun, T. G. Rowan and A. D. Jernigan, Safety of selamectin in dogs, *Vet. Parasitol.*, 2000, **91**, 377–391.
164. P. A. Karns and D. G. Luther, A survey of adverse effects associated with ivermectin use in Louisiana horses, *J. Am. Vet. Med. Assoc.*, 1984, **185**, 782–783.
165. L. M. Godber, F. Derksen, R. Williams and B. Mahmoud, Ivermectin toxicosis in a mule foal, *Austr. Vet. J.*, 1995, **72**, 191–192.

166. T. M. Swor, J. L. Whittenburg and M. K. Chaffin, Ivermectin toxicosis in three adult horses, *J. Am. Vet. Med. Assoc.*, 2009, **235**, 558–562.
167. C. E. Plummer, M. E. Kallberg, F. J. Ollivier, D. E. Brooks and K. N. Gelatt, Suspected ivermectin toxicoses in a miniature mule foal causing blindness, *Vet. Ophthalmol.*, 2006, **9**, 29–32.
168. L. A. Hautekeete, S. A. Khan and W. S. Hales, Ivermectin toxicosis in a zebra, *Vet. Human Toxicol.*, 1998, **40**, 29–31.
169. S. A. Khan, D. E. Kuster and S. R. Hansen, A review of moxidectin overdose cases in equines from 1998 through 2000, *Vet. Human Toxicol.*, 2002, **44**, 232–235.
170. L. S. Goehring and M. M. Sloet van Oldruitenborgh-Oosterbann, Moxidectin overdose in a foal, *Tijdschr. Diergeneeskd.*, 1999, **124**, 412–414.
171. P. J. Johnson, D. R. Mrad, A. J. Schwarz and L. Kellam, Presumed moxidectin toxicosis in 3 foals, *J. Am. Vet, Med. Assoc.*, 1999, **214**, 678–680.
172. J. K. Thomas, Suspect ivermectin toxicity, *Vet. Rec.*, 1988, **123**, 631–632.
173. C. Button, R. Barton, P. Honey and P. Rickford, Avermectin toxicity in calves and an evaluation of picrotoxin as an antidote, *Austr. Vet. J.*, 1988, **65**, 157–158.
174. J. T. Seaman, J. S. Eagleson, M. J. Carrigan and R. F. Webb, Avermectin B_1 toxicity in a herd of Murray Grey cattle, *Austr. Vet. J.*, 1987, **64**, 284–285.
175. J. M. Rose, S. L. Peckham, J. L. Scism and K. L. Audus, Evaluation of the role of P-glycoprotein in ivermectin uptake by primary cultures of bovine brain microvessel endothelial cells, *Neurochem. Res.*, 1998, **23**, 203–209.
176. R. B. Richards and J. R. Edwards, A progressive spinal myelinopathy in beef cattle, *Vet. Pathol.*, 1986, **23**, 35–41.
177. J. R. Edwards, R. B. Richards and M. J. Carrick, Inherited progressive spinal myelinopathy in Murray Grey cattle, *Aust. Vet. J.*, 1988, **5**, 108–109.
178. R. D. Jolly, Mannosidosis and its control in Angus and Murray Grey cattle, *N. Z. Vet. J.*, 1978, **26**, 194–198.
179. K. N. Woodward, Toxicity in animals: target species. *Current Pharm. Biotechnol.,* 2011, in press.
180. B. Skopets, R. F. Wilson, J. W. Griffith and C. M. Lang, Ivermectin toxicity in young mice, *Lab. Anim. Sci.*, 1996, **46**, 111–112.
181. S. A. Iliff-Sizemore, M. R. Partlow and S. T. Kelley, Ivermectin toxicology in a rhesus macaque, *Vet. Human Toxicol.*, 1990, **32**, 530–532.
182. J. S. Kim and E. C. Crichlow, Clinical signs of ivermectin toxicity and efficacy of antigabaergic convulsants as antidotes for ivermectin poisoning in epileptic chickens, *Vet. Human Toxicol.*, 1995, **37**, 122–126.
183. J. A. Teare and M. Bush, Toxicity and efficacy of ivermectin in chelonians, *J. Am. Vet. Med. Assoc.*, 1983, **183**, 1195–1197.

184. Z. Széll, T. Sréter and I. Varga, Ivermectin toxicosis in a chameleon (*Chamaeleo senegalensis*) infected with *Foleyella furcata, J. Zoo Wildl. Med.*, 2001, **32**, 115–117.
185. J. H. DeMarco, D. J. Heard, G. J. Fleming, B. A. Lock and T. J. Scase, Ivermectin toxicosis after topical administration in dog-faced fruit bats (*Cynopterus brachyotis*), *J. Zoo Wildl. Med.*, 2002, **33**, 147–150.
186. M. J. Costello, Review of methods to control sea lice (Caligidae: Crustacea) infestations on salmon (*Salmo salar*) farms, in *Pathogens of Wild and Farmed Fish*, ed, G. A. Boxall and D. Defaye, Ellis Horwood, London, 1993, pp. 219–252.
187. R. Palmer, H. Rodger, E. Drinan, C. Dwyer and P. R. Smith, Preliminary trials on the efficacy of ivermectin against parasitic copepods of Atlantic salmon, *Bull. Eur. Assoc. Fish Pathol.*, 1987, **7**, 47–54.
188. J. O'Halloran, J. Carpenter, D. Ogden, W. E. Hogans and M. Jansen, Atlantic Canada. *Ergasilus labracis* on Atlantic salmon, *Can. Vet. J.*, 1992, **33**, 75.

Subject Index